Radon Series on Computational and Applied Mathematics 9

Radon Series on Computational and Applied Mathematics

1 *Lectures on Advanced Computational Methods in Mechanics*
Johannes Kraus and Ulrich Langer (eds.), 2007

2 *Gröbner Bases in Symbolic Analysis*
Markus Rosenkranz and Dongming Wang (eds.), 2007

3 *Gröbner Bases in Control Theory and Signal Processing*
Hyungju Park and Georg Regensburger (eds.), 2007

4 *A Posteriori Estimates for Partial Differential Equations*
Sergey Repin, 2008

5 *Robust Algebraic Multilevel Methods and Algorithms*
Johannes Kraus and Svetozar Margenov, 2009

6 *Iterative Regularization Methods for Nonlinear Ill-Posed Problems*
Barbara Kaltenbacher, Andreas Neubauer and Otmar Scherzer, 2008

7 *Robust Static Super-Replication of Barrier Options*
Jan H. Maruhn, 2009

8 *Advanced Financial Modelling*
Hansjörg Albrecher, Wolfgang J. Runggaldier and Walter Schachermayer (eds.), 2009

9 *Theoretical Foundations and Numerical Methods for Sparse Recovery*
Massimo Fornasier (ed.), 2010

Theoretical Foundations and Numerical Methods for Sparse Recovery

Edited by

Massimo Fornasier

De Gruyter

Mathematics Subject Classification 2010
49-01, 65-01, 15B52, 26B30, 42A61, 49M29, 49N45, 65K10, 65K15, 90C90, 90C06

ISBN 978-3-11-022614-0
e-ISBN 978-3-11-022615-7
ISSN 1865-3707

Library of Congress Cataloging-in-Publication Data

Theoretical foundations and numerical methods for sparse recovery / edited by Massimo Fornasier.
p. cm. – (Radon series on computational and applied mathematics ; 9)
Includes bibliographical references and index.
ISBN 978-3-11-022614-0 (alk. paper)
1. Sparse matrices. 2. Equations – Numerical solutions. 3. Differential equations, Partial – Numerical solutions. I. Fornasier, Massimo.
QA297.T486 2010
512.9'434–dc22
2010018230

Bibliographic information published by the Deutsche Nationalbibliothek

The Deutsche Nationalbibliothek lists this publication in the Deutsche Nationalbibliografie; detailed bibliographic data are available in the Internet at http://dnb.d-nb.de.

Typesetting: Da-TeX Gerd Blumenstein, Leipzig, www.da-tex.de
Printing: Hubert & Co. GmbH & Co. KG, Göttingen
∞ Printed on acid-free paper

Printed in Germany

www.degruyter.com

Preface

Sparsity has become an important concept in recent years in applied mathematics, especially in mathematical signal and image processing, in the numerical treatment of partial differential equations, and in inverse problems. The key idea is that many types of functions arising naturally in these contexts can be described by only a small number of significant degrees of freedom. This feature allows the exact recovery of solutions from a minimal amount of information. The theory of sparse recovery exhibits fundamental and intriguing connections with several mathematical fields, such as probability, geometry of Banach spaces, harmonic analysis, calculus of variations and geometric measure theory, theory of computability, and information-based complexity. The link to convex optimization and the development of efficient and robust numerical methods make sparsity a concept concretely useful in a broad spectrum of natural science and engineering applications.

The present collection of four lecture notes is the very first contribution of this type in the field of sparse recovery and aims at describing the novel ideas that have emerged in the last few years. Emphasis is put on theoretical foundations and numerical methodologies. The lecture notes have been prepared by the authors on the occasion of the Summer School "Theoretical Foundations and Numerical Methods for Sparse Recovery" held at the Johann Radon Institute for Computational and Applied Mathematics (RICAM) of the Austrian Academy of Sciences on August 31 – September 4, 2009. The aim of organizing the school and editing this book was to provide a systematic and self-contained presentation of the recent developments. Indeed, there seemed to be a high demand of a friendly guide to this rapidly emerging field. In particular, our intention is to provide a useful reference which may serve as a textbook for graduate courses in applied mathematics and engineering. Differently from a unique monograph, the chapters of this book are already in the form of self-contained lecture notes and collect a selection of topics on specific facets of the field. We tried to keep the presentation simple, and always start from basic facts. However, we did not neglect to present also more advanced techniques which are at the core of sparse recovery from probability, nonlinear approximation, and geometric measure theory as well as tools from nonsmooth convex optimization for the design of efficient recovery algorithms. Part of the material presented in the book comes from the research work of the authors. Hence, it might also be of interest for advanced researchers who may find useful details and use the book as a reference for their work. An outline of the content of the book is as follows.

The first chapter by Holger Rauhut introduces the theoretical foundations of compressive sensing. It focuses on ℓ_1-minimization as a recovery method and on structured random measurement matrices, such as the random partial Fourier matrix and partial random circulant matrices. A detailed presentation of the basic tools of probability and more advanced techniques, such as scalar and noncommutative Khintchine inequalities, Dudley's inequality, and Rudelson's lemma, is provided. This chapter contains some improvements and generalizations of existing results, which have not yet appeared elsewhere in the literature.

The second chapter by Massimo Fornasier starts by addressing numerical methods for ℓ_1-minimization for compressive sensing applications. The analysis of the homotopy method, the iteratively re-weighted least squares method, and iterative hard thresholding is carefully outlined. Numerical methods for sparse optimization in Hilbert spaces are also provided, starting from the analysis of iterative soft-thresholding and continuing with a discussion on several improvements and acceleration methods. Numerical techniques for large scale computing based on domain decomposition methods are discussed in detail.

The focus of the third chapter by Ronny Ramlau and Gerd Teschke is on the regularization theory and on numerical methods for inverse and ill-posed problems with sparse solutions. The emphasis is on regularization properties of iterative algorithms, in particular convergence rates to exact solutions. Adaptive techniques in soft-shrinkage and projected gradient methods are carefully analyzed. The chapter starts with a friendly guide to classical linear inverse problems and then addresses more advanced techniques for nonlinear inverse problems with sparsity constraints. The results are illustrated also by numerical examples inspired by single photon emission computed tomography (SPECT) and in nonlinear sensing.

Besides sparsity with respect to classical bases, e.g., wavelets or curvelets, one may facilitate the robust reconstruction of images from only partial linear or nonlinear measurements by imposing that the interesting solution is the one which matches the given data and also has few discontinuities localized on sets of lower dimension, i.e., it is sparse in terms of its derivatives. As described in the mentioned chapters, the minimization of ℓ_1-norms occupies a fundamental role for the promotion of sparsity. This understanding furnishes an important interpretation of *total variation minimization*, i.e., the minimization of the L^1-norm of derivatives, as a regularization technique for image restoration. The fourth and last chapter by Antonin Chambolle, Vicent Caselles, Daniel Cremers, Matteo Novaga and Thomas Pock, addresses various theoretical and practical topics related to total variation based image reconstruction. The chapter first focuses on basic theoretical results on functions of bounded variation, and on more advanced results on fine properties of minimizers of the total variation. Standard and more sophisticated minimization algorithms for the total variation in a finite-difference setting are carefully presented. A series of applications from simple denoising to stereo, or deconvolution issues are discussed. More exotic applications of total variation minimization are also considered, like the solution of minimal partition problems.

Linz, April 2010 Massimo Fornasier

Table of Contents

Radon Series Comp. Appl. Math **9**, 1–92

Compressive Sensing and Structured Random Matrices

Holger Rauhut

Abstract. These notes give a mathematical introduction to compressive sensing focusing on recovery using ℓ_1-minimization and structured random matrices. An emphasis is put on techniques for proving probabilistic estimates for condition numbers of structured random matrices. Estimates of this type are key to providing conditions that ensure exact or approximate recovery of sparse vectors using ℓ_1-minimization.

Keywords. Compressive sensing, ℓ_1-minimization, basis pursuit, structured random matrices, condition numbers, random partial Fourier matrix, partial random circulant matrix, Khintchine inequalities, bounded orthogonal systems.

2010 Mathematics Subject Classification. 15A12, 15A60, 15B05, 15B10, 15B52, 42A05, 42A61, 60B20, 60G50, 90C05, 90C25, 90C90, 94A12, 94A20.

Table of Contents.

1 Introduction

Compressive sensing is a recent theory that predicts that sparse vectors in high dimensions can be recovered from what was previously believed to be incomplete information. The seminal papers by E. Candès, J. Romberg and T. Tao [19, 23] and by D. Donoho [38] have caught significant attention and have triggered enormous research activities after their appearance. These notes make an attempt to introduce to some mathematical aspects of this vastly growing field. In particular, we focus on

The author acknowledges support by the Hausdorff Center for Mathematics and by the WWTF project SPORTS (MA 07-004).

ℓ_1-minimization as recovery method and on structured random measurement matrices such as the random partial Fourier matrix and partial random circulant matrices. We put emphasis on methods for showing probabilistic condition number estimates for structured random matrices. Among the main tools are scalar and noncommutative Khintchine inequalities. It should be noted that modified parts of these notes together with much more material will appear in a monograph on compressive sensing [55] that is currently under preparation by the author and Simon Foucart.

The main motivation for compressive sensing is that many real-world signals can be well-approximated by sparse ones, that is, they can be approximated by an expansion in terms of a suitable basis, which has only a few non-vanishing terms. This is the key why many (lossy) compression techniques such as JPEG or MP3 work so well. To obtain a compressed representation one computes the coefficients in the basis (for instance a wavelet basis) and then keeps only the largest coefficients. Only these will be stored while the rest of them will be put to zero when recovering the compressed signal.

When complete information on the signal or image is available this is certainly a valid strategy. However, when the signal has to be acquired first with a somewhat costly, difficult, or time-consuming measurement process, this seems to be a waste of resources: First one spends huge efforts to collect complete information on the signal and then one throws away most of the coefficients to obtain its compressed version. One might ask whether there is a more clever way of obtaining somewhat more directly the compressed version of the signal. It is not obvious at first sight how to do this: measuring directly the large coefficients is impossible since one usually does not know *a-priori*, which of them are actually the large ones. Nevertheless, compressive sensing provides a way of obtaining the compressed version of a signal using only a small number of linear and non-adaptive measurements. Even more surprisingly, compressive sensing predicts that recovering the signal from its undersampled measurements can be done with computationally efficient methods, for instance convex optimization, more precisely, ℓ_1-minimization.

Of course, arbitrary undersampled linear measurements – described by the so-called measurement matrix – will not succeed in recovering sparse vectors. By now, necessary and sufficient conditions are known for the matrix to recover sparse vectors using ℓ_1-minimization: the null space property and the restricted isometry property. Basically, the restricted isometry property requires that all column submatrices of the measurement matrix of a certain size are well-conditioned. It turns out to be quite difficult to check this condition for deterministic matrices – at least when one aims to work with the minimal amount of measurements. Indeed, the seminal papers [19, 38] obtained their breakthrough by actually using random matrices. While the use of random matrices in sparse signal processing was rather uncommon before the advent of compressive sensing, we note that they were used quite successfully already much earlier, for instance in the very related problem from Banach space geometry of estimating Gelfand widths of ℓ_1^N-balls [54, 57, 74].

Introducing randomness allows to show optimal (or at least near-optimal) conditions on the number of measurements in terms of the sparsity that allow recovery of sparse vectors using ℓ_1-minimization. To this end, often Gaussian or Bernoulli matrices are used, that is, random matrices with stochastically independent entries having a standard normal or Bernoulli distribution. Tanner in this volume.

Applications, however, often do not allow the use of "completely" random matrices, but put certain physical constraints on the measurement process and limit the amount of randomness that can be used. For instance, when sampling a trigonometric polynomial having sparse coefficients one might only have the freedom to choose the sampling points at random. This leads then to a structured random measurement matrix, more precisely, a random partial Fourier type matrix. Indeed, such type of matrices were already investigated in the initial papers [19, 23] on compressive sensing. These notes will give an introduction on recovery results for ℓ_1-minimization that can be obtained using such structured random matrices. A focus is put on methods for probabilistic estimates of condition numbers such as the noncommutative Khintchine inequalities and Dudley's inequality.

Although we will not cover specific applications in these notes, let us mention that compressive sensing may be applied in imaging [44, 109], A/D conversion [133], radar [69, 49] and wireless communication [126, 95], to name a few.

These notes contain some improvements and generalizations of existing results, that have not yet appeared elsewhere in the literature. In particular, we generalize from random sampling of sparse trigonometric polynomials to random sampling of functions having sparse expansions in terms of bounded orthonormal systems. The probability estimate for the so-called restricted isometry constants for the corresponding matrix is slightly improved. Further, also the sparse recovery result for partial random circulant and Toeplitz matrices presented below is an improvement over the one in [105].

These lecture notes only require basic knowledge of analysis, linear algebra and probability theory, as well as some basic facts about vector and matrix norms.

2 Recovery via ℓ_1-Minimization

2.1 Preliminaries and Notation

Let us first introduce some notation. For a vector $\mathbf{x} = (x_1, \ldots, x_N) \in \mathbb{C}^N$, the usual p-norm is denoted

$$\|\mathbf{x}\|_p := \left(\sum_{\ell=1}^{N} |x_\ell|^p \right)^{1/p}, \quad 1 \le p < \infty,$$

$$\|\mathbf{x}\|_\infty := \max_{\ell \in [N]} |x_\ell|,$$

where $[N] := \{1, 2, \dots, N\}$. For a matrix $A = (a_{jk}) \in \mathbb{C}^{m \times N}$ we denote $A^* = (\overline{a_{kj}})$ its conjugate transpose. The operator norm of a matrix from ℓ_p into ℓ_p is defined as

$$\|A\|_{p \to p} := \max_{\|\mathbf{x}\|_p = 1} \|A\mathbf{x}\|_p.$$

For the cases $p = 1, 2, \infty$ an explicit expression for the operator norm of A is given by

$$\|A\|_{1 \to 1} = \max_{k \in [N]} \sum_{j=1}^{m} |a_{jk}|, \tag{2.1}$$

$$\|A\|_{\infty \to \infty} = \max_{j \in [m]} \sum_{k=1}^{N} |a_{jk}|,$$

$$\|A\|_{2 \to 2} = \sigma_{\max}(A) = \sqrt{\lambda_{\max}(A^* A)},$$

where $\sigma_{\max}(A)$ denotes the largest singular value of A and $\lambda_{\max}(A^* A) \geq 0$ is the largest eigenvalue of $A^* A$. Clearly, $\|A\|_{1 \to 1} = \|A^*\|_{\infty \to \infty}$. It follows from the Riesz–Thorin interpolation theorem [118, 7] that

$$\|A\|_{2 \to 2} \leq \max\{\|A\|_{1 \to 1}, \|A\|_{\infty \to \infty}\}. \tag{2.2}$$

The above inequality is sometimes called the Schur test, and it can also be derived using Hölder's inequality, see for instance [64]; or alternatively using Gershgorin's disc theorem [8, 71, 135]. In particular, if $A = A^*$ is hermitian, then

$$\|A\|_{2 \to 2} \leq \|A\|_{1 \to 1}. \tag{2.3}$$

All eigenvalues of a hermitian matrix $A = A^* \in \mathbb{C}^{n \times n}$ are contained in

$$\{\langle A\mathbf{x}, \mathbf{x} \rangle : \mathbf{x} \in \mathbb{C}^n, \|\mathbf{x}\|_2 = 1\} \subset \mathbb{R}.$$

In particular, for hermitian $A = A^*$,

$$\|A\|_{2 \to 2} = \sup_{\|\mathbf{x}\|_2 = 1} |\langle A\mathbf{x}, \mathbf{x} \rangle|. \tag{2.4}$$

For real scalars $\alpha_1, \dots, \alpha_n \in \mathbb{R}$ and vectors $\mathbf{z}_1, \dots, \mathbf{z}_n \in \mathbb{C}^m$ the matrix $\sum_{j=1}^n \alpha_j \mathbf{z}_j \mathbf{z}_j^*$ is hermitian and we have

$$\begin{aligned} \left\| \sum_{j=1}^{n} \alpha_j \mathbf{z}_j \mathbf{z}_j^* \right\|_{2 \to 2} &= \sup_{\|\mathbf{x}\|_2 = 1} \left| \left\langle \sum_{j=1}^{n} \alpha_j \mathbf{z}_j \mathbf{z}_j^* \mathbf{x}, \mathbf{x} \right\rangle \right| = \sup_{\|\mathbf{x}\|_2 = 1} \left| \sum_{j=1}^{n} \alpha_j |\langle \mathbf{z}_j, \mathbf{x} \rangle|^2 \right| \\ &\leq \max_{k \in [n]} |\alpha_k| \sup_{\|\mathbf{x}\|_2 = 1} \sum_{j=1}^{n} |\langle \mathbf{z}_j, \mathbf{x} \rangle|^2 = \left\| \sum_{j=1}^{n} \mathbf{z}_j \mathbf{z}_j^* \right\|_{2 \to 2} \max_{k \in [n]} |\alpha_k|. \end{aligned} \tag{2.5}$$

Also the Frobenius norm will be of importance. For a matrix $A = (a_{jk})$ it is defined as

$$\|A\|_F := \sqrt{\sum_{j,k} |a_{jk}|^2} = \sqrt{\mathrm{Tr}(A^*A)},$$

where Tr denotes the trace. The Frobenius norm is induced by the inner product $\langle A, B\rangle_F = \mathrm{Tr}(B^*A)$. The Cauchy–Schwarz inequality for the trace states that

$$|\langle A, B\rangle_F| = |\mathrm{Tr}\,(B^*A)| \leq \|A\|_F \|B\|_F. \tag{2.6}$$

The null space of a matrix $A \in \mathbb{C}^{m\times N}$ is denoted by $\ker A = \{\mathbf{x} \in \mathbb{C}^N, A\mathbf{x} = 0\}$. We usually write $\mathbf{a}_\ell \in \mathbb{C}^m$, $\ell = 1, \ldots, N$, for the columns of a matrix $A \in \mathbb{C}^{m\times N}$. The column submatrix of A consisting of the columns indexed by S will be written $A_S = (\mathbf{a}_j)_{j\in S}$. If $S \subset [N]$, then for $\mathbf{x} \in \mathbb{C}^N$ we denote by $\mathbf{x}_S \in \mathbb{C}^N$ the vector that coincides with $\mathbf{x}$ on S and is set to zero on $S^c = [N] \setminus S$. Similarly, $\mathbf{x}^S \in \mathbb{C}^S$ denotes the vector $\mathbf{x}$ restricted to the entries in S. The support of a vector is defined as $\mathrm{supp}\,\mathbf{x} = \{\ell, x_\ell \neq 0\}$. We write Id for the identity matrix. The complement of a set $S \subset [N]$ is denoted $S^c = [N] \setminus S$, while $|S|$ is its cardinality.

If $A \in \mathbb{C}^{m\times n}$, $m \geq n$, is of full rank (i.e. injective), then its Moore–Penrose pseudo-inverse is given by

$$A^\dagger = (A^*A)^{-1}A^*. \tag{2.7}$$

In this case, it satisfies $A^\dagger A = \mathrm{Id} \in \mathbb{C}^{n\times n}$. We refer to [8, 71, 59] for more information on the pseudo inverse.

All the constants appearing in this note – usually denoted by C or D – are universal, which means that they do not depend on any of the involved quantities.

2.2 Sparse Recovery

Let $\mathbf{x} \in \mathbb{C}^N$ be a (high-dimensional) vector that we will sometimes call signal. It is called s-sparse if

$$\|\mathbf{x}\|_0 := |\mathrm{supp}\,\mathbf{x}| \leq s. \tag{2.8}$$

The quantity $\|\cdot\|_0$ is often called ℓ_0-norm although it is actually not a norm, not even a quasi-norm.

In practice it is generally not realistic that a signal $\mathbf{x}$ is exactly s-sparse, but rather that its error of best s-term approximation $\sigma_s(\mathbf{x})_p$ is small,

$$\sigma_s(\mathbf{x})_p := \inf\{\|\mathbf{x} - \mathbf{z}\|_p, \mathbf{z} \text{ is } s\text{-sparse}\}. \tag{2.9}$$

(This is the standard notation in the literature, and we hope that no confusion with the singular values of a matrix will arise.)

Taking linear measurements of $\mathbf{x}$ is modeled as the application of a measurement matrix $A \in \mathbb{C}^{m\times N}$,

$$\mathbf{y} = A\mathbf{x}. \tag{2.10}$$

The vector $\mathbf{y} \in \mathbb{C}^m$ is called the measurement vector. We are interested in the case of undersampled measurements, that is, $m \ll N$. Reconstructing $\mathbf{x}$ amounts to solving (2.10). By basic linear algebra, this system of equations has infinitely many solutions (at least if A has full rank). Hence, it seems impossible at first sight to guess the correct $\mathbf{x}$ among these solutions. If, however, we impose the additional requirement (2.8) that $\mathbf{x}$ is s-sparse, the situation changes, as we will see. Intuitively, it is natural to search then for the solution with smallest support, that is, to solve the ℓ_0-minimization problem

$$\min_{\mathbf{z} \in \mathbb{C}^N} \|\mathbf{z}\|_0 \quad \text{subject to } A\mathbf{z} = \mathbf{y}. \tag{2.11}$$

The hope is that the solution $\mathbf{x}^{\#}$ of this optimization problem coincides with $\mathbf{x}$. Indeed, rather easy recovery conditions on $A \in \mathbb{C}^{m \times N}$ and on the sparsity s can be shown, see for instance [28]. There exist matrices $A \in \mathbb{C}^{m \times N}$ such that $2s \leq m$ suffices to always ensure recovery; choose the columns of A in general position.

Unfortunately, the combinatorial optimization problem (2.11) is NP hard in general [35, 88]. In other words, an algorithm that solves (2.11) for any matrix A and any vector $\mathbf{y}$ must be intractable (unless maybe the famous Millennium problem $P = NP$ is solved in the affirmative, on which we will not rely here). Therefore, (2.11) is completely impractical for applications and tractable alternatives have to be found. Essentially two approaches have mainly been pursued: greedy algorithms and convex relaxation. We will concentrate here on the latter and refer the reader to the literature [40, 58, 78, 90, 91, 103, 131, 127] for further information concerning greedy methods.

The ℓ_1-minimization problem

$$\min_{\mathbf{z} \in \mathbb{C}^N} \|\mathbf{z}\|_1 \quad \text{subject to } A\mathbf{z} = \mathbf{y} \tag{2.12}$$

can be understood as convex relaxation of (2.11). Sometimes (2.12) is also referred to as basis pursuit [25]. In contrast to (2.11), the ℓ_1-minimization problem can be solved with efficient convex optimization methods. In the real-valued case (2.12) can be rewritten as a linear program and can be solved with linear programming techniques, while in the complex-valued case (2.12) is equivalent to a second order cone program (SOCP), for which also efficient solvers exist [15]. We refer the interested reader to [32, 33, 34, 43, 47, 76] for further efficient algorithms for ℓ_1-minimization.

Of course, our hope is that the solution of (2.12) coincides with the one of (2.11). One purpose of these notes is to provide an understanding under which conditions this is actually guaranteed.

2.3 Null Space Property and Restricted Isometry Property

In this section we present conditions on the matrix A that ensure exact reconstruction of all s-sparse vectors using ℓ_1-minimization. Our first notion is the so-called null space property.

Definition 2.1. A matrix $A \in \mathbb{C}^{m \times N}$ satisfies the null space property of order s if for all subsets $S \subset [N]$ with $|S| = s$ it holds

$$\|\mathbf{v}_S\|_1 < \|\mathbf{v}_{S^c}\|_1 \quad \text{for all } \mathbf{v} \in \ker A \setminus \{0\}. \tag{2.13}$$

Remark 2.2. We deal here with the complex case. For real-valued matrices one might restrict the kernel to the real-valued vectors and define an obvious real-valued analogue of the null space property above. However, it is not obvious that the real and the complex null space property are the same for real-valued matrices. Nevertheless this fact can be shown [52].

Based on this notion we have the following recovery result concerning ℓ_1-minimization.

Theorem 2.3. *Let $A \in \mathbb{C}^{m \times N}$. Then every s-sparse vector $\mathbf{x} \in \mathbb{C}^N$ is the unique solution of the ℓ_1-minimization problem* (2.12) *with $\mathbf{y} = A\mathbf{x}$ if and only if A satisfies the null space property of order s.*

Proof. Assume first that every s-sparse vector $\mathbf{x} \in \mathbb{C}^N$ is the unique minimizer of $\|\mathbf{z}\|_1$ subject to $A\mathbf{z} = A\mathbf{x}$. Then, in particular, for any $\mathbf{v} \in \ker A \setminus \{0\}$ and any $S \subset [N]$ with $|S| = s$, the s-sparse vector $\mathbf{v}_S$ is the unique minimizer of $\|\mathbf{z}\|_1$ subject to $A\mathbf{z} = A\mathbf{v}_S$. Observe that $A(-\mathbf{v}_{S^c}) = A\mathbf{v}_S$ and $-\mathbf{v}_{S^c} \neq \mathbf{v}_S$, because $A(\mathbf{v}_{S^c} + \mathbf{v}_S) = A\mathbf{v} = 0$ and because $\mathbf{v} \neq 0$. Therefore we must have $\|\mathbf{v}_S\|_1 < \|\mathbf{v}_{S^c}\|_1$. This establishes the null space property.

For the converse, let us assume that the null space property of order s holds. Then, given an s-sparse vector $\mathbf{x} \in \mathbb{C}^N$ and a vector $\mathbf{z} \in \mathbb{C}^N$, $\mathbf{z} \neq \mathbf{x}$, satisfying $A\mathbf{z} = A\mathbf{x}$, we consider $\mathbf{v} := \mathbf{x} - \mathbf{z} \in \ker A \setminus \{0\}$ and $S := \operatorname{supp}(\mathbf{x})$. In view of the null space property we obtain

$$\begin{aligned} \|\mathbf{x}\|_1 &\leq \|\mathbf{x} - \mathbf{z}_S\|_1 + \|\mathbf{z}_S\|_1 = \|\mathbf{x}_S - \mathbf{z}_S\| + \|\mathbf{z}_S\|_1 = \|\mathbf{v}_S\|_1 + \|\mathbf{z}_S\|_1 \\ &< \|\mathbf{v}_{S^c}\|_1 + \|\mathbf{z}_S\|_1 = \|-\mathbf{z}_{S^c}\|_1 + \|\mathbf{z}_S\|_1 = \|\mathbf{z}\|_1. \end{aligned}$$

This establishes the required minimality of $\|\mathbf{x}\|_1$. □

This theorem seems to have first appeared explicitly in [60], although it was used implicitly already in [41, 48, 97]. The term null space property was coined by A. Cohen, W. Dahmen, and R. DeVore in [28]. One may obtain also a stable version of the above theorem by passing from sparse vectors to compressible ones, for which $\sigma_s(\mathbf{x})_1$ is small. Then the condition (2.13) has to be strengthened to $\|\mathbf{v}_S\|_1 < \gamma\|\mathbf{v}_{S^c}\|_1$ for some $\gamma \in (0, 1)$.

The null space property is usually somewhat difficult to show directly. Instead, the so-called restricted isometry property [22], which was introduced by E. Candès and T. Tao in [23] under the term uniform uncertainty principle (UUP), has become very popular in compressive sensing.

Definition 2.4. The restricted isometry constant δ_s of a matrix $A \in \mathbb{C}^{m \times N}$ is defined as the smallest δ_s such that

$$(1 - \delta_s)\|\mathbf{x}\|_2^2 \leq \|A\mathbf{x}\|_2^2 \leq (1 + \delta_s)\|\mathbf{x}\|_2^2 \tag{2.14}$$

for all s-sparse $\mathbf{x} \in \mathbb{C}^N$.

We say that a matrix A satisfies the restricted isometry property (RIP) if δ_s is small for reasonably large s (whatever "small" and "reasonably large" might mean in a concrete situation).

Before relating the restricted isometry property with the null space property let us first provide some simple properties of the restricted isometry constants.

Proposition 2.5. *Let $A \in \mathbb{C}^{m \times N}$ with isometry constants δ_s.*

(a) *The restricted isometry constants are ordered, $\delta_1 \leq \delta_2 \leq \delta_3 \leq \cdots$.*

(b) *It holds*

$$\begin{aligned}\delta_s &= \max_{S \subset [N], |S| \leq s} \|A_S^* A_S - \mathrm{Id}\|_{2 \to 2} \\ &= \sup_{\mathbf{x} \in T_s} |\langle (A^* A - \mathrm{Id})\mathbf{x}, \mathbf{x}\rangle|,\end{aligned}$$

where $T_s = \{\mathbf{x} \in \mathbb{C}^N, \|\mathbf{x}\|_2 = 1, \|\mathbf{x}\|_0 \leq s\}$.

(c) *Let $\mathbf{u}, \mathbf{v} \subset \mathbb{C}^N$ with disjoint supports, $\operatorname{supp} \mathbf{u} \cap \operatorname{supp} \mathbf{v} = \emptyset$. Let $s = |\operatorname{supp} \mathbf{u}| + |\operatorname{supp} \mathbf{v}|$. Then*

$$|\langle A\mathbf{u}, A\mathbf{v}\rangle| \leq \delta_s \|\mathbf{u}\|_2 \|\mathbf{v}\|_2.$$

Proof. Since an s-sparse vector is also $s+1$-sparse the statement (a) is immediate.

The definition (2.14) is equivalent to

$$|\|A\mathbf{x}\|_2^2 - \|\mathbf{x}\|_2^2| \leq \delta_s \|\mathbf{x}\|_2^2 \quad \text{for all } S \subset [N], |S| \leq s, \text{ for all } \mathbf{x} \in \mathbb{C}^N, \operatorname{supp} \mathbf{x} \subset S.$$

The term on the left hand side can be rewritten as $|\langle (A^*A - \mathrm{Id})\mathbf{x}, \mathbf{x}\rangle|$. Taking the supremum over all $\mathbf{x} \in \mathbb{C}^N$ with $\operatorname{supp} \mathbf{x} \subset S$ and unit norm $\|\mathbf{x}\|_2 = 1$ yields the operator norm $\|A_S^* A_S - \mathrm{Id}\|_{2 \to 2}$ by (2.4). Taking also the maximum over all subsets S of cardinality at most s completes the proof of (b).

For (c) we denote $S = \operatorname{supp} \mathbf{u}$, $\Xi = \operatorname{supp} \mathbf{v}$ and let $\tilde{\mathbf{u}}, \tilde{\mathbf{v}}$ denote the vectors $\mathbf{u}, \mathbf{v}$ restricted to their supports. Then we write

$$\begin{aligned}\langle A\mathbf{u}, A\mathbf{v}\rangle &= \tilde{\mathbf{u}}^* A_S^* A_\Xi \tilde{\mathbf{v}} = (\tilde{\mathbf{u}}^*, 0_\Xi^*) A_{S \cup \Xi}^* A_{S \cup \Xi} (0_S^*, \tilde{\mathbf{v}}^*)^* \\ &= (\tilde{\mathbf{u}}^*, 0_\Xi^*)(A_{S \cup \Xi}^* A_{S \cup \Xi} - \mathrm{Id})(0_S^*, \tilde{\mathbf{v}}^*)^*,\end{aligned}$$

where 0_S is the zero-vector on the indices in S. Therefore, one may estimate

$$|\langle A\mathbf{u}, A\mathbf{v}\rangle| \leq \|A_{S \cup \Xi}^* A_{S \cup \Xi} - \mathrm{Id}\|_{2 \to 2} \|\mathbf{u}\|_2 \|\mathbf{v}\|_2.$$

Applying part (b) completes the proof. □

Part (b) shows that the restricted isometry property requires in particular that all column submatrices of A of size s are well-conditioned. Indeed, all eigenvalues of $A_S^* A_S$ should be contained in the interval $[1-\delta_s, 1+\delta_s]$, which bounds the condition number of $A_S^* A_S$ by $\frac{1+\delta_s}{1-\delta_s}$ and therefore the one of A_S by $\sqrt{\frac{1+\delta_s}{1-\delta_s}}$.

The restricted isometry property implies the null space property as stated in the next theorem.

Theorem 2.6. *Suppose the restricted isometry constants δ_{2s} of a matrix $A \in \mathbb{C}^{m\times N}$ satisfies*

$$\delta_{2s} < \frac{1}{3}, \tag{2.15}$$

then the null space property of order s is satisfied. In particular, every s-sparse vector $\mathbf{x} \in \mathbb{C}^N$ is recovered by ℓ_1-minimization.

Proof. Let $\mathbf{v} \in \ker A$ be given. It is enough to consider an index set S_0 of s largest modulus entries of the vector $\mathbf{v}$. We partition the complement of S_0 as $S_0^c = S_1 \cup S_2 \cup \cdots$, where S_1 is an index set of s largest absolute entries of $\mathbf{v}$ in $[N] \setminus S_0$, S_2 is an index set of s largest absolute entries of $\mathbf{v}$ in $[N] \setminus (S_0 \cup S_1)$ etc. In view of $\mathbf{v} \in \ker A$, we have $A(\mathbf{v}_{S_0}) = -A(\mathbf{v}_{S_1} + \mathbf{v}_{S_2} + \cdots)$, so that

$$\begin{aligned}\|\mathbf{v}_{S_0}\|_2^2 &\le \frac{1}{1-\delta_{2s}}\|A(\mathbf{v}_{S_0})\|_2^2 = \frac{1}{1-\delta_{2s}}\langle A(\mathbf{v}_{S_0}), A(-\mathbf{v}_{S_1}) + A(-\mathbf{v}_{S_2}) + \cdots\rangle \\ &= \frac{1}{1-\delta_{2s}}\sum_{k\ge 1}\langle A(\mathbf{v}_{S_0}), A(-\mathbf{v}_{S_k})\rangle. \end{aligned}\tag{2.16}$$

Proposition 2.5(c) yields then

$$\langle A(\mathbf{v}_{S_0}), A(-\mathbf{v}_{S_k})\rangle \le \delta_{2s}\|\mathbf{v}_{S_0}\|_2\|\mathbf{v}_{S_k}\|_2. \tag{2.17}$$

Substituting (2.17) into (2.16) and dividing by $\|\mathbf{v}_{S_0}\|_2$ gives

$$\|\mathbf{v}_{S_0}\|_2 \le \frac{\delta_{2s}}{1-\delta_{2s}}\sum_{k\ge 1}\|\mathbf{v}_{S_k}\|_2.$$

Since the s entries of $\mathbf{v}_{S_k}$ do not exceed the s entries of $\mathbf{v}_{S_{k-1}}$ for $k \ge 1$, we have

$$|v_j| \le \frac{1}{s}\sum_{\ell\in S_{k-1}}|v_\ell| \quad \text{for all } j \in S_k$$

and therefore

$$\|\mathbf{v}_{S_k}\|_2 = \Big(\sum_{j\in S_k}|v_j|^2\Big)^{1/2} \le \frac{1}{\sqrt{s}}\|\mathbf{v}_{S_{k-1}}\|_1.$$

We obtain by the Cauchy–Schwarz inequality

$$\begin{aligned}\|\mathbf{v}_{S_0}\|_1 \leq \sqrt{s}\|\mathbf{v}_{S_0}\|_2 &\leq \frac{\delta_{2s}}{1-\delta_{2s}} \sum_{k\geq 1} \|\mathbf{v}_{S_{k-1}}\|_1 \\ &\leq \frac{\delta_{2s}}{1-\delta_{2s}}(\|\mathbf{v}_{S_0}\|_1 + \|\mathbf{v}_{S^c}\|_1) \end{aligned} \tag{2.18}$$

as announced. Since $\frac{\delta_{2s}}{1-\delta_{2s}} < 1/2$ by assumption, the null space property follows. □

The restricted isometry property also implies stable recovery by ℓ_1-minimization for vectors that can be well-approximated by sparse ones, and it further implies robustness under noise on the measurements. This fact was first noted in [23, 21]. The sufficient condition on the restricted isometry constants was successively improved in [18, 28, 53, 51]. We present without proof the so far best known result [51, 55] concerning recovery using a noise aware variant of ℓ_1-minimization.

Theorem 2.7. *Assume that the restricted isometry constant δ_{2s} of the matrix $A \in \mathbb{C}^{m\times N}$ satisfies*

$$\delta_{2s} < \frac{2}{3+\sqrt{7/4}} \approx 0.4627. \tag{2.19}$$

Then the following holds for all $\mathbf{x} \in \mathbb{C}^N$. Let noisy measurements $\mathbf{y} = A\mathbf{x} + \mathbf{e}$ be given with $\|\mathbf{e}\|_2 \leq \eta$. Let $\mathbf{x}^\#$ be the solution of

$$\min_{\mathbf{z}\in\mathbb{C}^N} \|\mathbf{z}\|_1 \quad \text{subject to } \|A\mathbf{z}-\mathbf{y}\|_2 \leq \eta. \tag{2.20}$$

Then

$$\|\mathbf{x}-\mathbf{x}^\#\|_2 \leq c\eta + d\frac{\sigma_s(\mathbf{x})_1}{\sqrt{s}}$$

for some constants $c, d > 0$ that depend only on δ_{2s}.

Note that the previous theorem ensures exact recovery of s-sparse signals using ℓ_1-minimization (2.12) under condition (2.19) in the noise-free case $\eta = 0$.

In contrast to the null space property, the restricted isometry property is not necessary for sparse recovery by ℓ_1-minimization. Indeed, the null space property of A is invariant under multiplication from the left with an invertible matrix $U \in \mathbb{C}^{m\times m}$ as this does not change the null space, while the restricted isometry property is certainly not invariant (simply take a matrix U with large condition number).

We will soon see examples of measurement matrices with small restricted isometry constants.

2.4 Recovery of Individual Vectors

We will later need also a condition ensuring sparse recovery which not only depends on the matrix A but also on the sparse vector $\mathbf{x} \in \mathbb{C}^N$ to be recovered. The following theorem is due to J. J. Fuchs [56] in the real-valued case and was extended to the complex-valued case by J. Tropp [128]. Its statement requires introducing the sign vector $\operatorname{sgn}(\mathbf{x}) \in \mathbb{C}^N$ having entries

$$\operatorname{sgn}(\mathbf{x})_j := \begin{cases} \frac{x_j}{|x_j|} & \text{if } x_j \neq 0, \\ 0 & \text{if } x_j = 0, \end{cases} \qquad j \in [N].$$

Theorem 2.8. *Let $A \in \mathbb{C}^{m \times N}$ and $\mathbf{x} \in \mathbb{C}^N$ with $S := \operatorname{supp}(\mathbf{x})$. Assume that A_S is injective and that there exists a vector $\mathbf{h} \in \mathbb{C}^m$ such that*

$$\begin{aligned} A_S^*\mathbf{h} &= \operatorname{sgn}(x^S), \\ |(A^*\mathbf{h})_\ell| &< 1, \quad \ell \in [N] \setminus S. \end{aligned} \tag{2.21}$$

Then $\mathbf{x}$ is the unique solution to the ℓ_1-minimization problem (2.12) *with $\mathbf{y} = A\mathbf{x}$.*

Proof. Let $\mathbf{h} \in \mathbb{C}^m$ be the vector with the described property. We have

$$\|\mathbf{x}\|_1 = \langle A^*\mathbf{h}, \mathbf{x}\rangle = \langle \mathbf{h}, A\mathbf{x}\rangle.$$

Thus, for $\mathbf{z} \in \mathbb{C}^N$, $\mathbf{z} \neq \mathbf{x}$, such that $A\mathbf{z} = \mathbf{y}$, we derive

$$\begin{aligned} \|\mathbf{x}\|_1 &= \langle \mathbf{h}, A\mathbf{z}\rangle = \langle A^*\mathbf{h}, \mathbf{z}\rangle = \langle A^*\mathbf{h}, \mathbf{z}_S\rangle + \langle A^*\mathbf{h}, \mathbf{z}_{S^c}\rangle \\ &\leq \|(A^*\mathbf{h})_S\|_\infty \|\mathbf{z}_S\|_1 + \|(A^*\mathbf{h})_{S^c}\|_\infty \|\mathbf{z}_{S^c}\|_1 < \|\mathbf{z}_S\|_1 + \|\mathbf{z}_{S^c}\|_1 = \|\mathbf{z}\|_1. \end{aligned}$$

The strict inequality follows from $\|\mathbf{z}_{S^c}\|_1 > 0$, which holds because otherwise the vector $\mathbf{z}$ would be supported on S and the equality $A\mathbf{z} = A\mathbf{x}$ would then be in contradiction with the injectivity of A_S. We have therefore shown that the vector $\mathbf{x}$ is the unique minimizer of $\|\mathbf{z}\|_1$ subject to $A\mathbf{z} = \mathbf{y}$, as desired. □

The above result makes clear that the success of sparse recovery by ℓ_1-minimization only depends on the support set S and on the sign pattern of the non-zero coefficients of $\mathbf{x}$.

Choosing the vector $h = (A_S^\dagger)^* \operatorname{sgn}(\mathbf{x}^S)$ leads to the following corollary, which will become a key tool later on.

Corollary 2.9. *Let $A \in \mathbb{C}^{m \times N}$ and $\mathbf{x} \in \mathbb{C}^N$ with $S := \operatorname{supp}(\mathbf{x})$. If the matrix A_S is injective and if*

$$|\langle A_S^\dagger \mathbf{a}_\ell, \operatorname{sgn}(\mathbf{x}^S)\rangle| < 1 \quad \textit{for all } \ell \in [N] \setminus S, \tag{2.22}$$

then the vector $\mathbf{x}$ is the unique solution to the ℓ_1-minimization problem (2.12) *with $\mathbf{y} = A\mathbf{x}$.*

Proof. The vector $h = (A_S^{\dagger})^* \mathrm{sgn}(\mathbf{x}^S)$ satisfies $A_S^* h = A_S^* A_S (A_S^* A_S)^{-1} \mathrm{sgn}(\mathbf{x}^S) = \mathrm{sgn}(\mathbf{x}^S)$, and the condition (2.22) translates into (2.21). Hence, the statement follows from Theorem 2.8. □

2.5 Coherence

A classical way to measure the quality of a measurement matrix A with normalized columns, $\|\mathbf{a}_j\|_2 = 1$, $j \in [N]$, is the coherence [39, 40, 60, 61, 127], defined by

$$\mu := \max_{j \neq k} |\langle \mathbf{a}_j, \mathbf{a}_k \rangle|.$$

If the coherence is small than the columns of A are almost mutually orthogonal. A small coherence is desired in order to have good sparse recovery properties.

A refinement of the coherence is the 1-coherence function or Babel function, defined by

$$\mu_1(s) := \max_{\ell \in [N]} \max_{\substack{S \subset [N] \setminus \{\ell\} \\ |S| \leq s}} \sum_{j \in S} |\langle \mathbf{a}_j, \mathbf{a}_\ell \rangle| \leq s\mu.$$

The following proposition lists simple properties of μ and μ_1 and relates the coherence to the restricted isometry constants.

Proposition 2.10. *Let $A \in \mathbb{C}^{m \times N}$ with unit norm columns, coherence μ, 1-coherence function $\mu_1(s)$ and restricted isometry constants δ_s. Then*

(a) $\mu = \delta_2$,

(b) $\mu_1(s) = \max_{S \subset [N], |S| \leq s+1} \|A_S^* A_S - \mathrm{Id}\|_{1 \to 1}$,

(c) $\delta_s \leq \mu_1(s-1) \leq (s-1)\mu$.

Proof. (a) If $S = \{j, \ell\}$ has cardinality two then

$$A_S^* A_S - \mathrm{Id} = \begin{pmatrix} 0 & \langle \mathbf{a}_j, \mathbf{a}_\ell \rangle \\ \langle \mathbf{a}_\ell, \mathbf{a}_j \rangle & 0 \end{pmatrix},$$

by the normalization $\|\mathbf{a}_j\|_2 = \|\mathbf{a}_\ell\|_2 = 1$. The operator norm of this matrix equals $|\langle \mathbf{a}_j, \mathbf{a}_\ell \rangle|$. Taking the maximum over all two element subsets S shows that $\delta_2 = \mu$ by Proposition 2.5(b).

(b) Again by normalization, the matrix $A_S^* A_S - \mathrm{Id}$ has zeros on the diagonal. The explicit expression (2.1) for the operator norm on ℓ_1 then yields

$$\|A_S^* A_S - \mathrm{Id}\|_{1 \to 1} = \max_{j \in S} \sum_{k \in S \setminus \{j\}} |\langle \mathbf{a}_j, \mathbf{a}_k \rangle|.$$

Taking also the maximum over all $S \subset [N]$ with $|S| \leq s+1$ gives

$$\begin{aligned}\max_{S\subset[N],|S|\leq s+1} \|A_S^* A_S - \mathrm{Id}\|_{1\to 1} &= \max_{S\subset[N],|S|\leq s+1} \max_{j\in S} \sum_{k\in S\setminus\{j\}} |\langle \mathbf{a}_j, \mathbf{a}_k\rangle| \\ &= \max_{j\in[N]} \max_{S\subset[N]\setminus\{j\},|S|\leq s} \sum_{k\in S} |\langle \mathbf{a}_j, \mathbf{a}_k\rangle| \\ &= \mu_1(s),\end{aligned}$$

which establishes (b).

For (c) observe that by Proposition 2.5(b) and inequality (2.3) for hermitian matrices

$$\begin{aligned}\delta_s = \max_{S\subset[N],|S|\leq s} \|A_S^* A_S - \mathrm{Id}\|_{2\to 2} &\leq \max_{S\subset[N],|S|\leq s} \|A_S^* A_S - \mathrm{Id}\|_{1\to 1} = \mu_1(s-1) \\ &\leq (s-1)\mu \qquad (2.23)\end{aligned}$$

by part (b). □

In combination with Theorem 2.6 (or Theorem 2.7) we see that $s - 1 \leq 1/(3\mu)$ or $\mu_1(s-1) \leq 1/3$ implies exact recovery (and also stable recovery) of all s-sparse vectors by ℓ_1-minimization. We note that the slightly weaker sufficient conditions

$$\mu_1(s-1) + \mu_1(s) < 1 \qquad (2.24)$$

or $(2s-1)\mu < 1$ ensuring recovery by ℓ_1-minimization can be shown by working directly with the coherence or the 1-coherence function [39, 60, 127, 129] instead of the restricted isometry constants. It is worth noting that (2.24) also implies recovery by the greedy algorithm (orthogonal) matching pursuit [127, 62].

A simple example of a matrix $A \in \mathbb{C}^{m\times 2m}$ with small coherence is a concatenation of the identity with a Fourier matrix $F \in \mathbb{C}^{m\times m}$, i.e., $A = (\mathrm{Id}|F)$, where the entries of F are given by

$$F_{j,k} = \frac{1}{\sqrt{m}} e^{2\pi i jk/m}.$$

It is well known that F is unitary and it is easy to see that $\mu = \frac{1}{\sqrt{m}}$ and $\mu_1(s) = \frac{s}{\sqrt{m}}$ for $s = 1, \dots, m-1$. It follows that

$$\delta_s \leq \frac{s-1}{\sqrt{m}}. \qquad (2.25)$$

Hence, if

$$s < \frac{\sqrt{m}}{6} + 1 \qquad (2.26)$$

then recovery by ℓ_1-minimization is ensured. There exist also matrices with many more columns still having coherence on the order $1/\sqrt{m}$. Indeed, [2, 121] give examples of matrices $A \in \mathbb{C}^{m \times m^2}$ satisfying

$$\mu = \frac{1}{\sqrt{m}}$$

(and one can also check that $\mu_1(s) = \frac{s}{\sqrt{m}}$ for $s = 1, \ldots, m-1$ for those matrices).

The drawback of these results is that the sparsity s is required to be tiny compared to the number m of measurements in (2.26). Or in other words, the number m of samples (measurements) required to recover an s-sparse vector scales quadratically in s. As we will see, there exists (random) matrices for which the quadratic scaling can be improved to a much better linear scaling (up to log-factors). However, such results cannot be obtained by analyzing the coherence or the 1-coherence function as follows from the lower bounds in the next theorem.

Theorem 2.11. *Let $A \in \mathbb{C}^{m \times N}$ with normalized columns, coherence μ and 1-coherence function $\mu_1(s)$. Then*

(a) $\mu \geq \sqrt{\frac{N-m}{m(N-1)}}$,

(b) $\mu_1(s) \geq s\sqrt{\frac{N-m}{m(N-1)}}$ *whenever* $s \leq \sqrt{N-1}$.

The inequality in part (a) is also called Welch bound and can be found in [121, 111]. The proof of Part (b) is contained in [119]. Note that the case $s > \sqrt{N-1}$ is of minor importance to us, since then $\mu_1(s) > \sqrt{N-1}\sqrt{\frac{N-m}{m(N-1)}} = \sqrt{\frac{N}{m} - 1}$ which will be larger than 1 provided $N \geq 2m$. The latter will be the case in all situations where compressive sensing is interesting. Then Proposition 2.10 implies only that $\delta_s \leq 1$, which does not allow any conclusion concerning ℓ_1-minimization.

For large enough N – say $N \geq 2m$ – the above lower bound for the coherence scales like $\frac{1}{\sqrt{m}}$, while the one for $\mu_1(s)$ scales like $\frac{s}{\sqrt{m}}$. Hence, those bounds explain to some extent why it is difficult to obtain significantly better recovery bounds than (2.26) for deterministic matrices. Indeed, the estimate (2.3) – or Gershgorin's theorem [8, 71, 135] that is often applied in the sparse approximation literature [127, 37] – which is used to establish Proposition 2.10(c), seems to be the optimal estimate one may obtain by taking into account only the absolute values of the Gramian matrix A^*A. In particular, it is *not* possible to improve on (2.25) by using Gershgorin's disc theorem, or by using Riesz–Thorin interpolation between $\|\cdot\|_{1\to 1}$ and $\|\cdot\|_{\infty\to\infty}$ (Schur's test).

Hence, to overcome the 'quadratic bottleneck' (2.25) or (2.26), that is, $m \geq Cs^2$, one should take into account cancellations that result from the signs of the entries of the Gramian A^*A. This task seems to be rather difficult, however, for deterministic matrices. The major breakthrough for beating the "quadratic bottleneck" was obtained using random matrices [19, 23, 38]. The problem of exploiting cancellations in the

Gramian matrix is handled much easier with probabilistic methods than with deterministic techniques. And indeed, it is presently still an open problem to come up with deterministic matrices offering the same performance guarantees for sparse recovery as the ones for random matrices we will see below.

2.6 Restricted Isometry Property of Gaussian and Bernoulli Random Matrices

By now, many papers deal with Gaussian or Bernoulli random matrices in connection with sparse recovery, or more generally, subgaussian random matrices, [5, 23, 38, 42, 87, 114, 116]. The entries of a random Bernoulli matrix take the value $+\frac{1}{\sqrt{m}}$ or $-\frac{1}{\sqrt{m}}$ with equal probability, while the entries of a Gaussian matrix are independent and follow a normal distribution with expectation 0 and variance $1/m$. With high probability such random matrices satisfy the restricted isometry property with a (near) optimal order in s, and therefore allow sparse recovery using ℓ_1-minimization.

Theorem 2.12. *Let $A \in \mathbb{R}^{m\times N}$ be a Gaussian or Bernoulli random matrix. Let $\epsilon, \delta \in (0,1)$ and assume*

$$m \geq C\delta^{-2}(s\ln(N/s) + \ln(\varepsilon^{-1})) \tag{2.27}$$

for a universal constant $C > 0$. Then with probability at least $1-\varepsilon$ the restricted isometry constant of A satisfies $\delta_s \leq \delta$.

There are by now several proofs of this result. In [5] a particularly nice and simple proof is given, which, however, yields an additional $\log(\delta^{-1})$-term. It shows in connection with Theorem 2.6 that with probability at least $1-\varepsilon$ all s-sparse vectors $\mathbf{x} \in \mathbb{C}^N$ can be recovered from $\mathbf{y} = A\mathbf{x}$ using ℓ_1-minimization (2.12) provided

$$m \geq C'(s\ln(N/s) + \ln(\varepsilon^{-1})). \tag{2.28}$$

Moreover, Theorem 2.7 predicts also stable and robust recovery under this condition. Note that choosing $\varepsilon = \exp(-cm)$ with $c = 1/(2C')$, we obtain that recovery by ℓ_1-minimization is successful with probability at least $1 - e^{-cm}$ provided

$$m \geq 2C's\ln(N/s). \tag{2.29}$$

This is the statement usually found in the literature.

The important point in the bound (2.29) is that the number of required samples only scales linearly in s up to the logarithmic factor $\ln(N/s)$ – in contrast to the quadratic scaling in the relation $m \geq 36(s-1)^2$ deduced from (2.26). Moreover, the ambient dimension N enters only very mildly into (2.29), and if N is large and s is rather small then m can be chosen significantly smaller than N and still allow for recovery by

ℓ_1-minimization. In particular, an s-sparse $\mathbf{x}$ can be reconstructed exactly although at first sight the available information seems highly incomplete.

Let us note that (2.29) is optimal as can be shown by using lower bounds for Gelfand widths of the ℓ_1^N ball [54, 57]. In particular, the factor $\ln(N/s)$ cannot be improved.

3 Structured Random Matrices

While Gaussian and Bernoulli matrices ensure sparse recovery via ℓ_1-minimization with the optimal bound (2.28) on the number of measurements, they are of somewhat limited use in applications for several reasons. Often the design of the measurement matrix is subject to physical or other constraints of the application, or it is actually given to us without having the freedom to design anything, and therefore it is often not justifiable that the matrix follows a Gaussian or Bernoulli distribution. Moreover, Gaussian or other unstructured matrices have the disadvantage that no fast matrix multiplication is available, which may speed up recovery algorithms significantly, so that large scale problems are not practicable with Gaussian or Bernoulli matrices. Even storing an unstructured matrix may be difficult.

From a computational and an application oriented view point it is desirable to have measurement matrices with structure. Since it is hard to rigorously prove good recovery conditions for deterministic matrices as outlined above, we will nevertheless allow randomness to come into play. This leads to the study of structured random matrices.

We will consider basically two types of structured random matrices. The larger part of these notes will be devoted to the recovery of randomly sampled functions that have a sparse expansion in terms of an orthonormal system $\{\psi_j, j = 1, \ldots, N\}$ with uniformly bounded L^∞-norm, $\sup_{j\in[N]} \|\psi_j\|_\infty = \sup_{j\in[N]} \sup_x |\phi_j(x)| \leq K$. The corresponding measurement matrix has entries $(\psi_j(t_\ell))_{\ell,j}$, where the t_ℓ are random sampling points. So the structure is determined by the function system ψ_j, while the randomness comes from the sampling locations.

The random partial Fourier matrix, which consists of randomly chosen rows of the discrete Fourier matrix can be seen as a special case of this setup and was studied already in the very first papers on compressive sensing [19, 23]. It is important to note that in this case the fast Fourier transform (FFT) algorithm can be used to compute a fast application of a partial Fourier matrix in $\mathcal{O}(N\log(N))$ operations [30, 59, 137] – to be compared with the usual $\mathcal{O}(mN)$ operations for a matrix vector multiply with an $m \times N$ matrix. Commonly, $m \geq Cs\log(N)$ in compressive sensing, so that an $\mathcal{O}(N\log(N))$ matrix multiply implies a substantial complexity gain.

The second type of structured random matrices we will study are partial random circulant and Toeplitz matrices. They arise in applications where convolutions are involved. Since circulant and Toeplitz matrices can be applied efficiently using again the FFT, they are also of interest for computationally efficient sparse recovery.

Other types of structured random matrices, that will not be discussed here in detail, are the following.

- **Random Gabor System.** On $\mathbb{C}^m$ a time-shift or translation is the circular shift operator $(T_k g)_j = g_{j-k \mod m}$, while a frequency shift is the modulation operator $(M_\ell g)_j = e^{2\pi i \ell j/m} g_j$. Now fix a vector g and construct a matrix $A = A_g \in \mathbb{C}^{m \times m^2}$ by selecting its columns as the time-frequency shifts $M_\ell T_k g \in \mathbb{C}^m$, $\ell, k \in [m]$. choose Here the entries of g are chosen independently and uniformly at random from the torus $\{\mathbf{z} \in \mathbb{C}, |\mathbf{z}| = 1\}$. Then $A = A_g$ is a structured random matrix called a random Gabor system. Corresponding sparse recovery results can be found in [95, 96].
- **Random Demodulator.** This type of random matrix is motivated by analog to digital conversion. We refer to [133] for details.

3.1 Nonuniform versus Uniform Recovery

Showing recovery results for ℓ_1-minimization in connection with structured random matrices is more delicate than for unstructured Gaussian matrices. Nevertheless, we will try to get as close to the recovery condition (2.28) as possible. We will not be able to obtain precisely this condition, but we will only suffer from a slightly larger log-term. Our recovery bounds will have the form

$$m \geq C s \log^\alpha(N/\varepsilon)$$

(or similar) for some $\alpha \geq 1$, where $\varepsilon \in (0, 1)$ corresponds to the probability of failure. In particular, the important linear scaling of m in s up to log-factors is retained.

We will pursue different strategies in order to come up with rigorous recovery results. In particular, we distinguish between uniform and nonuniform recovery guarantees. A uniform recovery guaranty means that once the random matrix is chosen, then with high probability all sparse signals can be recovered. A nonuniform recovery result states only that each fixed sparse signal can be recovered with high probability using a random draw of the matrix. In particular, such weaker results allow in principle that the small exceptional set of matrices for which recovery may fail is dependent on the signal, in contrast to a uniform statement. Clearly, uniform recovery implies nonuniform recovery, but the converse is not true.

It is usually easier to obtain nonuniform recovery results for structured random matrices, and the provable bounds on the maximal allowed sparsity (or on the minimal number of measurements) are usually slightly worse for uniform recovery.

Uniform recovery is clearly guaranteed once we prove that the restricted isometry property of a random matrix holds with high probability. Indeed, the corresponding Theorems 2.6 or 2.7 are purely deterministic and guarantee recovery of all s-sparse signals once the restricted isometry constant δ_{2s} of the measurement matrix is small enough.

In order to obtain nonuniform recovery results we will use the recovery condition for individual vectors, Corollary 2.9. If the signal is fixed then also its support is fixed, and hence, applying Corollary 2.9 means in the end that only a weaker property than the restricted isometry property has to be checked for the random matrix. In order to simplify arguments even further we can also choose the signs of the non-zero coefficients of the sparse vector at random.

4 Random Sampling in Bounded Orthonormal Systems

An important class of structured random matrices is connected with random sampling of functions in certain finite dimensional function spaces. We require an orthonormal basis of functions which are uniformly bounded in the L^∞-norm. The most prominent example consists of the trigonometric system [19, 102, 104, 78]. In a discrete setup, the resulting matrix is a random partial Fourier matrix, which actually was the first structured random matrix investigated in connection with compressive sensing [19, 23, 116].

4.1 Bounded Orthonormal Systems

Let $\mathcal{D} \subset \mathbb{R}^d$ be endowed with a probability measure ν. Further, let $\psi_1, \dots, \psi_N$ be an orthonormal system of complex-valued functions on $\mathcal{D}$, that is, for $j, k \in [N]$,

$$\int_{\mathcal{D}} \psi_j(t)\overline{\psi_k(t)}d\nu(t) = \delta_{j,k} = \begin{cases} 0 & \text{if } j \neq k, \\ 1 & \text{if } j = k. \end{cases} \tag{4.1}$$

The orthonormal system will be assumed to be uniformly bounded in L^∞,

$$\|\psi_j\|_\infty = \sup_{t \in \mathcal{D}} |\psi_j(t)| \leq K \quad \text{for all } j \in [N]. \tag{4.2}$$

The smallest value that the constant K can take is $K = 1$. Indeed,

$$1 = \int_{\mathcal{D}} |\psi_j(t)|^2 d\nu(x) \leq \sup_{t \in \mathcal{D}} |\psi_j(t)|^2 \int_{\mathcal{D}} d\nu(t) = K^2.$$

In the extreme case $K = 1$ we necessarily have $|\psi_j(t)| = 1$ for ν-almost all $t \in \mathcal{D}$.

Remark 4.1. (a) Note that *some* bound K can be found for most reasonable sets of functions ψ_j, $j \in [N]$. The crucial point of the boundedness condition (4.2) is that $K = \sup_{j \in [N]} \|\psi_j\|_\infty$ should ideally be independent of N, or at least depend only mildly on N, such as $K \leq C \ln^\alpha(N)$ for some $\alpha > 0$. Such a condition excludes for instance that the functions ψ_j are very localized in small regions of $\mathcal{D}$.

Expressed differently, the quotients $\|\psi_j\|_\infty / \|\psi_j\|_2$ should be uniformly bounded in j (in case that the functions ψ_j are not yet normalized); or at least grow only very slowly.

(b) It is not essential that $\mathcal{D}$ is a (measurable) subset of $\mathbb{R}^d$. This assumption was only made for convenience. In fact, $\mathcal{D}$ can be any measure space endowed with a probability measure ν.

We consider functions of the form

$$f(t) = \sum_{k=1}^{N} x_k \psi_k(t), \quad t \in \mathcal{D} \tag{4.3}$$

with coefficients $x_1, \dots, x_N \in \mathbb{C}$.

Let $t_1, \dots, t_m \in \mathcal{D}$ be some points and suppose we are given the sample values

$$y_\ell = f(t_\ell) = \sum_{k=1}^{N} x_k \psi_k(t_\ell), \quad \ell = 1, \dots, m.$$

Introducing the sampling matrix $A \in \mathbb{C}^{m \times N}$ with entries

$$A_{\ell,k} = \psi_k(t_\ell), \quad \ell = 1, \dots, m, \ k = 1, \dots, N, \tag{4.4}$$

the vector $\mathbf{y} = (y_1, \dots, y_m)^T$ of sample values (measurements) can be written in the form

$$\mathbf{y} = A\mathbf{x}, \tag{4.5}$$

where $\mathbf{x}$ is the vector of coefficients in (4.3).

Our task is to reconstruct the polynomial f – or equivalently its vector $\mathbf{x}$ of coefficients – from the vector of samples $\mathbf{y}$. We wish to perform this task with as few samples as possible. Without further knowledge this is clearly impossible if $m < N$. As common in compressive sensing we therefore assume sparsity.

A polynomial f of the form (4.3) is called s-sparse if its coefficient vector $\mathbf{x}$ is s-sparse. The problem of recovering an s-sparse polynomial from m sample values reduces then to solving (4.5) with a sparsity constraint, where A is the matrix in (4.4). We consider ℓ_1-minimization for this task.

Now we introduce randomness. We assume to this end that the sampling points $t_1, \dots, t_m$ are selected independently at random according to the probability measure ν. This means in particular that $\mathbb{P}(t_\ell \in B) = \nu(B)$, $\ell = 1, \dots, m$, for a measurable subset $B \subset \mathcal{D}$. The matrix A in (4.4) becomes then a structured random matrix.

Let us give examples of bounded orthonormal systems.

(i) **Trigonometric Polynomials.** Let $\mathcal{D} = [0, 1]$ and for $k \in \mathbb{Z}$ set

$$\psi_k(t) = e^{2\pi i k t}, \quad t \in [0, 1].$$

The probability measure ν is taken to be the Lebesgue measure on [0, 1]. Then for all $j, k \in \mathbb{Z}$,

$$\int_0^1 \psi_k(t)\overline{\psi_j(t)}dt = \delta_{j,k}. \tag{4.6}$$

The constant in (4.2) is clearly $K = 1$. For a subset $\Gamma \subset \mathbb{Z}$ of size N we then consider the trigonometric polynomials of the form

$$f(t) = \sum_{k \in \Gamma} x_k \psi_k(t) = \sum_{k \in \Gamma} x_k e^{2\pi i k t}.$$

A common choice is $\Gamma = \{-q, -q+1, \ldots, q-1, q\}$ resulting in trigonometric polynomials of degree at most q (then $N = 2q+1$). We emphasize, however, that an arbitrary choice of $\Gamma \subset \mathbb{Z}$ of size $|\Gamma| = N$ is possible. Introducing sparsity on the coefficient vector $\mathbf{x} \in \mathbb{C}^N$ then leads to the notion of s-sparse trigonometric polynomials.

The sampling points $t_1, \ldots, t_m$ will be chosen independently and uniformly at random from [0, 1]. The entries of the associated structured random matrix A are given by

$$A_{\ell,k} = e^{2\pi i k t_\ell}, \quad \ell = 1, \ldots, m, \ k \in \Gamma. \tag{4.7}$$

Such A is a Fourier type matrix, sometimes also called a nonequispaced Fourier matrix.

This example extends to multivariate trigonometric polynomials on $[0, 1]^d$, $d \in \mathbb{N}$. Indeed, the monomials $\psi_{\mathbf{k}}(t) = e^{2\pi i \langle \mathbf{k}, t \rangle}$, $\mathbf{k} \in \mathbb{Z}^d$, $t \in [0, 1]^d$, form an orthonormal system. For readers familiar with abstract harmonic analysis we mention that this example can be further generalized to characters of a compact commutative group. The corresponding measure will be the Haar measure of the group [50, 117].

The matrix A in (4.7) has a fast (approximate) matrix multiplication algorithm, called the nonequispaced fast Fourier transform (NFFT) [46, 101]. Similarly to the FFT, it has complexity $\mathcal{O}(N \log(N))$.

(ii) **Real Trigonometric Polynomials.** Instead of the complex exponentials above we may also take the real functions

$$\begin{aligned}
\psi_{2k}(t) &= \sqrt{2}\cos(2\pi k t), \quad k \in \mathbb{N}_0, \ \psi_0(t) = 1, \\
\psi_{2k+1}(t) &= \sqrt{2}\sin(2\pi k t), \quad k \in \mathbb{N}.
\end{aligned} \tag{4.8}$$

They also form an orthonormal system on [0, 1] with respect to the Lebesgue measure and the constant in (4.2) is $K = \sqrt{2}$. The samples $t_1, \ldots, t_m$ are chosen again according to the uniform distribution on [0, 1].

(iii) **Discrete Orthonormal Systems.** Let $U = (U_{tk}) \in \mathbb{C}^{N \times N}$ be a unitary matrix. The normalized columns $\sqrt{N}\mathbf{u}_k \in \mathbb{C}^N$, $k \in [N]$, then form an orthonormal system with respect to the discrete uniform probability measure on $[N]$, $\nu(B) = |B|/N$ for $B \subset [N]$; written out, this means

$$\frac{1}{N}\sum_{t=1}^{N} \sqrt{N}\mathbf{u}_k(t)\overline{\sqrt{N}\mathbf{u}_\ell(t)} = \langle \mathbf{u}_k, \mathbf{u}_\ell \rangle = \delta_{k,\ell}, \quad k, \ell \in [N].$$

Here, $\mathbf{u}_k(t) = U_{tk}$ denotes the tth entry of the kth column of U. The boundedness condition (4.2) requires that the normalized entries of U are bounded, i.e.,

$$\sqrt{N} \max_{k,t\in[N]} |U_{tk}| = \max_{k,t\in[N]} |\sqrt{N}\mathbf{u}_k(t)| \leq K. \tag{4.9}$$

Choosing the points $t_1, \ldots, t_m$ independently and uniformly at random from $[N]$ corresponds then to creating the random matrix A by selecting its rows independently and uniformly at random from the rows of $\sqrt{N}U$, that is,

$$A = \sqrt{N} R_T U,$$

where $R_T : \mathbb{C}^N \to \mathbb{C}^m$ denotes the random subsampling operator

$$(R_T \mathbf{z})_\ell = \mathbf{z}_{t_\ell}, \quad \ell = 1, \ldots, m. \tag{4.10}$$

Compressive sensing in this context yields the situation that only a small portion of the entries of $\tilde{\mathbf{y}} = \sqrt{N}U\mathbf{x} \in \mathbb{C}^N$ are observed of a sparse vector $\mathbf{x} \in \mathbb{C}^N$. In other words, $\mathbf{y} = R_T\tilde{\mathbf{y}} \in \mathbb{C}^m$, and we wish to recover $\mathbf{x}$ from the undersampled $\mathbf{y}$.

Note that it may happen with non-zero probability that a row of $\sqrt{N}U$ is selected more than once because the probability measure is discrete in this example. Hence, A is allowed to have repeated rows. One can avoid this effect by passing to a different probability model where the subset $\{t_1, \ldots, t_m\} \subset [N]$ is selected uniformly at random among all subsets of $[N]$ of cardinality m. This probability model requires a slightly different analysis than the model described above, and we refer to [19, 23, 20, 55, 116, 130] for more information. The difference between the two models, however, is very slight in practice and the corresponding recovery results are almost the same.

(iv) **Partial Discrete Fourier Transform.** Our next example uses the discrete Fourier matrix $F \in \mathbb{C}^{N\times N}$ with entries

$$F_{\ell,k} = \frac{1}{\sqrt{N}} e^{2\pi i \ell k/N}, \quad \ell, k = 1, \ldots, N. \tag{4.11}$$

It is well known (and easy to see) that F is unitary. The constant in (4.2) or (4.9) is clearly $K = 1$. The result $\hat{\mathbf{x}} = F\mathbf{x}$ of a applying F to a vector is called the Fourier transform of $\mathbf{x}$. Applying the setup of the previous example to this situation results in the problem of reconstructing a sparse vector $\mathbf{x}$ from m random entries of its Fourier transform $\hat{\mathbf{x}}$, that are independent and uniformly distributed on $\mathbb{Z}_N := \{\frac{k}{N}, k = 1, \ldots, N\}$. The resulting matrix A is called random partial Fourier matrix. Such a matrix can also be seen as a special case of the nonequispaced Fourier type matrix in (4.7) with the points t_ℓ being chosen from the grid $\mathbb{Z}_N$ instead of from the whole interval $[0, 1]$. Note that the discrete Fourier matrix in (4.11) can also be extended to higher dimensions, i.e., to grids $\mathbb{Z}_N^d$ for $d \in \mathbb{N}$.

A crucial point for applications is that the Fourier transform has a fast algorithm for matrix-vector multiplication, the so-called fast Fourier transform (FFT) [30, 137]. It computes the Fourier transform of a vector $\mathbf{x} \in \mathbb{C}^N$ in complexity $\mathcal{O}(N \log(N))$.

(v) **Incoherent Bases.** Let $V, W \in \mathbb{C}^{N\times N}$ be two unitary matrices. Their columns $(\mathbf{v}_\ell)_{\ell=1}^N$ and $(\mathbf{w}_\ell)_{\ell=1}^N$ form two orthonormal bases of $\mathbb{C}^N$. Assume that a vector $\mathbf{z} \in \mathbb{C}^N$ is sparse with respect to the basis $(\mathbf{v}_\ell)$ rather than the canonical basis, that is, $\mathbf{z} = V\mathbf{x}$ for a sparse vector $\mathbf{x}$. Further, assume that $\mathbf{z}$ is sampled with respect to the basis $(\mathbf{w}_\ell)$, i.e., we obtain measurements

$$y_k = \langle z, \mathbf{w}_{t_k}\rangle, \quad k = 1, \ldots, m$$

with $T := \{t_1, \ldots, t_m\} \subset [N]$. In matrix vector form this can be written as

$$\mathbf{y} = R_T W^* \mathbf{z} = R_T W^* V \mathbf{x},$$

where R_T is again the random sampling operator (4.10). Defining the unitary matrix $U = W^* V \in \mathbb{C}^{N\times N}$ we are back to the situation of the third example. The condition (4.9) now reads

$$\sqrt{N} \max_{\ell,k\in[N]} |\langle \mathbf{v}_\ell, \mathbf{w}_k\rangle| \leq K. \tag{4.12}$$

The quantity on the left hand side (without the $\sqrt{N}$) is known as the mutual coherence of the bases $(\mathbf{v}_\ell)$, $(\mathbf{w}_\ell)$, and they are called incoherent if K can be chosen small. The two previous examples also fall into this setting by choosing one of the bases as the canonical basis, $W = \mathrm{Id} \in \mathbb{C}^N$. The Fourier basis and the canonical basis are actually maximally incoherent, since then $K = 1$.

(vi) **Haar-Wavelets and Noiselets.** This example is a special case of the previous one, which is potentially useful for image processing applications. It is convenient to start with a continuous description of Haar-wavelets and noiselets [29], and then pass to the discrete setup via sampling. The Haar scaling function on $\mathbb{R}$ is defined as the characteristic function of the interval $[0, 1)$,

$$\phi(x) = \chi_{[0,1)}(x) = \begin{cases} 1 & \text{if } x \in [0, 1), \\ 0 & \text{otherwise.} \end{cases} \tag{4.13}$$

The Haar wavelet is then defined as

$$\psi(x) = \phi(2x) - \phi(2x-1) = \begin{cases} 1 & \text{if } x \in [0, 1/2), \\ -1 & \text{if } x \in [1/2, 1), \\ 0 & \text{otherwise.} \end{cases} \tag{4.14}$$

Further, denote

$$\psi_{j,k}(x) = 2^{j/2}\psi(2^j x - k), \quad \phi_k(x) = \phi(x-k), \quad x \in \mathbb{R},\ j \in \mathbb{Z},\ k \in \mathbb{Z}. \tag{4.15}$$

It is well known [138] (and can easily be seen) that, for $n \in \mathbb{N}$, the Haar-wavelet system

$$\Psi_n := \{\phi_k, k \in \mathbb{Z}\} \cup \{\psi_{j,k},\ k = 0, \ldots, 2^j - 1, j = 0, \ldots, n-1\} \tag{4.16}$$

forms an orthonormal basis of

$$V_n = \{f \in L^2([0,1]) : f \text{ is constant on } [k2^{-n}, (k+1)2^{-n}), k = 0, \dots, 2^n - 1\}.$$

Now let $N = 2^n$ for some $n \in \mathbb{N}$. Since the functions $\psi_{j,k}$, $j \le n-1$, are constant on intervals of the form $[2^{-n}k, 2^{-n}(k+1))$ we conclude that the vectors $\tilde{\phi}, \tilde{\psi}^{(j,k)} \in \mathbb{C}^N$, $j = 0, \dots, n-1, k = 0, \dots, 2^j - 1$, with entries

$$\tilde{\phi}_t = 2^{-n/2}\phi(t/N), \quad t = 0, \dots, N-1$$
$$\tilde{\psi}_t^{(j,k)} = 2^{-n/2}\psi_{j,k}(t/N), \quad t = 0, \dots, N-1$$

form an orthonormal basis of $\mathbb{C}^N$. We collect these vectors as the columns of a unitary matrix $\Psi \in \mathbb{C}^{N \times N}$.

Next we introduce the noiselet system on [0, 1]. Let $g_1 = \phi = \chi_{[0,1)}$ be the Haar scaling function and define, for $r \ge 1$, recursively the complex-valued functions

$$g_{2r}(x) = (1-i)g_r(2x) + (1+i)g_r(2x-1),$$
$$g_{2r+1}(x) = (1+i)g_r(2x) + (1-i)g_r(2x-1).$$

It is shown in [29] that the functions $\{2^{-n/2}g_r, r = 2^n, \dots, 2^{n+1} - 1\}$ form an orthonormal basis of V_n. The key property for us consists in the fact that they are maximally incoherent with respect to the Haar basis. Indeed, Lemma 10 in [29] states that

$$\left| \int_0^1 g_r(x)\psi_{j,k}(x)dx \right| = 1 \text{ provided } r \ge 2^j - 1, \quad 0 \le k \le 2^j - 1. \tag{4.17}$$

For the discrete noiselet basis on $\mathbb{C}^N$, $N = 2^n$, we take the vectors

$$\tilde{g}_t^{(r)} = 2^{-n} g_{N+r}(t/N), \quad r = 0, \dots, N-1, \ t = 0, \dots, N-1.$$

Again, since the functions $g_{N+r}, r = 0, \dots, N-1$, are constant on intervals of the form $[2^{-n}k, 2^{-n}(k+1))$ it follows that the vectors $\tilde{g}^{(r)}$, $r = 0, \dots, N-1$, form an orthonormal basis of $\mathbb{C}^N$. We collect these as columns into a unitary matrix $G \in \mathbb{C}^{N \times N}$. Due to (4.17) the unitary matrix $U = G^*\Psi \in \mathbb{C}^{N \times N}$ satisfies (4.9) with $K = 1$ – or in other words, the incoherence condition (4.12) for the Haar basis and the noiselet basis holds with the minimal constant $K = 1$.

Due to the their recursive definition, both the Haar wavelet transform and the noiselet transform, that is, the application of Ψ and G and their adjoints, come with a fast algorithm that computes a matrix vector multiply in $\mathcal{O}(N \log(N))$ time.

As a simple signal model, images or other types of signals are sparse in the Haar wavelet basis. The described setup corresponds to randomly sampling such functions with respect to noiselets. For more information on wavelets we refer to [27, 31, 83, 138].

(vii) **Legendre Polynomials**. The Legendre polynomials P_j are a system of orthogonal polynomials, where P_j is a polynomial of precise degree j, and orthonormality is with respect to the normalized Lebesgue measure $dx/2$ on $[-1,1]$. Their supremum norm is given by $\|P_j\|_\infty = \sqrt{2j+1}$, so considering the polynomials P_j, $j = 0,\ldots,N-1$, yields the constant $K = \sqrt{2N-1}$. Unfortunately, K grows therefore rather quickly with N. This problem can be avoided with a trick. One takes sampling points with respect to the "Chebyshev" measure $d\nu(x) = \pi^{-1}(1-x^2)^{-1/2}dx$ and uses a preconditioned measurement matrix. We refer to [106] for details.

Figure 1 shows an example of exact recovery of a 10-sparse vector in dimension 300 from 30 Fourier samples (example (iv) above) using ℓ_1-minimization. For comparison the reconstruction via ℓ_2-minimization is also shown.

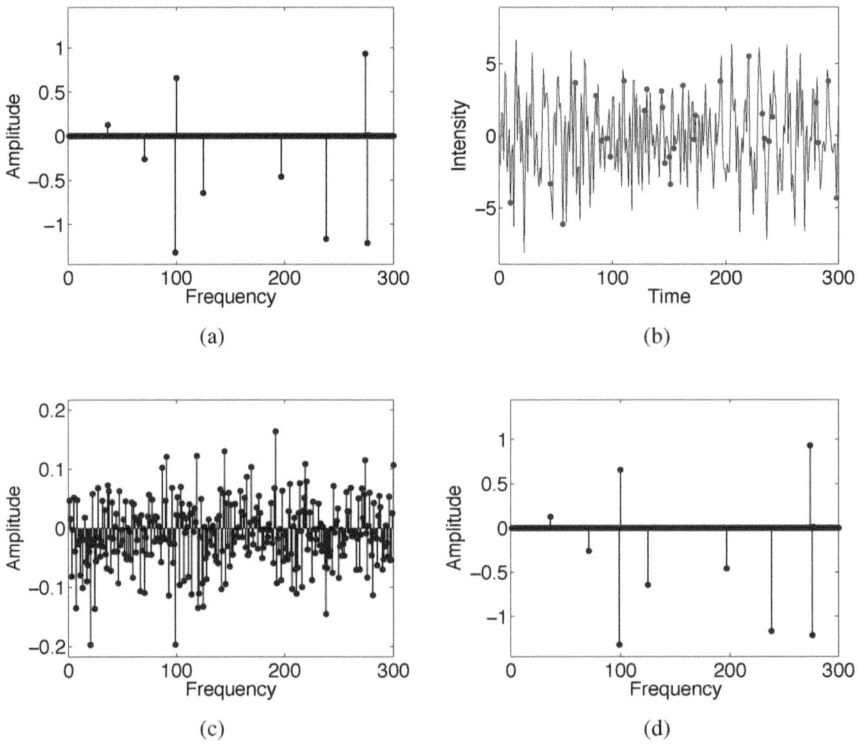

Figure 1. (a) 10-sparse Fourier spectrum, (b) time domain signal of length 300 with 30 samples, (c) reconstruction via ℓ_2-minimization, (d) exact reconstruction via ℓ_1-minimization.

4.2 Nonuniform Recovery

We start with a nonuniform recovery result that additionally assumes that the signs of the non-zero entries of the signal $\mathbf{x}$ are chosen at random.

Theorem 4.2. *Let $S \subset [N]$ be of cardinality $|S| = s$ and let $\boldsymbol{\epsilon} = (\epsilon_\ell)_{\ell \in S} \in \mathbb{C}^s$ be a sequence of independent random variables that take the values ± 1 with equal probability. Alternatively, the ϵ_ℓ may be uniformly distributed on the torus $\{\mathbf{z} \in \mathbb{C}, |\mathbf{z}| = 1\}$. Let $\mathbf{x}$ be an s-sparse vector with support S and $\operatorname{sgn}(\mathbf{x}^S) = \boldsymbol{\epsilon}$.*

Let $A \in \mathbb{C}^{m \times N}$ be the sampling matrix (4.4) *associated to an orthonormal system that satisfies the boundedness condition* (4.2) *for some constant $K \geq 1$. Assume that the random sampling points $t_1, \ldots, t_m$ are chosen independently and distributed according to the orthogonalization measure ν. Assume that*

$$m \geq C K^2 s \ln^2(6N/\varepsilon), \tag{4.18}$$

where $C \approx 26.25$. Set $\mathbf{y} = A\mathbf{x}$. Then with probability at least $1 - \varepsilon$ the vector $\mathbf{x}$ is the unique solution to the ℓ_1-minimization problem (2.12).

The proof will be contained in Chapter 7. With more effort (which we will not do here), the exponent 2 at the log-term in (4.18) can be replaced by 1. More precisely, one may obtain also the following sufficient recovery condition [55]

$$m \geq C_1 K^2 \max\{s, C_2 \ln(6N/\varepsilon)\} \ln(6N/\varepsilon) \tag{4.19}$$

with (reasonable) constants $C_1, C_2 > 0$. In the special case of a discrete orthonormal system (see example (iii) in the previous section), a version of Theorem 4.2 with recovery condition (4.19) was shown in [20] under a slightly different probability model.

The constants provided in (4.18) and (4.19) are likely not optimal. Numerical experiments suggest much better values. In the special case of the Fourier matrix (examples (i) and (iv) in the previous section) indeed slightly better constants are available [65, 102, 55]. However, we note that condition (4.19) gives an estimate that is valid for any possible support set S of size $|S| \leq s$. Clearly, it is impossible to test all such subsets numerically. So only limited conclusions on the constants in (4.19) and (4.18) can be drawn from numerical experiments.

In case of random sampling in the Fourier system (examples (i) and (iv) in the previous section) the assumption of randomness in the sign pattern of the non-zero entries of $\mathbf{x}$ can be removed [19, 102].

Theorem 4.3. *Let $\mathbf{x} \in \mathbb{C}^N$ be s-sparse. Assume A is the random Fourier type matrix* (4.7) *or the random partial Fourier matrix of example* (iv) *above. If*

$$m \geq C s \log(N/\varepsilon),$$

then $\mathbf{x}$ is the unique solution of the ℓ_1-minimization problem (2.12) *with probability at least $1 - \varepsilon$.*

The techniques of the proof of this theorem [19, 102] heavily use the algebraic structure of the Fourier system and do not easily extend to general bounded orthonormal systems. In fact, the general case is still open.

4.3 Uniform Recovery

Our main theorem concerning the recovery of sparse polynomials in bounded orthonormal systems from random samples reads as follows.

Theorem 4.4. *Let $A \in \mathbb{C}^{m \times N}$ be the sampling matrix* (4.4) *associated to an orthonormal system that satisfies the boundedness condition* (4.2) *for some constant $K \geq 1$. Assume that the random sampling points $t_1, \ldots, t_m$ are chosen independently at random according to the orthogonalization measure ν. Suppose*

$$\frac{m}{\ln(m)} \geq CK^2 s \ln^2(s) \ln(N), \tag{4.20}$$

$$m \geq DK^2 s \ln(\varepsilon^{-1}), \tag{4.21}$$

where $C, D > 0$ are some universal constants. Then with probability at least $1 - \varepsilon$ every s-sparse vector $\mathbf{x} \in \mathbb{C}^N$ is recovered from the samples

$$\mathbf{y} = A\mathbf{x} = \left(\sum_{j=1}^{N} x_j \phi_j(t_\ell) \right)_{\ell=1}^{m}$$

by ℓ_1-minimization (2.12).

Moreover, with probability at least $1 - \varepsilon$ the following holds for every $\mathbf{x} \in \mathbb{C}^N$. Let noisy samples $\mathbf{y} = A\mathbf{x} + \mathbf{e}$ with

$$\|\mathbf{e}\|_2 = \sqrt{\sum_{\ell=1}^{m} |e_\ell|^2} \leq \eta\sqrt{m}$$

be given and let $\mathbf{x}^*$ be the solution of the ℓ_1-minimization problem (2.20), *where η is replaced by $\eta\sqrt{m}$. Then*

$$\|\mathbf{x} - \mathbf{x}^*\|_2 \leq c\frac{\sigma_s(\mathbf{x})_1}{\sqrt{s}} + d\eta$$

for suitable constants $c, d > 0$.

This result is proven in Chapter 8 by estimating the restricted isometry constants δ_s of A. Thereby, also explicit constants are provided, see Theorem 8.4. The reader will notice that its proof is considerably more involved than the one of Theorem 4.2.

Remark 4.5. We may choose ε such that there is equality in (4.21). Then condition (4.20) implies recovery with probability at least

$$1 - N^{-\gamma \ln(m) \ln^2(s)}$$

where $\gamma = C/D$. A condition that is easier to remember is derived by noting that $s \le N$ and $m \le N$ (otherwise, we are not in the range of interest for compressive sensing). Indeed,

$$m \ge CK^2 s \ln^4(N) \tag{4.22}$$

implies recovery by ℓ_1-minimization with probability at least $1 - N^{-\gamma \ln^3(N)}$.

E. Candès and T. Tao [23] obtained the sufficient condition (4.22) in case of the random partial Fourier matrix with an exponent 6 instead of 4 at the $\ln(N)$ term. M. Rudelson and R. Vershynin [116] improved this to an exponent 5 at the $\ln(N)$-term; or alternatively to an exponent 4 for constant probability ε, see also Theorem 8.1 below. The condition (4.22) with exponent 4 and super-polynomially decreasing failure probability $N^{-\gamma \ln(N)^3}$ is presently the best known result. (In the Fourier case this is already contained in the proof of the main result in [104], but the author did not realize at that time that this was actually a slight improvement over the estimate of Rudelson and Vershynin in [116].) Our proof in Chapter 8 follows the ideas of Rudelson and Vershynin in [116] with some modifications and the mentioned improvements.

5 Partial Random Circulant Matrices

This chapter will be devoted to a different type of structured random matrices, which are important in applications such as wireless communications and radar, see [4, 68, 110]. We will study partial random circulant matrices and partial random Toeplitz matrices. Presently, there are less recovery results available than for the structured random matrices in the preceding chapter. In particular, a good estimate for the restricted isometry constants is still under investigation at the time of writing. (The estimates in [4, 68] only provide a quadratic scaling of the number of measurements in terms of the sparsity, similar to (2.26).) Therefore, we will only be able to present a nonuniform recovery result in the spirit of Theorem 4.2, which is a slight improvement of the main result in [105]. We believe that the mathematical approach to its proof should be of interest on its own.

We consider the following measurement matrices. For $\mathbf{b} = (b_0, b_1, \dots, b_{N-1}) \in \mathbb{C}^N$ we define its associated circulant matrix $\Phi = \Phi(\mathbf{b}) \in \mathbb{C}^{N \times N}$ by setting

$$\Phi_{k,j} = b_{j-k \bmod N}, \quad k, j = 1, \dots, N.$$

Note that the application of Φ to a vector is the convolution,

$$(\Phi \mathbf{x})_j = (\mathbf{x} * \tilde{\mathbf{b}})_j = \sum_{\ell=1}^{N} x_\ell \tilde{b}_{j-\ell \bmod N},$$

where $\tilde{b}_j = b_{N-j}$. Similarly, for a vector $\mathbf{c} = (c_{-N+1}, c_{-N+2}, \ldots, c_{N-1})$ its associated Toeplitz matrix $\Psi = \Psi(\mathbf{c}) \in \mathbb{C}^{N \times N}$ has entries $\Psi_{k,j} = c_{j-k}$, $k, j = 1, \ldots, N$.

Now we choose an arbitrary subset $\Theta \subset [N]$ of cardinality $m < N$ and let the partial circulant matrix $\Phi^\Theta = \Phi^\Theta(\mathbf{b}) \in \mathbb{C}^{m \times N}$ be the submatrix of Φ consisting of the rows indexed by Θ. The partial Toeplitz matrix $\Psi^\Theta = \Psi^\Theta(\mathbf{c}) \in \mathbb{C}^{m \times N}$ is defined similarly. For the purpose of this exposition we will choose the vectors $\mathbf{b}$ and $\mathbf{c}$ as Rademacher sequences, that is, the entries of $\mathbf{b}$ and $\mathbf{c}$ are independent random variables that take the value ± 1 with equal probability. Standard Gaussian vectors or Steinhaus sequences (independent random variables that are uniformly distributed on the complex torus) are possible as well.

It is important from a computational viewpoint that circulant matrices can be diagonalized using the discrete Fourier transform [59]. Therefore, there is a fast matrix vector multiplication algorithm for partial circulant matrices of complexity $\mathcal{O}(N \log(N))$ that uses the FFT. Since Toeplitz matrices can be seen as submatrices of circulant matrices [59], this remark applies to partial Toeplitz matrices as well.

Of particular interest is the case $N = mL$ with $L \in \mathbb{N}$ and $\Theta = \{L, 2L, \ldots, mL\}$. Then the application of $\Phi^\Theta(\mathbf{b})$ and $\Psi^\Theta(\mathbf{c})$ corresponds to (periodic or non-periodic) convolution with the sequence $\mathbf{b}$ (or $\mathbf{c}$, respectively) followed by a downsampling by a factor of L. This setting was studied numerically in [132] by J. Tropp et al. (using orthogonal matching pursuit instead of ℓ_1-minimization). Also of interest is the case $\Theta = [m]$ which was investigated in [4, 68].

Since Toeplitz matrices can be embedded into circulant matrices as just mentioned, we will deal only with the latter in the following. The result below (including its proof) holds without a difference (and even with the same constants) for Toeplitz matrices as well. Similarly to Theorem 4.2 we deal with nonuniform recovery, where additionally the signs $x_j/|x_j|$ of the non-zero coefficients of the vector x to be recovered are chosen at random.

Theorem 5.1. *Let $\Theta \subset [N]$ be an arbitrary (deterministic) set of cardinality m. Let $\mathbf{x} \in \mathbb{C}^N$ be s-sparse such that the signs of its non-zero entries form a Rademacher or Steinhaus sequence. Choose $\mathbf{b} = \boldsymbol{\epsilon} \in \mathbb{R}^N$ to be a Rademacher sequence. Let $\mathbf{y} = \Phi^\Theta(\boldsymbol{\epsilon})\mathbf{x} \in \mathbb{C}^m$. Then*

$$m \geq 57 s \ln^2(17N^2/\varepsilon) \tag{5.1}$$

implies that with probability at least $1 - \varepsilon$ the vector $\mathbf{x}$ is the unique solution to the ℓ_1-minimization problem (2.12).

The proof of Theorem 5.1 will be presented in Chapter 9. We note that the exponent 2 of the log-term in (5.1) is a slight improvement over an exponent 3 present in the main result of [105]. The constant 57 is very likely not optimal. With the much more technical (and combinatorial) approach of [19, 102, 95] we expect that the randomness in the signs can be removed and the exponent 2 at the log-factor can be improved to 1.

6 Tools from Probability Theory

The proofs of the results presented in the two previous chapters will require tools from probability theory that might not be part of an introductory course on probability. This chapter collects the necessary background. We will only assume familiarity of the reader with basic probability theory that can be found in most textbooks on the subject, see for instance [63, 112].

In the following we discuss the relation of moments and tail estimates, symmetrization, decoupling, and scalar and noncommutative Khintchine inequalities. The latter represent actually a very powerful tool that presently does not seem to be widely acknowledged. Furthermore, we present Dudley's inequality on the expectation of the supremum of a subgaussian process. Much more material of a similar flavor can be found in the monographs [36, 73, 79, 80, 125, 134].

6.1 Basics on Probability

In this section we recall some important facts from basic probability theory. Let $(\Omega, \Sigma, \mathbb{P})$ be a probability space, where Σ denotes a σ-algebra on the sample space Ω and $\mathbb{P}$ a probability measure on (Ω, Σ). The probability of an event $B \in \Sigma$ is denoted by

$$\mathbb{P}(B) = \int_B d\mathbb{P}(\omega) = \int_\Omega I_B(\omega) d\mathbb{P}(\omega),$$

where the characteristic function $I_B(\omega)$ takes the value 1 if $\omega \in B$ and 0 otherwise. The union bound (or Bonferroni's inequality, or Boole's inequality) states that for a collection of events $B_\ell \in \Sigma$, $\ell = 1, \ldots, n$, we have

$$\mathbb{P}\left(\bigcup_{\ell=1}^{n} B_\ell\right) \le \sum_{\ell=1}^{n} \mathbb{P}(B_\ell). \tag{6.1}$$

We assume knowledge of basic facts on random variables. The expectation or mean of a random variable X is denoted by

$$\mathbb{E}X = \int_\Omega X(\omega) d\mathbb{P}(\omega).$$

The quantities $\mathbb{E}|X|^p$, $0 < p < \infty$, are called (absolute) moments. For $1 \le p < \infty$, $(\mathbb{E}|\mathbf{X}|^p)^{1/p}$ defines a norm on the $L^p(\Omega, \mathbb{P})$-space of all p-integrable random variables, in particular, the triangle inequality

$$(\mathbb{E}|X + Y|^p)^{1/p} \le (\mathbb{E}|X|^p)^{1/p} + (\mathbb{E}|Y|^p)^{1/p} \tag{6.2}$$

holds for $X, Y \in L^p(\Omega, \mathbb{P}) = \{X \text{ measurable}\,, \mathbb{E}|X|^p < \infty\}$.

Let $p, q \ge 1$ with $1/p + 1/q = 1$, Hölder's inequality states that $|\mathbb{E}XY| \le (\mathbb{E}|X|^p)^{1/p}(\mathbb{E}|Y|^q)^{1/q}$ for random variables X, Y. The special case $p = q = 2$

is the Cauchy–Schwarz inequality. It follows from Hölder's inequality that for all $0 < p \le q < \infty$,

$$(\mathbb{E}|X|^p)^{1/p} \le (\mathbb{E}|X|^q)^{1/q}. \tag{6.3}$$

Absolute moments can be computed by means of the following formula.

Proposition 6.1. *The absolute moments of a random variable X can be expressed as*

$$\mathbb{E}|X|^p = p\int_0^\infty \mathbb{P}(|X| \ge t)\, t^{p-1}dt, \quad p > 0.$$

Proof. Recall that $I_{\{|X|^p \ge x\}}$ is the random variable that takes the value 1 on the event $|X|^p \ge x$ and 0 otherwise. Using Fubini's theorem we derive

$$\begin{aligned}
\mathbb{E}|X|^p &= \int_\Omega |X|^p d\mathbb{P} = \int_\Omega \int_0^{|X|^p} dx d\mathbb{P} = \int_\Omega \int_0^\infty I_{\{|X|^p \ge x\}} dx d\mathbb{P} \\
&= \int_0^\infty \int_\Omega I_{\{|X|^p \ge x\}} d\mathbb{P} dx = \int_0^\infty \mathbb{P}(|X|^p \ge x) dx \\
&= p\int_0^\infty \mathbb{P}(|X|^p \ge t^p) t^{p-1} dt = p\int_0^\infty \mathbb{P}(|X| \ge t) t^{p-1} dt,
\end{aligned}$$

where we also applied a change of variables. □

The function $t \mapsto \mathbb{P}(|X| \ge t)$ is called the tail of X. The Markov inequality is a simple way of estimating a tail.

Theorem 6.2 (Markov inequality). *Let X be a random variable. Then*

$$\mathbb{P}(|X| \ge t) \le \frac{\mathbb{E}|X|}{t} \quad \textit{for all } t > 0.$$

Proof. Note that $\mathbb{P}(|X| \ge t) = \mathbb{E}I_{\{|X| \ge t\}}$ and $tI_{\{|X| \ge t\}} \le |X|$. Hence, $t\mathbb{P}(|X| \ge t) = \mathbb{E}tI_{\{|X| \ge t\}} \le \mathbb{E}|X|$ and the proof is complete. □

A random vector $\mathbf{X} = (X_1, \ldots, X_n)^T \in \mathbb{R}^n$ is a collection of n random variables X_ℓ on a common probability space. Its expectation is the vector

$$\mathbb{E}\mathbf{X} = (\mathbb{E}X_1, \ldots, \mathbb{E}X_n)^T \in \mathbb{R}^n.$$

A complex random vector $\mathbf{Z} = \mathbf{X} + i\mathbf{Y} \in \mathbb{C}^n$ is a special case of a $2n$-dimensional real random vector $(\mathbf{X}, \mathbf{Y}) \in \mathbb{R}^{2n}$.

A collection of random vectors $\mathbf{X}_1, \ldots, \mathbf{X}_N \in \mathbb{C}^n$ is called (stochastically) independent if for all measurable subsets $B_1, \ldots, B_N \subset \mathbb{C}^n$,

$$\mathbb{P}(\mathbf{X}_1 \in B_1, \mathbf{X}_2 \in B_2, \ldots, \mathbf{X}_N \in B_N) = \mathbb{P}(\mathbf{X}_1 \in B_1)\mathbb{P}(\mathbf{X}_2 \in B_2) \cdots \mathbb{P}(\mathbf{X}_N \in B_N).$$

Functions of independent random vectors are again independent. A random vector $\mathbf{X}'$ in $\mathbb{C}^n$ will be called an independent copy of $\mathbf{X}$ if $\mathbf{X}$ and $\mathbf{X}'$ are independent and have the same distribution, that is, $\mathbb{P}(\mathbf{X} \in B) = \mathbb{P}(\mathbf{X}' \in B)$ for all measurable $B \subset \mathbb{C}^n$.

Jensen's inequality reads as follows.

Theorem 6.3 (Jensen's inequality). *Let $f : \mathbb{C}^n \to \mathbb{R}$ be a convex function, and let $\mathbf{X} \in \mathbb{C}^n$ be a random vector. Then*

$$f(\mathbb{E}\mathbf{X}) \le \mathbb{E} f(\mathbf{X}). \tag{6.4}$$

Finally, we state the Borel–Cantelli lemma.

Lemma 6.4 (Borel–Cantelli). *Let $A_1, A_2, \ldots \in \Sigma$ be events and let*

$$A^* = \limsup_{n\to\infty} A_n = \cap_{n=1}^{\infty} \cup_{m=n}^{\infty} A_m.$$

If $\sum_{n=1}^{\infty} \mathbb{P}(A_n) < \infty$ then $\mathbb{P}(A^*) = 0$.

Proof. Since $A^* \subset \bigcup_{m=n}^{\infty} A_m$ for all n, it holds $\mathbb{P}(A^*) \le \sum_{m=n}^{\infty} \mathbb{P}(A_m) \to 0$ as $n \to \infty$ whenever $\sum_{n=1}^{\infty} \mathbb{P}(A_n) < \infty$. □

6.2 Moments and Tails

It will be very crucial for us that tails of random variables can be estimated by means of their moments. The next statement is rather simple but very effective, see also [130].

Proposition 6.5. *Suppose Z is a random variable satisfying*

$$(\mathbb{E}|Z|^p)^{1/p} \le \alpha\beta^{1/p} p^{1/\gamma} \quad \textit{for all } p \ge p_0$$

for some constants $\alpha, \beta, \gamma, p_0 > 0$. Then

$$\mathbb{P}(|Z| \ge e^{1/\gamma}\alpha u) \le \beta e^{-u^\gamma/\gamma}$$

for all $u \ge p_0^{1/\gamma}$.

Proof. By Markov's inequality, Theorem 6.2, we obtain for an arbitrary $\kappa > 0$

$$\mathbb{P}(|Z| \ge e^\kappa \alpha u) \le \frac{\mathbb{E}|Z|^p}{(e^\kappa \alpha u)^p} \le \beta\left(\frac{\alpha p^{1/\gamma}}{e^\kappa \alpha u}\right)^p.$$

Choose $p = u^\gamma$ and the optimal value $\kappa = 1/\gamma$ to obtain the claim. □

Also a converse of the above proposition can be shown [55, 80]. Important special cases are $\gamma = 1, 2$. In particular, if $(\mathbb{E}|Z|^p)^{1/p} \le \alpha\beta^{1/p}\sqrt{p}$ for all $p \ge 2$ then Z satisfies the subgaussian tail estimate.

$$\mathbb{P}(|Z| \ge e^{1/2}\alpha u) \le \beta e^{-u^2/2} \quad \text{for all } u \ge \sqrt{2}. \tag{6.5}$$

For random variables satisfying a subgaussian tail estimate, the following useful estimate of the expectation of their maximum can be shown [80].

Lemma 6.6. *Let $X_1, \ldots, X_M$ be random variables satisfying*

$$\mathbb{P}(|X_\ell| \geq u) \leq \beta e^{-u^2/2} \quad \text{for } u \geq \sqrt{2},\ \ell = 1, \ldots, M,$$

for some $\beta \geq 1$. Then

$$\mathbb{E} \max_{\ell=1,\ldots,M} |X_\ell| \leq C_\beta \sqrt{\ln(4\beta M)}$$

with $C_\beta \leq \sqrt{2} + \frac{1}{4\sqrt{2}\ln(4\beta)}$.

Proof. According to Proposition 6.1 we have, for some $\alpha \geq \sqrt{2}$,

$$\begin{aligned}
\mathbb{E} \max_{\ell=1,\ldots,M} |X_\ell| &= \int_0^\infty \mathbb{P}\Big(\max_{\ell=1,\ldots,M} |X_\ell| > u\Big) du \\
&\leq \int_0^\alpha 1 du + \int_\alpha^\infty \mathbb{P}\Big(\max_{\ell=1,\ldots,M} |X_\ell| > u\Big) du \\
&\leq \alpha + \int_\alpha^\infty \sum_{\ell=1}^M \mathbb{P}(|X_\ell| > u) du \\
&\leq \alpha + M\beta \int_\alpha^\infty e^{-u^2/2} du.
\end{aligned}$$

In the second line we have applied the union bound. Using Proposition 10.2 in the Appendix we obtain

$$\mathbb{E} \max_{\ell=1,\ldots,M} |X_\ell| \leq \alpha + \frac{M\beta}{\alpha} e^{-\alpha^2/2}.$$

Now we choose $\alpha = \sqrt{2\ln(4\beta M)} \geq \sqrt{2\ln(4)} \geq \sqrt{2}$. This yields

$$\begin{aligned}
\mathbb{E} \max_{\ell=1,\ldots,M} |X_\ell| &\leq \sqrt{2\ln(4\beta M)} + \frac{1}{4\sqrt{2\ln(4\beta M)}} \\
&= \left(\sqrt{2} + \frac{1}{4\sqrt{2}\ln(4\beta M)}\right) \sqrt{\ln(4\beta M)} \leq C_\beta \sqrt{\ln(4\beta M)}
\end{aligned}$$

by our choice of C_β. The proof is completed. □

6.3 Rademacher Sums and Symmetrization

A Rademacher variable is presumably the simplest random variable. It takes the values $+1$ or -1, each with probability $1/2$. A sequence $\boldsymbol{\epsilon}$ of independent Rademacher variables ϵ_j, $j = 1, \ldots, M$, is called a Rademacher sequence. The technique of symmetrization leads to so-called Rademacher sums $\sum_{j=1}^M \epsilon_j x_j$ where the x_j are scalars, vectors or matrices. Although quite simple, symmetrization is very powerful because there are nice estimates for Rademacher sums available – the so-called Khintchine inequalities to be treated later on.

A random vector $\mathbf{X}$ is called symmetric, if $\mathbf{X}$ and $-\mathbf{X}$ have the same distribution. In this case $\mathbf{X}$ and $\epsilon\mathbf{X}$, where ϵ is a Rademacher variable independent of $\mathbf{X}$, have the same distribution as well.

The following lemma, see also [80, 36], is the key to symmetrization.

Lemma 6.7 (Symmetrization). *Assume that* $\boldsymbol{\xi} = (\xi_j)_{j=1}^M$ *is a sequence of independent random vectors in* $\mathbb{C}^n$ *equipped with a (semi-)norm* $\|\cdot\|$, *having expectations* $\mathbf{x}_j = \mathbb{E}\xi_j$. *Then for* $1 \le p < \infty$

$$\left(\mathbb{E}\left\|\sum_{j=1}^M (\xi_j - \mathbf{x}_j)\right\|^p\right)^{1/p} \le 2\left(\mathbb{E}\left\|\sum_{j=1}^M \epsilon_j \xi_j\right\|^p\right)^{1/p}, \tag{6.6}$$

where $\boldsymbol{\epsilon} = (\epsilon_j)_{j=1}^N$ *is a Rademacher sequence independent of* $\boldsymbol{\xi}$.

Proof. Let $\boldsymbol{\xi}' = (\xi'_1, \dots, \xi'_M)$ denote an independent copy of the sequence of random vectors $(\xi_1, \dots, \xi_M)$. Since $\mathbb{E}\xi'_j = \mathbf{x}_j$ an application of Jensen's inequality (6.4) yields

$$E := \mathbb{E}\left\|\sum_{j=1}^M (\xi_j - \mathbf{x}_j)\right\|^p = \mathbb{E}\left\|\sum_{j=1}^M (\xi_j - \mathbb{E}\xi'_j)\right\|^p \le \mathbb{E}\left\|\sum_{j=1}^M (\xi_j - \xi'_j)\right\|^p.$$

Now observe that $(\xi_j - \xi'_j)_{j=1}^M$ is a vector of independent symmetric random variables; hence, it has the same distribution as $(\epsilon_j(\xi_j - \xi'_j))_{j=1}^M$. The triangle inequality gives

$$\begin{aligned} E^{1/p} &\le \left(\mathbb{E}\left\|\sum_{j=1}^M \epsilon_j(\xi_j - \xi'_j)\right\|^p\right)^{1/p} \le \left(\mathbb{E}\left\|\sum_{j=1}^M \epsilon_j \xi_j\right\|^p\right)^{1/p} + \left(\mathbb{E}\left\|\sum_{j=1}^M \epsilon_j \xi'_j\right\|^p\right)^{1/p} \\ &= 2\left(\mathbb{E}\left\|\sum_{j=1}^M \epsilon_j \xi_j\right\|^p\right)^{1/p}. \end{aligned}$$

The last equality is due to the fact that $\boldsymbol{\xi}'$ is an independent copy of $\boldsymbol{\xi}$. □

Note that this lemma holds also in infinite-dimensional spaces [80]. Since it is rather technical to introduce random vectors in infinite dimensions we stated the lemma only for the finite-dimensional case. Further, also a converse inequality to (6.6) can be shown [36, 80].

6.4 Scalar Khintchine Inequalities

Khintchine inequalities provide estimates of the moments of Rademacher and related sums. In this section we present the scalar Khintchine inequalities, while in the next section we concentrate on the noncommutative (matrix-valued) Khintchine inequalities.

Theorem 6.8 (Khintchine's inequality). *Let* $\mathbf{b} \in \mathbb{C}^M$ *and* $\boldsymbol{\epsilon} = (\epsilon_1, \ldots, \epsilon_M)$ *be a Rademacher sequence. Then, for all* $n \in \mathbb{N}$,

$$\mathbb{E}\left|\sum_{j=1}^{M} \epsilon_j b_j\right|^{2n} \leq \frac{(2n)!}{2^n n!}\|\mathbf{b}\|_2^{2n}. \tag{6.7}$$

Proof. First assume that the b_j are real-valued. Expanding the expectation on the left hand side of (6.7) with the multinomial theorem, which states that

$$\left(\sum_{j=1}^{M} x_j\right)^n = \sum_{\substack{k_1+\cdots+k_M=n \\ k_i \in \{0,1,\ldots,n\}}} \frac{n!}{k_1!\cdots k_M!} x_1^{k_1} \cdots x_M^{k_M}, \tag{6.8}$$

yields

$$\begin{aligned} E := \mathbb{E}\left|\sum_{j=1}^{M} \epsilon_j b_j\right|^{2n} \\ &= \sum_{\substack{j_1+\cdots+j_M=n \\ j_i \in \{0,1,\ldots,n\}}} \frac{(2n)!}{(2j_1)!\cdots(2j_M)!} |b_1|^{2j_1} \cdots |b_M|^{2j_M} \mathbb{E}\epsilon_1^{2j_1} \cdots \mathbb{E}\epsilon_M^{2j_M} \\ &= \sum_{\substack{j_1+\cdots+j_M=n \\ j_i \in \{0,1,\ldots,n\}}} \frac{(2n)!}{(2j_1)!\cdots(2j_M)!} |b_1|^{2j_1} \cdots |b_M|^{2j_M}. \end{aligned}$$

Hereby we used the independence of the ϵ_j and the fact that $\mathbb{E}\epsilon_j^k = 0$ if k is an odd integer. For integers satisfying $j_1 + \cdots + j_M = n$ it holds

$$2^n j_1! \cdots j_M! = 2^{j_1} j_1! \cdots 2^{j_M} j_M! \leq (2j_1)! \cdots (2j_M)!.$$

This implies

$$\begin{aligned} E &\leq \frac{(2n)!}{2^n n!} \sum_{\substack{j_1+\cdots+j_M=n \\ j_i \in \{0,1,\ldots,n\}}} \frac{n!}{j_1!\cdots j_n!} |b_1|^{2j_1} \cdots |b_M|^{2j_M} \\ &= \frac{(2n)!}{2^n n!}\left(\sum_{j=1}^{M} |b_j|^2\right)^n = \frac{(2n)!}{2^n n!}\|\mathbf{b}\|_2^{2n}. \end{aligned}$$

The general complex case is derived by splitting into real and imaginary parts as follows

$$\left(\mathbb{E}\left|\sum_{j=1}^{M}\epsilon_j(\mathrm{Re}(b_j)+i\,\mathrm{Im}(b_j))\right|^{2n}\right)^{1/2n}$$
$$=\left(\mathbb{E}\left(\left|\sum_{j=1}^{M}\epsilon_j\,\mathrm{Re}(b_j)\right|^2+\left|\sum_{j=1}^{M}\epsilon_j\,\mathrm{Im}(b_j)\right|^2\right)^n\right)^{1/2n}$$
$$\leq\left(\left(\mathbb{E}\left|\sum_{j=1}^{M}\epsilon_j\,\mathrm{Re}(b_j)\right|^{2n}\right)^{1/n}+\left(\mathbb{E}\left|\sum_{j=1}^{M}\epsilon_j\,\mathrm{Im}(b_j)\right|^{2n}\right)^{1/n}\right)^{1/2}$$
$$\leq\left(\left(\frac{(2n)!}{2^n n!}\right)^{1/n}(\|\mathrm{Re}(\mathbf{b})\|_2^2+\|\mathrm{Im}(\mathbf{b})\|_2^2)\right)^{1/2}=\left(\frac{(2n)!}{2^n n!}\right)^{1/2n}\|\mathbf{b}\|_2.$$

This concludes the proof. □

Except that we allowed the coefficient vector $\mathbf{b}$ to be complex valued, the above formulation and the proof is due to Khintchine [75]. Using the central limit theorem, one can show that the constants in (6.7) are optimal. Based on Theorem 6.8 we can also estimate the general absolute pth moment of a Rademacher sum.

Corollary 6.9 (Khintchine's inequality). *Let $\mathbf{b} \in \mathbb{C}^M$ and $\boldsymbol{\epsilon} = (\epsilon_1, \dots, \epsilon_M)$ be a Rademacher sequence. Then, for all $p \geq 2$,*

$$\left(\mathbb{E}\left|\sum_{j=1}^{M}\epsilon_j b_j\right|^p\right)^{1/p} \leq 2^{3/(4p)}e^{-1/2}\sqrt{p}\|\mathbf{b}\|_2. \tag{6.9}$$

Proof. Without loss of generality we assume that $\|\mathbf{b}\|_2 = 1$. Stirling's formula for the factorial,

$$n! = \sqrt{2\pi n}\, n^n e^{-n} e^{\lambda_n}, \tag{6.10}$$

where $\frac{1}{12n+1} \leq \lambda_n \leq \frac{1}{12n}$, gives

$$\frac{(2n)!}{2^n n!} = \frac{\sqrt{2\pi 2n}(2n/e)^{2n}e^{\lambda_{2n}}}{2^n\sqrt{2\pi n}(n/e)^n e^{\lambda_n}} \leq \sqrt{2}\,(2/e)^n n^n. \tag{6.11}$$

An application of Hölder's inequality yields for $\theta \in [0,1]$ and an arbitrary random variable Z,

$$\mathbb{E}|Z|^{2n+2\theta} = \mathbb{E}[|Z|^{(1-\theta)2n}|Z|^{\theta(2n+2)}] \leq (\mathbb{E}|Z|^{2n})^{1-\theta}(\mathbb{E}|Z|^{2n+2})^{\theta}. \tag{6.12}$$

Combine the two estimates above and the Khintchine inequality (6.7) to get

$$
\begin{aligned}
\mathbb{E}\Big|\sum_{j=1}^{M}\epsilon_j b_j\Big|^{2n+2\theta} &\le \left(\mathbb{E}\Big|\sum_{j=1}^{M}\epsilon_j b_j\Big|^{2n}\right)^{1-\theta}\left(\mathbb{E}\Big|\sum_{j=1}^{M}\epsilon_j b_j\Big|^{2n+2}\right)^{\theta}\\
&\le (\sqrt{2}(2/e)^n n^n)^{1-\theta}(\sqrt{2}(2/e)^{n+1}(n+1)^{n+1})^{\theta}\\
&= \sqrt{2}(2/e)^{n+\theta} n^{n(1-\theta)}(n+1)^{\theta(n+1)}\\
&= \sqrt{2}(2/e)^{n+\theta}(n^{1-\theta}(n+1)^{\theta})^{n+\theta}\left(\frac{n+1}{n}\right)^{\theta(1-\theta)}\\
&\le \sqrt{2}(2/e)^{n+\theta}(n+\theta)^{n+\theta}\left(\frac{n+1}{n}\right)^{\theta(1-\theta)}\\
&\le 2^{3/4}(2/e)^{n+\theta}(n+\theta)^{n+\theta}. \qquad (6.13)
\end{aligned}
$$

In the second line from below the inequality of the geometric and arithmetic mean was applied. The last step used that $(n+1)/n \le 2$ and $\theta(1-\theta) \le 1/4$. Replacing $n+\theta$ by $p/2$ completes the proof. □

The optimal constants $C_p = (2^{\frac{p-1}{2}}\frac{\Gamma(p/2)}{\Gamma(3/2)})^{1/p}$, $p \ge 2$, instead of $2^{3/(4p)}e^{-1/2}\sqrt{p}$ for Khintchine's inequality (6.9) are actually slightly better than the ones computed above, but deriving these requires much more effort [67, 89].

Combining Corollary 6.9 with Proposition 6.5 yields the following special case of Hoeffding's inequality [70], also known as Chernoff's bound [26].

Corollary 6.10 (Hoeffding's inequality for Rademacher sums). *Let* $\mathbf{b} = (b_1, \ldots, b_M) \in \mathbb{C}^M$ *and* $\boldsymbol{\epsilon} = (\epsilon_1, \ldots, \epsilon_M)$ *be a Rademacher sequence. Then, for* $u \ge \sqrt{2}$,

$$\mathbb{P}\left(\Big|\sum_{j=1}^{M}\epsilon_j b_j\Big| \ge \|\mathbf{b}\|_2 u\right) \le 2^{3/4}\exp(-u^2/2). \qquad (6.14)$$

For completeness we also give the standard version and proof of Hoeffding's inequality for Rademacher sums.

Proposition 6.11 (Hoeffding's inequality for Rademacher sums). *Let* $\mathbf{b} = (b_1, \ldots, b_M) \in \mathbb{C}^M$ *and* $\boldsymbol{\epsilon} = (\epsilon_1, \ldots, \epsilon_M)$ *be a Rademacher sequence. Then, for* $u > 0$,

$$\mathbb{P}\left(\sum_{j=1}^{M}\epsilon_j b_j \ge \|\mathbf{b}\|_2 u\right) \le \exp(-u^2/2) \qquad (6.15)$$

and consequently,

$$\mathbb{P}\left(\Big|\sum_{j=1}^{M}\epsilon_j b_j\Big| \ge \|\mathbf{b}\|_2 u\right) \le 2\exp(-u^2/2). \qquad (6.16)$$

Proof. Without loss of generality we may assume $\|\mathbf{b}\|_2 = 1$. By Markov's inequality (Theorem 6.2) and independence we have, for $\lambda > 0$,

$$\mathbb{P}\Big(\sum_{j=1}^{M} \epsilon_j b_j \geq u\Big) = \mathbb{P}\Big(\exp\Big(\lambda \sum_{j=1}^{M} \epsilon_j b_j\Big) \geq e^{\lambda u}\Big) \leq e^{-\lambda u}\mathbb{E}\Big[\exp\Big(\lambda \sum_{j=1}^{M} \epsilon_j b_j\Big)\Big]$$
$$= e^{-\lambda u} \prod_{j=1}^{M} \mathbb{E}[\exp(\epsilon_j \lambda b_j)].$$

Note that, for $s \in \mathbb{R}$,

$$\mathbb{E}[\exp(\epsilon_j s)] = \frac{1}{2}(e^{-s} + e^{s}) = \frac{1}{2}\Big(\sum_{k=0}^{\infty} \frac{(-s)^k}{k!} + \sum_{k=0}^{\infty} \frac{s^k}{k!}\Big) = \sum_{k=0}^{\infty} \frac{s^{2k}}{(2k)!}$$
$$\leq \sum_{k=0}^{\infty} \frac{s^{2k}}{2^k k!} = e^{s^2/2}.$$

This yields

$$\mathbb{P}\Big(\sum_{j=1}^{M} \epsilon_j b_j \geq u\Big) \leq e^{-\lambda u} \prod_{j=1}^{M} e^{\lambda^2 b_j^2/2} = e^{-\lambda u + \lambda^2 \|\mathbf{b}\|_2^2/2}.$$

Choosing $\lambda = u$ and recalling that $\|\mathbf{b}\|_2 = 1$ yields (6.15). Finally,

$$\mathbb{P}\Big(\Big|\sum_{j=1}^{M} \epsilon_j b_j\Big| \geq \|\mathbf{b}\|_2 u\Big) = \mathbb{P}\Big(\sum_{j=1}^{M} \epsilon_j b_j \geq \|\mathbf{b}\|_2 u\Big) + \mathbb{P}\Big(\sum_{j=1}^{M} \epsilon_j b_j \leq -\|\mathbf{b}\|_2 u\Big)$$
$$\leq 2e^{-u^2/2},$$

since $-\epsilon_j$ has the same distribution as ϵ_j. □

As mentioned earlier, a complex random variable which is uniformly distributed on the torus $\mathbb{T} = \{z \in \mathbb{C}, |z| = 1\}$ is called a Steinhaus variable. A sequence $\boldsymbol{\epsilon} = (\epsilon_1, \ldots, \epsilon_N)$ of independent Steinhaus variables is called a Steinhaus sequence. There is also a version of Khintchine's inequality for Steinhaus sequences.

Theorem 6.12 (Khintchine's inequality for Steinhaus sequences). *Let $\boldsymbol{\epsilon} = (\epsilon_1, \ldots, \epsilon_M)$ be a Steinhaus sequence and $\mathbf{b} = (b_1, \ldots, b_M) \in \mathbb{C}^M$. Then*

$$\mathbb{E}\Big|\sum_{j=1}^{M} \epsilon_j b_j\Big|^{2n} \leq n! \|\mathbf{b}\|_2^{2n} \quad \textit{for all } n \in \mathbb{N}.$$

Proof. Expand the moment of the Steinhaus sum using the multinomial theorem (6.8),

$$\mathbb{E}\left|\sum_{j=1}^{M} \epsilon_j b_j\right|^{2n} = \mathbb{E}\left(\sum_{j=1}^{M} \epsilon_j b_j\right)^n \left(\sum_{k=1}^{M} \overline{\epsilon_k b_k}\right)^n$$
$$= \mathbb{E}\left(\sum_{\substack{j_1+\cdots+j_M=n \\ j_\ell \geq 0}} \frac{n!}{j_1!\cdots j_M!} b_1^{j_1}\cdots b_M^{j_M} \epsilon_1^{j_1}\cdots \epsilon_M^{j_M}\right)$$
$$\times \left(\sum_{\substack{k_1+\cdots k_M=n \\ k_\ell \geq 0}} \frac{n!}{k_1!\cdots k_M!} \overline{b_1^{k_1}\cdots b_M^{k_M} \epsilon_1^{k_1}\cdots \epsilon_M^{k_M}}\right)$$
$$= \sum_{\substack{j_1+\cdots+j_M=n \\ k_1+\cdots+k_M=n \\ j_\ell,k_\ell\geq 0}} \frac{n!}{j_1!\cdots j_M!}\frac{n!}{k_1!\cdots k_M!} b_1^{j_1}\overline{b_1^{k_1}}\cdots b_M^{j_M}\overline{b_M^{k_M}}\,\mathbb{E}[\epsilon_1^{j_1}\overline{\epsilon_1^{k_1}}\cdots \epsilon_M^{j_M}\overline{\epsilon_M^{k_M}}].$$

Since the ϵ_j are independent and uniformly distributed on the torus it holds

$$\mathbb{E}[\epsilon_1^{j_1}\overline{\epsilon_1^{k_1}}\cdots \epsilon_M^{j_M}\overline{\epsilon_M^{k_M}}] = \mathbb{E}[\epsilon_1^{j_1-k_1}]\cdots \mathbb{E}[\epsilon_M^{j_M-k_M}] = \delta_{j_1,k_1}\cdots\delta_{j_M,k_M}.$$

This yields

$$\mathbb{E}\left|\sum_{j=1}^{M} \epsilon_j b_j\right|^{2n} = \sum_{\substack{k_1+\cdots+k_M=n \\ k_\ell\geq 0}} \left(\frac{n!}{k_1!\cdots k_M!}\right)^2 |b_1|^{2k_1}\cdots |b_M|^{2k_M}$$
$$\leq n! \sum_{\substack{k_1+\cdots+k_M=n \\ k_\ell\geq 0}} \frac{n!}{k_1!\cdots k_M!}|b_1|^{2k_1}\cdots |b_M|^{2k_M}$$
$$= n!\left(\sum_{j=1}^{M} |b_j|^2\right)^{2n},$$

where the multinomial theorem (6.8) was applied once more in the last step. □

The above moment estimate leads to a Hoeffding type inequality for Steinhaus sums.

Corollary 6.13 (Hoeffding's inequality for Steinhaus sequences). *Let* $\boldsymbol{\epsilon} = (\epsilon_1,\ldots,\epsilon_M)$ *be a Steinhaus sequence,* $\mathbf{b} = (b_1,\ldots,b_M) \in \mathbb{C}^M$ *and* $0 < \lambda < 1$. *Then*

$$\mathbb{P}\left(\left|\sum_{j=1}^{M} \epsilon_j b_j\right| \geq u\|\mathbf{b}\|_2\right) \leq \frac{1}{1-\lambda} e^{-\lambda u^2} \quad \text{for all } u \geq 0. \tag{6.17}$$

In particular, using the optimal choice $\lambda = 1 - u^{-2}$,

$$\mathbb{P}\left(\left|\sum_{j=1}^{M} \epsilon_j b_j\right| \geq u\|\mathbf{b}\|_2\right) \leq \exp(-u^2 + \log(u^2) + 1) \quad \textit{for all } u \geq 1. \tag{6.18}$$

Note that the argument of the exponential in (6.18) is always negative for $u > 1$.

Proof. Without loss of generality assume $\|\mathbf{b}\|_2 = 1$. Markov's inequality gives

$$\begin{aligned}
\mathbb{P}\left(\left|\sum_{j=1}^{M} \epsilon_j b_j\right| \geq u\right) &= \mathbb{P}\left(\exp\left(\lambda\left|\sum_{j=1}^{M} \epsilon_j b_j\right|^2\right) \geq \exp(\lambda u^2)\right) \\
&\leq \mathbb{E}\left[\exp\left(\lambda\left|\sum_{j=1}^{M} \epsilon_j b_j\right|^2\right)\right]\exp(-\lambda u^2) \\
&= \exp(-\lambda u^2)\sum_{n=0}^{\infty} \frac{\lambda^n \mathbb{E}|\sum_{j=1}^{M} \epsilon_j b_j|^{2n}}{n!} \\
&\leq \exp(-\lambda u^2)\sum_{n=0}^{\infty} \lambda^n = \frac{1}{1-\lambda}e^{-\lambda u^2}.
\end{aligned}$$

In the second line the Taylor expansion of the exponential function was used together with Fubini's theorem in order to interchange the expectation and the series. In the third line Theorem 6.12 was applied. □

For more information and extensions of scalar Khintchine inequalities we refer the interested reader to [94, 93].

6.5 Noncommutative Khintchine Inequalities

The scalar Khintchine inequalities above can be generalized to the case where the coefficients are matrices. Combined with symmetrization the resulting noncommutative Khintchine inequalities are a very powerful tool in the theory of random matrices. Schatten class norms have to be introduced to formulate them.

For a matrix A we let $\sigma(A) = (\sigma_1(A), \ldots, \sigma_n(A))$ be its sequence of singular values. Then the Schatten p-norm is defined as

$$\|A\|_{S_p} := \|\sigma(A)\|_p, \quad 1 \leq p \leq \infty. \tag{6.19}$$

It is actually nontrivial to show the triangle inequality for Schatten p-norms. We refer the interested reader to [8, 72, 120].

The hermitian matrix AA^* can be diagonalized using a unitary matrix U,

$$AA^* = U^* D^2 U$$

where $D = \operatorname{diag}(\sigma_1(A), \dots, \sigma_n(A))$ (possibly filled up with zeros). As the trace is cyclic, that is $\operatorname{Tr}(AB) = \operatorname{Tr}(BA)$, and since $UU^* = \operatorname{Id}$, we get for $n \in \mathbb{N}$

$$\begin{aligned}\|A\|_{S_{2n}}^{2n} &= \|\sigma(A)\|_{2n}^{2n} = \operatorname{Tr}(D^{2n}) = \operatorname{Tr}(D^{2n}UU^*) = \operatorname{Tr}(U^*D^{2n}U)\\ &= \operatorname{Tr}((U^*D^2U)^n) = \operatorname{Tr}((AA^*)^n).\end{aligned} \tag{6.20}$$

As a special case, the Frobenius norm is the Schatten 2-norm, $\|A\|_F = \|A\|_{S_2}$. The operator norm is also a Schatten norm,

$$\|A\|_{2\to 2} = \sigma_1(A) = \|\sigma(A)\|_\infty = \|A\|_{S_\infty}.$$

By the analogous property of the vector p-norm we have $\|A\|_{S_q} \le \|A\|_{S_p}$ for $q \ge p$. In particular, the following estimate will be very useful,

$$\|A\|_{2\to 2} \le \|A\|_{S_p} \quad \text{for all } 1 \le p \le \infty. \tag{6.21}$$

If A has rank r then it follows from the corresponding property of ℓ_p-norms that

$$\|A\|_{S_p} \le r^{1/p}\|A\|_{2\to 2}. \tag{6.22}$$

Let us now state the noncommutative Khintchine inequality for matrix-valued Rademacher sums, which was first formulated by F. Lust-Piquard [82] with unspecified constants. The optimal constants were provided by A. Buchholz in [16, 17], although it is not obvious at first sight that the results of his paper [16] allow to deduce our next theorem, see also [130]. The proof follows the ideas of Buchholz [16].

Theorem 6.14. *Let $\epsilon = (\epsilon_1, \dots, \epsilon_M)$ be a Rademacher sequence, and let B_j, $j = 1, \dots, M$, be complex matrices of the same dimension. Choose $n \in \mathbb{N}$. Then*

$$\mathbb{E}\left\|\sum_{j=1}^M \epsilon_j B_j\right\|_{S_{2n}}^{2n} \le \frac{(2n)!}{2^n n!} \max\left\{\left\|\left(\sum_{j=1}^M B_j B_j^*\right)^{1/2}\right\|_{S_{2n}}^{2n}, \left\|\left(\sum_{j=1}^M B_j^* B_j\right)^{1/2}\right\|_{S_{2n}}^{2n}\right\}. \tag{6.23}$$

Note that the matrices $B_j B_j^*$ and $B_j^* B_j$ in (6.23) are self-adjoint and positive, so that the square-roots in (6.23) are well-defined.

In order to prove the noncommutative Khintchine inequalities we need to introduce the notion of pairings.

Definition 6.15. (a) A pairing is a partition of the set $[2n]$ into two-element subsets, called blocks. The set $\mathbb{P}_{2n}$ denotes the set of all pairings of $[2n]$.

(b) The canonical pairing $\mathbb{1} = \{D_1, \dots, D_n\}$ has blocks $D_j = \{2j-1, 2j\}$.

(c) Let $\pi = \{D_1, \dots, D_n\}$ be a pairing. Then its cyclic shift $T\pi$ is the pairing with blocks TD_ℓ, where $T\{j,k\} = \{j+1, k+1\}$ with addition understood modulo $2n$.

(d) The "symmetrized" pairing $\overleftarrow{\pi}$ contains all blocks $\{j,k\}$ of a pairing π satisfying $j,k \leq n$ and in addition the "reflected" blocks $\{2n+1-j, 2n+1-k\}$. The blocks of π with both elements being larger than n are omitted in $\overleftarrow{\pi}$ and the blocks $\{j,k\}$ with $j \leq n$ and $k > n$ are replaced by the "symmetric" block $\{j, 2n+1-j\}$.

(e) Similarly, the pairing $\overrightarrow{\pi}$ contains all blocks $\{j,k\}$ of the pairing π satisfying $j,k > n$ and in addition the "reflected" blocks $\{2n+1-j, 2n+1-k\}$. The blocks of π with both elements smaller than $n+1$ are omitted in $\overrightarrow{\pi}$ and the blocks $\{j,k\}$ with $j \leq n$ and $k > n$ are replaced by the "symmetric" block $\{2n+1-k, k\}$.

Let $\mathcal{B} = (B_1, \dots, B_M)$ be a sequence of matrices of the same dimension and $\pi = \{D_1, \dots, D_n\} \in \mathbb{P}_{2n}$. We define the mapping $\alpha = \alpha_\pi : [2n] \to [n]$ such that $\alpha(j) = \ell$ iff $j \in D_\ell$. Using this notation we introduce

$$\pi(\mathcal{B}) = \sum_{k_1,\dots,k_n=1}^{M} B_{k_{\alpha(1)}} B^*_{k_{\alpha(2)}} B_{k_{\alpha(3)}} B^*_{k_{\alpha(4)}} \cdots B_{k_{\alpha(2n-1)}} B^*_{k_{\alpha(2n)}}. \tag{6.24}$$

Note that $\pi(\mathcal{B})$ is independent of the chosen numbering of $D_1, \dots, D_n$. The following lemma will be the key to the proof of the noncommutative Khintchine inequalities.

Lemma 6.16. *Let $\pi \in \mathbb{P}_{2n}$ and $\mathcal{B} = (B_1, \dots, B_M)$ a sequence of complex matrices of the same dimension. Then there is $\gamma \geq 1/(4n)$ and non-negative numbers $p_\rho = p_\rho(\pi)$, $\rho \in \mathbb{P}_{2n}$, satisfying $\gamma + \sum_{\rho \in \mathbb{P}_{2n}} p_\rho = 1$, such that*

$$|\mathrm{Tr}\, \pi(\mathcal{B})| \tag{6.25}$$
$$\leq \max\left\{ \left\| \left(\sum_{k=1}^{M} B_k B_k^* \right)^{1/2} \right\|_{S_{2n}}^{2n}, \left\| \left(\sum_{k=1}^{M} B_k^* B_k \right)^{1/2} \right\|_{S_{2n}}^{2n} \right\}^{\gamma} \prod_{\rho \in \mathbb{P}_{2n}} |\mathrm{Tr}\, \rho(\mathcal{B})|^{p_\rho}.$$

Proof. First observe that

$$\mathbb{1}(\mathcal{B}) = \sum_{k_1,\dots,k_n=1}^{N} \prod_{j=1}^{n} B_{k_j} B^*_{k_j} = \left(\sum_{k=1}^{M} B_k B_k^* \right)^n.$$

Since the matrix inside the bracket is self-adjoint and positive semi-definite we can take its square root and (6.20) yields

$$\mathrm{Tr}\, \mathbb{1}(\mathcal{B}) = \left\| \left(\sum_{k=1}^{M} B_k B_k^* \right)^{1/2} \right\|_{S_{2n}}^{2n}. \tag{6.26}$$

Since the trace is cyclic we similarly obtain

$$\operatorname{Tr} T\mathbb{1}(\mathcal{B}) = \left\| \Big(\sum_{k=1}^{M} B_k^* B_k \Big)^{1/2} \right\|_{S_{2n}}^{2n}. \tag{6.27}$$

The idea of the proof is to successively provide estimates of $|\operatorname{Tr}\pi(\mathcal{B})|$ in terms of traces of operators $\rho(\mathcal{B})$ which become more and more 'similar' to $\mathbb{1}(\mathcal{B})$ or $T\mathbb{1}(\mathcal{B})$.

Let $t \in \{0, 1, \ldots, n\}$ be the maximal number such that, for some p, $\{p, p+1\}, \{p+2, p+3\}, \ldots, \{p+2t-2, p+2t-1\}$ are blocks of the partition π. If $t = n$ then $\pi = \mathbb{1}$ or $\pi = T\mathbb{1}$ and we are done. We postpone the case $t = 0$ to later and assume $t \in [n-1]$.

By cyclicity of the trace, it holds $\operatorname{Tr}\pi(\mathcal{B}) = \operatorname{Tr}(T^{n-p-2t+1}\pi)(\mathcal{B})$ if $n-p$ is odd and $\operatorname{Tr}\pi(\mathcal{B}) = \operatorname{Tr}(T^{n-p-2t+1}\pi)(\mathcal{B}^*)$ if $n-p$ is even, where $\mathcal{B}^* = (B_1^*, \ldots, B_M^*)$. Note that the blocks $\{n-2t+1, n-2t+2\}, \{n-2t+3, n-2t+4\}, \ldots, \{n-1, n\}$ (with addition modulo $2n$) are part of the partition $T^{n-p-2t+1}\pi$. Assume n even and p odd for the moment. Denote the blocks of $T^{n-p-2t+1}\pi$ by $D_1, \ldots, D_n$ and let $\alpha = \alpha_\pi : [2n] \to [n]$ be the mapping defined by $\alpha(j) = \ell$ iff $j \in D_\ell$. Divide $[n]$ into three sets L, R, U. The subset L (resp. R) contains the indices ℓ, for which both elements of D_ℓ are in $\{1, \ldots, n\}$ (resp. $\{n+1, \ldots, 2n\}$), while U contains the remaining indices for which the blocks have elements in both $\{1, \ldots, n\}$ and $\{n+1, \ldots, 2n\}$.

The Cauchy–Schwarz inequality for the trace (2.6) and for the usual Euclidean inner product yields

$$\begin{aligned}
&|\operatorname{Tr}\pi(\mathcal{B})| = |\operatorname{Tr}(T^{n-p-2t+1}\pi)(\mathcal{B})| \\
&= \Bigg| \sum_{k_i\in[M], i\in U} \operatorname{Tr}\Bigg(\Big(\sum_{k_i\in[M], i\in L} B_{k_{\alpha(1)}} \cdots B^*_{k_{\alpha(n)}} \Big) \Big(\sum_{k_i\in[M], i\in R} B_{k_{\alpha(n+1)}} \cdots B^*_{k_{\alpha(2n)}} \Big) \Bigg) \Bigg| \\
&\le \sum_{k_i, i\in U} \sqrt{ \operatorname{Tr}\Bigg(\Big(\sum_{k_i, i\in L} B_{k_{\alpha(1)}} \cdots B^*_{k_{\alpha(n)}} \Big) \Big(\sum_{k_i, i\in L} B_{k_{\alpha(n)}} \cdots B^*_{k_{\alpha(1)}} \Big) \Bigg) } \\
&\qquad \times \sqrt{ \operatorname{Tr}\Bigg(\Big(\sum_{k_i, i\in R} B_{k_{\alpha(n+1)}} \cdots B^*_{k_{\alpha(2n)}} \Big) \Big(\sum_{k_i, i\in R} B_{k_{\alpha(2n)}} \cdots B^*_{k_{\alpha(n+1)}} \Big) \Bigg) } \\
&\le \sqrt{ \sum_{k_i, i\in U} \operatorname{Tr}\Bigg(\Big(\sum_{k_i, i\in L} B_{k_{\alpha(1)}} \cdots B^*_{k_{\alpha(n)}} \Big) \Big(\sum_{k_i, i\in L} B_{k_{\alpha(n)}} \cdots B^*_{k_{\alpha(1)}} \Big) \Bigg) } \\
&\qquad \times \sqrt{ \sum_{k_i, i\in U} \operatorname{Tr}\Bigg(\Big(\sum_{k_i, i\in R} B_{k_{\alpha(n+1)}} \cdots B^*_{k_{\alpha(2n)}} \Big) \Big(\sum_{k_i, i\in R} B_{k_{\alpha(2n)}} \cdots B^*_{k_{\alpha(n+1)}} \Big) \Bigg) } \\
&= |\operatorname{Tr} \overleftarrow{(T^{n-p-2t+1}\pi)}(\mathcal{B})|^{1/2} |\operatorname{Tr}\rho(\mathcal{B})|^{1/2}
\end{aligned} \tag{6.28}$$

with $\rho = \overrightarrow{T^{n-p-2t+1}\pi} \in \mathbb{P}_{2n}$. If $t \geq n/2$ then $\overleftarrow{T^{n-p-2t+1}\pi}$ equals $\mathbb{1}$ or $T\mathbb{1}$ and we are done. If $t < n/2$ then $\overleftarrow{T^{n-p-2t+1}\pi}$ contains the blocks $\{n-2t+1, n-2t+2\}, \dots, \{n-1, n\}, \{n+1, n+2\}, \dots, \{n+2t-1, n+2t\}$. Apply the same estimates with $t' = 2t$ as above to $T^{-2t}(\overleftarrow{T^{n-p-2t+1}\pi})$ to obtain

$$|\operatorname{Tr}\pi(\mathcal{B})| \leq |\operatorname{Tr}(\overleftarrow{T^{-2t}(\overleftarrow{T^{n-p-2t+1}\pi})}(\mathcal{B})|^{1/4} |\operatorname{Tr}\tilde{\rho}(\mathcal{B})|^{1/4} |\operatorname{Tr}\rho(\mathcal{B})|^{1/2}$$

for suitable $\tilde{\rho}, \rho \in \mathbb{P}_{2n}$. Similarly, as above if $t \geq n/4$ then $\overleftarrow{T^{-2t}(\overleftarrow{T^{n-p-2t}\pi})}$ equals $\mathbb{1}$ or $T\mathbb{1}$ and we are done. If $t < n/4$ then we continue in this way, and after at most $\lceil \log_2(n) \rceil$ estimation steps of the form (6.28) inequality (6.25) is obtained with $\gamma \geq 1/2^{\lceil \log_2(n) \rceil} \geq 1/(2n)$.

If initially $t = 0$, then we apply the above method to $T^q\pi$ where q was chosen such that $\{n, p\}$ for some $p > n$ is a block of $T^q\pi$. Using the same estimates as in (6.28) yields $|\operatorname{Tr}\pi(\mathcal{B})| \leq |\operatorname{Tr}\tilde{\pi}(\mathcal{B})|^{1/2} |\operatorname{Tr}\rho(\mathcal{B})|^{1/2}$ for some partition ρ, where $\tilde{\pi}$ contains the block $\{n, n+1\}$. Then invoke the above method to obtain (6.25) with $\gamma \geq 1/(4n)$.

If n is odd and p even, then an obvious modification of the chain of inequalities (6.28) applies. If $n - p$ is even then $\mathcal{B}^*$ instead of $\mathcal{B}$ appears after the first equality in (6.28). Noting that $\operatorname{Tr}\mathbb{1}(\mathcal{B}^*) = \operatorname{Tr}T\mathbb{1}(\mathcal{B})$ and $\operatorname{Tr}T\mathbb{1}(\mathcal{B}^*) = \operatorname{Tr}\mathbb{1}(\mathcal{B})$ by cyclicity concludes the proof. □

Corollary 6.17. *Under the same assumptions as in Lemma* 6.16, *for all* $\pi \in \mathbb{P}_{2n}$,

$$|\operatorname{Tr}\pi(\mathcal{B})| \leq \max\left\{ \left\| \left(\sum_{k=1}^{M} B_k B_k^* \right)^{1/2} \right\|_{S_{2n}}^{2n}, \left\| \left(\sum_{k=1}^{M} B_k^* B_k \right)^{1/2} \right\|_{S_{2n}}^{2n} \right\}. \tag{6.29}$$

Proof. Denote the right hand side of (6.29) by D. The constant γ in Lemma 6.16 may be chosen the same for all partitions $\pi \in \mathbb{P}_{2n}$, for instance $\gamma = \gamma_1 = 1/(4n)$. Indeed, if γ is initially larger, then by (6.26) and (6.27) simply move some weight from D^γ to $|\operatorname{Tr}\mathbb{1}(\mathcal{B})|^{p_{\mathbb{1}}(\pi)}$ or to $|\operatorname{Tr}T\mathbb{1}(\mathcal{B})|^{p_{T\mathbb{1}}(\pi)}$, whichever term is larger.

Apply Lemma 6.16 to itself to obtain

$$\begin{aligned} |\operatorname{Tr}\pi(\mathcal{B})| &\leq D^{\gamma_1} \prod_{\kappa \in \mathbb{P}_{2n}} |\operatorname{Tr}\kappa(\mathcal{B})|^{p_\kappa(\pi)} \\ &\leq D^{\gamma_1} \prod_{\kappa \in \mathbb{P}_{2n}} D^{\gamma_1 p_\kappa(\pi)} \prod_{\rho \in \mathbb{P}_{2n}} |\operatorname{Tr}\rho(\mathcal{B})|^{p_\kappa(\pi) p_\rho(\kappa)} \\ &= D^{\gamma_1 + \gamma_1(1-\gamma_1)} \prod_{\rho \in \mathbb{P}_{2n}} |\operatorname{Tr}\rho(\mathcal{B})|^{\sum_{\kappa \in \mathbb{P}_{2n}} p_\rho(\kappa) p_\kappa(\pi)}. \end{aligned}$$

This yields (6.25) with new constants

$$\gamma_2 = \gamma_1 + \gamma_1(1-\gamma_1), \quad p_\rho^{(2)}(\pi) = \sum_{\kappa \in \mathbb{P}_{2n}} p_\rho(\kappa) p_\kappa(\pi).$$

Since $\gamma_1 = 1/(4n)$, in particular, $0 < \gamma_1 < 1$, the new constant γ_2 is larger than γ_1. Iterating this process yields increasingly larger constants γ_ℓ defined recursively by

$$\gamma_{\ell+1} = \gamma_\ell + \gamma_\ell(1 - \gamma_\ell).$$

Elementary calculus shows that $\lim_{\ell\to\infty} \gamma_\ell = 1$. Since the corresponding constants $p_\rho^{(\ell)}(\pi)$ satisfy $\gamma_\ell + \sum_{\rho\in\mathbb{P}_{2n}} p_\rho^{(\ell)}(\pi) = 1$ for all ℓ one concludes $\lim_{\ell\to\infty} p_\rho^{(\ell)}(\pi) = 0$ for all $\rho \in \mathbb{P}_{2n}$. This completes the proof. □

Now we are in the position to complete the proof of the noncommutative Khintchine inequalities.

Proof of Theorem 6.14. By (6.20)

$$\begin{aligned} E := \mathbb{E}\left\| \sum_{k=1}^{M} \epsilon_k B_k \right\|_{S_{2n}}^{2n} &= \mathbb{E}\,\mathrm{Tr}\left(\left(\sum_{k=1}^{M} \epsilon_k B_k \sum_{j=1}^{M} \epsilon_j B_j^*\right)^n\right) \\ &= \sum_{k_1,\ldots,k_{2n}=1}^{M} \mathbb{E}[\epsilon_{k_1}\cdots\epsilon_{k_{2n}}]\,\mathrm{Tr}(B_{k_1} B_{k_2}^* B_{k_3}\cdots B_{k_{2n}}^*). \end{aligned}$$

Observe that $\mathbb{E}[\epsilon_{k_1}\cdots\epsilon_{k_{2n}}] = 1$ if and only if each $j \in [2n]$ can be paired with an $\ell \in [2n]$ such that $k_j = k_\ell$ and $\mathbb{E}[\epsilon_{k_1}\cdots\epsilon_{k_{2n}}] = 0$ otherwise. Therefore, denoting $\mathcal{B} = (B_1, \ldots, B_M)$, Corollary 6.17 yields (recall also the definition in (6.24))

$$\begin{aligned} E &= \sum_{\pi\in\mathbb{P}_{2n}} \mathrm{Tr}\,\pi(\mathcal{B}) \\ &\le |\mathbb{P}_{2n}| \max\left\{ \left\| \left(\sum_{k=1}^{M} B_k B_k^*\right)^{1/2} \right\|_{S_{2n}}^{2n}, \left\| \left(\sum_{k=1}^{M} B_k^* B_k\right)^{1/2} \right\|_{S_{2n}}^{2n} \right\}. \end{aligned}$$

Elementary considerations show that the number $|\mathbb{P}_{2n}|$ of pairings of a set with $2n$ elements equals $\frac{(2n)!}{2^n n!}$. □

The noncommutative Khintchine inequalities may be extended to general $p \ge 2$, similarly to Corollary 6.9 in the scalar case, see also [130]. For our purposes the present version will be sufficient.

6.6 Rudelson's Lemma

Rudelson's lemma [113] is a very useful estimate for the operator norm of a Rademacher sum of rank one matrices. The statement below is slightly different from the formulation in [113], but allows to draw the same conclusions, and makes constants explicit. Its proof is a nice application of the noncommutative Khintchine inequality.

Lemma 6.18. *Let $A \in \mathbb{C}^{m \times M}$ of rank r with columns $\mathbf{a}_1, \ldots, \mathbf{a}_M$. Let $\epsilon = (\epsilon_1, \ldots, \epsilon_M)$ be a Rademacher sequence. Then, for $2 \le p < \infty$,*

$$\left(\mathbb{E}\left\|\sum_{j=1}^{M} \epsilon_j \mathbf{a}_j \mathbf{a}_j^*\right\|_{2\to 2}^{p}\right)^{1/p} \le 2^{3/(4p)} r^{1/p} \sqrt{p} e^{-1/2} \|A\|_{2\to 2} \max_{j=1,\ldots,M} \|\mathbf{a}_j\|_2. \tag{6.30}$$

Proof. Write $p = 2n + 2\theta$ with $n \in N$ and $\theta \in [0,1)$. Denote $C_n = (\frac{(2n)!}{2^n n!})^{1/(2n)}$. Note that $(\mathbf{a}_j \mathbf{a}_j^*)^*(\mathbf{a}_j \mathbf{a}_j^*) = (\mathbf{a}_j \mathbf{a}_j^*)(\mathbf{a}_j \mathbf{a}_j^*)^* = \|\mathbf{a}_j\|_2^2 \mathbf{a}_j \mathbf{a}_j^*$. Therefore, the noncommutative Khintchine inequality (6.23) yields

$$\begin{aligned} E := \left(\mathbb{E}\left\|\sum_j \epsilon_j \mathbf{a}_j \mathbf{a}_j^*\right\|_{2\to 2}^{2n}\right)^{1/(2n)} &\le \left(\mathbb{E}\left\|\sum_j \epsilon_j \mathbf{a}_j \mathbf{a}_j^*\right\|_{S_{2n}}^{2n}\right)^{1/(2n)} \\ &\le C_n \left\|\left(\sum_j \|\mathbf{a}_j\|_2^2 \mathbf{a}_j \mathbf{a}_j^*\right)^{1/2}\right\|_{S_{2n}}. \end{aligned}$$

The operator $\sum_j \|\mathbf{a}_j\|_2^2 \mathbf{a}_j \mathbf{a}_j^*$ has rank at most r. The estimate (6.22) of the Schatten norm by the operator norm together with (2.5) gives therefore

$$\begin{aligned} E &\le C_n r^{1/(2n)} \left\|\left(\sum_j \|\mathbf{a}_j\|_2^2 \mathbf{a}_j \mathbf{a}_j^*\right)^{1/2}\right\|_{2\to 2} \\ &\le C_n r^{1/(2n)} \left\|\sum_j \mathbf{a}_j \mathbf{a}_j^*\right\|_{2\to 2}^{1/2} \max_{k=1,\ldots,M} \|\mathbf{a}_k\|_2. \end{aligned}$$

Observing that $\|\sum_j \mathbf{a}_j \mathbf{a}_j^*\|_{2\to 2}^{1/2} = \|AA^*\|_{2\to 2}^{1/2} = \|A\|_{2\to 2}$ yields

$$E \le C_n r^{1/(2n)} \|A\|_{2\to 2} \max_{j=1,\ldots,M} \|\mathbf{a}_j\|_2.$$

With the estimate (6.21) of the operator norm by the Schatten norm together with (6.12) we obtain

$$\begin{aligned} \mathbb{E}\left\|\sum_{j=1}^{M} \epsilon_j \mathbf{a}_j \mathbf{a}_j^*\right\|_{2\to 2}^{2n+2\theta} &\le \left(\mathbb{E}\left\|\sum_j \epsilon_j \mathbf{a}_j \mathbf{a}_j^*\right\|_{S_{2n}}^{2n}\right)^{1-\theta} \left(\mathbb{E}\left\|\sum_j \epsilon_j \mathbf{a}_j \mathbf{a}_j^*\right\|_{S_{2n+2}}^{2n+2}\right)^{\theta} \\ &\le \left(\frac{(2n)!}{2^n n!}\right)^{1-\theta} \left(\frac{(2n+2)!}{2^{n+1}(n+1)!}\right)^{\theta} r\left(\|A\|_{2\to 2} \max_{j=1,\ldots,M} \|\mathbf{a}_j\|_2\right)^{2n+2\theta} \\ &\le 2^{3/4} (2/e)^{n+\theta} (n+\theta)^{n+\theta} r\left(\|A\|_{2\to 2} \max_{j=1,\ldots,M} \|\mathbf{a}_j\|_2\right)^{2n+2\theta}. \end{aligned}$$

Hereby, we applied (6.11) and the same chain of inequalities as in (6.13). Substituting $p/2 = n + \theta$ completes the proof. □

Proposition 6.5 leads to the following statement.

Corollary 6.19. *Let $A \in \mathbb{C}^{m\times M}$ of rank r with columns $\mathbf{a}_1, \ldots, \mathbf{a}_M$. Let $\boldsymbol{\epsilon} \in \mathbb{R}^M$ be a Rademacher sequence. Then for all $u \geq \sqrt{2}$*

$$\mathbb{P}\left(\left\| \sum_{j=1}^{M} \epsilon_j \mathbf{a}_j \mathbf{a}_j^* \right\|_{2\to 2} \geq u \|A\|_{2\to 2} \max_{j=1,\ldots,M} \|\mathbf{a}_j\|_2 \right) \leq 2^{3/4} r e^{-u^2/2}. \tag{6.31}$$

The formulation of Rudelson's lemma which is most commonly used follows then from an application of Lemma 6.6 (where the "maximum" is taken only over one random variable) after estimating $2^{3/4} < 2$.

Corollary 6.20. *Let $A \in \mathbb{C}^{m\times M}$ of rank r with columns $\mathbf{a}_1, \ldots, \mathbf{a}_M$. Let $\boldsymbol{\epsilon} \in \mathbb{R}^M$ be a Rademacher sequence. Then*

$$\mathbb{E}\left\| \sum_{j=1}^{M} \epsilon_j \mathbf{a}_j \mathbf{a}_j^* \right\|_{2\to 2} \leq C \sqrt{\ln(8r)} \|A\|_{2\to 2} \max_{j=1,\ldots,M} \|\mathbf{a}_j\|_2$$

with $C \leq \sqrt{2} + \frac{1}{4\sqrt{2\ln(8)}} \approx 1.499 < 1.5$.

6.7 Decoupling

Decoupling is a technique that reduces stochastic dependencies in certain sums of random variables, called chaos variables. A typical example is a sum of the form

$$\sum_{j \neq \ell} \epsilon_j \epsilon_\ell \mathbf{x}_{j\ell}$$

where $\mathbf{x}_{j\ell}$ are some vectors and $\boldsymbol{\epsilon} = (\epsilon_j)$ is a Rademacher series. Such a sum is called Rademacher chaos of order 2. The following statement, taken from [13], provides a way of "decoupling" the sum. Many more results concerning decoupling can be found in the monograph [36].

Lemma 6.21. *Let $\boldsymbol{\xi} = (\xi_1, \ldots, \xi_M)$ be a sequence of independent random variables with $\mathbb{E}\xi_j = 0$ for all $j = 1, \ldots, M$. Let $B_{j,k}$, $j,k = 1, \ldots, M$, be a double sequence of elements in a vector space with norm $\|\cdot\|$, where $B_{j,j} = 0$ for all $j = 1, \ldots, M$. Then for $1 \leq p < \infty$*

$$\mathbb{E}\left\| \sum_{j,k=1}^{M} \xi_j \xi_k B_{j,k} \right\|^p \leq 4^p \mathbb{E}\left\| \sum_{j,k=1}^{M} \xi_j \xi_k' B_{j,k} \right\|^p, \tag{6.32}$$

where $\boldsymbol{\xi}'$ denotes an independent copy of $\boldsymbol{\xi}$.

Proof. Introduce a sequence $\delta = (\delta_j)_{j=1}^M$, of independent random variables δ_j taking only the values 0 and 1 with probability $1/2$. Then for $j \neq k$

$$\mathbb{E}\delta_j(1-\delta_k) = 1/4.$$

Since $B_{j,j} = 0$ this gives

$$\begin{aligned} E := \mathbb{E}\left\| \sum_{j,k=1}^M \xi_j \xi_k B_{j,k} \right\|^p &= 4^p\, \mathbb{E}_\xi \left\| \sum_{j,k=1}^M \mathbb{E}_\delta[\delta_j(1-\delta_k)]\xi_j\xi_k B_{j,k} \right\|^p \\ &\leq 4^p\, \mathbb{E}\left\| \sum_{j,k=1}^M \delta_j(1-\delta_k)\xi_j\xi_k B_{j,k} \right\|^p, \end{aligned}$$

where Jensen's inequality was applied in the last step. Now let

$$\sigma(\delta) := \{j = 1, \ldots, M : \delta_j = 1\}.$$

Then, by Fubini's theorem,

$$E \leq 4^p\, \mathbb{E}_\delta \mathbb{E}_\xi \left\| \sum_{j\in\sigma(\delta)} \sum_{k\notin\sigma(\delta)} \xi_j\xi_k B_{j,k} \right\|^p.$$

For a fixed δ the sequences $(\xi_j)_{j\in\sigma(\delta)}$ and $(\xi_k)_{k\notin\sigma(\delta)}$ are independent, and hence,

$$E \leq 4^p\, \mathbb{E}_\delta \mathbb{E}_\xi \mathbb{E}_{\xi'} \left\| \sum_{j\in\sigma(\delta)} \sum_{k\notin\sigma(\delta)} \xi_j\xi'_k B_{j,k} \right\|^p.$$

This implies the existence of a δ_0, and hence a $\sigma = \sigma(\delta_0)$ such that

$$E \leq 4^p\, \mathbb{E}_\xi \mathbb{E}_{\xi'} \left\| \sum_{j\in\sigma} \sum_{k\notin\sigma} \xi_j\xi'_k B_{j,k} \right\|^p.$$

Since $\mathbb{E}\xi_j = \mathbb{E}\xi'_j = 0$, an application of Jensen's inequality yields

$$\begin{aligned} E &\leq 4^p\, \mathbb{E}\left\| \sum_{j\in\sigma}\Big(\sum_{k\notin\sigma} \xi_j\xi'_k B_{j,k} + \sum_{k\in\sigma} \xi_j \mathbb{E}[\xi'_k] B_{j,k} \Big) + \sum_{j\notin\sigma} \mathbb{E}[\xi_j] \sum_{k=1}^M \xi'_k B_{j,k} \right\|^p \\ &\leq 4^p\, \mathbb{E}\left\| \sum_{j=1}^M \sum_{k=1}^M \xi_j\xi'_k B_{j,k} \right\|^p, \end{aligned}$$

and the proof is completed. □

We note that the mean-zero assumption $\mathbb{E}\xi_j = 0$ may be removed by introducing a larger constant 8 instead of 4, see Theorem 3.1.1 in [36] and its proof. The sum $\sum_{j,k} \xi_j \xi'_k B_{j,k}$ on the right hand side of (6.32) is called a decoupled chaos.

6.8 Noncommutative Khintchine Inequalities for Decoupled Rademacher Chaos

The previous section showed the usefulness of studying decoupled chaoses. Next we state the noncommutative Khintchine inequality for decoupled Rademacher chaos [105], see also [100, p. 111] for a slightly more general inequality (without explicit constants). A scalar version can be found, for instance, in [86].

Theorem 6.22. *Let $B_{j,k} \in \mathbb{C}^{r\times t}$, $j,k = 1,\ldots,M$, be complex matrices of the same dimension. Let $\boldsymbol{\epsilon},\boldsymbol{\epsilon}'$ be independent Rademacher sequences. Then, for $n \in \mathbb{N}$,*

$$\left[\mathbb{E}\left\|\sum_{j,k=1}^{M}\epsilon_j\epsilon'_k B_{j,k}\right\|_{S_{2n}}^{2n}\right]^{1/2n} \leq 2^{1/(2n)}\left(\frac{(2n)!}{2^n n!}\right)^{1/n} \tag{6.33}$$

$$\times\max\left\{\left\|\left(\sum_{j,k=1}^{M}B_{j,k}B_{j,k}^*\right)^{1/2}\right\|_{S_{2n}},\left\|\left(\sum_{j,k=1}^{M}B_{j,k}^*B_{j,k}\right)^{1/2}\right\|_{S_{2n}},\|F\|_{S_{2n}}\right\},$$

where F is the block matrix $F = (B_{j,k})_{j,k=1}^{M}$.

We note that the factor $2^{1/(2n)}$ may be removed with a more technical proof that uses the same strategy as the proof of the (ordinary) noncommutative Khintchine inequality (6.25) above. Our proof below rather proceeds by applying (6.25) twice. Taking scalars instead of matrices $B_{j,k}$ results in a scalar Khintchine inequality for decoupled Rademacher chaos. In the scalar case the first two terms in the maximum in (6.33) coincide and the third one is always dominated by the first.

Proof of Theorem 6.22. Denote $C_n = \frac{(2n)!}{2^n n!}$. Fubini's theorem and an application of the noncommutative Khintchine inequality (6.23) yields

$$E := \mathbb{E}\left\|\sum_{j,k=1}^{M}\epsilon_j\epsilon'_k B_{j,k}\right\|_{S_{2n}}^{2n} \tag{6.34}$$

$$\leq C_n\mathbb{E}_\epsilon\max\left\{\left\|\left(\sum_{k=1}^{M}H_k(\epsilon)^*H_k(\epsilon)\right)^{1/2}\right\|_{S_{2n}}^{2n},\left\|\left(\sum_{k=1}^{M}H_k(\epsilon)H_k(\epsilon)^*\right)^{1/2}\right\|_{S_{2n}}^{2n}\right\},$$

where $H_k(\epsilon) := \sum_{j=1}^{N}\epsilon_j B_{j,k}$. We define

$$\widehat{B}_{j,k} = (0|\ldots|0|B_{j,k}|0|\ldots|0) \in \mathbb{C}^{r\times tM},$$

and similarly

$$\tilde{B}_{j,k} = (0|\ldots|0|B_{j,k}^*|0|\ldots|0)^* \in \mathbb{C}^{rM\times t},$$

where in both cases the non-zero block $B_{j,k}$ is the kth one. Then

$$\widehat{B}_{j,k}\widehat{B}^*_{j',k'} = \begin{cases} 0 & \text{if } k \neq k', \\ B_{j,k}B^*_{j',k} & \text{if } k = k', \end{cases} \tag{6.35}$$

$$\tilde{B}^*_{j,k}\tilde{B}_{j',k'} = \begin{cases} 0 & \text{if } k \neq k', \\ B^*_{j,k}B_{j',k} & \text{if } k = k'. \end{cases}$$

Since the singular values obey $\sigma_k(A) = \sigma_k((AA^*)^{1/2})$, the Schatten class norm satisfies $\|A\|_{S_{2n}} = \|(AA^*)^{1/2}\|_{S_{2n}}$. This allows us to verify that

$$\begin{aligned}
\Big\|\sum_{j=1}^M \epsilon_j \sum_{k=1}^M \widehat{B}_{j,k}\Big\|_{S_{2n}} &= \Big\|\Big(\sum_{j,j'} \epsilon_j\epsilon_{j'} \sum_{k,k'} \widehat{B}_{j,k}\widehat{B}^*_{j',k'}\Big)^{1/2}\Big\|_{S_{2n}} \\
&= \Big\|\Big(\sum_{j,j'} \epsilon_j\epsilon_{j'} \sum_{k} B_{j,k}B^*_{j',k}\Big)^{1/2}\Big\|_{S_{2n}} \\
&= \Big\|\Big(\sum_{k} H_k(\epsilon)H_k(\epsilon)^*\Big)^{1/2}\Big\|_{S_{2n}}.
\end{aligned}$$

Similarly, we also get

$$\Big\|\Big(\sum_{k} H_k(\epsilon)^*H_k(\epsilon)\Big)^{1/2}\Big\|_{S_{2n}} = \Big\|\sum_{j=1}^M \epsilon_j \sum_{k=1}^M \tilde{B}_{j,k}\Big\|_{S_{2n}}.$$

Plugging the above expressions into (6.34) we can further estimate

$$E \leq C_n\Big(\mathbb{E}\Big\|\sum_{j=1}^M \epsilon_j \sum_{k=1}^M \widehat{B}_{j,k}\Big\|^{2n}_{S_{2n}} + \mathbb{E}\Big\|\sum_{j=1}^M \epsilon_j \sum_{k=1}^M \tilde{B}_{j,k}\Big\|^{2n}_{S_{2n}}\Big).$$

Using Khintchine's inequality (6.23) once more we obtain

$$\begin{aligned}
E_1 &:= \mathbb{E}\Big\|\sum_{j=1}^M \epsilon_j \sum_{k=1}^M \widehat{B}_{j,k}\Big\|^{2n}_{S_{2n}} \\
&\leq C_n \max\left\{\Big\|\Big(\sum_{j} \tilde{H}_j\tilde{H}_j^*\Big)^{1/2}\Big\|^{2n}_{S_{2n}}, \Big\|\Big(\sum_{j} \tilde{H}_j^*\tilde{H}_j\Big)^{1/2}\Big\|^{2n}_{S_{2n}}\right\},
\end{aligned}$$

where $\tilde{H}_j = \sum_{k=1}^M \widehat{B}_{j,k}$. Using (6.35) we see that

$$\sum_{j} \tilde{H}_j\tilde{H}_j^* = \sum_{k,j} B_{j,k}B^*_{j,k}.$$

Furthermore, noting that

$$F = \begin{pmatrix} B_{1,1} & B_{1,2} & \dots & B_{1,M} \\ B_{2,1} & B_{2,2} & \dots & B_{2,M} \\ \vdots & \vdots & \vdots & \vdots \\ B_{M,1} & B_{M,2} & \dots & B_{M,M} \end{pmatrix} = \begin{pmatrix} \tilde{H}_1 \\ \tilde{H}_2 \\ \vdots \\ \tilde{H}_M \end{pmatrix},$$

we have

$$\Big\| \Big(\sum_j \tilde{H}_j^* \tilde{H}_j \Big)^{1/2} \Big\|_{S_{2n}}^{2n} = \|(F^*F)^{1/2}\|_{S_{2n}}^{2n} = \|F\|_{S_{2n}}^{2n}.$$

Hence,

$$E_1 \le C_n \max\left\{ \Big\| \Big(\sum_{j,k=1}^{M} B_{j,k} B_{j,k}^* \Big)^{1/2} \Big\|_{S_{2n}}^{2n}, \|F\|_{S_{2n}}^{2n} \right\}.$$

As $\tilde{B}_{j,k}$ differs from $\widehat{B}_{j,k}$ only by interchanging $B_{j,k}$ with $B_{j,k}^*$ we obtain similarly

$$E_2 := \mathbb{E} \Big\| \sum_{j=1}^{M} \epsilon_j \sum_{k=1}^{M} \tilde{B}_{j,k} \Big\|_{S_{2n}}^{2n} \le C_n \max\left\{ \Big\| \Big(\sum_{j,k=1}^{M} B_{j,k}^* B_{j,k} \Big)^{1/2} \Big\|_{S_{2n}}^{2n}, \|F\|_{S_{2n}}^{2n} \right\}.$$

Finally,

$$\begin{aligned} E &\le C_n(E_1 + E_2) \\ &\le 2 \cdot C_n^2 \max\left\{ \Big\| \Big(\sum_{j,k=1}^{M} B_{j,k}^* B_{j,k} \Big)^{1/2} \Big\|_{S_{2n}}^{2n}, \Big\| \Big(\sum_{j,k=1}^{M} B_{j,k} B_{j,k}^* \Big)^{1/2} \Big\|_{S_{2n}}^{2n}, \|F\|_{S_{2n}}^{2n} \right\}. \end{aligned}$$

This concludes the proof. □

6.9 Dudley's Inequality

A stochastic process is a collection X_t, $t \in \hat{T}$, of complex-valued random variables indexed by some set $\hat{T}$. We are interested in bounding the moments of its supremum. In order to avoid measurability issues (in general, the supremum of an uncountable number of random variables might not be measurable any more) we define, for a subset $T \subset \hat{T}$, the so-called lattice supremum as

$$\mathbb{E} \sup_{t \in T} |X_t| := \sup \left\{ \mathbb{E} \sup_{t \in F} |X_t|, F \subset T, F \text{ finite} \right\}. \tag{6.36}$$

Note that for a countable set T, where no measurability problems can arise, $\mathbb{E}(\sup_{t \in T} |X_t|)$ equals the right hand side above. Dudley's inequality, which was originally formulated and shown in [45] for the expectation, bounds the moments

$\mathbb{E}\sup_{t\in T}|X_t|^p$ from above by a geometric quantity involving the covering numbers of T.

We endow $\hat{T}$ with the pseudometric

$$d(s,t) = (\mathbb{E}|X_t - X_s|^2)^{1/2}. \tag{6.37}$$

Recall that in contrast to a metric a pseudometric does not need to separate points, i.e., $d(s,t) = 0$ does not necessarily imply $s = t$. We assume that the increments of the process satisfy,

$$\mathbb{P}(|X_t - X_s| \geq u d(t,s)) \leq 2\exp(-u^2/2), \quad t,s \in \hat{T},\ u > 0. \tag{6.38}$$

We will later apply Dudley's inequality for the special case of Rademacher processes of the form

$$X_t = \sum_{j=1}^{M} \epsilon_j x_j(t), \quad t \in \hat{T}, \tag{6.39}$$

where $\boldsymbol{\epsilon} = (\epsilon_1, \ldots, \epsilon_M)$ is a Rademacher sequence and the $x_j : \hat{T} \to \mathbb{C}$ are some deterministic functions. Observe that

$$\begin{aligned} d(t,s)^2 = \mathbb{E}|X_t - X_s|^2 &= \mathbb{E}\left|\sum_{j=1}^{M} \epsilon_j (x_j(t) - x_j(s))\right|^2 \\ &= \sum_{j=1}^{M} (x_j(t) - x_j(s))^2 = \|\mathbf{x}(t) - \mathbf{x}(s)\|_2^2, \end{aligned}$$

where $\mathbf{x}(t)$ denotes the vector with components $x_j(t)$, $j = 1, \ldots, M$. Therefore, we define the (pseudo-)metric

$$d(s,t) = (\mathbb{E}|X_t - X_s|^2)^{1/2} = \|\mathbf{x}(t) - \mathbf{x}(s)\|_2. \tag{6.40}$$

Hoeffding's inequality (Proposition 6.11) shows that the Rademacher process (6.39) satisfies (6.38). Although we will need Dudley's inequality only for Rademacher processes here, we note that the original formulation was for Gaussian processes, see also [3, 79, 80, 99, 125].

For a subset $T \subset \hat{T}$, the covering number $N(T,d,\varepsilon)$ is defined as the smallest integer N such that there exists a subset $E \subset \hat{T}$ with cardinality $|E| = N$ satisfying

$$T \subset \bigcup_{t\in E} B_d(t,\varepsilon),$$

where $B_d(t,\varepsilon) = \{s \in \hat{T}, d(t,s) \leq \varepsilon\}$. In words, T can be covered with N balls of radius ε in the metric d. Note that some authors additionally require that $E \subset T$. For us $E \subset \hat{T}$ will be sufficient. Denote the diameter of T in the metric d by

$$\Delta(T) := \sup_{s,t\in T} d(s,t).$$

With these concepts at hand our version of Dudley's inequality reads as follows.

Theorem 6.23. *Let $X_t, t \in \hat{T}$, be a complex-valued process indexed by a pseudometric space $(\hat{T}, d)$ with pseudometric defined by* (6.37) *which satisfies* (6.38). *Then, for a subset $T \subset \hat{T}$ and any point $t_0 \in T$ it holds*

$$\mathbb{E} \sup_{t \in T} |X_t - X_{t_0}| \le C_1 \int_0^{\Delta(T)} \sqrt{\ln(N(T,d,u))} du + D_1 \Delta(T) \tag{6.41}$$

with constants $C_1 = 16.51$ and $D_1 = 4.424$. Furthermore, for $p \ge 2$,

$$\left(\mathbb{E} \sup_{t \in T} |X_t - X_{t_0}|^p\right)^{1/p} \le \beta^{1/p} \sqrt{p} \left(C \int_0^{\Delta(T)} \sqrt{\ln(N(T,d,u))} du + D\Delta(T)\right) \tag{6.42}$$

with constants $C = 14.372$, $D = 5.818$ and $\beta = 6.028$.

We note that the estimate (6.42) also holds for $1 \le p \le 2$ with possibly slightly different constants (this can be seen, for instance, via interpolation between $p = 1$ and $p = 2$). Further, the theorem and its proof easily extend to Banach space valued processes satisfying $\mathbb{P}(\|X_t - X_s\| > ud(t,s)) \le 2e^{-u^2/2}$. Inequality (6.42) for the increments of the process can be used in the following way to bound the supremum,

$$\begin{aligned} \left(\mathbb{E} \sup_{t \in T} |X_t|^p\right)^{1/p} &\le \inf_{t_0 \in T} \left(\left(\mathbb{E} \sup_{t \in T} |X_t - X_{t_0}|^p\right)^{1/p} + (\mathbb{E}|X_{t_0}|^p)^{1/p}\right) \\ &\le \beta^{1/p} \sqrt{p} \left(\int_0^{\Delta(T)} \sqrt{\log(N(T,d,u))} du + D\Delta(T)\right) + \inf_{t_0 \in T} (\mathbb{E}|X_{t_0}|^p)^{1/p}. \end{aligned}$$

The second term is usually easy to estimate. Further, note that for a centered real-valued process, that is, $\mathbb{E} X_t = 0$ for all $t \in \hat{T}$, we have

$$\mathbb{E} \sup_{t \in T} X_t = \mathbb{E} \sup_{t \in T} (X_t - X_{t_0}) \le \mathbb{E} \sup_{t \in T} |X_t - X_{t_0}|. \tag{6.43}$$

For completeness we also state the usual version of Dudley's inequality.

Corollary 6.24. *Let $X_t, t \in T$, be a real-valued centered process indexed by a pseudometric space (T, d) such that* (6.38) *holds. Then*

$$\mathbb{E} \sup_{t \in T} X_t \le 30 \int_0^{\Delta(T)} \sqrt{\ln(N(T,d,u))} du. \tag{6.44}$$

Proof. As in the proof of Theorem 6.23 below, we may assume without loss of generality that $\Delta(T) = 1$. Then it follows that $N(T,d,u) \ge 2$ for all $u < 1/2$. Indeed, if $N(T,d,u) = 1$ for some $u < 1/2$ then, for any $\epsilon > 0$, there would be two points of distance at least $1 - \epsilon$ that are covered by one ball of radius u, a contradiction to the triangle inequality. Therefore,

$$\int_0^{\Delta(T)} \sqrt{\ln(N(T,d,u))} du \ge \int_0^{1/2} \sqrt{\ln(2)} du = \frac{\sqrt{\ln 2}}{2} \Delta(T).$$

Therefore, (6.44) follows from (6.41) and (6.43) and the estimate

$$C_1 + \frac{2D_1}{\sqrt{\ln 2}} < 30. \qquad \square$$

Generalizations of Dudley's inequality are contained in [80, 125]; in particular, extensions to generic chaining inequalities, or bounds of suprema of random processes by means of majorizing measure conditions.

Proof of Theorem 6.23. Without loss of generality we may assume that the right hand sides of (6.41) and (6.42) are finite and non-vanishing. Otherwise, the statement becomes trivial. In particular, $0 < \Delta(T) < \infty$ and $N(T, d, u) < \infty$ for all $u > 0$. By eventually passing to a rescaled process $X'_t = X_t/\Delta(T)$ we may assume $\Delta(T) = 1$.

Now let $b > 1$ to be specified later. According to the definition of the covering numbers, there exist finite subsets $E_j \subset \hat{T}$, $j \in \mathbb{N} \setminus \{0\}$, of cardinality $|E_j| = N(T, d, b^{-j})$ such that

$$T \subset \bigcup_{t \in E_j} B_d(t, b^{-j}).$$

For each $t \in T$ and $j \in \mathbb{N} \setminus \{0\}$ we can therefore define $\pi_j(t) \in E_j$ such that

$$d(t, \pi_j(t)) \leq b^{-j}.$$

Further set $\pi_0(t) = t_0$. Then by the triangle inequality

$$d(\pi_j(t), \pi_{j-1}(t)) \leq d(\pi_j(t), t) + d(\pi_{j-1}(t), t) \leq (1+b) \cdot b^{-j} \quad \text{for all } j \geq 2$$

and $d(\pi_1(t), \pi_0(t)) \leq \Delta(T) = 1$. Therefore,

$$d(\pi_j(t), \pi_{j-1}(t)) \leq (1+b) \cdot b^{-j}, \quad \text{for all } j \geq 1. \tag{6.45}$$

Now we claim the chaining identity

$$X_t - X_{t_0} = \sum_{j=1}^{\infty} (X_{\pi_j(t)} - X_{\pi_{j-1}(t)}) \quad \text{almost surely.} \tag{6.46}$$

Indeed, by (6.38) we have

$$\mathbb{P}(|X_{\pi_j(t)} - X_{\pi_{j-1}(t)}| \geq b^{-j/2})$$
$$\leq \mathbb{P}\left(|X_{\pi_j(t)} - X_{\pi_{j-1}(t)}| \geq \frac{b^{j/2}}{1+b} d(\pi_j(t), \pi_{j-1}(t)) \right) \leq 2 \exp\left(-\frac{1}{2(1+b)^2} b^j \right).$$

This implies that $\sum_{j=1}^{\infty} \mathbb{P}(|X_{\pi_j(t)} - X_{\pi_{j-1}(t)}| \geq b^{-j/2}) < \infty$. It follows from the Borel–Cantelli lemma (Lemma 6.4) that the event that there exists an increasing sequence $j_\ell, \ell = 1, 2, \ldots$ of integers with $j_\ell \to \infty$ as $\ell \to \infty$ such that $|X_{\pi_{j_\ell}(t)} - X_{\pi_{j_\ell - 1}(t)}| \geq b^{-j/2}$ has zero probability. In conclusion, $|X_{\pi_j(t)} - X_{\pi_{j-1}(t)}| < b^{-j/2}$ for all $j \geq j_0$ and some j_0 holds almost surely. Consequently, the series on the right hand side of (6.46) converges almost surely. Furthermore,

$$\mathbb{E}\left|X_t - X_{t_0} - \sum_{j=1}^{J}(X_{\pi_j(t)} - X_{\pi_{j-1}(t)})\right|^2 = \mathbb{E}|X_t - X_{\pi_J(t)}|^2$$
$$= d(t, \pi_J(t))^2 \to 0 \quad (J \to \infty)$$

by definition (6.37) of the metric d and construction of the $\pi_j(t)$. The chaining identity (6.46) follows.

Now let F be a finite subset of T. Let $a_j > 0$, $j > 0$, be numbers to be determined later. For brevity of notation we write $N(T, d, b^{-j}) = N(b^{-j})$. Then

$$\begin{aligned}
&\mathbb{P}\left(\max_{t \in F} |X_t - X_{t_0}| > u \sum_{j=1}^{\infty} a_j\right) \\
&\leq \mathbb{P}\left(\max_{t \in F} \sum_{j=1}^{\infty} |X_{\pi_j(t)} - X_{\pi_{j-1}(t)}| > u \sum_{j=1}^{\infty} a_j\right) \\
&\leq \sum_{j=1}^{\infty} \mathbb{P}\left(\max_{t \in F} |X_{\pi_j(t)} - X_{\pi_{j-1}(t)}| > u a_j\right) \\
&\leq \sum_{j=1}^{\infty} N(b^{-j}) N(b^{-(j-1)}) \max_{t \in F} \mathbb{P}(|X_{\pi_j(t)} - X_{\pi_{j-1}(t)}| \geq u a_j) \\
&\leq \sum_{j=1}^{\infty} N(b^{-j}) N(b^{-(j-1)}) \max_{t \in F} \mathbb{P}\left(|X_{\pi_j(t)} - X_{\pi_{j-1}(t)}| \geq \frac{u a_j d(\pi_j(t), \pi_{j-1}(t))}{(1+b) \cdot b^{-j}}\right) \\
&\leq 2 \sum_{j=1}^{\infty} N(b^{-j}) N(b^{-(j-1)}) \exp\left(-\frac{u^2 (b^j a_j)^2}{2(1+b)^2}\right).
\end{aligned} \tag{6.47}$$

Hereby we used that the number of possible increments $|X_{\pi_j(t)} - X_{\pi_{j-1}(t)}|$ is bounded by the product $N(b^{-j})N(b^{-(j-1)})$ of the cardinalities of the sets E_j and E_{j-1}. Further, we have applied (6.45) and (6.38). Now for a number $\alpha > 0$ to be determined later, we choose

$$a_j = \sqrt{2}\alpha^{-1}(1+b) \cdot b^{-j} \sqrt{\ln(b^j N(b^{-j}) N(b^{-(j-1)}))}, \quad j \geq 1.$$

Continuing the chain of inequalities (6.47) yields, for $u \geq \alpha$,

$$\mathbb{P}\Big(\max_{t\in F}|X_t - X_{t_0}| > u\sum_{j=1}^{\infty} a_j\Big)$$
$$\leq 2\sum_{j=1}^{\infty} N(b^{-j})N(b^{-(j-1)})(b^j N(b^{-j})N(b^{-(j-1)}))^{-u^2/\alpha^2}$$
$$\leq 2\sum_{j=1}^{\infty} b^{-ju^2/\alpha^2} \leq 2b^{-u^2/\alpha^2}\sum_{j=0}^{\infty} b^{-j} = \frac{2b}{b-1}b^{-u^2/\alpha^2}.$$

Using $N(b^{-(j-1)}) \leq N(b^{-j})$ we further obtain

$$\Theta := \sum_{j=1}^{\infty} a_j \leq \sqrt{2}\alpha^{-1}(b+1)\sum_{j=1}^{\infty} b^{-j}\sqrt{j\ln(b) + 2\ln(N(b^{-j}))}$$
$$\leq \sqrt{2}\alpha^{-1}(b+1)\sum_{j=1}^{\infty} b^{-j}\sqrt{j\ln(b)} + 2\alpha^{-1}(b+1)\sum_{j=1}^{\infty} b^{-j}\sqrt{\ln(N(b^{-j}))}. \tag{6.48}$$

By comparing sums and integrals, the second sum in (6.48) is upperbounded by

$$\sum_{j=1}^{\infty} b^{-j}\sqrt{\ln(N(b^{-j}))} = \frac{b}{b-1}\sum_{j=1}^{\infty}\sqrt{\ln(N(b^{-j}))}\int_{b^{-(j+1)}}^{b^{-j}} du$$
$$\leq \frac{b}{b-1}\int_0^{b^{-1}} \sqrt{\ln(N(T,d,u))}du$$
$$\leq \frac{b}{b-1}\int_0^{\Delta(T)} \sqrt{\ln(N(T,d,u))}du,$$

where we additionally used that $N(T,d,b^{-j}) \leq N(T,d,u)$ for all $u \in [b^{-(j+1)}, b^{-j}]$. Plugging into (6.48) shows that

$$\Theta \leq C(b,\alpha)\Delta(T) + \frac{2b(b+1)}{\alpha(b-1)}\int_0^{\Delta(T)} \sqrt{\ln(N(T,d,u))}du \tag{6.49}$$

with

$$C(b,\alpha) := \sqrt{2}\alpha^{-1}(b+1)\sqrt{\ln(b)}\sum_{j=1}^{\infty} b^{-j}\sqrt{j}, \tag{6.50}$$

while

$$\mathbb{P}\Big(\max_{t\in F}|X_t - X_{t_0}| > u\Theta\Big) \leq \frac{2b}{b-1}b^{-u^2/\alpha^2}, \quad u \geq \alpha.$$

Using that any probability is bounded by 1, Proposition 6.1 yields for the moments

$$\begin{aligned}
\mathbb{E} \sup_{t\in F} |X_t - X_{t_0}|^p &= p \int_0^\infty \mathbb{P}\Big(\sup_{t\in F} |X_t - X_{t_0}| \geq v \Big) v^{p-1} dv \\
&= p\Theta^p \int_0^\infty \mathbb{P}\Big(\sup_{t\in F} |X_t - X_{t_0}| \geq u\Theta \Big) u^{p-1} du \\
&\leq p\Theta^p \Big(\int_0^\alpha u^{p-1} du + \frac{2b}{b-1} \int_\alpha^\infty b^{-u^2/\alpha^2} u^{p-1} du \Big) \\
&= p\Theta^p \Big(\frac{\alpha^p}{p} + \frac{2b}{b-1} \int_\alpha^\infty b^{-u^2/\alpha^2} u^{p-1} du \Big).
\end{aligned}$$

Taking the supremum over all finite subsets $F \subset T$ yields

$$\begin{aligned}
\Big(\mathbb{E} \sup_{t\in T} |X_t - X_{t_0}|^p \Big)^{1/p} \leq K_1(p,b,\alpha) \int_0^{\Delta(T)} & \sqrt{\ln(N(T,d,u))} du \\
&+ K_2(p,b,\alpha)\Delta(T),
\end{aligned}$$

where

$$K_1(p,b,\alpha) = p^{1/p} \frac{2b(b+1)}{\alpha(b-1)} \Big(\frac{\alpha^p}{p} + \frac{2b}{b-1} \int_\alpha^\infty b^{-u^2/\alpha^2} u^{p-1} du \Big)^{1/p},$$

and

$$K_2(p,b,\alpha) = p^{1/p} C(b,\alpha) \Big(\frac{\alpha^p}{p} + \frac{2b}{b-1} \int_\alpha^\infty b^{-u^2/\alpha^2} u^{p-1} du \Big)^{1/p}.$$

(Readers who do not care about the values of the constants may be satisfied at this point.) We choose $\alpha = \sqrt{2\ln(b)}$. Consider first $p = 1$. Lemma 10.2 in the Appendix yields

$$\int_\alpha^\infty b^{-u^2/\alpha^2} du = \int_{\sqrt{2\ln(b)}}^\infty e^{-u^2/2} du \leq \frac{1}{b\sqrt{2\ln(b)}}.$$

Hence,

$$\hat{K}_1(b) := K_1(1, b, \sqrt{2\ln(b)}) \leq \frac{2b(b+1)}{b-1} + \frac{2b(b+1)}{\ln(b)(b-1)^2}.$$

In order to estimate K_2 we note that, for $x < 1$,

$$\sum_{j=1}^\infty x^j \sqrt{j} \leq \sum_{j=1}^\infty x^j j = x \frac{d}{dx} \Big(\sum_{j=1}^\infty x^j \Big) = \frac{x}{(1-x)^2}.$$

Therefore,

$$C(b,\alpha) \leq \sqrt{2\ln(b)} \alpha^{-1} \frac{b(b+1)}{(b-1)^2}, \tag{6.51}$$

and

$$\hat{K}_2(b) := K_2(1, b, \sqrt{2\ln(b)}) \leq \frac{b(b+1)}{(b-1)^2}\sqrt{2\ln(b)} + \frac{b(b+1)}{(b-1)^3\ln(b)}.$$

The choice $b = 3.8$ yields $\hat{K}_1(3.8) \leq 16.51 = C_1$ and $\hat{K}_2(3.8) \leq 4.424 = D_1$. This yields the claim for $p = 1$.

Now assume $p \geq 2$. We use the Gamma function $\Gamma(x) = \int_0^\infty e^{-t}t^{x-1}dt$ and the inequality $\Gamma(x) \leq \frac{x^{x-1/2}}{e^{x-1}}$, for $x \geq 1$, see [81], to estimate (recall $\alpha = \sqrt{2\ln(b)}$)

$$\begin{aligned}
\int_\alpha^\infty b^{-u^2/\alpha^2}u^{p-1}du &= \int_{\sqrt{2\ln(b)}}^\infty e^{-u^2/2}u^{p-1}du \leq \int_0^\infty e^{-u^2/2}u^{p-1}du \\
&= 2^{p/2-1}\int_0^\infty e^{-t}t^{p/2-1}dt = 2^{p/2-1}\Gamma(p/2) \\
&\leq (2/e)^{p/2-1}(p/2)^{p/2-1/2} = \frac{e}{\sqrt{2p}}(p/e)^{p/2}.
\end{aligned}$$

This yields, recalling that $p \geq 2$ and using that $p^{1/(2p)} \leq e^{1/(2e)}$,

$$\begin{aligned}
&\hat{K}_1(p) := K_1(p, b, \sqrt{2\ln b}) \\
&\leq p^{1/p}\frac{2b(b+1)}{\sqrt{2\ln b}(b-1)}\left(\frac{(2\ln b)^{p/2}}{p} + \frac{2b}{b-1}\frac{e}{\sqrt{2p}}(p/e)^{p/2}\right)^{1/p} \\
&\leq \frac{2b(b+1)}{b-1} + p^{1/(2p)}\frac{2b(b+1)}{\sqrt{2\ln b}(b-1)}\left(\frac{\sqrt{2}eb}{b-1}\right)^{1/p}e^{-1/2}\sqrt{p} \\
&\leq \frac{\sqrt{2}b(b+1)}{b-1}\left(\frac{\sqrt{2}eb}{b-1}\right)^{1/p}\sqrt{p} + \frac{2e^{1/(2e)}e^{-1/2}b(b+1)}{\sqrt{2\ln b}(b-1)}\left(\frac{\sqrt{2}eb}{b-1}\right)^{1/p}\sqrt{p} \\
&= \sqrt{2}b(b+1)\left(\frac{1}{b-1} + \frac{e^{1/(2e)-1/2}}{\sqrt{\ln b}(b-1)}\right)\left(\frac{\sqrt{2}eb}{b-1}\right)^{1/p}\sqrt{p}.
\end{aligned}$$

Using (6.51) we estimate similarly

$$\begin{aligned}
\hat{K}_2(p) &:= K_2(p, b, \sqrt{2\ln b}) \\
&\leq p^{1/p}\frac{b(b+1)}{(b-1)^2}\left(\frac{(2\ln b)^{p/2}}{p} + \frac{2b}{b-1}\frac{e}{\sqrt{2p}}(p/e)^{p/2}\right)^{1/p} \\
&\leq \frac{b(b+1)\sqrt{2\ln b}}{(b-1)^2} + p^{1/(2p)}e^{-1/2}\frac{b(b+1)}{(b-1)^2}\left(\frac{\sqrt{2}eb}{b-1}\right)^{1/p}\sqrt{p} \\
&\leq \left(\frac{b(b+1)\sqrt{\ln b}}{(b-1)^2} + \frac{e^{1/(2e)-1/2}b(b+1)}{(b-1)^2}\right)\left(\frac{\sqrt{2}eb}{b-1}\right)^{1/p}\sqrt{p}.
\end{aligned}$$

In conclusion, inequality (6.42) holds with

$$\beta = \frac{\sqrt{2}eb}{b-1}, \quad C = \sqrt{2}b(b+1)\left(\frac{1}{b-1} + \frac{e^{1/(2e)-1/2}}{\sqrt{\ln b}(b-1)}\right),$$

$$D = \frac{b(b+1)\sqrt{\ln b}}{(b-1)^2} + \frac{e^{1/(2e)-1/2}b(b+1)}{(b-1)^2}.$$

Now we choose $b = 2.76$ to obtain $\beta = 6.028$, $C = 14.372$ and $D = 5.818$. This completes the proof. □

6.10 Deviation Inequalities for Suprema of Empirical Processes

The strong probability estimate of Theorem 4.4 depends on a deviation inequality for suprema of empirical processes that we present in this section. Let $\mathbf{Y}_1, \ldots, \mathbf{Y}_M$ be independent random vectors in $\mathbb{C}^n$ and let $\mathcal{F}$ be a countable collection of functions from $\mathbb{C}^n$ into $\mathbb{R}$. We are interested in the random variable

$$Z = \sup_{f \in \mathcal{F}} \sum_{\ell=1}^{M} f(\mathbf{Y}_\ell), \tag{6.52}$$

that is, the supremum of an empirical process. The next theorem estimates the probability that Z deviates much from its mean.

Theorem 6.25. *Let $\mathcal{F}$ be a countable set of functions $f : \mathbb{C}^n \to \mathbb{R}$. Let $\mathbf{Y}_1, \ldots, \mathbf{Y}_M$ be independent copies of a random vector $\mathbf{Y}$ on $\mathbb{C}^n$ such that $\mathbb{E} f(\mathbf{Y}) = 0$ for all $f \in \mathcal{F}$, and assume $f(\mathbf{Y}) \leq 1$ almost surely. Let Z be the random variable defined in (6.52) and $\mathbb{E}Z$ its expectation. Let $\sigma^2 > 0$ such that $\mathbb{E}[f(\mathbf{Y})^2] \leq \sigma^2$ for all $f \in \mathcal{F}$. Set $v_M = M\sigma^2 + 2\mathbb{E}Z$. Then, for all $t > 0$,*

$$\mathbb{P}(Z \geq \mathbb{E}Z + t) \leq \exp(-v_M h(t/v_M)) \leq \exp\left(-\frac{t^2}{2v_M + 2t/3}\right), \tag{6.53}$$

where $h(t) = (1+t)\ln(1+t) - t$.

This theorem, in particular, the left-hand inequality (6.53), is taken from [14]. The second inequality in (6.53) follows from $h(t) \geq \frac{t^2}{2+2t/3}$ for all $t > 0$. If $\mathcal{F}$ consists only of a single function f, then Theorem 6.25 reduces to the ordinary Bernstein or Bennett inequality [6, 134]. Hence, (6.53) can be viewed as a far reaching generalization of these inequalities.

The proof of (6.53), which uses the concept of entropy, is beyond the scope of these notes. We refer the interested reader to [14]. Deviation inequalities for suprema of empirical processes were already investigated in the 1980s by P. Massart and others, see e.g. [84, 1]. M. Talagrand obtained major breakthroughs in [122, 123], in particular, he obtained also a concentration inequality of the following type: Let $\mathbf{Y}_1, \ldots, \mathbf{Y}_M$

be independent random vectors and $|f(\mathbf{Y}_\ell)| \leq 1$ almost surely for all $f \in \mathcal{F}$ and all $\ell = 1, \ldots, M$. Then

$$\mathbb{P}(|Z - \mathbb{E}Z| \geq t) \leq 3 \exp\Big(- \frac{t}{C} \log \Big(1 + \frac{t}{M\sigma^2}\Big)\Big), \tag{6.54}$$

where $C > 0$ is a universal constant. The constants in the deviation and concentration inequalities were successfully improved in [85, 107, 108, 14, 77]. Extensions of deviation and concentration inequalities can be found in [10, 9, 79].

7 Proof of Nonuniform Recovery Result for Bounded Orthonormal Systems

In this chapter we prove Theorem 4.2.

7.1 Nonuniform Recovery with Coefficients of Random Signs

In order to obtain nonuniform recovery results we use the recovery condition for individual vectors, Corollary 2.9. In order to simplify arguments we also choose the signs of the non-zero coefficients of the sparse vector at random. A general recovery result reads as follows.

Proposition 7.1. *Let $A = (\mathbf{a}_1, \ldots, \mathbf{a}_N) \in \mathbb{C}^{m \times N}$ and let $S \subset [N]$ of size $|S| = s$. Assume A_S is injective and*

$$\|A_S^\dagger \mathbf{a}_\ell\|_2 \leq \alpha < 1/\sqrt{2} \quad \textit{for all } \ell \notin S, \tag{7.1}$$

where $A^\dagger$ is the Moore–Penrose pseudo-inverse of A_S. Let $\boldsymbol{\epsilon} = (\epsilon_j)_{j \in S} \in \mathbb{C}^s$ be a (random) Rademacher or Steinhaus sequence. Then with probability at least

$$1 - 2^{3/4}(N - s) \exp(-\alpha^{-2}/2)$$

every vector $\mathbf{x} \in \mathbb{C}^N$ with support S and $\operatorname{sgn}(\mathbf{x}^S) = \boldsymbol{\epsilon}$ is the unique solution to the ℓ_1-minimization problem (2.12).

Proof. In the Rademacher case the union bound and Hoeffding's inequality, Corollary 6.10, yield

$$\begin{aligned}\mathbb{P}\Big(\max_{\ell \notin S} |\langle A_S^\dagger \mathbf{a}_\ell, \operatorname{sgn}(\mathbf{x}_S)\rangle| \geq 1\Big) &\leq \sum_{\ell \notin S} \mathbb{P}(|\langle A_S^\dagger \mathbf{a}_\ell, \operatorname{sgn}(\mathbf{x}_S)\rangle| \geq \|A_S^\dagger \mathbf{a}_\ell\|_2 \alpha^{-1}) \\ &\leq (N - s) 2^{3/4} \exp(-\alpha^{-2}/2).\end{aligned}$$

In the Steinhaus case we even obtain a better estimate from Corollary 6.13. An application of Corollary 2.9 finishes the proof. □

In view of the previous proposition it is enough to show that $\|A_S^\dagger \mathbf{a}_\ell\|_2$ is small. The next statement indicates a way how to pursue this task.

Proposition 7.2. *Let $A \in \mathbb{C}^{m\times N}$ with coherence μ and let $S \subset [N]$ of size s. Assume that*

$$\|A_S^* A_S - \mathrm{Id}\|_{2\to 2} \le \delta \tag{7.2}$$

for some $\delta \in (0,1)$. Then

$$\|A_S^\dagger \mathbf{a}_\ell\|_2 \le \frac{\sqrt{s}\mu}{1-\delta} \quad \textit{for all } \ell \notin S.$$

Proof. By definition of the operator norm

$$\|A_S^\dagger \mathbf{a}_\ell\|_2 = \|(A_S^* A_S)^{-1} A_S^* \mathbf{a}_\ell\|_2 \le \|(A_S^* A_S)^{-1}\|_{2\to 2} \|A_S^* \mathbf{a}_\ell\|_2. \tag{7.3}$$

The Neumann series yields

$$\|(A_S^* A_S)^{-1}\|_{2\to 2} = \left\| \sum_{k=0}^\infty (\mathrm{Id} - A_S^* A_S)^k \right\|_{2\to 2} \le \sum_{k=0}^\infty \|\mathrm{Id} - A_S^* A_S\|_{2\to 2}^k$$
$$\le \sum_{k=0}^\infty \delta^k = \frac{1}{1-\delta}$$

by the geometric series formula. The second term in (7.3) can be estimated using the coherence,

$$\|A_S^* \mathbf{a}_\ell\|_2 = \sqrt{\sum_{j\in S} |\langle \mathbf{a}_\ell, \mathbf{a}_j\rangle|^2} \le \sqrt{s}\mu.$$

Combining the two estimates completes the proof. □

We note that in contrast to the usual definition of coherence, we do not require the columns of A to be normalized in the previous statement. Condition (7.2) is a different way of saying that the eigenvalues of $A_S^* A_S$ are contained in $[1-\delta, 1+\delta]$, or that the singular values of A_S are contained in $[\sqrt{1-\delta}, \sqrt{1+\delta}]$.

7.2 Condition Number Estimate for Column Submatrices

Let us return now to the situation of Theorem 4.4. Proposition 7.2 requires to provide an estimate on the coherence of A and on $\|A_S^* A_S - \mathrm{Id}\|_{2\to 2}$. The latter corresponds to a probabilistic condition number estimate of a column submatrix of the structured random matrix A of the form (4.4). The estimate of the coherence will follow as a corollary. (Note, however, that the coherence alone might be estimated with simpler tools, see for instance [78].) The main theorem of this section reads as follows.

Theorem 7.3. *Let $A \in \mathbb{C}^{m \times N}$ be the sampling matrix* (4.4) *associated to an orthonormal system that satisfies the boundedness condition* (4.2) *for some constant $K \geq 1$. Let $S \subset [N]$ be of cardinality $|S| = s \geq 2$. Assume that the random sampling points $t_1, \ldots, t_m$ are chosen independently according to the orthogonalization measure ν. Let $\delta \in (0, 1/2]$. Then with probability at least*

$$1 - 2^{3/4} s \exp\left(-\frac{m\delta^2}{\tilde{C} K^2 s}\right), \tag{7.4}$$

where $\tilde{C} = 9 + \sqrt{17} \approx 13.12$, the normalized matrix $\tilde{A} = \frac{1}{\sqrt{m}} A$ satisfies

$$\|\tilde{A}_S^* \tilde{A}_S - \mathrm{Id}\|_{2\to 2} \leq \delta.$$

We note that the theorem also holds for $1/2 \leq \delta < 1$ with a slightly larger constant $\tilde{C}$.

Proof. Denote by $\mathbf{X}_\ell = (\overline{\psi_j(t_\ell)})_{j \in S} \in \mathbb{C}^s$ a column vector of A_S^*. By independence of the t_ℓ these are i.i.d. random vectors. Their 2-norm is bounded by

$$\|\mathbf{X}_\ell\|_2 = \sqrt{\sum_{j \in S} |\psi_j(t_\ell)|^2} \leq K\sqrt{s}. \tag{7.5}$$

Furthermore,

$$\mathbb{E}(\mathbf{X}_\ell \mathbf{X}_\ell^*)_{j,k} = \mathbb{E}[\psi_k(t_\ell)\overline{\psi_j(t_\ell)}] = \int_{\mathcal{D}} \psi_j(t)\overline{\psi_k(t)} d\nu(t) = \delta_{j,k}, \quad j,k \in S,$$

or in other words, $\mathbb{E}\mathbf{X}_\ell \mathbf{X}_\ell^* = \mathrm{Id}$. Using symmetrization, Lemma 6.7, we estimate, for $p \geq 2$,

$$\begin{aligned} E_p := \mathbb{E}\|\tilde{A}_S^* \tilde{A}_S - \mathrm{Id}\|_{2\to 2}^p &= \mathbb{E}\left\|\frac{1}{m}\sum_{\ell=1}^m (\mathbf{X}_\ell \mathbf{X}_\ell^* - \mathbb{E}\mathbf{X}_\ell \mathbf{X}_\ell^*)\right\|_{2\to 2}^p \\ &\leq \left(\frac{2}{m}\right)^p \mathbb{E}\left\|\sum_{\ell=1}^m \epsilon_\ell \mathbf{X}_\ell \mathbf{X}_\ell^*\right\|_{2\to 2}^p, \end{aligned}$$

where $\boldsymbol{\epsilon} = (\epsilon_1, \ldots, \epsilon_m)$ is a Rademacher sequence, independent of $\mathbf{X}_1, \ldots, \mathbf{X}_m$. Now, we are in the position to apply Rudelson's lemma 6.18. To this end we note that A_S has rank at most s. Using Fubini's theorem and applying Rudelson's lemma conditional on $(\mathbf{X}_1, \ldots, \mathbf{X}_m)$ yields

$$\begin{aligned} E_p &\leq \left(\frac{2}{m}\right)^p 2^{3/4} s p^{p/2} e^{-p/2} \mathbb{E}\Big[\|A_S\|_{2\to 2}^p \max_{\ell \in [m]} \|\mathbf{X}_\ell\|_2^p\Big] \\ &\leq \left(\frac{2}{\sqrt{m}}\right)^p 2^{3/4} s p^{p/2} e^{-p/2} \sqrt{\mathbb{E}\|\tilde{A}_S^* \tilde{A}_S\|_{2\to 2}^p \mathbb{E}\Big[\max_{\ell \in [m]} \|\mathbf{X}_\ell\|_2^{2p}\Big]}. \end{aligned} \tag{7.6}$$

In the last step we applied the Cauchy–Schwarz inequality. Using the bound (7.5), which holds for all realizations of $\mathbf{X}_1, \dots, \mathbf{X}_m$, inserting the identity Id into the operator norm and applying the triangle inequality yields

$$E_p \le \left(\frac{2}{\sqrt{m}}\right)^p 2^{3/4} s p^{p/2} e^{-p/2} s^{p/2} K^p \sqrt{\mathbb{E}[(\|\tilde{A}_S^*\tilde{A}_S - \mathrm{Id}\|_{2\to 2} + 1)^p]}$$
$$\le \left(2K\sqrt{\frac{s}{m}}\right)^p 2^{3/4} s e^{-p/2} p^{p/2} ((\mathbb{E}\|\tilde{A}_S^*\tilde{A}_S - \mathrm{Id}\|_{2\to 2}^p)^{1/p} + 1)^{p/2}.$$

Denoting

$$D_{p,m,s} = 2K\sqrt{\frac{s}{m}} 2^{3/(4p)} s^{1/p} e^{-1/2}\sqrt{p}$$

we have deduced

$$E_p^{1/p} \le D_{p,m,s}\sqrt{E_p^{1/p} + 1}.$$

Squaring this inequality and completing the squares yields

$$(E_p^{1/p} - D_{p,m,s}^2/2)^2 \le D_{p,m,s}^2 + D_{p,m,s}^4/4,$$

which gives

$$E_p^{1/p} \le \sqrt{D_{p,m,s}^2 + D_{p,m,s}^4/4} + D_{p,m,s}^2/2. \tag{7.7}$$

Assuming $D_{p,m,s} \le 1/2$ this yields

$$E_p^{1/p} \le \sqrt{1 + \frac{1}{16}} D_{p,m,s} + \frac{1}{4} D_{p,m,s} = \kappa D_{p,m,s} \tag{7.8}$$

with $\kappa = \frac{\sqrt{17}+1}{4}$. Hence,

$$(\mathbb{E}\min\{1/2, \|\tilde{A}_S^*\tilde{A}_S - \mathrm{Id}\|_{2\to 2}^p\})^{1/p} \le \min\{1/2, (\mathbb{E}\|\tilde{A}_S^*\tilde{A}_S - \mathrm{Id}\|_{2\to 2}^p)^{1/p})\}$$
$$\le \kappa D_{p,M,s}.$$

It follows from Proposition 6.5 that for $u \ge \sqrt{2}$,

$$\mathbb{P}\left(\min\{1/2, \|\tilde{A}_S^*\tilde{A}_S - \mathrm{Id}\|_{2\to 2}\} \ge 2\kappa K\sqrt{\frac{s}{m}}u\right) \le 2^{3/4} s \exp(-u^2/2),$$

hence, for $2\kappa K\sqrt{\frac{2s}{m}} \le \delta \le 1/2$

$$\mathbb{P}(\|A_S^* A_S - \mathrm{Id}\|_{2\to 2} \ge \delta) \le 2^{3/4} s \exp\left(-\frac{m\delta^2}{8\kappa^2 K^2 s}\right). \tag{7.9}$$

The right hand side in (7.9) is less than ε provided

$$m \geq \frac{\tilde{C} K^2 s}{\delta^2} \ln(2^{3/4} s/\varepsilon) \tag{7.10}$$

with $\tilde{C} = 8\kappa^2 = (\sqrt{17}+1)^2/2 = 9 + \sqrt{17} \approx 13.12$. In order to have a non-trivial statement we must have $\varepsilon < 1$. In fact, for $s \geq 2$ condition (7.10) then implies that $\delta \geq 2\kappa K\sqrt{2s/m}$. We conclude that (7.9) holds trivially also for $0 < \delta < 2\kappa K\sqrt{2s/m}$, which finishes the proof. □

The above proof followed ideas contained in [115, 136]. Similar techniques were used in [92]. We remark that in the special case of the trigonometric system (examples (i) and (iv) in Section 4.1), the constant 13.12 in (7.4) can be essentially improved to $3e \approx 8.15$ (see [65] for the precise statement) by exploiting the algebraic structure of the Fourier system [102, 65]. Indeed, one may estimate $\mathbb{E}\|\frac{1}{m}A_S^* A_S - \mathrm{Id}\|_{S_{2n}}^{2n} = \mathbb{E}\,\mathrm{Tr}((\frac{1}{m}A_S A_S - \mathrm{Id})^n)$ directly in this case, i.e., without Rudelson's lemma or the Khintchine inequality. This approach, however, is more technical and uses elements from combinatorics.

Note furthermore that the conclusion of the theorem can be reformulated as follows: If for $\varepsilon \in (0,1), \delta \in (0,1/2]$ condition (7.10) holds, then with probability at least $1-\varepsilon$ the normalized matrix $\tilde{A} = \frac{1}{\sqrt{m}}A$ satisfies

$$\|\tilde{A}_S^* \tilde{A}_S - \mathrm{Id}\|_{2\to 2} \leq \delta.$$

The above proof also indicates how the boundedness condition (4.2) may be weakened. Indeed, the term $\mathbb{E}\max_{\ell\in[m]}\|\mathbf{X}_\ell\|_2^{2p}$ in (7.6) was estimated by $K^{2p}s^p$ using the boundedness condition (4.2). Instead, we might actually impose also finite moment conditions of the form

$$\sup_{j\in[N]} \int_{\mathcal{D}} |\psi_j(t)|^p d\nu(t) \leq K_p, \quad 2 \leq p < \infty.$$

A suitable growth condition on the constants K_p should then still allow a probabilistic estimate of $\|\tilde{A}_S^* \tilde{A}_S - \mathrm{Id}\|_{2\to 2}$ – possibly with a worse probability decay than in (7.4). Details remain to be worked out.

Let us now turn to the probabilistic estimate of the coherence of the matrix A in (4.4).

Corollary 7.4. *Let $A \in \mathbb{C}^{m\times N}$ be the sampling matrix* (4.4) *associated to an orthonormal system that satisfies the boundedness condition* (4.2) *for some constant $K \geq 1$. Then the coherence of the renormalized matrix $\tilde{A} = \frac{1}{\sqrt{m}}A$ satisfies*

$$\mu \leq \sqrt{\frac{2\tilde{C} K^2 \ln(2^{3/4} N^2/\varepsilon)}{m}}$$

with probability at least $1-\varepsilon$ *– provided the right hand side is at most* $1/2$. *The constant is the same as in the previous statement,* $\tilde{C} = 9+\sqrt{17} \approx 13.12$.

Proof. Let $S = \{j,k\}$ be a two element set. Then the matrix $\tilde{A}_S^*\tilde{A}_S - \mathrm{Id}$ contains $\langle \tilde{\mathbf{a}}_j, \tilde{\mathbf{a}}_k\rangle$ as a matrix entry. Since the absolute value of any entry of a matrix is bounded by the operator norm of the matrix on ℓ_2, we have

$$|\langle \tilde{\mathbf{a}}_j, \tilde{\mathbf{a}}_k\rangle| \le \|\tilde{A}_S^*\tilde{A}_S - \mathrm{Id}\|_{2\to 2}.$$

By Theorem 7.3 the probability that the operator norm on the right is *not* bounded by $\delta \in (0,1/2]$ is at most

$$2^{3/4}\cdot 2\exp\left(-\frac{m\delta^2}{\tilde{C}K^2\cdot 2}\right).$$

Taking the union bound over all $N(N-1)/2 \le N^2/2$ two element sets $S \subset [N]$ shows that

$$\mathbb{P}(\mu \ge \delta) \le 2^{3/4}N^2\exp\left(-\frac{m\delta^2}{2\tilde{C}K^2}\right).$$

Requiring that the right hand side is at most ε leads to the desired conclusion. □

7.3 Finishing the Proof

Now we complete the proof of Theorem 4.2. Set $\alpha = \frac{\sqrt{st}}{1-\delta}$ for some $t,\delta \in (0,1/2]$ to be chosen later. Let μ be the coherence of $\tilde{A} = \frac{1}{\sqrt{m}}A$. By Proposition 7.1 and Proposition 7.2 the probability that recovery by ℓ_1-minimization fails is bounded from above by

$$2^{3/4}(N-s)e^{-\alpha^{-2}/2}\mathbb{P}\Big(\max_{\ell\in[N]\setminus S}\|\tilde{A}_S^\dagger\tilde{\mathbf{a}}_\ell\|_2 \ge \alpha\Big)$$
$$\le 2^{3/4}(N-s)e^{-\alpha^{-2}/2} + \mathbb{P}(\|\tilde{A}_S^*\tilde{A}_S - \mathrm{Id}\|_{2\to 2} > \delta) + \mathbb{P}(\mu > t). \tag{7.11}$$

By Theorem 7.3 we have $\mathbb{P}(\|\tilde{A}_S^*\tilde{A}_S - \mathrm{Id}\|_{2\to 2} > \delta) \le \varepsilon$ provided

$$m \ge \frac{\tilde{C}K^2}{\delta^2}s\ln(2^{3/4}s/\varepsilon), \tag{7.12}$$

while Corollary 7.4 asserts that $\mathbb{P}(\mu > t) \le \varepsilon$ provided

$$m \ge \frac{2\tilde{C}K^2}{t^2}\ln(2^{3/4}N^2/\varepsilon). \tag{7.13}$$

Set $t = \delta\sqrt{\frac{2}{s}}$. Then (7.13) implies (7.12), and $\alpha = \frac{\delta\sqrt{2}}{1-\delta}$. The first term in (7.11) is then bounded by ε if

$$\delta^{-2} = 4\ln(2^{3/4}N/\varepsilon).$$

Plugging this into the definition of t and then into (7.13) we find that recovery by ℓ_1-minimization fails with probability at most 3ε provided

$$\begin{aligned} m &\geq \tilde{C} K^2 s \ln(2^{3/4} N/\varepsilon) \ln(2^{3/4} N^2/\varepsilon) \\ &= \tilde{C} K^2 s \ln(2^{3/4} N/\varepsilon)(\ln(N) + \ln(2^{3/4} N/\varepsilon)). \end{aligned}$$

Replacing ε by $\varepsilon/3$, this is satisfied if (4.18) holds with $C = 2\tilde{C}$. □

8 Proof of Uniform Recovery Result for Bounded Orthonormal Systems

In this chapter we first prove the theorem below concerning the restricted isometry constants δ_s of $\tilde{A} = \frac{1}{\sqrt{m}} A$, associated to random sampling in bounded orthogonal system, see (4.4). Rudelson and Vershynin have shown an analog result for discrete orthonormal systems in [116]. Later in Section 8.6 we strengthen Theorem 8.1 to Theorem 8.4, which ultimately shows Theorem 4.4.

Theorem 8.1. *Let $A \in \mathbb{C}^{m \times N}$ be the sampling matrix* (4.4) *associated to an orthonormal system that satisfies the boundedness condition* (4.2) *for some constant $K \geq 1$. Assume that the random sampling points $t_1, \ldots, t_m$ are chosen independently at random according to the orthogonalization measure ν. Suppose, for some $\varepsilon \in (0,1), \delta \in (0, 1/2]$, that*

$$\frac{m}{\ln(10m)} \geq D K^2 \delta^{-2} s \ln^2(100s) \ln(4N) \ln(7\varepsilon^{-1}) \tag{8.1}$$

where the constant $D \leq 163\,932$, then with probability at least $1 - \varepsilon$ the restricted isometry constant of the renormalized matrix $\frac{1}{\sqrt{m}} A$ satisfies $\delta_s \leq \delta$.

8.1 Start of Proof

We use the characterization of the restricted isometry constants in Proposition 2.5(b),

$$\delta_s = \max_{S \subset N, |S| \leq s} \|\tilde{A}_S^* \tilde{A}_S - \mathrm{Id}\|_{2 \to 2}.$$

Let us introduce the set

$$D_{s,N}^2 := \{\mathbf{z} \in \mathbb{C}^N, \|\mathbf{z}\|_2 = 1, \|\mathbf{z}\|_0 \leq s\} = \bigcup_{S \subset [N], |S| = s} \mathcal{S}_S^2,$$

where $\mathcal{S}_S^2 = \{\mathbf{z} \in \mathbb{C}^N, \|\mathbf{z}\|_2 = 1, \operatorname{supp} \mathbf{z} \subset S\}$. The quantity

$$|\!|\!| B |\!|\!|_s := \sup_{\mathbf{z} \in D_{s,N}^2} |\langle B\mathbf{z}, \mathbf{z} \rangle|$$

defines a norm on self-adjoint matrices $B = B^* \in \mathbb{C}^{N\times N}$ (a semi-norm on all of $\mathbb{C}^{N\times N}$), and

$$\delta_s = |||\tilde{A}^*\tilde{A} - \mathrm{Id}|||_s.$$

Let $\mathbf{X}_\ell = (\overline{\psi_j(t_\ell)})_{j=1}^N \in \mathbb{C}^N$ be the random column vector associated to the sampling point t_ℓ, $\ell \in [m]$. Then $\mathbf{X}_\ell^*$ is a row of A. Observe that $\mathbb{E}\mathbf{X}_\ell\mathbf{X}_\ell^* = \mathrm{Id}$ by the orthogonality relation 4.1. We can express the restricted isometry constant of $\tilde{A}$ as

$$\delta_s = \left|\left|\left|\frac{1}{m}\sum_{\ell=1}^m \mathbf{X}_\ell\mathbf{X}_\ell^* - \mathrm{Id}\right|\right|\right|_s = \frac{1}{m}\left|\left|\left|\sum_{\ell=1}^m (\mathbf{X}_\ell\mathbf{X}_\ell^* - \mathbb{E}\mathbf{X}_\ell\mathbf{X}_\ell^*)\right|\right|\right|_s. \tag{8.2}$$

Let us first consider the moments of δ_s. Using symmetrization (Lemma 6.7) we estimate, for $p \geq 1$,

$$\left(\mathbb{E}\left|\left|\left|\sum_{\ell=1}^m (\mathbf{X}_\ell\mathbf{X}_\ell^* - \mathbb{E}\mathbf{X}_\ell\mathbf{X}_\ell^*)\right|\right|\right|_s^p\right)^{1/p} \leq 2\left(\mathbb{E}\left|\left|\left|\sum_{\ell=1}^m \epsilon_\ell\mathbf{X}_\ell\mathbf{X}_\ell^*\right|\right|\right|_s^p\right)^{1/p}. \tag{8.3}$$

where $\boldsymbol{\epsilon} = (\epsilon_1, \ldots, \epsilon_m)$ is a Rademacher sequence, which is independent of the random sampling points t_ℓ, $\ell \in [m]$.

8.2 The Crucial Lemma

The following lemma, which heavily relies on Dudley's inequality, is key to the estimate of the moments in (8.3).

Lemma 8.2. *Let $\mathbf{x}_1, \ldots, \mathbf{x}_m$ be vectors in $\mathbb{C}^N$ with $\|\mathbf{x}_\ell\|_\infty \leq K$ for $\ell \in [m]$ and assume $s \leq m$. Then,*

$$\mathbb{E}\left|\left|\left|\sum_{\ell=1}^m \epsilon_\ell\mathbf{x}_\ell\mathbf{x}_\ell^*\right|\right|\right|_s \leq \tilde{C}_1 K\sqrt{2}\ln(100s)\sqrt{\ln(4N)\ln(10m)}\sqrt{\left|\left|\left|\sum_{\ell=1}^m \mathbf{x}_\ell\mathbf{x}_\ell^*\right|\right|\right|_s} \tag{8.4}$$

where $\tilde{C}_1 = 78.04$. Furthermore, for $p \geq 2$,

$$\left(\mathbb{E}\left|\left|\left|\sum_{\ell=1}^m \epsilon_\ell\mathbf{x}_\ell\mathbf{x}_\ell^*\right|\right|\right|_s^p\right)^{1/p}$$

$$\leq \beta^{1/p}\tilde{C}_2\sqrt{p}K\sqrt{s}\ln(100s)\sqrt{\ln(4N)\ln(10m)}\sqrt{\left|\left|\left|\sum_{\ell=1}^m \mathbf{x}_\ell\mathbf{x}_\ell^*\right|\right|\right|_s}, \tag{8.5}$$

where $\tilde{C}_2 \approx 67.97$ and $\beta = 6.028$ is the constant in Dudley's inequality (6.42).

Proof. We introduce the norm

$$\|\mathbf{z}\|_1^* := \sum_{j=1}^{N} (|\operatorname{Re}(z_j)| + |\operatorname{Im}(z_j)|), \quad \mathbf{z} \in \mathbb{C}^N,$$

which is the usual ℓ_1-norm after identification of $\mathbb{C}^N$ with $\mathbb{R}^{2N}$. By the Cauchy–Schwarz inequality we have $\|\mathbf{z}\|_1^* \leq \sqrt{2s}\|\mathbf{z}\|_2 = \sqrt{2s}$ for $\mathbf{z} \in D_{s,N}^2$.

Now we write out the norm on the left hand side of (8.4)

$$\begin{aligned}\left\|\sum_{\ell=1}^{m} \epsilon_\ell \mathbf{x}_\ell \mathbf{x}_\ell^*\right\|_s &= \sup_{\mathbf{z}\in D_{s,N}^2} \left|\sum_{\ell=1}^{m} \epsilon_\ell |\langle \mathbf{x}_\ell, \mathbf{z}\rangle|^2\right| \\ &\leq \sqrt{2s} \sup_{\mathbf{z}\in D_{s,N}^2} \left|\sum_{\ell=1}^{m} \epsilon_\ell |\langle \mathbf{x}_\ell, \mathbf{z}/\sqrt{\|\mathbf{z}\|_1^*}\rangle|^2\right|. \qquad (8.6)\end{aligned}$$

Introduce

$$X_z := \sum_{\ell=1}^{m} \epsilon_\ell |\langle \mathbf{x}_\ell, \mathbf{z}/\sqrt{\|\mathbf{z}\|_1^*}\rangle|^2, \quad \mathbf{z} \in \mathbb{C}^N \setminus \{0\}.$$

Then (8.6) is actually the supremum of this Rademacher process over $D_{s,N}^2$. For technical reasons we introduce $X_0 := 0$, so that then X_z is defined on all of $\mathbb{C}^N$. Additionally, we enlarge our original set of interest, $D_{s,N}^{2,0} = D_{s,N}^2 \cup \{0\}$. Then Dudley's inequality, Theorem 6.23, yields

$$\begin{aligned}E_p &:= \left(\mathbb{E}\left\|\sum_{\ell=1}^{m} \epsilon_\ell \mathbf{x}_\ell \mathbf{x}_\ell^*\right\|_s^p\right)^{1/p} = \left(\mathbb{E} \sup_{\mathbf{z}\in D_{s,N}^2} |X_z|^p\right)^{1/p} = \left(\mathbb{E} \sup_{\mathbf{z}\in D_{s,N}^{2,0}} |X_z - X_0|^p\right)^{1/p} \\ &\leq \beta^{1/p}\sqrt{p}\left(C\int_0^\infty \sqrt{\ln(N(D_{s,N}^{2,0}, d, u))}du + D\Delta(D_{s,N}^{2,0})\right),\end{aligned}$$

where $N(D_{s,N}^{2,0}, d, t)$ denotes the covering numbers and $\Delta(D_{s,N}^{2,0})$ the diameter of $D_{s,N}^{2,0} \subset \mathbb{C}^N$ with respect to the (pseudo-)metric d defined in (6.37), that is,

$$d(\mathbf{u}, \mathbf{v}) = \sqrt{\sum_{\ell=1}^{m} \left(\frac{|\langle \mathbf{x}_\ell, \mathbf{u}\rangle|^2}{\|\mathbf{u}\|_1^*} - \frac{|\langle \mathbf{x}_\ell, \mathbf{v}\rangle|^2}{\|\mathbf{v}\|_1^*}\right)^2}, \quad \mathbf{u}, \mathbf{v} \in \mathbb{C}^N \setminus \{0\},$$

and

$$d(\mathbf{u}, 0) = \sqrt{\sum_{\ell=1}^{m} \frac{|\langle \mathbf{x}_\ell, \mathbf{u}\rangle|^4}{(\|\mathbf{u}\|_1^*)^2}}.$$

Let $\mathbf{u}, \mathbf{v} \in D^2_{s,N}$ and assume without loss of generality that $\|\mathbf{u}\|_1^* \leq \|\mathbf{v}\|_1^*$. Then we can estimate

$$d(\mathbf{u},\mathbf{v}) = \Bigg(\sum_{\ell=1}^m \left(\frac{|\langle \mathbf{x}_\ell, \mathbf{u}\rangle|}{\|\mathbf{u}\|_1^*} - \frac{|\langle \mathbf{x}_\ell, \mathbf{v}\rangle|}{\|\mathbf{v}\|_1^*} \right)^2$$

$$\times \left(|\langle \mathbf{x}_\ell, \mathbf{u}\rangle| + |\langle \mathbf{x}_\ell, \mathbf{v}\rangle| - |\langle \mathbf{x}_\ell, \mathbf{u}\rangle \langle \mathbf{x}_\ell, \mathbf{v}\rangle| (1/\|\mathbf{u}\|_1^* - 1/\|\mathbf{v}\|_1^*)\right)^2 \Bigg)^{1/2}$$

$$\leq \max_{\ell\in[m]} \left| \frac{|\langle \mathbf{x}_\ell, \mathbf{u}\rangle|}{\|\mathbf{u}\|_1^*} - \frac{|\langle \mathbf{x}_\ell, \mathbf{v}\rangle|}{\|\mathbf{v}\|_1^*} \right| \sup_{\mathbf{u},\mathbf{v}\in D^2_{s,N}} \sqrt{\sum_{\ell=1}^m (|\langle \mathbf{x}_\ell, \mathbf{u}\rangle| + |\langle \mathbf{x}_\ell, \mathbf{v}\rangle|)^2}$$

$$\leq 2R \max_{\ell\in[m]} \left| \left\langle \mathbf{x}_\ell, \frac{\mathbf{u}}{\|\mathbf{u}\|_1^*} - \frac{\mathbf{v}}{\|\mathbf{v}\|_1^*} \right\rangle \right|,$$

where

$$R = \sup_{\mathbf{u}\in D^2_{s,N}} \sqrt{\sum_{\ell=1}^m |\langle \mathbf{x}_\ell, \mathbf{u}\rangle|^2} = \sqrt{\left\| \sum_{\ell=1}^m \mathbf{x}_\ell \mathbf{x}_\ell^* \right\|_s}.$$

It is even simpler to deduce $d(\mathbf{u}, 0) \leq R \max_{\ell\in[m]} |\langle \mathbf{x}_\ell, \frac{\mathbf{u}}{\|\mathbf{u}\|_1^*}\rangle|$. We further introduce the auxiliary pseudo-metric

$$\tilde{d}(\mathbf{u},\mathbf{v}) := \max_{\ell\in[m]} \left| \left\langle \mathbf{x}_\ell, \frac{\mathbf{u}}{\|\mathbf{u}\|_1^*} - \frac{\mathbf{v}}{\|\mathbf{v}\|_1^*} \right\rangle \right|, \quad \mathbf{u},\mathbf{v} \in \mathbb{C}^N \setminus \{0\}.$$

Using basic monotonicity and rescaling properties of the covering numbers and the diameter we obtain

$$E_p \leq 2\sqrt{2s}\beta^{1/p} R \Big(C \int_0^\infty \sqrt{\ln(N(D^{2,0}_{s,N}, \tilde{d}, t))} du + D\Delta(D^{2,0}_{s,N}, \tilde{d}) \Big), \tag{8.7}$$

where $\Delta(D^{2,0}_{s,N}, \tilde{d})$ denotes the diameter with respect to the pseudo-metric $\tilde{d}$.

The mapping $\mathbf{u} \mapsto h(\mathbf{u}) = \frac{\mathbf{u}}{\|\mathbf{u}\|_1^*}$ is a bijection between $D^2_{s,N}$ and

$$D^1_{s,N} := \{\mathbf{z} \in \mathbb{C}^N : \|\mathbf{z}\|_1^* = 1, \|\mathbf{z}\|_0 \leq s\},$$

which transforms the pseudo-metric $\tilde{d}$ into the pseudo-metric induced by the semi-norm

$$\|\mathbf{u}\|_X = \max_{\ell\in[m]} |\langle \mathbf{x}_\ell, \mathbf{u}\rangle|, \tag{8.8}$$

i.e., $\|h(\mathbf{u}) - h(\mathbf{v})\|_X = \tilde{d}(\mathbf{u},\mathbf{v})$. Distances to 0 are also left invariant under this transformation. This implies

$$N(D^{2,0}_{s,N}, \tilde{d}, t) = N(D^{1,0}_{s,N}, \|\cdot\|_X, t) \quad \text{for all } t > 0,$$

where $D^{1,0}_{s,N} = D^1_{s,N} \cup \{0\}$. Our task consists in estimating the latter covering numbers. To this end we distinguish between small and large values of t. For small values, we use the embedding $D^1_{s,N} \subset \mathcal{S}_1^N = \{\mathbf{x} \in \mathbb{C}^N, \|\mathbf{x}\|_1^* = 1\}$. □

8.3 Covering Number Estimate

The next lemma provides an estimate of the covering numbers of an arbitrary subset of $\mathcal{S}_1^N$.

Lemma 8.3. *Let U be a subset of $\mathcal{S}_1^N \cup \{0\}$ and $0 < t < \sqrt{2}K$. Then*

$$\sqrt{\ln(N(U, \|\cdot\|_X, t))} \leq 3K\sqrt{\ln(10m)\ln(4N)}t^{-1}.$$

Proof. Fix $\mathbf{x} \in U \setminus \{0\}$. The idea is to approximate $\mathbf{x}$ by a finite set of very sparse vectors. In order to find a vector $\mathbf{z}$ from this finite set that is close to $\mathbf{x}$ we use the so-called empirical method of Maurey [24, 98]. To this end we define a random vector $\mathbf{Z}$ that takes the value $\operatorname{sgn}(\operatorname{Re}(x_j))\mathbf{e}_j$ with probability $|\operatorname{Re}(x_j)|$ and the value $i\operatorname{sgn}(\operatorname{Im}(x_j))\mathbf{e}_j$ with probability $|\operatorname{Im}(x_j)|$ for $j = 1, \ldots, N$. Here, $\mathbf{e}_j$ denotes the jth canonical unit vector, $(\mathbf{e}_j)_k = \delta_{j,k}$. Since $\|\mathbf{x}\|_1^* = 1$ this is a valid probability distribution. Note that

$$\mathbb{E}\mathbf{Z} = \sum_{j=1}^N \operatorname{sgn}(\operatorname{Re}(x_j))|\operatorname{Re}(x_j)|\mathbf{e}_j + i\sum_{j=1}^N \operatorname{sgn}(\operatorname{Im}(x_j))|\operatorname{Im}(x_j)|\mathbf{e}_j = \mathbf{x}.$$

Let $\mathbf{Z}_1, \ldots, \mathbf{Z}_M$ be independent copies of $\mathbf{Z}$, where M is a number to be determined later. We attempt to approximate $\mathbf{x}$ with the M-sparse vector

$$\mathbf{z} = \frac{1}{M}\sum_{k=1}^M \mathbf{Z}_k.$$

We estimate the expected distance of $\mathbf{z}$ to $\mathbf{x}$ in $\|\cdot\|_X$ by first using symmetrization (Lemma 6.7),

$$\begin{aligned}\mathbb{E}\|\mathbf{z}-\mathbf{x}\|_X &= \mathbb{E}\left\|\frac{1}{M}\sum_{k=1}^M(\mathbf{Z}_k - \mathbb{E}\mathbf{Z}_k)\right\|_X \leq \frac{2}{M}\mathbb{E}\left\|\sum_{k=1}^M \epsilon_k\mathbf{Z}_k\right\|_X \\ &= \frac{2}{M}\mathbb{E}\max_{\ell\in[m]}\left|\sum_{k=1}^M \epsilon_k\langle \mathbf{x}_\ell, \mathbf{Z}_k\rangle\right|,\end{aligned}$$

where $\boldsymbol{\epsilon}$ is a Rademacher sequence, which is independent of $(\mathbf{Z}_1, \ldots, \mathbf{Z}_M)$. Now we fix a realization of $(\mathbf{Z}_1, \ldots, \mathbf{Z}_M)$ and consider only expectation and probability with respect to $\boldsymbol{\epsilon}$ for the moment (that is, conditional on $(\mathbf{Z}_1, \ldots, \mathbf{Z}_M)$). Since $\|\mathbf{x}_\ell\|_\infty \leq K$

and $\mathbf{Z}_k$ has only a single non-zero component of magnitude 1, we have $|\langle \mathbf{x}_\ell, \mathbf{Z}_k\rangle| \le K$. It follows that

$$\|(\langle \mathbf{x}_\ell, \mathbf{Z}_k\rangle)_{k=1}^M\|_2 \le \sqrt{M}K, \quad \ell \in [m].$$

Using Hoeffding's inequality (Proposition 6.11) we obtain

$$\mathbb{P}_\epsilon\left(\left|\sum_{k=1}^M \epsilon_k \langle \mathbf{x}_\ell, \mathbf{Z}_k\rangle\right| > K\sqrt{M}u\right) \le \mathbb{P}_\epsilon\left(\left|\sum_{k=1}^M \epsilon_k \langle \mathbf{x}_\ell, \mathbf{Z}_k\rangle\right| > \|(\langle \mathbf{x}_\ell, \mathbf{Z}_k\rangle)_{k=1}^M\|_2 u\right)$$
$$\le 2e^{-u^2/2}, \quad \text{for all } u > 0, \ell \in [m].$$

Lemma 6.6 yields

$$\mathbb{E} \max_{\ell\in[m]} \left|\sum_{k=1}^M \epsilon_k \langle \mathbf{x}_\ell, \mathbf{Z}_k\rangle\right| \le CK\sqrt{M}\sqrt{\ln(8m)} \tag{8.9}$$

with $C = \sqrt{2} + \frac{1}{4\sqrt{2}\ln(8)} \approx 1.499 < 1.5$. By Fubini's theorem we finally obtain

$$\mathbb{E}\|\mathbf{z}-\mathbf{x}\|_X \le \frac{2}{M}\mathbb{E}_\mathbf{Z}\mathbb{E}_\epsilon \max_{\ell\in[m]} \left|\sum_{k=1}^M \epsilon_k \langle \mathbf{x}_\ell, \mathbf{Z}_k\rangle\right| \le \frac{3K}{\sqrt{M}}\sqrt{\ln(8m)}.$$

This implies that there exists a vector of the form

$$\mathbf{z} = \frac{1}{M}\sum_{k=1}^M \mathbf{z}_k, \tag{8.10}$$

where each $\mathbf{z}_k$ is one of the vectors in $\{\pm\mathbf{e}_j, \pm i\mathbf{e}_j : j \in [N]\}$, such that

$$\|\mathbf{z}-\mathbf{x}\|_X \le \frac{3K}{\sqrt{M}}\sqrt{\ln(8m)}. \tag{8.11}$$

(Note that $\mathbf{z}$ has sparsity at most M.) In particular,

$$\|\mathbf{z}-\mathbf{x}\|_X \le t \tag{8.12}$$

provided

$$\frac{3K}{\sqrt{M}}\sqrt{\ln(8m)} \le t. \tag{8.13}$$

Each $\mathbf{z}_k$ takes $4N$ values, so that $\mathbf{z}$ can take at most $(4N)^M$ values. (Actually, it takes strictly less than $(4N)^M$ values, since if some $\mathbf{e}_j$ appears more than once in the sum, then it always appears with the same sign.) For each $\mathbf{x} \in U$ we can therefore find a vector $\mathbf{z}$ of the form (8.10) such that $\|\mathbf{x}-\mathbf{z}\|_X \le t$. Further, the vector 0 is covered as well, when simply adding it to the covering. (This enlarges the size of the covering only by one, and therefore it still has cardinality at most $(4N)^M$.) Hence, we provided a covering of U, also when $0 \in U$. The choice

$$M = \left\lfloor \frac{9K^2}{t^2} \ln(10m) \right\rfloor$$

satisfies (8.13). Indeed, then

$$\begin{aligned} M &\geq \frac{9K^2}{t^2} \ln(10m) - 1 = \frac{9K^2}{t^2} \ln(8m) + \frac{9K^2 \ln(10/8)}{t^2} - 1 \\ &\geq \frac{9K^2}{t^2} \ln(8m) + \frac{9 \ln(10/8)}{2} - 1 \geq \frac{9K^2}{t^2} \ln(8m) \end{aligned}$$

since $t \leq \sqrt{2}K$ and $\frac{9\ln(10/8)}{2} > 1$. Therefore, (8.13) is satisfied. We deduce that the covering numbers can be estimated by

$$\begin{aligned} \sqrt{\ln(N(U, \|\cdot\|_X, t))} &\leq \sqrt{\ln((4N)^M)} \leq \sqrt{\left\lfloor \frac{9K^2}{t^2} \ln(10m) \right\rfloor \ln(4N)} \\ &\leq 3K\sqrt{\ln(10m)\ln(4N)}\, t^{-1}. \end{aligned}$$

This completes the proof of the lemma. □

8.4 Finishing the Proof of the Crucial Lemma

The estimate of the covering number in the lemma of the previous section will be good for larger values of t. For small values of t we use a volumetric argument. First note that by duality of ℓ_1 and ℓ_∞ we have

$$\|\mathbf{x}\|_X = \max_{\ell \in [m]} |\langle \mathbf{x}_\ell, \mathbf{x} \rangle| \leq \|\mathbf{x}\|_1 \max_{\ell \in [m]} \|\mathbf{x}_\ell\|_\infty \leq \sqrt{2}\|\mathbf{x}\|_1^* K = \sqrt{2}K$$

for all $\mathbf{x} \in \mathbb{C}^N$ with $\|\mathbf{x}\|_1^* = 1$. Hence, $\mathcal{S}_S^1 = \{\mathbf{x} \in \mathbb{C}^N, \|\mathbf{x}\|_1^* = 1, \operatorname{supp} \mathbf{x} \subset S\}$ is contained in $\sqrt{2}KB_X$ where $B_X = \{\mathbf{x} \in \mathbb{C}^N, \|\mathbf{x}\|_X \leq 1\}$. Using Proposition 10.1 on covering numbers of subsets of unit spheres we obtain (identifying $\mathcal{S}_S^1$ with a subset of $\mathbb{R}^{2|S|}$)

$$N(\mathcal{S}_S^1 \cup \{0\}, \|\cdot\|_X, t) \leq \left(1 + \frac{2\sqrt{2}K}{t}\right)^{2|S|}.$$

Our set of interest can be written as

$$D_{s,N}^{1,0} = \bigcup_{S \subset [N], |S| \leq s} \mathcal{S}_S^1 \cup \{0\}.$$

There are

$$\binom{N}{s} = \frac{N(N-1)\cdots(N-s+1)}{s!} \leq \frac{N^s}{s!} = \frac{s^s}{s!}\frac{N^s}{s^s} \leq e^s \frac{N^s}{s^s} = \left(\frac{eN}{s}\right)^s$$

subsets of $[N]$ of cardinality s. Hence, by subadditivity of the covering numbers we obtain

$$N(D_{s,N}^{1,0}, \|\cdot\|_X, t) \leq (eN/s)^s \left(1 + \frac{2\sqrt{2}K}{t}\right)^{2s}. \tag{8.14}$$

Next we note that $D_{s,N}^{1,0} \subset \sqrt{2}K B_{\mathbf{X}}$, so that

$$\Delta(D_{s,N}^{1,0}, \|\cdot\|_X) \leq \sqrt{2}K. \tag{8.15}$$

We obtain, for $\kappa \in (0, \sqrt{2}K)$,

$$\begin{aligned}
I &:= \int_0^{\Delta(D_{s,N}^{2,0})} \sqrt{\ln(N(D_{s,N}^{2,0}, \tilde{d}, t))}dt = \int_0^{\sqrt{2}K} \sqrt{\ln(N(D_{s,N}^{1,0}, \|\cdot\|_X, t))}dt \\
&\leq \sqrt{s}\int_0^{\kappa} \sqrt{\ln(eN/s) + 2\ln(1 + 2\sqrt{2}Kt^{-1})}dt \\
&\quad + 3K\sqrt{\ln(10m)\ln(4N)} \int_{\kappa}^{\sqrt{2}K} t^{-1}dt \\
&\leq \kappa\sqrt{s}\sqrt{\ln(eN/s)} + 4K\sqrt{s}\int_0^{\kappa/(2\sqrt{2}K)} \sqrt{\ln(1+t^{-1})}dt \\
&\quad + 3K\sqrt{\ln(10m)\ln(4N)}\ln(\sqrt{2}K/\kappa) \\
&\leq \kappa\sqrt{s}\left(\sqrt{\ln(eN/s)} + \frac{4}{2\sqrt{2}}\sqrt{\ln(e(1+2\sqrt{2}K\kappa^{-1}))}\right) \\
&\quad + 3K\sqrt{\ln(10m)\ln(4N)}\ln(\sqrt{2}K/\kappa).
\end{aligned} \tag{8.16}$$

In the last step we have applied Lemma 10.3. The choice $\kappa = \frac{2\sqrt{2}}{20}K/\sqrt{s}$ yields

$$\begin{aligned}
I &\leq \frac{2\sqrt{2}K}{20}\sqrt{\ln(eN/s)} + \frac{K}{5}\sqrt{\ln(e(1+20\sqrt{s}))} \\
&\quad + 3K\sqrt{\ln(10m)\ln(4N)}\ln(\sqrt{100s}) \\
&\leq C_0 K\sqrt{\ln(10m)\ln(4N)}\ln(100s),
\end{aligned}$$

where $C_0 = \frac{\sqrt{2}}{10} + \frac{1}{5\sqrt{2}\ln(100)} + 3/2 \approx 1.67$. Hereby, we applied the inequality

$$\begin{aligned}\sqrt{\ln(e(1+20\sqrt{s}))} &\leq \sqrt{\ln(100s)/2 + \ln(21e/10)}\\ &\leq \frac{1}{\sqrt{2\ln(100)}} \ln(100s)\sqrt{\ln(21e/10)}\\ &\leq \frac{1}{\sqrt{2\ln(100)}} \ln(100s)\sqrt{\ln(10m)\ln(4N)}.\end{aligned}$$

Plugging the above estimate and (8.15) into (8.7) yields

$$\begin{aligned}E_p &\leq \beta^{1/p}\sqrt{s}2\sqrt{2}(C_0CK\sqrt{\ln(10m)\ln(4N)}\ln(100s) + \sqrt{2}DK)R\\ &\leq \beta^{1/p}\tilde{C}_2\sqrt{s}\sqrt{\ln(10m)\ln(4N)}\ln(100s)R,\end{aligned}$$

where, for $p \geq 2$, (and $N, m \geq 2$),

$$\tilde{C} = \tilde{C}_2 = 2\sqrt{2}\Big(C_0C + \frac{\sqrt{2}D}{C_0C\sqrt{\ln(20)\ln(8)\ln(100)}}\Big) \approx 67.97.$$

For the case $p = 1$ we can use the slight better constants C_1 and D_1 in Dudley's inequality (6.41) to obtain

$$\tilde{C} = \tilde{C}_1 = 2\sqrt{2}\Big(C_0C_1 + \frac{\sqrt{2}D_1}{C_1C\sqrt{\ln(20)\ln(8)\ln(100)}}\Big) \approx 78.04.$$

The proof of Lemma 8.2 is completed.

8.5 Completing the Proof of Theorem 8.1

We proceed similarly as in Section 7.2. Denote, for $p \geq 2$,

$$E_p := (\mathbb{E}\delta_s^p)^{1/p} = \left(\mathbb{E}\left|\!\left|\!\left|\frac{1}{m}\sum_{\ell=1}^m \mathbf{X}_\ell\mathbf{X}_\ell^* - \mathrm{Id}\right|\!\right|\!\right|_s^p\right)^{1/p}.$$

Then (8.3) together with Lemma 8.2 yields

$$\begin{aligned}E_p^p &\leq \left(\frac{2D_{N,m,s,p}}{\sqrt{m}}\right)^p \mathbb{E}\left|\!\left|\!\left|\frac{1}{m}\sum_{\ell=1}^m \mathbf{X}_\ell\mathbf{X}_\ell^* - \mathrm{Id}\right|\!\right|\!\right|_s^{p/2}\\ &\leq \left(\frac{2D_{N,m,s,p}}{\sqrt{m}}\right)^p \mathbb{E}\left(\left|\!\left|\!\left|\frac{1}{m}\sum_{\ell=1}^m \mathbf{X}_\ell\mathbf{X}_\ell^* - \mathrm{Id}\right|\!\right|\!\right|_s + 1\right)^{p/2},\end{aligned} \tag{8.17}$$

where

$$D_{N,m,s,p} = \beta^{1/p}\tilde{C}_2\sqrt{p}K\sqrt{s}\ln(100s)\sqrt{\ln(4N)\ln(10m)}.$$

Using the triangle inequality we conclude that

$$E_p \leq \frac{2D_{N,m,s,p}}{\sqrt{m}} \sqrt{E_p + 1}.$$

Proceeding in the same way as in Section 7.2, see (7.8), and setting $\kappa = \frac{\sqrt{17}+1}{4}$ yields

$$(\mathbb{E} \min\{1/2, \delta_s\}^p)^{1/p} \leq \frac{2\kappa D_{N,m,s,p}}{\sqrt{m}}$$
$$= \beta^{1/p} 2\kappa \tilde{C}_2 \sqrt{p} \sqrt{\frac{s}{m}} \sqrt{s} \ln(100s) \sqrt{\ln(4N)\ln(10m)}, \quad p \geq 2. \tag{8.18}$$

Proposition 6.5 shows that for all $u \geq 2$,

$$\mathbb{P}\left(\min\{1/2, \delta_s\} \geq 2\kappa e^{1/2} \tilde{C}_2 \sqrt{\frac{s}{m}} \ln(100s) \sqrt{\ln(4N)\ln(10m)}\, u \right) < 7e^{-u^2/2},$$

where we used that $\beta < 7$. Expressed differently, $\delta_s \leq \delta \leq 1/2$ with probability at least $1 - \varepsilon$ provided

$$m \geq D\delta^{-2} s \ln^2(100s) \ln(4N) \ln(10m) \ln(7\varepsilon^{-1})$$

with $D = 2(2\kappa e^{1/2}\tilde{C}_2)^2 \approx 163\,931.48 < 163\,932$.

8.6 Strengthening the Probability Estimate

In this section we slightly improve on Theorem 8.1. The next theorem immediately implies Theorem 4.4 by noting Theorems 2.6 and 2.7. Its proof uses the deviation inequality of Section 6.10.

Theorem 8.4. *Let A be the random sampling matrix* (4.4) *associated to random sampling in a bounded orthonormal system obeying* (4.2) *with some constant $K \geq 1$. Let $\varepsilon \in (0,1), \delta \in (0,1/2]$. If*

$$\frac{m}{\ln(10m)} \geq C\delta^{-2} K^2 s \ln^2(100s) \ln(4N), \tag{8.19}$$
$$m \geq D\delta^{-2} K^2 s \ln(\varepsilon^{-1}),$$

then with probability at least $1-\varepsilon$ the restricted isometry constant δ_s of $\frac{1}{\sqrt{m}}A$ satisfies $\delta_s \leq \delta$. The constants satisfy $C \leq 17\,190$ and $D \leq 456$.

Proof. Set $E = \mathbb{E}\delta_s$. Using Lemma 8.2 for $p = 1$ and proceeding similarly as in the preceding section we obtain

$$E \leq \frac{2D_{N,m,s,1}}{\sqrt{m}} \sqrt{E+1} = G_{N,m,s}\sqrt{E+1}$$

with

$$G_{N,m,s} = C' \sqrt{\frac{s}{m}} \ln(100s) \sqrt{\ln(10m)\ln(4N)}$$

and $C' = 2\tilde{C}_1$. It follows from (7.7) that, if

$$G_{N,m,s} \leq \sigma\delta, \quad \text{with } \sigma := 0.84 \tag{8.20}$$

for $\delta \leq 1/2$, then

$$E \leq \mathbb{E}\delta_s < 8\delta/9.$$

It remains to show that δ_s does not deviate much from its expectation with high probability. To this end we use the deviation inequality of Theorem 6.25. By definition of the norm $\|\!|\cdot|\!\|_s$ we can write

$$\begin{aligned} m\delta_s &= \left|\!\left|\!\left| \sum_{\ell=1}^m (\mathbf{X}_\ell \mathbf{X}_\ell^* - \mathrm{Id}) \right|\!\right|\!\right|_s = \sup_{S\subset[N],|S|\leq s} \left\| \sum_{\ell=1}^m (\mathbf{X}_\ell^S (\mathbf{X}_\ell^S)^* - \mathrm{Id}_S) \right\|_{2\to2} \\ &= \sup_{(z,w)\in Q_{s,N}^2} \mathrm{Re}\left(\left\langle \sum_{\ell=1}^m (\mathbf{X}_\ell \mathbf{X}_\ell^* - \mathrm{Id})\mathbf{z}, \mathbf{w} \right\rangle \right) \\ &= \sup_{(z,w)\in Q_{s,N}^{2,*}} \sum_{\ell=1}^m \mathrm{Re}\left(\left\langle \sum_{\ell=1}^m (\mathbf{X}_\ell \mathbf{X}_\ell^* - \mathrm{Id})\mathbf{z}, \mathbf{w} \right\rangle \right), \end{aligned}$$

where $\mathbf{X}_\ell^S$ denotes the vector $\mathbf{X}_\ell$ restricted to the entries in S, and

$$Q_{s,N}^2 = \bigcup_{S\subset[N],|S|\leq s} Q_{S,N},$$

where $Q_{S,N} = \{(\mathbf{z},\mathbf{w}) : \mathbf{z},\mathbf{w} \in \mathbb{C}^N, \|\mathbf{z}\|_2 = \|\mathbf{w}\|_2 = 1, \operatorname{supp}\mathbf{z}, \operatorname{supp}\mathbf{w} \subset S\}$. Further, let $Q_{s,N}^{2,*}$ denote a dense countable subset of $Q_{s,N}^2$. Introducing $f_{\mathbf{z},\mathbf{w}}(\mathbf{X}) = \mathrm{Re}(\langle(\mathbf{X}\mathbf{X}^* - \mathrm{Id})\mathbf{z}, \mathbf{w}\rangle)$ we therefore have

$$m^{-1}\delta_s = \sup_{(\mathbf{z},\mathbf{w})\in Q_{s,N}^{2,*}} \sum_{\ell=1}^m f_{\mathbf{z},\mathbf{w}}(\mathbf{X}_\ell).$$

Since $\mathbb{E}\mathbf{X}\mathbf{X}^* = \mathrm{Id}$ it follows that $f_{\mathbf{z},\mathbf{w}}(\mathbf{X}) = 0$. Let us check the boundedness of $f_{\mathbf{z},\mathbf{w}}$ for $(\mathbf{z},\mathbf{w}) \in Q_{S,N}$ with $|S| \leq s$,

$$\begin{aligned} |f_{\mathbf{z},\mathbf{w}}(\mathbf{X})| &\leq |\langle(\mathbf{X}\mathbf{X}^* - \mathrm{Id})\mathbf{z}, \mathbf{w}\rangle| \leq \|\mathbf{z}\|_2 \|\mathbf{w}\|_2 \|\mathbf{X}^S (\mathbf{X}^S)^* - \mathrm{Id}_S\|_{2\to2} \\ &\leq \|\mathbf{X}^S (\mathbf{X}^S)^* - \mathrm{Id}_S\|_{1\to1} = \max_{j\in S} \sum_{k\in S} |\psi_j(t)\overline{\psi_k(t)} - \delta_{j,k}| \\ &\leq sK^2 \end{aligned}$$

by the boundedness condition (4.2). Hereby, we used that the operator norm on ℓ_2 is bounded by the one on ℓ_1 for self-adjoint matrices, see (2.3) as well as the explicit expression (2.1) for $\|\cdot\|_{1\to1}$. For the variance term σ^2 we estimate

$$\begin{aligned}\mathbb{E}|f_{\mathbf{z},\mathbf{w}}(\mathbf{X}_\ell)|^2 &\le \mathbb{E}|\langle(\mathbf{X}\mathbf{X}^* - \mathrm{Id})\mathbf{z},\mathbf{w}\rangle|^2 = \mathbb{E}\mathbf{w}^*(\mathbf{X}\mathbf{X} - \mathrm{Id})\mathbf{z}((\mathbf{X}\mathbf{X}^* - \mathrm{Id})\mathbf{z})^*\mathbf{w}\\ &\le \|\mathbf{w}\|_2^2\mathbb{E}\|(\mathbf{X}\mathbf{X} - \mathrm{Id})\mathbf{z}((\mathbf{X}\mathbf{X}^* - \mathrm{Id})\mathbf{z})^*\|_{2\to2} = \mathbb{E}\|(\mathbf{X}\mathbf{X}^* - \mathrm{Id})\mathbf{z}\|_2^2\\ &= \mathbb{E}[\|\mathbf{X}\|_2^2|\langle\mathbf{X},\mathbf{z}\rangle|^2] - 2\mathbb{E}|\langle\mathbf{X},\mathbf{z}\rangle|^2 + 1.\end{aligned}$$

Hereby we used that $\|\mathbf{u}\mathbf{u}^*\|_{2\to2} = \|\mathbf{u}\|_2^2$. Observe that $\|X\|_2 \le \sqrt{s}K$ by the Cauchy–Schwarz inequality and the boundedness condition (4.2). Furthermore,

$$\mathbb{E}|\langle\mathbf{X},z\rangle|^2 = \sum_{j,k\in S} z_j\overline{z_k}\mathbb{E}[\psi_k(t)\overline{\psi_j(t)}] = \|\mathbf{z}\|_2^2 = 1$$

by orthogonality (4.1). Hence,

$$\begin{aligned}\mathbb{E}|f_{\mathbf{z},\mathbf{w}}(\mathbf{X}_\ell)|^2 &\le \mathbb{E}[\|X\|_2^2|\langle\mathbf{X},\mathbf{z}\rangle|^2] - 2\mathbb{E}|\langle\mathbf{X},\mathbf{z}\rangle|^2 + 1 \le (sK^2-2)\mathbb{E}|\langle\mathbf{X},\mathbf{z}\rangle|^2 + 1\\ &= sK^2 - 1 < sK^2.\end{aligned}$$

Now we are prepared to apply Theorem 6.25. Under condition (8.20) it gives

$$\begin{aligned}\mathbb{P}(\delta_s \ge \delta) &\le \mathbb{P}(\delta_s \ge \mathbb{E}\delta_s + \delta/9)\\ &= \mathbb{P}\left(\left\|\sum_{\ell=1}^m(\mathbf{X}_\ell\mathbf{X}_\ell^* - \mathrm{Id})\right\|_s \ge \mathbb{E}\left\|\sum_{\ell=1}^m(\mathbf{X}_\ell\mathbf{X}_\ell^* - \mathrm{Id})\right\|_s + \delta m/9\right)\\ &\le \exp\left(-\frac{(\delta m/9)^2}{2msK^2 + 4(8\delta/9)m + 2(\delta m/9)/3}\right)\\ &= \exp\left(-\frac{m\delta^2}{sK^2}\frac{1}{9^2(2 + 4\frac{8\delta}{9sK^2} + \frac{2\delta}{3\cdot9sK^2})}\right)\\ &\le \exp\left(-\frac{m\delta^2}{sK^2}\frac{1}{162 + 9\cdot32 + 6}\right)\\ &= \exp\left(-\frac{m\delta^2}{456sK^2}\right).\end{aligned}$$

In the third line, it was used a second time that $\mathbb{E}\|\sum_{\ell=1}^m(\mathbf{X}_\ell\mathbf{X}_\ell^* - \mathrm{Id}\|_s = m\mathbb{E}\delta_s \le m8\delta/9$. Also, note that $\delta/(sK^2) < 1$. It follows that $\delta_s \le \delta$ with probability at least $1-\varepsilon$ provided

$$m \ge 456\,\delta^{-2}K^2 s\ln(\varepsilon^{-1}).$$

Taking also (8.20) into account, we proved that $\delta_s \le \delta$ with probability at least $1 - \varepsilon$ provided that m satisfies the two conditions

$$\frac{m}{\ln(10m)} \ge C\delta^{-2}K^2 s \ln^2(100s)\ln(4N),$$
$$m \ge 456\,\delta^{-2}K^2 s \ln(\varepsilon^{-1}),$$

with $C = \sigma^{-2}(C')^2 = 4\sigma^{-2}\tilde{C}_1^2 \approx 17\,189.09 < 17\,190$. □

8.7 Notes

The estimates of the restricted isometry constants are somewhat related to the Λ_1-set problem [11, 12], where one aims at selecting a subset of characters (or bounded orthonormal functions), such that all their linear combinations have comparable L_1 and L_2-norms, up to a logarithmic factor, see [124, 66]. The paper [66] considers also the more involved problem of providing a Kashin splitting of a set of bounded orthonormal functions. It is interesting to note that the analysis in [66] also uses the norm $\|\cdot\|_X$ introduced in (8.8).

9 Proof of Recovery Theorem for Partial Circulant Matrices

The proof of Theorem 5.1 is based on Proposition 7.2, which requires to estimate the coherence of $A = \frac{1}{\sqrt{m}}\Phi^{\Theta}(\mathbf{b})$ and to provide a probabilistic estimate of $\|A_S^*A_S - \mathrm{Id}\|_{2\to 2}$, where $S = \mathrm{supp}(\mathbf{x})$. We start with the coherence estimate.

9.1 Coherence

The proof of the following coherence bound uses similar ideas as the one of Theorem 5.1 in [96].

Proposition 9.1. *Let μ be the coherence of the partial random circulant matrix $A = \frac{1}{\sqrt{m}}\Phi^{\Theta}(\boldsymbol{\epsilon}) \in \mathbb{R}^{m\times N}$, where $\boldsymbol{\epsilon}$ is a Rademacher sequence and $\Theta \subset [N]$ has cardinality m. For convenience assume that m is divisible by three. Then with probability at least $1 - \varepsilon$ the coherence satisfies*

$$\mu \le \sqrt{\frac{6\log(3N^2/\varepsilon)}{m}}. \tag{9.1}$$

Proof. The inner product between two different columns $\mathbf{a}_\ell, \mathbf{a}_k$, $\ell \ne k$, of A can be written

$$\langle \mathbf{a}_\ell, \mathbf{a}_k\rangle = \frac{1}{m}\sum_{j\in\Theta} \epsilon_{\ell-j}\epsilon_{k-j},$$

where here and in the following $\ell - j$ and $k - j$ is understood modulo N. The random variables $\tilde{\epsilon}_j = \epsilon_{\ell-j}\epsilon_{k-j}$, $j \in \Theta$, are again Rademacher variables by independence of the ϵ_j and since $\ell \neq k$. We would like to apply Hoeffding's inequality, but unfortunately the $\tilde{\epsilon}_j$, $j \in \Theta$, are not independent in general. Nevertheless, we claim that we can always partition the $\tilde{\epsilon}_j$, $j \in \Theta$ into three sets $\Theta_1, \Theta_2, \Theta_3$ of cardinality $m/3$, such that for each Θ_i the corresponding family $\{\tilde{\epsilon}_j, j \in \Theta_i\}$ forms a sequence of independent Rademacher variables. To this end consider the sets $G_j = \{\ell - j, k - j\}$, $j \in \Theta$, and let

$$H = \{j \in \Theta : \exists j' \in \Theta \text{ such that } G_j \cap G_{j'} \neq \emptyset\}.$$

The random variables $\tilde{\epsilon}_j$, $j \in H$, are not independent. In order to construct the desired splitting into three sets, consider the graph with vertices $j \in \Theta$. The graph contains an edge between j and j' if and only if $G_j \cap G_{j'} \neq \emptyset$. Since any $r \in \Theta$ can be contained in at most two sets G_j (once as $\ell - j$ and once as $k - j$), this graph has degree at most 2. The independence problems are caused by the connected components of the graph. In order to start with the construction of the three sets $\Theta_1, \Theta_2, \Theta_3$ we choose a connected component of the graph, and then one of its endpoints (that is, a vertex that is only connected to one other vertex). If the connected component is a cycle then we choose an arbitrary vertex as starting point. We then move along the connected component, and add the starting vertex j to Θ_1, the second vertex to Θ_2, the third to Θ_3, the fourth to Θ_1 etc. If the connected component is actually a cycle, and if the last vertex was to be added to Θ_1, then we add it to Θ_2 instead. It is easily seen that after dealing with the first connected component of the graph by this process, the random variables $\{\tilde{\epsilon}_j, j \in \Theta_i\}$, are independent for each $i = 1, 2, 3$. All random variables $\tilde{\epsilon}_j$ with j not being a member of the first connected component are independent of the already treated ones. So we may repeat this process with the next connected component in the same way, starting now with a Θ_i satisfying $|\Theta_i| \leq |\Theta_j|$, $j \in \{1, 2, 3\} \setminus \{i\}$. After going through all connected components in this way, we add each element of $\Theta \setminus H$ arbitrarily to one of the Θ_i, such that at the end $|\Theta_i| = m/3$ for $i = 1, 2, 3$. By construction, the random variables $\tilde{\epsilon}_j$, $j \in \Theta_i$, are independent for each $i = 1, 2, 3$. By the triangle inequality, the union bound, and Hoeffding's inequality (6.16) we obtain

$$\begin{aligned}
\mathbb{P}(|\langle \mathbf{a}_\ell, \mathbf{a}_k \rangle| \geq u) &\leq \sum_{i=1}^{3} \mathbb{P}\left(\frac{1}{m} \left| \sum_{j \in \Theta_i} \tilde{\epsilon}_j \right| \geq u/3 \right) \\
&= \sum_{i=1}^{3} \mathbb{P}\left(\left| \sum_{j \in \Theta_i} \tilde{\epsilon}_j \right| \geq \sqrt{|\Theta_i|} \frac{um}{3\sqrt{|\Theta_i|}} \right) \leq 2 \sum_{i=1}^{3} \exp\left(-\frac{u^2 m^2}{18|\Theta_i|} \right) \\
&\leq 6 \exp\left(-\frac{u^2 m}{6} \right).
\end{aligned} \tag{9.2}$$

Taking the union bound over all $N(N-1)/2$ possible pairs $\{\ell, k\} \subset [N]$ we get the coherence bound

$$\mathbb{P}(\mu \geq u) \leq 3N(N-1)e^{-u^2 m/6}.$$

This implies that the coherence satisfies $\mu \leq u$ with probability at least $1-\varepsilon$ provided

$$m \geq \frac{6}{u^2} \ln(3N^2/\varepsilon).$$

Yet another reformulation is the statement of the proposition. □

Note that (9.1) is a slight improvement with respect to Proposition III.2 in [105]. It implies a non-optimal estimate for the restricted isometry constants of A.

Corollary 9.2. *The restricted isometry constant of the renormalized partial random circulant matrix $A \in \mathbb{R}^{m\times N}$ (with m divisible by three) satisfies $\delta_s \leq \delta$ with probability exceeding $1-\varepsilon$ provided*

$$m \geq 6\delta^{-2} s^2 \ln(3N^2/\varepsilon).$$

Proof. Combine Proposition 9.1 with Proposition 2.10(c). □

9.2 Conditioning of Submatrices

Our key estimate, which will be presented next, is mainly based on the noncommutative Khintchine inequality for Rademacher chaos, Theorem 6.22.

Theorem 9.3. *Let $\Theta, S \subset [N]$ with $|\Theta| = m$ and $|S| = s \in \mathbb{N}$. Let $\boldsymbol{\epsilon} \in \mathbb{R}^N$ be a Rademacher sequence. Denote $A = \frac{1}{\sqrt{m}}\Phi^\Theta(\boldsymbol{\epsilon})$ and assume, for $\varepsilon \in (0, 1/2]$, $\delta \in (0,1)$,*

$$m \geq 16\delta^{-2} s \ln^2(2^{5/2} s/\varepsilon), \tag{9.3}$$

Then with probability at least $1-\varepsilon$ it holds $\|A_S^* A_S - \mathrm{Id}\|_{2\to 2} \leq \delta$.

Proof. Let us denote $H_S = A_S^* A_S - \mathrm{Id}_S$. We introduce the elementary shift operators on $\mathbb{R}^N$,

$$(T_j \mathbf{x})_\ell = x_{\ell - j \bmod N}, \quad j = 1, \ldots, N.$$

Further, denote by $R_\Theta : \mathbb{C}^N \to \mathbb{C}^\Theta$ the operator that restricts a vector to the indices in Θ. Then we can write

$$\Phi^\Theta(\boldsymbol{\epsilon}) = R_\Theta \sum_{j=1}^N \epsilon_j T_j. \tag{9.4}$$

We introduce $R_S^*: \mathbb{C}^S \to \mathbb{C}^N$ to be the extension operator that fills up a vector in $\mathbb{C}^S$ with zeros outside S. Observe that

$$A_S^* A_S = \frac{1}{m} \sum_{j=1}^N \epsilon_j R_S T_j^* R_\Theta^* \sum_{k=1}^N \epsilon_k R_\Theta T_k R_S^*$$

$$= \frac{1}{m} \sum_{\substack{j,k=1 \\ j \neq k}}^N \epsilon_j \epsilon_k R_S T_j^* P_\Theta T_k R_S^* + \frac{1}{m} R_S \Big(\sum_{j=1}^N T_j^* P_\Theta T_j \Big) R_S^*,$$

where $P_\Theta = R_\Theta^* R_\Theta$ denotes the projection operator which cancels all components of a vector outside Θ. It is straightforward to check that

$$\sum_{j=1}^N T_j^* P_\Theta T_j = m \mathrm{Id}_N, \tag{9.5}$$

where Id_N is the identity on $\mathbb{C}^N$. Since $R_S R_S^* = \mathrm{Id}_S$ we obtain

$$H_S = \frac{1}{m} \sum_{j \neq k} \epsilon_j \epsilon_k R_S T_j^* P_\Theta T_k R_S^* = \frac{1}{m} \sum_{j \neq k} \epsilon_j \epsilon_k B_{j,k}$$

with $B_{j,k} = R_S T_j^* P_\Theta T_k R_S^*$. Our goal is to apply the noncommutative Khintchine inequality for decoupled Rademacher chaos, Theorem 6.22. To this end we first observe that by (9.5)

$$\sum_{j=1}^N B_{j,k}^* B_{j,\ell} = R_S T_k^* P_\Theta \Big(\sum_{j=1}^N T_j P_S T_j^* \Big) P_\Theta T_\ell R_S^* = s R_S T_k^* P_\Theta T_\ell R_S^*.$$

Using (9.5) once more this yields

$$\sum_{j,k=1}^N B_{j,k}^* B_{j,k} = s R_S \Big(\sum_{k=1}^N T_k^* P_\Theta T_k \Big) R_S^* = s m R_S R_S^* = s m \mathrm{Id}_S.$$

Since the entries of all matrices $B_{j,k}$ are non-negative we get

$$\Big\| \Big(\sum_{j \neq k} B_{j,k}^* B_{j,k} \Big)^{1/2} \Big\|_{S_{2n}}^{2n} = \mathrm{Tr} \Big(\sum_{j \neq k} B_{j,k}^* B_{j,k} \Big)^n$$

$$\leq \mathrm{Tr} \Big(\sum_{j,k} B_{j,k}^* B_{j,k} \Big)^n = \mathrm{Tr}(s m \mathrm{Id}_S)^n = s^{n+1} m^n.$$

Furthermore, since $B_{j,k}^* = B_{k,j}$ we have $\sum_{j \neq k} B_{j,k}^* B_{j,k} = \sum_{j \neq k} B_{j,k} B_{j,k}^*$. Let F denote the block matrix $F = (\tilde{B}_{j,k})_{j,k}$ where $\tilde{B}_{j,k} = B_{j,k}$ if $j \neq k$ and $\tilde{B}_{j,j} = 0$.

Expressing the product $(F^*F)^n$ as multiple sums over the block-components $\tilde{B}_{j,k}$ and applying the trace yields

$$\begin{aligned}
\|F\|_{S_{2n}}^{2n} &= \mathrm{Tr}[(F^*F)^n] \\
&= \mathrm{Tr}\Big[\sum_{\substack{j_1,j_2,\dots,j_n=1\\ k_1,k_2,\dots,k_n=1}}^{N} \tilde{B}_{j_1,k_1}^* \tilde{B}_{j_1,k_2} \tilde{B}_{j_2,k_2}^* \tilde{B}_{j_2,k_3} \cdots \tilde{B}_{j_n,k_n}^* \tilde{B}_{j_n,k_1}\Big] \\
&\le \mathrm{Tr} \sum_{k_1,\dots,k_n=1}^{N} \Big[\sum_{j_1=1}^{N} B_{j_1,k_1}^* B_{j_1,k_2} \cdots \sum_{j_n=1}^{N} B_{j_n,k_n}^* B_{j_n,k_1}\Big] \\
&= s^n\, \mathrm{Tr} \sum_{k_1,\dots,k_n=1}^{N} [R_S T_{k_1}^* P_\Theta T_{k_2} R_S^* R_S T_{k_2}^* P_\Theta T_{k_3} R_S^* \cdots R_S T_{k_n}^* P_\Theta T_{k_1} R_S^*],
\end{aligned}$$

where we applied also (9.5) once more. In the inequality step we used again that the entries of all matrices are non-negative. Using the cyclicity of the trace and applying (9.5) another time, together with the fact that $T_k = T_{-k \bmod N}^*$, gives

$$\begin{aligned}
\|F\|_{S_{2n}}^{2n} &\le s^n\, \mathrm{Tr}\Big[\sum_{k_1=1}^{N} T_{k_1} P_S T_{k_1}^* P_\Theta \sum_{k_2=1}^{N} T_{k_2} P_S T_{k_2}^* P_\Theta \cdots \sum_{k_n=1}^{N} T_{k_n} P_S T_{k_n}^* P_\Theta\Big] \\
&= s^{2n}\, \mathrm{Tr}[P_\Theta] = m s^{2n}.
\end{aligned}$$

Since by assumption (9.3) $s \le m$ it follows that

$$\max\left\{ \Big\|\Big(\sum_{j\neq k} B_{j,k}^* B_{j,k}\Big)^{1/2}\Big\|_{S_{2n}}^{2n}, \Big\|\Big(\sum_{j\neq k} B_{j,k} B_{j,k}^*\Big)^{1/2}\Big\|_{S_{2n}}^{2n}, \|F\|_{S_{2n}}^{2n} \right\} \le m^n s^{n+1}.$$

Using $\|H_S\|_{2\to 2} = \|H_S\|_{S_\infty} \le \|H_S\|_{S_p}$ and applying the decoupling Lemma 6.21 and the Khintchine inequality in Theorem 6.22 we obtain for an integer n

$$\begin{aligned}
\mathbb{E}\|H_S\|_{2\to 2}^{2n} &= \mathbb{E}\|A_S^* A_S - \mathrm{Id}_S\|_{2\to 2}^{2n} \le \mathbb{E}\|A_S^* A_S - \mathrm{Id}_S\|_{S_{2n}}^{2n} \\
&= \frac{1}{m^{2n}} \mathbb{E}\Big\|\sum_{j\neq k} \epsilon_j \epsilon_k B_{j,k}\Big\|_{S_{2n}}^{2n} \\
&\le \frac{4^{2n}}{m^{2n}} \mathbb{E}\Big\|\sum_{j\neq k} \epsilon_j \epsilon_k' B_{j,k}\Big\|_{S_{2n}}^{2n} \le 2 \cdot 4^{2n} \left(\frac{(2n)!}{2^n n!}\right)^2 \frac{s^{n+1}}{m^n}.
\end{aligned}$$

Here $\boldsymbol{\epsilon}'$ denotes a Rademacher sequence, inpendent of $\boldsymbol{\epsilon}$. Let $p = 2n + 2\theta = (1-\theta)2n + \theta(2n+2)$ with $\theta \in [0,1]$. Applying Hölder's inequality, see also (6.12), and the series of inequalities in (6.13) yields

$$\begin{aligned}\mathbb{E}\|H_S\|_{2\to 2}^{2n+2\theta} &\le (\mathbb{E}\|H_S\|_{2\to 2}^{2n})^{1-\theta}(\mathbb{E}\|H_S\|_{2\to 2}^{2n+2})^{\theta}\\ &\le 2\cdot 4^{2n+2\theta}\left(\left(\frac{(2n)!}{2^n n!}\right)^{1-\theta}\left(\frac{(2(n+1))!}{2^{n+1}(n+1)!}\right)^{\theta}\right)^2\frac{s^{n+\theta+1}}{m^{n+\theta}}\\ &\le 2\cdot 2^{3/2}4^{2n+2\theta}(2/e)^{2n+2\theta}(n+\theta)^{2n+2\theta}\frac{s^{n+\theta+1}}{m^{n+\theta}}.\end{aligned}$$

In other words, for $p \ge 2$,

$$(\mathbb{E}\|H_S\|_{2\to 2}^{p})^{1/p} \le 4e^{-1}\sqrt{\frac{s}{m}}(2^{5/2}s)^{1/p}p.$$

An application of Proposition 6.5 yields

$$\mathbb{P}\left(\|H_S\|_{2\to 2} \ge 4\sqrt{\frac{s}{m}}u\right) \le 2^{5/2}se^{-u} \tag{9.6}$$

for all $u \ge 2$. Note that $s \ge 1$ implies $2^{5/2}se^{-u} \ge 1/2$ for $u < 2$. Therefore, setting the right hand side equal $\varepsilon \le 1/2$ yields $u \ge 2$. In particular, $\|H_S\| \le \delta$ with probability at least $1-\varepsilon$ provided (9.3) holds true. □

9.3 Completing the Proof

Let us now complete the proof of Theorem 5.1. Set

$$\alpha = \frac{\sqrt{st}}{1-\delta} \tag{9.7}$$

for some $t, \delta \in (0,1)$ to be chosen later such that $\alpha < 1/\sqrt{2}$. According to Propositions 7.1 and 7.2, the probability that recovery fails is bounded from above by

$$2^{3/4}(N-s)e^{-\alpha^{-2}/2} + \mathbb{P}(\|A_S^*A_S - \mathrm{Id}\|_{2\to 2} > \delta) + \mathbb{P}(\mu > t). \tag{9.8}$$

By Theorem 9.3 we have $\mathbb{P}(\|A_S^*A_S - \mathrm{Id}\|_{2\to 2} > \delta) \le \varepsilon/3$ provided

$$m \ge 16\delta^{-2}s\ln^2(3\cdot 2^{5/2}s/\varepsilon), \tag{9.9}$$

and Proposition 9.1 yields $\mathbb{P}(\mu > t) \le \varepsilon/3$ if

$$m \ge 6t^{-2}\ln(9N^2/\varepsilon). \tag{9.10}$$

The first term of (9.8) equals $\varepsilon/3$ for

$$\alpha = \frac{1}{\sqrt{2\ln(2^{3/4}\cdot 3(N-s)/\varepsilon)}} < \frac{1}{\sqrt{2}}.$$

Solving for t in (9.7) gives

$$t = \frac{1-\delta}{\sqrt{2s \ln(2^{3/4} \cdot 3(N-s)/\varepsilon)}},$$

and plugging into (9.10) yields the condition

$$m \geq \frac{12s}{(1-\delta)^2} \ln(9N^2/\varepsilon) \ln(2^{3/4} \cdot 3(N-s)/\varepsilon). \tag{9.11}$$

Choose $\delta = 8/15$. Then (5.1) implies both (9.9) and (9.11). □

10 Appendix

Here we show some lemmas that are needed in some of the proofs.

10.1 Covering Numbers for the Unit Ball

Proposition 10.1. *Let $\|\cdot\|$ be some semi-norm on $\mathbb{R}^n$ and let U be a subset of the unit ball $B = \{\mathbf{x} \in \mathbb{R}^n, \|\mathbf{x}\| \leq 1\}$. Then the covering numbers satisfy, for $t > 0$,*

$$N(U, \|\cdot\|, t) \leq \left(1 + \frac{2}{t}\right)^n. \tag{10.1}$$

Proof. If $\|\cdot\|$ fails to be a norm, we consider the quotient space $X = \mathbb{R}^n/\mathcal{N}$ where $\mathcal{N} = \{\mathbf{x} \in \mathbb{R}^n, \|\mathbf{x}\| = 0\}$ is the kernel of $\|\cdot\|$. Then $\|\cdot\|$ is a norm on X, and the latter is isomorphic to $\mathbb{R}^{n-d}$, where d is the dimension of $\mathcal{N}$. Hence, we may assume without loss of generality that $\|\cdot\|$ is actually a norm.

Let $\{\mathbf{x}_1, \ldots, \mathbf{x}_N\} \subset U$ be a maximal t packing of U, that is, a maximal set satisfying $d(\mathbf{x}_i, \mathbf{x}_j) > t$ for all $i \neq j$. Then the balls $B(\mathbf{x}_\ell, t/2) = \{\mathbf{x} \in \mathbb{R}^n, \|\mathbf{x} - \mathbf{x}_\ell\| \leq t/2\}$ do not intersect and they are contained in the scaled unit ball $(1 + t/2)B$. By comparing volumes (that is, Lebesgue measures) of the involved balls we get

$$\mathrm{vol}\left(\bigcup_{\ell=1}^{N} B(\mathbf{x}_\ell, t/2)\right) = N \,\mathrm{vol}\left((t/2)B\right) \leq \mathrm{vol}\left((1 + t/2)B\right).$$

(Note that $\mathrm{vol}(B) < \infty$ since $\|\cdot\|$ is a norm.) On $\mathbb{R}^n$ the volume satisfies $\mathrm{vol}(tB) = t^n \,\mathrm{vol}(B)$, hence, $N(t/2)^n \,\mathrm{vol}(B) \leq (1 + t/2)^n \,\mathrm{vol}(B)$ or $N \leq (1 + 2/t)^n$. To conclude the proof, observe that the balls $B(\mathbf{x}_\ell, t)$, $\ell = 1, \ldots, N$ form a covering of U. Indeed, if there were an $\mathbf{x} \in U$ that is not covered, then $d(\mathbf{x}_\ell, \mathbf{x}) > t$, so that $\mathbf{x}$ could be added to the packing. But this is a contradiction to the maximality of the packing. □

10.2 Integral Estimates

This section contains estimates for two integrals.

Lemma 10.2. *For $u > 0$ it holds*

$$\int_u^\infty e^{-t^2/2}dt \le \min\left\{\sqrt{\frac{\pi}{2}}, \frac{1}{u}\right\}\exp(-u^2/2).$$

Proof. A change of variables yields

$$\int_u^\infty e^{-t^2/2}dt = \int_0^\infty e^{-\frac{(t+u)^2}{2}}dt = e^{-u^2/2}\int_0^\infty e^{-tu}e^{-t^2/2}dt.$$

On the one hand, using that $e^{-tu} \le 1$ for $t, u \ge 0$, we get

$$\int_u^\infty e^{-t^2/2}dt \le e^{-u^2/2}\int_0^\infty e^{-t^2/2}dt = \sqrt{\frac{\pi}{2}}e^{-u^2/2}.$$

On the other hand, using that $e^{-t^2} \le 1$ for $t \ge 0$ yields

$$\int_u^\infty e^{-t^2/2}dt \le e^{-u^2/2}\int_0^\infty e^{-tu}dt = \frac{1}{u}e^{-u^2/2}. \tag{10.2}$$

This shows the desired estimate. □

Lemma 10.3. *For $\alpha > 0$ it holds*

$$\int_0^\alpha \sqrt{\ln(1+t^{-1})}dt \le \alpha\sqrt{\ln(e(1+\alpha^{-1}))}. \tag{10.3}$$

Proof. First apply the Cauchy–Schwarz inequality to obtain

$$\int_0^\alpha \sqrt{\ln(1+t^{-1})}dt \le \sqrt{\int_0^\alpha 1dt \int_0^\alpha \ln(1+t^{-1})dt}.$$

A change of variables and integration by parts yields

$$\begin{aligned}
\int_0^\alpha \ln(1+t^{-1})dt &= \int_{\alpha^{-1}}^\infty u^{-2}\ln(1+u)du \\
&= -u^{-1}\ln(1+u)|_{\alpha^{-1}}^\infty + \int_{\alpha^{-1}}^\infty u^{-1}\frac{1}{1+u}du \\
&\le \alpha\ln(1+\alpha^{-1}) + \int_{\alpha^{-1}}^\infty \frac{1}{u^2}du \\
&= \alpha\ln(1+\alpha^{-1}) + \alpha.
\end{aligned}$$

Combining the above estimates concludes the proof. □

Acknowledgments. I would like to thank Massimo Fornasier for organizing the summer school "Theoretical Foundations and Numerical Methods for Sparse Recovery" and for inviting me to present this course. The time in Linz was very enjoyable and fruitful. Also I would like to thank Simon Foucart for the joint adventure of writing the monograph [55], which influenced very much these notes, and for sharing his insights and ideas on compressive sensing. Further, I greatly acknowledge RICAM and the START grant "Sparse Approximation and Optimization in High Dimensions" for hosting the summer school. I would further like to thank several people for nice and interesting discussions on the subject: Joel Tropp, Roman Vershynin, Rachel Ward, Stefan Kunis, Ingrid Daubechies, Karlheinz Gröchenig, Ron DeVore, Thomas Strohmer, Emmanuel Candès, Justin Romberg, Götz Pfander, Jared Tanner, Karin Schnass, Rémi Gribonval, Pierre Vandergheynst, Tino Ullrich, Albert Cohen, Alain Pajor. Also, I thank the following people for identifying errors and providing comments on previous versions of these notes: Jan Vybíral, Jan Haskovec, Silvia Gandy, Felix Krahmer, Ulas Ayaz, Tino Ullrich, Deanna Needell, Rachel Ward, Thomas Strohmer, Dirk Lorenz, and Pasc Gavruta. My work on this topic started when being a PostDoc at NuHAG in Vienna. I would like to thank Hans Feichtinger and the whole NuHAG group for providing a very nice and productive research environment. I enjoyed my time there very much. Also I am very grateful to the Hausdorff Center for Mathematics (funded by the DFG) and to the Institute for Numerical Simulation at the University of Bonn for providing excellent working conditions and financial support. Last but not least, my greatest thanks go to Daniela and to our little children Niels and Paulina for making my math-free time so enjoyable.

Bibliography

[1] K. Alexander, Probability inequalities for empirical processes and a law of the iterated logarithm, *Ann. Probab.* **12** (1984), 1041–1067.

[2] W. O. Alltop, Complex sequences with low periodic correlations, *IEEE Trans. Inform. Theory* **26** (1980), 350–354.

[3] J.-M. Azaïs and M. Wschebor, *Level Sets and Extrema of Random Processes and Fields*, John Wiley & Sons Inc., 2009.

[4] W. Bajwa, J. Haupt, G. Raz, S. J. Wright and R. Nowak, Toeplitz-structured compressed sensing matrices., 2007, IEEE Workshop SSP.

[5] R. G. Baraniuk, M. Davenport, R. A. DeVore and M. Wakin, A simple proof of the restricted isometry property for random matrices, *Constr. Approx.* **28** (2008), 253–263.

[6] G. Bennett, Probability inequalities for the sum of independent random variables., *J. Am. Statist. Assoc.* **57** (1962), 33–45.

[7] J. Bergh and J. Löfström, *Interpolation Spaces. An Introduction*, Springer, 1976.

[8] R. Bhatia, *Matrix Analysis*, Graduate Texts in Mathematics 169, Springer, New York, 1997.

[9] S. Boucheron, O. Bousquet, G. Lugosi and P. Massart, Moment inequalities for functions of independent random variables, *Ann. Probab.* **33** (2005), 514–560.

[10] S. Boucheron, G. Lugosi and P. Massart, Concentration inequalities using the entropy method, *Ann. Probab.* **31** (2003), 1583–1614.

[11] J. Bourgain, Bounded orthogonal systems and the $\Lambda(p)$-set problem, *Acta Math.* **162** (1989), 227–245.

[12] ———, *Λ_p-sets in analysis: results, problems and related aspects*, Handbook of the Geometry of Banach Spaces, Vol I, North-Holland, 2001, pp. 195–232.

[13] J. Bourgain and L. Tzafriri, Invertibility of 'large' submatrices with applications to the geometry of Banach spaces and harmonic analysis, *Israel J. Math.* **57** (1987), 137–224.

[14] O. Bousquet, *Concentration inequalities for sub-additive functions using the entropy method*, Stochastic Inequalities and Applications, Progr. Probab. 56, Birkhäuser, Basel, 2003, pp. 213–247.

[15] S. Boyd and L. Vandenberghe, *Convex Optimization*, Cambridge Univ. Press, 2004.

[16] A. Buchholz, Operator Khintchine inequality in non-commutative probability, *Math. Ann.* **319** (2001), 1–16.

[17] ———, Optimal constants in Khintchine type inequalities for fermions, Rademachers and q-Gaussian operators, *Bull. Pol. Acad. Sci. Math.* **53** (2005), 315–321.

[18] E. J. Candès, The restricted isometry property and its implications for compressed sensing, *C. R. Acad. Sci. Paris Sér. I Math.* **346** (2008), 589–592.

[19] E. J. Candès, J. T. Tao and J. Romberg, Robust uncertainty principles: exact signal reconstruction from highly incomplete frequency information, *IEEE Trans. Inform. Theory* **52** (2006), 489–509.

[20] E. J. Candès and J. Romberg, Sparsity and incoherence in compressive sampling, *Inverse Problems* **23** (2007), 969–985.

[21] E. J. Candès, J. Romberg and T. Tao, Stable signal recovery from incomplete and inaccurate measurements, *Comm. Pure Appl. Math.* **59** (2006), 1207–1223.

[22] E. J. Candès and T. Tao, Decoding by linear programming, *IEEE Trans. Inform. Theory* **51** (2005), 4203–4215.

[23] ———, Near optimal signal recovery from random projections: universal encoding strategies?, *IEEE Trans. Inform. Theory* **52** (2006), 5406–5425.

[24] B. Carl, Inequalities of Bernstein–Jackson-type and the degree of compactness of operators in Banach spaces, *Ann. Inst. Fourier (Grenoble)* **35** (1985), 79–118.

[25] S. S. Chen, D. L. Donoho and M. A. Saunders, Atomic decomposition by Basis Pursuit, *SIAM J. Sci. Comput.* **20** (1999), 33–61.

[26] H. Chernoff, A measure of asymptotic efficiency of tests of a hypothesis based on the sum of observations, *Ann. Math. Statist.* **23** (1952), 493–507.

[27] A. Cohen, *Numerical Analysis of Wavelet Methods*, North-Holland, 2003.

[28] A. Cohen, W. Dahmen and R. DeVore, Compressed sensing and best k-term approximation, *J. Am. Math. Soc.* **22** (2009), 211–231.

[29] R. Coifman, F. Geshwind and Y. Meyer, Noiselets, *Appl. Comput. Harmon. Anal.* **10** (2001), 27–44.

[30] J. Cooley and J. Tukey, An algorithm for the machine calculation of complex Fourier series, *Math. Comp.* **19** (1965), 297–301.

[31] I. Daubechies, *Ten Lectures on Wavelets*, CBMS-NSF Regional Conference Series in Applied Mathematics 61, SIAM, Society for Industrial and Applied Mathematics, 1992.

[32] I. Daubechies, M. Defrise and C. De Mol, An iterative thresholding algorithm for linear inverse problems with a sparsity constraint, *Comm. Pure Appl. Math.* **57** (2004), 1413–1457.

[33] I. Daubechies, R. DeVore, M. Fornasier and S. Güntürk, Iteratively re-weighted least squares minimization for sparse recovery, *Comm. Pure Appl. Math.* **63** (2010), 1–38.

[34] I. Daubechies, M. Fornasier and I. Loris, Accelerated projected gradient methods for linear inverse problems with sparsity constraints, *J. Fourier Anal. Appl.* **14** (2008), 764–792.

[35] G. Davis, S. Mallat and M. Avellaneda, Adaptive greedy approximations, *Constr. Approx.* **13** (1997), 57–98.

[36] V. de la Peña and E. Giné, *Decoupling. From Dependence to Independence*, Probability and its Applications (New York), Springer, New York, 1999.

[37] R. A. DeVore, Deterministic constructions of compressed sensing matrices, *J. Complexity* **23** (2007), 918–925.

[38] D. L. Donoho, Compressed sensing, *IEEE Trans. Inform. Theory* **52** (2006), 1289–1306.

[39] D. L. Donoho and M. Elad, Optimally sparse representations in general (non-orthogonal) dictionaries via ℓ^1 minimization, *Proc. Nat. Acad. Sci.* **100** (2002), 2197–2202.

[40] D. L. Donoho, M. Elad and V. N. Temlyakov, Stable recovery of sparse overcomplete representations in the presence of noise, *IEEE Trans. Inform. Theory* **52** (2006), 6–18.

[41] D. L. Donoho and X. Huo, Uncertainty principles and ideal atomic decompositions, *IEEE Trans. Inform. Theory* **47** (2001), 2845–2862.

[42] D. L. Donoho and J. Tanner, Counting faces of randomly-projected polytopes when the projection radically lowers dimension, *J. Am. Math. Soc.* **22** (2009), 1–53.

[43] D. L. Donoho and Y. Tsaig, Fast solution of l1-norm minimization problems when the solution may be sparse, *IEEE Trans. Inform. Theory* **54** (2008), 4789–4812.

[44] M. F. Duarte, M. A. Davenport, D. Takhar, J. N. Laska, S. Ting, K. F. Kelly and R. G. Baraniuk, Single-Pixel Imaging via Compressive Sampling, *IEEE Signal Processing Magazine* **25** (2008), 83–91.

[45] R. M. Dudley, The sizes of compact subsets of Hilbert space and continuity of Gaussian processes, *J. Functional Analysis* **1** (1967), 290–330.

[46] A. Dutt and V. Rokhlin, Fast Fourier transforms for nonequispaced data, *SIAM J. Sci. Comput.* **14** (1993), 1368–1393.

[47] B. Efron, T. Hastie, I. Johnstone and R. Tibshirani, Least angle regression, *Ann. Statist.* **32** (2004), 407–499.

[48] M. Elad and A. M. Bruckstein, A generalized uncertainty principle and sparse representation in pairs of bases., *IEEE Trans. Inform. Theory* **48** (2002), 2558–2567.

[49] A. Fannjiang, P. Yan and T. Strohmer, Compressed Remote Sensing of Sparse Objects, preprint (2009).

[50] G. B. Folland, *A Course in Abstract Harmonic Analysis*, CRC Press, 1995.

[51] S. Foucart, A note on ensuring sparse recovery via ℓ_1-minimization, *Appl. Comput. Harm. Anal.* (to appear).

[52] S. Foucart and R. Gribonval, Real vs. complex null space properties for sparse vector recovery, preprint (2009).

[53] S. Foucart and M. Lai, Sparsest solutions of underdetermined linear systems via ℓ_q-minimization for $0 < q \leq 1$, *Appl. Comput. Harmon. Anal.* **26** (2009), 395–407.

[54] S. Foucart, A. Pajor, H. Rauhut and T. Ullrich, The Gelfand widths of ℓ_p-balls for $0 < p \leq 1$, *J. Complexity* (to appear).

[55] S. Foucart and H. Rauhut, *A Mathematical Introduction to Compressive Sensing*, Appl. Numer. Harmon. Anal., Birkhäuser, Boston, in preparation.

[56] J. J. Fuchs, On sparse representations in arbitrary redundant bases, *IEEE Trans. Inform. Theory* **50** (2004), 1341–1344.

[57] A. Y. Garnaev and E. D. Gluskin, On widths of the Euclidean ball, *Sov. Math., Dokl.* **30** (1984), 200–204.

[58] A. C. Gilbert and J. A. Tropp, Signal recovery from random measurements via orthogonal matching pursuit, *IEEE Trans. Inform. Theory* **53** (2007), 4655–4666.

[59] G. Golub and C. F. van Loan, *Matrix Computations*, 3rd ed, The Johns Hopkins University Press, 1996.

[60] R. Gribonval and M. Nielsen, Sparse representations in unions of bases, *IEEE Trans. Inform. Theory* **49** (2003), 3320–3325.

[61] ———, Highly sparse representations from dictionaries are unique and independent of the sparseness measure, *Appl. Comput. Harmon. Anal.* **22** (2007), 335–355.

[62] R. Gribonval and P. Vandergheynst, On the exponential convergence of matching pursuits in quasi-incoherent dictionaries, *IEEE Trans. Inform. Theory* **52** (2006), 255–261.

[63] G. Grimmett and D. Stirzaker, *Probability and random processes*, Third ed, Oxford University Press, New York, 2001.

[64] K. Gröchenig, *Foundations of Time-Frequency Analysis*, Appl. Numer. Harmon. Anal., Birkhäuser, Boston, 2001.

[65] K. Gröchenig, B. Pötscher and H. Rauhut, Learning trigonometric polynomials from random samples and exponential inequalities for eigenvalues of random matrices, preprint (2007).

[66] O. Guédon, S. Mendelson, A. Pajor and N. Tomczak Jaegermann, Majorizing measures and proportional subsets of bounded orthonormal systems, *Rev. Mat. Iberoam.* **24** (2008), 1075–1095.

[67] U. Haagerup, The best constants in the Khintchine inequality, *Studia Math.* **70** (1981), 231–283 (1982).

[68] J. Haupt, W. Bajwa, G. Raz and R. Nowak, Toeplitz compressed sensing matrices with applications to sparse channel estimation, preprint (2008).

[69] M. Herman and T. Strohmer, High-resolution radar via compressed sensing, *IEEE Trans. Signal Process.* **57** (2009), 2275–2284.

[70] W. Hoeffding, Probability inequalities for sums of bounded random variables, *J. Am. Statist. Assoc.* **58** (1963), 13–30.

[71] R. Horn and C. Johnson, *Matrix Analysis*, Cambridge University Press, 1990.

[72] ———, *Topics in Matrix Analysis*, Cambridge University Press, Cambridge, 1994, Corrected reprint of the 1991 original.

[73] W. B. Johnson and J. Lindenstrauss (eds.), *Handbook of the Geometry of Banach Spaces*, Vol I, North-Holland Publishing Co., Amsterdam, 2001.

[74] B. S. Kashin, Diameters of some finite-dimensional sets and classes of smooth functions., *Math. USSR, Izv.* **11** (1977), 317–333.

[75] A. Khintchine, Über dyadische Brüche, *Math. Z.* **18** (1923), 109–116.

[76] S. Kim, K. Koh, M. Lustig, S. Boyd and D. Gorinevsky, A method for large-scale l1-regularized least squares problems with applications in signal processing and statistics, *IEEE J. Sel. Top. Signal Process.* **4** (2007), 606–617.

[77] T. Klein and E. Rio, Concentration around the mean for maxima of empirical processes, *Ann. Probab.* **33** (2005), 1060–1077.

[78] S. Kunis and H. Rauhut, Random sampling of sparse trigonometric polynomials II – orthogonal matching pursuit versus basis pursuit, *Found. Comput. Math.* **8** (2008), 737–763.

[79] M. Ledoux, *The Concentration of Measure Phenomenon*, AMS, 2001.

[80] M. Ledoux and M. Talagrand, *Probability in Banach Spaces*, Springer, Berlin, Heidelberg, NewYork, 1991.

[81] X. Li and C.-P. Chen, Inequalities for the Gamma function, *JIPAM. J. Inequal. Pure Appl. Math.* **8** (2007), Article 28, 3 pp. (electronic).

[82] F. Lust-Piquard, Inégalites de Khintchine dans $C_p(1 < p < \infty)$, *C. R. Acad. Sci. Paris Sér. I Math.* **303** (1986), 289–292.

[83] S. Mallat, *A Wavelet Tour of Signal Processing*, Academic Press, 1998.

[84] P. Massart, Rates of convergence in the central limit theorem for empirical processes, *Ann. Inst. H. Poincaré Probab. Statist.* **22** (1986), 381–423.

[85] ______, About the constants in Talagrand's concentration inequalities for empirical processes, *Ann. Probab.* **28** (2000), 863–884.

[86] T. McConnell and M. Taqqu, Decoupling inequalities for multilinear forms in independent symmetric random variables, *Annals Prob.* **11** (1986), 943–951.

[87] S. Mendelson, A. Pajor and N. Tomczak Jaegermann, Uniform uncertainty principle for Bernoulli and subgaussian ensembles, *Constr. Approx.* **28** (2009), 277–289.

[88] B. K. Natarajan, Sparse approximate solutions to linear systems., *SIAM J. Comput.* **24** (1995), 227–234.

[89] F. Nazarov and A. Podkorytov, *Ball, Haagerup, and distribution functions*, Complex Analysis, Operators, and related Topics, Oper. Theory Adv. Appl. 113, Birkhäuser, Basel, 2000, pp. 247–267.

[90] D. Needell and R. Vershynin, Uniform uncertainty principle and signal recovery via regularized orthogonal matching pursuit, *Found. Comput. Math.* **9** (2009), 317–334.

[91] ______, Signal recovery from incomplete and inaccurate measurements via regularized orthogonal matching pursuit, *IEEE J. Sel. Topics Sig. Process.* **4** (2010), 310–316.

[92] A. Pajor and S. Mendelson, On singular values of matrices with independent rows, *Bernoulli* **12** (2006), 761–773.

[93] G. Peškir, Best constants in Kahane–Khintchine inequalities for complex Steinhaus functions, *Proc. Amer. Math. Soc.* **123** (1995), 3101–3111.

[94] G. Peškir and A. N. Shiryaev, The Khintchine inequalities and martingale expanding sphere of their action, *Russian Math. Surveys* **50** (1995), 849–904.

[95] G. Pfander and H. Rauhut, Sparsity in time-frequency representations, *J. Fourier Anal. Appl.* **16** (2010), 233–260.

[96] G. E. Pfander, H. Rauhut and J. Tanner, Identification of matrices having a sparse representation, *IEEE Trans. Signal Process.* **56** (2008), 5376–5388.

[97] A. Pinkus, *On L^1-Approximation*, Cambridge Tracts in Mathematics 93, Cambridge University Press, Cambridge, 1989.

[98] G. Pisier, *Remarques sur un résultat non publié de B. Maurey*, Seminar on Functional Analysis, 1980–1981, École Polytech., 1981, pp. Exp. No. V, 13.

[99] ______, *The Volume of Convex Bodies and Banach Space Geometry*, Cambridge Tracts in Mathematics 94, Cambridge University Press, Cambridge, 1989.

[100] ______, Non-commutative vector valued L_p-spaces and completely p-summing maps, *Astérisque* (1998).

[101] D. Potts, G. Steidl and M. Tasche, *Fast Fourier transforms for nonequispaced data: A tutorial*, Modern Sampling Theory: Mathematics and Applications (Eds. J. J. Benedetto and P. J. S. G. Ferreira), Birkhäuser, 2001, pp. 247–270.

[102] H. Rauhut, Random sampling of sparse trigonometric polynomials, *Appl. Comput. Harmon. Anal.* **22** (2007), 16–42.

[103] ______, On the impossibility of uniform sparse reconstruction using greedy methods, *Sampl. Theory Signal Image Process.* **7** (2008), 197–215.

[104] ______, Stability results for random sampling of sparse trigonometric polynomials, *IEEE Trans. Information Theory* **54** (2008), 5661–5670.

[105] ______, Circulant and Toeplitz matrices in compressed sensing, in: *Proc. SPARS'09*, Saint-Malo, France, 2009.

[106] H. Rauhut and R. Ward, Sparse Legendre expansions via ℓ_1-minimization, preprint (2010).

[107] E. Rio, Inégalités de concentration pour les processus empiriques de classes de parties, *Probab. Theory Related Fields* **119** (2001), 163–175.

[108] ______, Une inégalité de Bennett pour les maxima de processus empiriques, *Ann. Inst. H. Poincaré Probab. Statist.* **38** (2002), 1053–1057, En l'honneur de J. Bretagnolle, D. Dacunha-Castelle, I. Ibragimov.

[109] J. Romberg, Imaging via compressive sampling, *IEEE Signal Process. Magazine* **25** (2008), 14–20.

[110] ______, Compressive sensing by random convolution, *SIAM J. Imaging Sci.* **2** (2009), 1098–1128.

[111] M. Rosenfeld, *In Praise of the Gram Matrix*, The Mathematics of Paul Erdős, II, Algorithms Combin. 14, Springer, 1997, pp. 318–323.

[112] S. Ross, *Introduction to Probability Models*, Nineth ed, Academic Press, 2006.

[113] M. Rudelson, Random vectors in the isotropic position, *J. Funct. Anal.* **164** (1999), 60–72.

[114] M. Rudelson and R. Vershynin, Geometric approach to error-correcting codes and reconstruction of signals, *Internat. Math. Res. Notices* (2005), 4019–4041.

[115] ______, Sampling from large matrices: an approach through geometric functional analysis, *J. ACM* **54** (2007), Art. 21, 19 pp. (electronic).

[116] ______, On sparse reconstruction from Fourier and Gaussian measurements, *Comm. Pure Appl. Math.* **61** (2008), 1025–1045.

[117] W. Rudin, *Fourier Analysis on Groups*, Interscience Publishers, 1962.

[118] ______, *Functional Analysis*, McGraw-Hill Book Company, 1973.

[119] K. Schnass and Pierre Vandergheynst, Dictionary preconditioning for greedy algorithms, *IEEE Trans. Signal Process.* **56** (2008), 1994–2002.

[120] B. Simon, *Trace Ideals and their Applications*, Cambridge University Press, Cambridge, 1979.

[121] T. Strohmer and R. W. jun. Heath, Grassmannian frames with applications to coding and communication., *Appl. Comput. Harmon. Anal.* **14** (2003), 257–275.

[122] M. Talagrand, Isoperimetry and integrability of the sum of independent Banach-space valued random variables, *Ann. Probab.* **17** (1989), 1546–1570.

[123] ———, New concentration inequalities in product spaces, *Invent. Math.* **126** (1996), 505–563.

[124] ———, Selecting a proportion of characters, *Israel J. Math.* **108** (1998), 173–191.

[125] ———, *The Generic Chaining*, Springer Monographs in Mathematics, Springer, Berlin, 2005.

[126] Georg Tauböck, Franz Hlawatsch, Daniel Eiwen and Holger Rauhut, Compressive Estimation of Doubly Selective Channels in Multicarrier Systems: Leakage Effects and Sparsity-Enhancing Processing, *IEEE J. Sel. Top. Sig. Process.* **4** (2010), 255–271.

[127] J. A. Tropp, Greed is good: Algorithmic results for sparse approximation, *IEEE Trans. Inform. Theory* **50** (2004), 2231–2242.

[128] ———, Recovery of short, complex linear combinations via l_1 minimization, *IEEE Trans. Inform. Theory* **51** (2005), 1568–1570.

[129] ———, Just relax: Convex programming methods for identifying sparse signals in noise, *IEEE Trans. Inform. Theory* **51** (2006), 1030–1051.

[130] ———, On the conditioning of random subdictionaries, *Appl. Comput. Harmon. Anal.* **25** (2008), 1–24.

[131] J. A. Tropp and D. Needell, CoSaMP: Iterative signal recovery from incomplete and inaccurate samples, *Appl. Comput. Harmon. Anal.* **26** (2008), 301–321.

[132] J. A. Tropp, M. Wakin, M. Duarte, D. Baron and R. G. Baraniuk, Random filters for compressive sampling and reconstruction, *Proc. 2006 IEEE Int. Conf. Acoustics, Speech, and Signal Processing* **3** (2006), 872–875.

[133] Joel A. Tropp, Jason N. Laska, Marco F. Duarte, Justin K. Romberg and Richard G. Baraniuk, Beyond Nyquist: Efficient sampling of sparse bandlimited signals, *IEEE Trans. Inform. Theory* **56** (2010), 520 –544.

[134] A. W. Van der Vaart and J. A. Wellner, *Weak Convergence and Empirical Processes*, Springer, 1996.

[135] R. Varga, *Gershgorin and his Circles*, Springer Series in Computational Mathematics, Springer, Berlin, 2004.

[136] R. Vershynin, Frame expansions with erasures: an approach through the non-commutative operator theory, *Appl. Comput. Harmon. Anal.* **18** (2005), 167–176.

[137] J. S. Walker, *Fast Fourier Transforms*, CRC Press, 1991.

[138] P. Wojtaszczyk, *A Mathematical Introduction to Wavelets*, Cambridge University Press, 1997.

Author information

Holger Rauhut, Hausdorff Center for Mathematics & Institute for Numerical Simulation, University of Bonn, Endenicher Allee 60, 53115 Bonn, Germany.
E-mail: `rauhut@hcm.uni-bonn.de`

Radon Series Comp. Appl. Math **9**, 93–200

Numerical Methods for Sparse Recovery

Massimo Fornasier

Abstract. These lecture notes address the analysis of numerical methods for performing optimizations with linear model constraints and additional sparsity conditions to solutions, i.e., we expect solutions which can be represented as sparse vectors with respect to a prescribed basis. In the first part of the manuscript we illustrate the theory of compressed sensing with emphasis on computational aspects. We present the analysis of the homotopy method, the iteratively re-weighted least squares method, and the iterative hard-thresholding. In the second part, starting from the analysis of iterative soft-thresholding, we illustrate several numerical methods for addressing sparse optimizations in Hilbert spaces. The third and final part is concerned with numerical techniques, based on domain decomposition methods, for dimensionality reduction in large scale computing. In the notes we are mainly focusing on the analysis of the algorithms, and we report a few illustrative numerical examples.

Keywords. Numerical methods for sparse optimization, calculus of variations, algorithms for nonsmooth optimization.

2010 Mathematics Subject Classification. 15A29, 65K10, 90C25, 52A41, 49M30.

Table of Contents.

1 Introduction

These lecture notes are an introduction to methods recently developed for performing numerical optimizations with linear model constraints and additional sparsity conditions to solutions, i.e., we expect solutions which can be represented as sparse vectors with respect to a prescribed basis. Such a type of problems has been recently greatly popularized by the development of the field of nonadaptive compressed acquisition of data, the so-called *compressed sensing*, and its relationship with ℓ_1-minimization. We start our presentation by recalling the mathematical setting of compressed sensing as a reference framework for developing further generalizations. In particular we focus on the analysis of algorithms for such problems and their performances. We introduce and analyse the homotopy method, the iteratively reweighted least squares method, and the iterative hard thresholding algorithm. We will see that the properties of convergence of these algorithms to solutions depend very much on special spectral properties, the Restricted Isometry Property or the Null Space Property, of the matrices which define the linear models. This provides a link to the analysis of random matrices which are typical examples of matrices with such properties. The concept of sparsity does not necessarily affect the entries of a vector only, but it can also be applied, for instance, to their variation. We will show that some of the algorithms proposed for compressed sensing are in fact useful for optimization problems with total variation constraints. These optimizations on continuous domains are related to the calculus of variations on bounded variation (BV) functions and to geometric measure theory.

In the second part of the lecture notes we address sparse optimizations in Hilbert spaces, and especially for situations where no Restricted Isometry Property or Null Space Property are assumed for the linear model. We will be able to formulate efficient algorithms based on so-called *iterative soft-thresholding* also for such situations, although their analysis will require different tools, typically from nonsmooth convex analysis.

A common feature of the illustrated algorithms will be their variational nature, in the sense that they are derived as minimization strategies of given energy functionals. Not only does the variational framework allow us to derive very precise statements about the convergence properties of these algorithms, but it also provides the algorithms with an intrinsic robustness.

We will finally address large scale computations, showing how we can define domain decomposition strategies for these nonsmooth optimizations, for problems coming from compressed sensing and ℓ_1-minimization as well as for total variation minimization problems.

The first part of the lecture notes is elementary and it does not require more than the basic knowledge of notions of linear algebra and standard inequalities. The second part of the course is slightly more advanced and addresses problems in Hilbert spaces, and we will make use of concepts of nonsmooth convex analysis. We refer the interested reader to the books [36,50] for an introduction to convex analysis, variational methods, and related numerical techniques.

1.1 Notations

In the following we collect general notations. More specific notations will be introduced and recalled in the following sections.

We will consider $X = \mathbb{R}^N$ as a Banach space endowed with different norms. In particular, later we use the ℓ_p-(quasi-)norms

$$\|x\|_p := \|x\|_{\ell_p} := \|x\|_{\ell_p^N} := \begin{cases} \left(\sum_{i=1}^N |x_j|^p\right)^{1/p}, & 0 < p < \infty, \\ \max_{j=1,\dots,N} |x_j|, & p = \infty. \end{cases} \tag{1.1}$$

Associated to these norms we denote their unit balls by $B_{\ell_p} := B_{\ell_p^N} := \{x \in \mathbb{R} : \|x\|_p \leq 1\}$ and the balls of radius R by $B_{\ell_p}(R) := B_{\ell_p^N}(R) := R \cdot B_{\ell_p^N}$. Associated to a closed convex body $0 \in \Omega \subset \mathbb{R}^N$, we define its polar set by $\Omega^\circ = \{y \in \mathbb{R}^N : \sup_{x\in\Omega}\langle x, y\rangle \leq 1\}$. This allow us to define an associated norm $\|x\|_\Omega = \sup_{y\in\Omega^\circ}\langle x, y\rangle$.

The index set $\mathcal{I}$ is supposed to be countable and we will consider the $\ell_p(\mathcal{I})$ spaces of p-summable sequences over the index set $\mathcal{I}$ as well. Their norm are defined as usual and similarly to the case of $\mathbb{R}^N$. We use the same notations B_{ℓ_p} for the $\ell_p(\mathcal{I})$-balls as for the ones in $\mathbb{R}^N$. With A we will usually denote a $m \times N$ real matrix, $m, N \in \mathbb{N}$ or an operator $A : \ell_2(\mathcal{I}) \to Y$. We denote with A^* the adjoint matrix or with K^* the adjoint of an operator K. We will always work on real vector spaces, hence, in finite dimensions, A^* usually coincides with the transposed matrix of A. The norm of an operator $K : X \to Y$ acting between two Banach spaces is denoted by $\|K\|_{X\to Y}$; for matrices the norm $\|A\|$ denotes the spectral norm. The support of a vector $u \in \ell_2(\mathcal{I})$, i.e., the set of coordinates which are not zero, is denoted by $\mathrm{supp}(u)$.

We will consider index sets $\Lambda \subset \mathcal{I}$ and their complements $\Lambda^c = \mathcal{I} \setminus \Lambda$. The symbols $|\Lambda|$ and $\#\Lambda$ are used indifferently for indicating the cardinality of Λ. With a slight abuse we will denote

$$\|u\|_0 := \|u\|_{\ell_0(\mathcal{I})} := \#\,\mathrm{supp}(u), \tag{1.2}$$

which is popularly called the "ℓ_0-norm" in the literature. When $\#\mathcal{I} = N$ then $\ell_2(\mathcal{I}) = \mathbb{R}^N$ and we may also denote $\|u\|_{\ell_0^N} := \|u\|_{\ell_0(\mathcal{I})}$ We use the notation A_Λ to indicate a submatrix extracted from A by retaining only the columns indexed in Λ as well as the restrictions u_Λ of vectors u to the index set Λ. Sometimes the notation A_Λ (resp. u_Λ) indicates also a matrix (resp. a vector) whose columns (resp. entries) indexed on Λ^c are set to zero. We also denote by $A^*A_{\Lambda\times\Lambda} := (A^*A)_{\Lambda\times\Lambda} := A^*_\Lambda A_\Lambda$ the submatrix extracted from A^*A by retaining only the entries indexed on $\Lambda \times \Lambda$.

Generic positive constants used in estimates are denoted as usual by

$$c, C, \tilde{c}, \tilde{C}, c_0, C_0, c_1, C_1, c_2, C_2, \ldots .$$

2 An Introduction to Sparse Recovery

2.1 A Toy Mathematical Model for Sparse Recovery

2.1.1 Adaptive and Compressed Acquisition

Let $k, N \in \mathbb{N}, k \leq N$ and

$$\Sigma_k := \{x \in \mathbb{R}^N : \|x\|_{\ell_0^N} := \#\,\mathrm{supp}(x) \leq k\}$$

be the set of vectors with at most k nonzero entries, which we will call *k-sparse vectors*. The *k-best approximation error* that we can achieve in this set to a vector

$x \in \mathbb{R}^N$ with respect to a suitable space quasi-norm $\|\cdot\|_X$ is defined by

$$\sigma_k(x)_X = \inf_{z \in \Sigma_k} \|x - z\|_X.$$

Example 2.1. Let $r(x)$ be the *nonincreasing rearrangement* of x, i.e.,

$$r(x) = (|x_{i_1}|, \ldots, |x_{i_N}|)^T \quad \text{and} \quad |x_{i_j}| \geq |x_{i_{j+1}}|, \quad \text{for } j = 1, \ldots, N-1.$$

Then it is straightforward to check that

$$\sigma_k(x)_{\ell_p^N} := \left(\sum_{j=k+1}^{N} r_j(x)^p \right)^{1/p}, \quad 1 \leq p < \infty.$$

In particular, the vector $x_{[k]}$ derived from x by setting to zero all the $N-k$ smallest entries in absolute value is called the *best k-term approximation* to x and it coincides with

$$x_{[k]} = \arg\min_{z \in \Sigma_k} \|x - z\|_{\ell_p^N} \tag{2.1}$$

for any $1 \leq p < \infty$.

Lemma 2.2. *Let $r = \frac{1}{q} - \frac{1}{p}$ and $x \in \mathbb{R}^N$. Then*

$$\sigma_k(x)_{\ell_p} \leq \|x\|_{\ell_q} k^{-r}, \quad k = 1, 2, \ldots, N.$$

Proof. Let Λ be the set of indexes of a k-largest entries of x in absolute value. If $\varepsilon = r_k(x)$, the kth-entry of the nonincreasing rearrangement $r(x)$, then

$$\varepsilon \leq \|x\|_{\ell_q} k^{-\frac{1}{q}}.$$

Therefore

$$\sigma_k(x)_{\ell_p}^p = \sum_{j \notin \Lambda} |x_j|^p \leq \sum_{j \notin \Lambda} \varepsilon^{p-q} |x_j|^q$$
$$\leq \|x\|_{\ell_q}^{p-q} k^{-\frac{p-q}{q}} \|x\|_{\ell_q}^q,$$

which implies

$$\sigma_k(x)_{\ell_p} \leq \|x\|_{\ell_q} k^{-r}.$$ □

The computation of the best k-term approximation of $x \in \mathbb{R}^N$ in general requires the search of the largest entries of x in absolute value, and therefore the testing of all the entries of the vector x. This procedure is *adaptive*, since it depends on the particular vector, and it is currently at the basis of lossy compression methods, such as JPEG [58].

2.1.2 Nonadaptive and Compressed Acquisition: Compressed Sensing

One would like to describe a *linear encoder* which allows us to compute approximately k measurements $(y_1, \dots, y_k)^T$ and a nearly optimal approximation of x in the following sense:

> Provided a set $K \subset \mathbb{R}^N$, there exists a linear map $A : \mathbb{R}^N \to \mathbb{R}^m$, with $m \approx k$ and a possibly nonlinear map $\Delta : \mathbb{R}^m \to \mathbb{R}^N$ such that
>
> $$\|x - \Delta(Ax)\|_X \le C\sigma_k(x)_X$$
>
> for all $x \in K$.

Note that the way we encode $y = Ax$ is via a prescribed map A which is independent of x. Also the decoding procedure Δ might depend on A, but not on x. This is why we call this strategy a *nonadaptive (or universal) and compressed acquisition* of x. Note further that we would like to recover an approximation to x from nearly k-linear measurements which is of the order of the k-best approximation error. In this sense we say that the performances of the encoder/decoder system (A, Δ) are nearly optimal.

2.1.3 Optimal Performances of Encoder/Decoder Pairs

Let us define $\mathcal{A}_{m,N}$ the set of all encoder/decoder pairs (A, Δ) with A a $m \times N$ matrix and $\Delta : \mathbb{R}^m \to \mathbb{R}^N$ any function. We wonder whether there exists such a nearly optimal pair as claimed above. Let us fix $m \le N$ two natural numbers, and $K \subset \mathbb{R}^N$. For $1 \le k \le N$ we denote

$$\sigma_k(K)_X := \sup_{x \in K} \sigma_k(x)_X, \quad \text{and} \quad E_m(K)_X := \inf_{(A,\Delta) \in \mathcal{A}_{m,N}} \sup_{x \in K} \|x - \Delta(Ax)\|_X.$$

We would like to find the largest k such that

$$E_m(K)_X \le C_0 \sigma_k(K)_X.$$

We answer this question in the particular case where $K = B_{\ell_1^N}$ and $X = \ell_2^N$. This setting will turn out to be particularly useful later on and it is already sufficient for showing that, unfortunately, it is impossible to reconstruct $x \in B_{\ell_1^N}$ with an accuracy asymptotically (for m, N larger and larger) of the order of the k-best approximation error in ℓ_2^M if $k = m$, but it is necessary to have a slightly larger number of measurements, i.e., $k = m - \varepsilon(m, N)$, $\varepsilon(m, N) > 0$.

The proper estimate of $E_m(K)_X$ turns out to be linked to the geometrical concept of *Gelfand width*.

Definition 2.3. Let K be a compact set in X. Then the *Gelfand width* of K of order m is

$$d^m(K)_X := \inf_{\substack{Y \leq X \\ \operatorname{codim}(Y) \leq m}} \sup\{\|x\|_X : x \in K \cap Y\}.$$

The infimum is taken over the set of linear subspaces Y of X with codimension less or equal than m.

We have the following fundamental equivalence.

Proposition 2.4. *Let $K \subset \mathbb{R}^N$ be any compact set for which $K = -K$ and such that there exists a constant $C_0 > 0$ for which $K + K \subset C_0 K$. If $X \subset \mathbb{R}^N$ is a normed space, then*

$$d^m(K)_X \leq E_m(K)_X \leq C_0 d^m(K)_X.$$

Proof. For a matrix $A \in \mathbb{R}^{m \times N}$, we denote $\mathcal{N} = \ker A$. Note that $Y = \mathcal{N}$ has codimension less or equal than m. Conversely, given any space $Y \subset \mathbb{R}^N$ of codimension less or equal than m, we can associate a matrix A whose rows are a basis for $Y^\perp$. With this identification we see that

$$d^m(K)_X = \inf_{A \in \mathbb{R}^{m \times N}} \sup\{\|\eta\|_X : \eta \in \mathcal{N} \cap K\}.$$

If (A, Δ) is an encoder/decoder pair in $\mathcal{A}_{m,N}$ and $z = \Delta(0)$, then for any $\eta \in \mathcal{N}$ we have also $-\eta \in \mathcal{N}$. It follows that either $\|\eta - z\|_X \geq \|\eta\|_X$ or $\|-\eta - z\|_X \geq \|\eta\|_X$. Indeed, if we assumed that both inequalities were false then

$$\|2\eta\|_X = \|\eta - z + z + \eta\|_X \leq \|\eta - z\|_X + \|-\eta - z\|_X < 2\|\eta\|_X,$$

a contradiction. Since $K = -K$ we conclude that

$$\begin{aligned} d^m(K)_X &= \inf_{A \in \mathbb{R}^{m \times N}} \sup\{\|\eta\|_X : \eta \in \mathcal{N} \cap K\} \\ &\leq \sup_{\eta \in \mathcal{N} \cap K} \|\eta - z\|_X \\ &= \sup_{\eta \in \mathcal{N} \cap K} \|\eta - \Delta(A\eta)\|_X \\ &\leq \sup_{x \in K} \|x - \Delta(Ax)\|_X. \end{aligned}$$

By taking the infimum over all $(A, \Delta) \in \mathcal{A}_{m,N}$ we obtain

$$d^m(K)_X \leq E_m(K)_X.$$

To prove the upper inequality, let us choose an optimal Y such that

$$d^m(K)_X = \sup\{\|x\|_X : x \in Y \cap K\}.$$

(Actually such an optimal subspace Y always exists [54].) As mentioned above, we can associate to Y a matrix A such that $\mathcal{N} = \ker A = Y$. Let us denote the affine solution space $\mathcal{F}(y) := \{x : Ax = y\}$. Let us now define a decoder as follows: If $\mathcal{F}(y) \cap K \neq \emptyset$ then we take $\bar{x}(y) \in \mathcal{F}(y) \cap K$ and $\Delta(y) = \bar{x}(y)$. If $\mathcal{F}(y) \cap K = \emptyset$ then $\Delta(y) \in \mathcal{F}(y)$. Hence, we can estimate

$$\begin{aligned} E_m(K)_X &= \inf_{(A,\Delta)\in\mathcal{A}_{m,N}} \sup_{x\in K} \|x - \Delta(Ax)\|_X \\ &\leq \sup_{x,x'\in K; Ax=Ax'} \|x - x'\|_X \\ &\leq \sup_{\eta\in C_0(\mathcal{N}\cap K)} \|\eta\|_X \leq C_0 d^m(K)_X. \end{aligned}$$ □

The following result was proven in the relevant work of Kashin, Garnaev, and Gluskin [46, 47, 51] already in the '70s and '80s. See [20, 33] for a description of the relationship between this result and the more modern point of view on compressed sensing.

Theorem 2.5. *The Gelfand widths of ℓ_q^N-balls in ℓ_p^N for $1 \leq q < p \leq 2$ are estimated by*

$$C_1 \Psi(m, N, p, q) \leq d^m(B_{\ell_q^N})_{\ell_p^N} \leq C_2 \Psi(m, N, p, q),$$

where

$$\Psi(m, N, p, q) = \min\{1, N^{1-\frac{1}{q}} m^{-\frac{1}{2}}\}^{\frac{1/q-1/p}{1/q-1/2}}, \quad 1 < q < p \leq 2,$$

$$\Psi(m, N, 2, 1) = \min\left\{1, \sqrt{\frac{\log(N/m)+1}{m}}\right\}, \quad q = 1 \text{ and } p = 2.$$

From Proposition 2.4 and Theorem 2.5 we obtain

$$\tilde{C}_1 \Psi(m, N, p, q) \leq E_m(B_{\ell_q^N})_{\ell_p^N} \leq \tilde{C}_2 \Psi(m, N, p, q).$$

In particular, for $q = 1$ and $p = 2$, we obtain, for m, N large enough, the estimate

$$\tilde{C}_1 \sqrt{\frac{\log(N/m)+1}{m}} \leq E_m(B_{\ell_1^N})_{\ell_2^N}.$$

If we wanted to enforce

$$E_m(B_{\ell_1^N})_{\ell_2^N} \leq C \sigma_k(B_{\ell_1^N})_{\ell_2^N},$$

then Lemma 2.2 would imply

$$\sqrt{\frac{\log(N/m)+1}{m}} \leq C_0 k^{-\frac{1}{2}}, \quad \text{or} \quad k \leq C_0 \frac{m}{\log\frac{N}{m}+1}.$$

Hence, we proved the following

Corollary 2.6. *For m, N fixed, there exists an optimal encoder/decoder pair $(A, \Delta) \in \mathcal{A}_{m,N}$, in the sense that*

$$E_m(B_{\ell_1^N})_{\ell_2^N} \leq C\sigma_k(B_{\ell_1^N})_{\ell_2^N},$$

only if

$$k \leq C_0 \frac{m}{\log \frac{N}{m} + 1}, \tag{2.2}$$

for some constant $C_0 > 0$ independent of m, N.

The next section is devoted to the construction of optimal encoder/decoder pairs $(A, \Delta) \in \mathcal{A}_{m,N}$ as stated in the latter corollary.

2.2 Survey on Mathematical Analysis of Compressed Sensing

In the following sections we want to show that under a certain property, called the *Restricted Isometry Property* (RIP) for a matrix A,

The decoder, which we call ℓ_1-*minimization*,

$$\Delta(y) = \arg \min_{Az=y=Ax} \|z\|_{\ell_1^N} \tag{2.3}$$

performs

$$\|x - \Delta(y)\|_{\ell_1^N} \leq C_1 \sigma_k(x)_{\ell_1^N}, \tag{2.4}$$

as well as

$$\|x - \Delta(y)\|_{\ell_2^N} \leq C_2 \frac{\sigma_k(x)_{\ell_1^N}}{k^{1/2}}, \tag{2.5}$$

for all $x \in \mathbb{R}^N$.

Note that by (2.5) we immediately obtain

$$E_m(B_{\ell_1^N})_{\ell_2^N} \leq C_0 k^{-1/2},$$

implying once again (2.2). Hence, the following question we will address is the existence of matrices A with RIP for which k is optimal, i.e.,

$$k \asymp \frac{m}{\log N/m + 1}.$$

2.2.1 An Intuition why ℓ_1-Minimization Works Well

In this section we would like to provide an intuitive explanation of the near-optimal error estimates (2.4) and (2.5) provided by ℓ_1-minimization (2.3) in recovering vectors from partial linear measurements. Equations (2.4) and (2.5) ensure in particular that if the vector x is k-*sparse*, then ℓ_1-minimization (2.3) will be able to recover it *exactly* from m linear measurements y obtained via the matrix A. This result is quite surprising because the problem of recovering a sparse vector, or the solution of the following optimization

$$\min \|x\|_{\ell_0^N} \quad \text{subject to } Ax = y, \tag{2.6}$$

is know to be *NP-complete*[1] [59, 61] whereas ℓ_1-minimization is a convex problem which can be solved at any prescribed accuracy in polynomial time. For instance interior-point methods are guaranteed to solve the ℓ_1-problem to a fixed precision in time $\mathcal{O}(m^2 N^{1.5})$ [63]. The first intuitive approach to this perhaps surprising result is by interpreting ℓ_1-minimization as the *convex relaxation* of the problem (2.6).

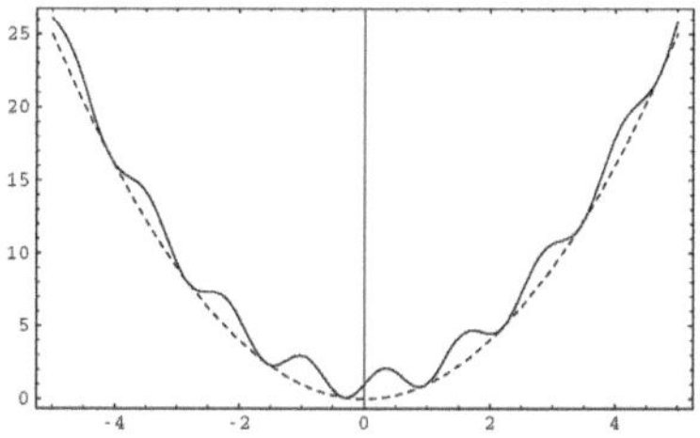

Figure 1. A non convex function f and a convex approximation $g \le f$ from below.

If we were interested in solving an optimization problem

$$\min f(x) \quad \text{subject to } x \in \mathcal{C},$$

where f is a nonconvex, lower-semincontinuous function, and $\mathcal{C}$ is a closed convex set, it might be convenient to recast the problem by considering its convexification, i.e.,

$$\min \bar{f}(x) \quad \text{subject to } x \in \mathcal{C},$$

where $\bar{f}$ is called the *convex relaxation* or the *convex envelope* of f and is given by

$$\bar{f}(x) := \sup\{g(x) \le f(x) : g \text{ is a convex function}\}.$$

The motivation of this choice is simply geometrical. While f can have many minimizers on $\mathcal{C}$, its convex envelop $\bar{f}$ has global minimizers, and such global minimizers are likely to be in a neighborhood of a global minimizer of f, see Figure 1. Actually if $\mathcal{C}$ is compact, then the global minima of f and $\bar{f}$ must necessarily coincide. Unfortunately, the precise computation of $\bar{f}$ is again a very difficult problem. In the case of

[1] NP stands for non-deterministic polynomial-time and indicates a class of problems for which the verification of their solution has a computational costs which is polynomial in the size of the input. However presently it is not known whether such problems can be solved with a polynomial complexity algorithm. This issue is the first in the list of the *Millennium Prize Problems* of the Clay Mathematics Institute.

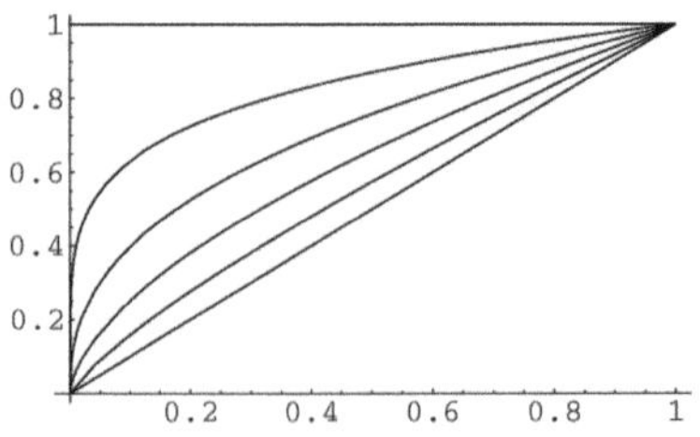

Figure 2. The absolute value function $|\cdot|$ is the convex relaxation of the function $|\cdot|_0$ on $[0, 1]$.

$\|x\|_{\ell_0^N}$, one rewrites

$$\|x\|_{\ell_0^N} := \sum_{j=1}^{N} |x_j|_0, \quad |t|_0 := \begin{cases} 0, & t = 0 \\ 1, & t \neq 0. \end{cases}$$

Its convex envelope in $B_{\ell_\infty^N}(R) \cap \{z : Az = y\}$ is bounded below by $\frac{1}{R}\|x\|_{\ell_1^N} := \frac{1}{R}\sum_{j=1}^{N} |x_j|$, see Figure 2. This observation gives already a first impression of the motivation why ℓ_1-minimization can help in approximating sparse solutions of $Ax = y$. However, it is not yet clear when a global minimizer of

$$\min \|x\|_{\ell_1^N} \quad \text{subject to } Ax = y, \tag{2.7}$$

really coincides with a solution to (2.6), since the ℓ_1-norm is not yet the precise convex envelope of $\|\cdot\|_{\ell_0^N}$ over the solution space $\{z : Az = y\}$. Again a simple geometrical reasoning can help us to get a feeling about more general principles which will be addressed more formally in the following sections.

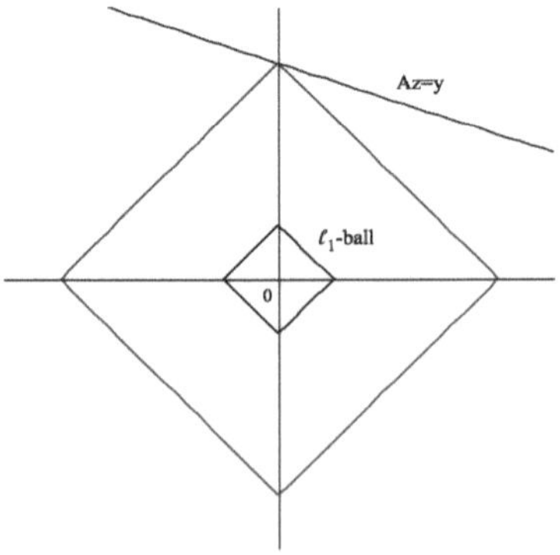

Figure 3. The ℓ_1-minimizer within the affine space of solutions of the linear system $Ax = y$ coincides with the sparsest solution.

Assume for a moment that $N = 2$ and $m = 1$. Hence we are dealing with an affine space of solutions $\mathcal{F}(y) = \{z : Az = y\}$ which is just a line in $\mathbb{R}^2$. When we search for the ℓ_1- norm minimizers among the elements $\mathcal{F}(y)$ (see Figure 3), we immediately realize that, except for pathological situations where $\mathcal{N} = \ker A$ is parallel to one of the faces of the polytope $B_{\ell_1^2}$, there is a unique solution which coincides also with a solution with a minimal number of nonzero entries. Therefore, if we exclude situations in which there exists $\eta \in \mathcal{N}$ such that $|\eta_1| = |\eta_2|$ or, equivalently, we assume that

$$|\eta_i| < |\eta_{\{1,2\}\setminus\{i\}}| \tag{2.8}$$

for all $\eta \in \mathcal{N}$ and for one $i = 1, 2$, then the solution to (2.7) is a solution to (2.6)! Note also that, if we give a uniform probability distribution to the angle in $[0, 2\pi]$ formed by $\mathcal{N}$ and any of the coordinate axes, then we realize that the pathological situation of violating (2.8) has zero probability. Of course, in higher dimension such simple reasoning becomes more involved, since the number of faces and edges of an ℓ_1^N-ball $B_{\ell_1^N}$ becomes larger and larger and one should cumulate the probabilities of different angles with respect to possible affine spaces of codimension $N-m$. However, condition (2.8) is the right prototype of a property (we call it the *Null Space Property* (NSP) and we describe it in detail in the next section) which guarantees, also in higher dimension, that the solution to (2.7) is a solution to (2.6).

2.2.2 Restricted Isometry Property and Null Space Property

Definition 2.7. One says that $A \in \mathbb{R}^{m\times N}$ has the *Null Space Property* (NSP) of order k for $0 < \gamma < 1$ if

$$\|\eta_\Lambda\|_{\ell_1^N} \le \gamma \|\eta_{\Lambda^c}\|_{\ell_1^N},$$

for all sets $\Lambda \subset \{1, \dots, N\}$, $\#\Lambda \le k$ and for all $\eta \in \mathcal{N} = \ker A$.

Note that this definition essentially generalizes condition (2.8) which we introduced by our simple and rough geometrical reasoning in $\mathbb{R}^2$. Further we need to introduce a related property for matrices.

Definition 2.8. One says that $A \in \mathbb{R}^{m\times N}$ has the *Restricted Isometry Property* (RIP) of order K if there exists $0 < \delta_K < 1$ such that

$$(1-\delta_K)\|z\|_{\ell_2^N} \le \|Az\|_{\ell_2^N} \le (1+\delta_K)\|z\|_{\ell_2^N},$$

for all $z \in \Sigma_K$.

The RIP turns out to be very useful in the analysis of stability of certain algorithms as we will show in Section 3.1.4. The RIP is also introduced because it implies the *Null Space Property*, and when dealing with random matrices (see Section 2.2.4) it is more easily addressed. In fact we have:

Lemma 2.9. *Assume that $A \in \mathbb{R}^{m \times N}$ has the RIP of order $K = k + h$ with $0 < \delta_K < 1$. Then A has the NSP of order k and constant $\gamma = \sqrt{\frac{k}{h}\frac{1+\delta_K}{1-\delta_K}}$.*

Proof. Let $\Lambda \subset \{1, \dots, N\}$, $\#\Lambda \leq k$. Define $\Lambda_0 = \Lambda$ and $\Lambda_1, \Lambda_2, \dots, \Lambda_s$ disjoint sets of indexes of size at most h, associated to a decreasing rearrangement of the entries of $\eta \in \mathcal{N} = \ker(A)$. Then, by using Cauchy–Schwarz inequality, the RIP twice, the fact that $A\eta = 0$, and eventually the triangle inequality, we have the following sequence of inequalities:

$$
\begin{aligned}
\|\eta_\Lambda\|_{\ell_1^N} &\leq \sqrt{k}\|\eta_\Lambda\|_{\ell_2^N} \leq \sqrt{k}\|\eta_{\Lambda_0 \cup \Lambda_1}\|_{\ell_2^N} \\
&\leq (1-\delta_K)^{-1}\sqrt{k}\|A\eta_{\Lambda_0 \cup \Lambda_1}\|_{\ell_2^N} = (1-\delta_K)^{-1}\sqrt{k}\|A\eta_{\Lambda_2 \cup \Lambda_3 \cup \dots \cup \Lambda_s}\|_{\ell_2^N} \\
&\leq (1-\delta_K)^{-1}\sqrt{k}\sum_{j=2}^{s}\|A\eta_{\Lambda_j}\|_{\ell_2^N} \leq \frac{1+\delta_K}{1-\delta_K}\sqrt{k}\sum_{j=2}^{s}\|\eta_{\Lambda_j}\|_{\ell_2^N}.
\end{aligned}
\tag{2.9}
$$

Note now that $i \in \Lambda_{j+1}$ and $\ell \in \Lambda_j$ imply by construction of Λ_j's by nonincreasing rearrangement of the entries of η

$$|\eta_i| \leq |\eta_\ell|.$$

By taking the sum over ℓ first and than the ℓ_2^N-norm over i we get

$$|\eta_i| \leq h^{-1}\|\eta_{\Lambda_j}\|_{\ell_1^N}, \quad \text{and} \quad \|\eta_{\Lambda_{j+1}}\|_{\ell_2^N} \leq h^{-1/2}\|\eta_{\Lambda_j}\|_{\ell_1^N}.$$

By using the latter estimates in (2.9) we obtain

$$\|\eta_\Lambda\|_{\ell_1^N} \leq \frac{1+\delta_K}{1-\delta_K}\sqrt{\frac{k}{h}}\sum_{j=1}^{s-1}\|\eta_{\Lambda_j}\|_{\ell_1^N} \leq \left(\frac{1+\delta_K}{1-\delta_K}\sqrt{\frac{k}{h}}\right)\|\eta_{\Lambda^c}\|_{\ell_1^N}. \qquad \square$$

The RIP property does imply the NSP, but the converse is not true. Actually the RIP is significantly more restrictive.

2.2.3 Performances of ℓ_1-Minimization as an Optimal Decoder

In this section we address the proofs of the approximation properties (2.4) and (2.5).

Theorem 2.10. *Let $A \in \mathbb{R}^{m \times N}$ satisfy the RIP of order $2k$ with $\delta_{2k} \leq \delta < \frac{\sqrt{2}-1}{\sqrt{2}+1}$ (or simply A satisfies the NSP of order k with constant $\gamma = \frac{1+\delta}{1-\delta}\sqrt{\frac{1}{2}} < 1$), then the decoder Δ as in (2.3) satisfies (2.4).*

Proof. By Lemma 2.9 we have

$$\|\eta_\Lambda\|_{\ell_1^N} \leq \frac{1+\delta}{1-\delta}\sqrt{\frac{1}{2}}\|\eta_{\Lambda^c}\|_{\ell_1^N},$$

for all $\Lambda \subset \{1,\dots,N\}$, $\#\Lambda \leq k$ and $\eta \in \mathcal{N} = \ker A$. Let $x^* = \Delta(Ax)$, so that $\eta = x^* - x \in \mathcal{N}$, and

$$\|x^*\|_{\ell_1^N} \leq \|x\|_{\ell_1^N}.$$

One denotes now with Λ the set of the k-largest entries of x in absolute value. One has

$$\|x^*_\Lambda\|_{\ell_1^N} + \|x^*_{\Lambda^c}\|_{\ell_1^N} \leq \|x_\Lambda\|_{\ell_1^N} + \|x_{\Lambda^c}\|_{\ell_1^N}.$$

It follows immediately by triangle inequality

$$\|x_\Lambda\|_{\ell_1^N} - \|\eta_\Lambda\|_{\ell_1^N} + \|\eta_{\Lambda^c}\|_{\ell_1^N} - \|x_{\Lambda^c}\|_{\ell_1^N} \leq \|x_\Lambda\|_{\ell_1^N} + \|x_{\Lambda^c}\|_{\ell_1^N}.$$

Hence

$$\|\eta_{\Lambda^c}\|_{\ell_1^N} \leq \|\eta_\Lambda\|_{\ell_1^N} + 2\|x_{\Lambda^c}\|_{\ell_1^N} \leq \frac{1+\delta}{1-\delta}\sqrt{\frac{1}{2}}\|\eta_{\Lambda^c}\|_{\ell_1^N} + 2\sigma_k(x)_{\ell_1^N},$$

or, equivalently,

$$\|\eta_{\Lambda^c}\|_{\ell_1^N} \leq \frac{2}{1-\frac{1+\delta}{1-\delta}\sqrt{\frac{1}{2}}}\sigma_k(x)_{\ell_1^N}. \tag{2.10}$$

In particular, note that for $\delta < \frac{\sqrt{2}-1}{\sqrt{2}+1}$ we have $\frac{1+\delta}{1-\delta}\sqrt{\frac{1}{2}} < 1$. Eventually we conclude the estimates

$$\begin{aligned}\|x - x^*\|_{\ell_1^N} &= \|\eta_\Lambda\|_{\ell_1^N} + \|\eta_{\Lambda^c}\|_{\ell_1^N} \\ &\leq \left(\frac{1+\delta}{1-\delta}\sqrt{\frac{1}{2}} + 1\right)\|\eta_{\Lambda^c}\|_{\ell_1^N} \\ &\leq C_1\sigma_k(x)_{\ell_1^N},\end{aligned}$$

where $C_1 := \left[\frac{2(\frac{1+\delta}{1-\delta}\sqrt{\frac{1}{2}}+1)}{1-\frac{1+\delta}{1-\delta}\sqrt{\frac{1}{2}}}\right]$. □

Similarly we address the second estimate (2.5).

Theorem 2.11. *Let $A \in \mathbb{R}^{m\times N}$ satisfy the RIP of order $3k$ with $\delta_{3k} \leq \delta < \frac{\sqrt{2}-1}{\sqrt{2}+1}$, then the decoder Δ as in* (2.3) *satisfies* (2.5).

Proof. Let $x^* = \Delta(Ax)$. As we proceeded in Lemma 2.9, we denote $\eta = x^* - x \in \mathcal{N}$, $\Lambda_0 = \Lambda$ the set of the $2k$-largest entries of η in absolute value, and Λ_j of size at most k composed of nonincreasing rearrangement entries of η. Then

$$\|\eta_\Lambda\|_{\ell_2^N} \leq \frac{1+\delta}{1-\delta} k^{-\frac{1}{2}} \|\eta_{\Lambda^c}\|_{\ell_1^N}.$$

Note now that by Lemma 2.2 and by Lemma 2.9

$$\begin{aligned}
\|\eta_{\Lambda^c}\|_{\ell_2^N} &\leq (2k)^{-\frac{1}{2}} \|\eta\|_{\ell_1^N} = (2k)^{-1/2}(\|\eta_\Lambda\|_{\ell_1^N} + \|\eta_{\Lambda^c}\|_{\ell_1^N}) \\
&\leq (2k)^{-1/2}(C\|\eta_{\Lambda^c}\|_{\ell_1^N} + \|\eta_{\Lambda^c}\|_{\ell_1^N}) \\
&= \frac{C+1}{\sqrt{2}} k^{-1/2} \|\eta_{\Lambda^c}\|_{\ell_1^N},
\end{aligned}$$

for a suitable constant $C > 0$. Note that, being Λ the set of the in absolute value $2k$-largest entries of η, one has also

$$\|\eta_{\Lambda^c}\|_{\ell_1^N} \leq \|\eta_{(\operatorname{supp} x_{[2k]})^c}\|_{\ell_1^N} \leq \|\eta_{(\operatorname{supp} x_{[k]})^c}\|_{\ell_1^N}, \tag{2.11}$$

where $x_{[h]}$ is the best h-term approximation to x. The use of this latter estimate, combined with inequality (2.10) finally gives

$$\begin{aligned}
\|x - x^*\|_{\ell_2^N} &\leq \|\eta_\Lambda\|_{\ell_2^N} + \|\eta_{\Lambda^c}\|_{\ell_2^N} \\
&\leq C_1 k^{-1/2} \|\eta_{\Lambda^c}\|_{\ell_1^N} \\
&\leq C_2 k^{-1/2} \sigma_k(x)_{\ell_1^N}.
\end{aligned}$$

□

We would like to conclude this section by mentioning a further stability property of ℓ_1-minimization as established in [12].

Theorem 2.12. *Let $A \in \mathbb{R}^{m \times N}$ which satisfies the RIP of order $4k$ with δ_{4k} sufficiently small. Assume further that $Ax + e = y$ where e is a measurement error. Then the decoder Δ has the further enhanced stability property:*

$$\|x - \Delta(y)\|_{\ell_2^N} \leq C_3 \left(\sigma_k(x)_{\ell_2^N} + \frac{\sigma_k(x)_{\ell_1^N}}{k^{1/2}} + \|e\|_{\ell_2^N} \right). \tag{2.12}$$

2.2.4 Random Matrices and Optimal RIP

In this section we would like to mention how, for different classes of random matrices, it is possible to show that the RIP property can hold with optimal constants, i.e.,

$$k \asymp \frac{m}{\log N/m + 1},$$

at least with high probability. This implies in particular, that such matrices exist, they are frequent, but they are given to us only with an uncertainty.

Gaussian and Bernoulli Random Matrices Let $(\Omega, \mathbb{P})$ be a probability space and $\mathcal{X}$ a random variable on $(\Omega, \mathbb{P})$. One can define a random matrix $A(\omega)$, $\omega \in \Omega^{mN}$, as the matrix whose entries are independent realizations of $\mathcal{X}$. We assume further that $\|A(\omega)x\|_{\ell_2^m}^2$ has expected value $\|x\|_{\ell_2^N}^2$ and

$$\mathbb{P}(|\|A(\omega)x\|_{\ell_2^m}^2 - \|x\|_{\ell_2^N}^2| \geq \varepsilon\|x\|_{\ell_2^N}^2) \leq 2e^{-mc_0(\varepsilon)}, \quad 0 < \varepsilon < 1. \tag{2.13}$$

Example 2.13. Here we collect two relevant examples for which the concentration property (2.13) holds:

1. One can choose, for instance, the entries of A as i.i.d. Gaussian random variables, $A_{ij} \sim \mathcal{N}(0, \frac{1}{m})$, and $c_0(\varepsilon) = \varepsilon^2/4 - \varepsilon^3/6$. This can be shown by using Chernoff inequalities and a comparison of the moments of a Bernoulli random variable with respect to those of a Gaussian random variable.
2. One can also use matrices where the entries are independent realizations of ± 1 Bernoulli random variables, i.e.,

$$A_{ij} = \begin{cases} +1/\sqrt{m}, & \text{with probability } \frac{1}{2} \\ -1/\sqrt{m}, & \text{with probability } \frac{1}{2}. \end{cases}$$

Then we have the following result, shown, for instance in [3].

Theorem 2.14. *Suppose that m, N and $0 < \delta < 1$ are fixed. If $A(\omega), \omega \in \Omega^{mN}$ is a random matrix of size $m \times N$ with the concentration property* (2.13)*, then there exist constants $c_1, c_2 > 0$ depending on δ such that the RIP holds for $A(\omega)$ with constant δ and $k \leq c_1 \frac{m}{\log(N/m)+1}$ with probability exceeding $1 - 2e^{-c_2 m}$.*

An extensive study on RIP properties of different types of matrices, for instance partial orthogonal matrices or random structured matrices, is provided in [70].

3 Numerical Methods for Compressed Sensing

The previous sections showed that ℓ_1-minimization performs very well in recovering sparse or approximately sparse vectors from undersampled measurements. In applications it is important to have fast methods for actually solving ℓ_1-minimization or at least with similar guarantees of stability. Three such methods – the homotopy (LARS) method introduced in [35,67], the iteratively reweighted least squares method (IRLS) [30], and the iterative hard thresholding algorithm [6,7] – will be explained in more detail below.

As a first remark, the ℓ_1-minimization problem

$$\min \|x\|_{\ell_1^N} \quad \text{subject to } Ax = y \tag{3.1}$$

is in the real case equivalent to the linear program

$$\min \sum_{j=1}^{2N} v_j \quad \text{subject to } v \geq 0,\ (A| - A)v = y. \tag{3.2}$$

The solution x^* to (3.1) is obtained from the solution v^* of (3.2) via $x^* = (I| - I)v^*$, for I the identity matrix. Any linear programming method may therefore be used for solving (3.1). The simplex method as well as interior point methods apply in particular [63], and standard software may be used. (In the complex case, (3.1) is equivalent to a second order cone program (SOCP) and can be solved with interior point methods as well.) However, such methods and software are of general purpose and one may expect that methods specialized to (3.1) outperform such existing standard methods. Moreover, standard software often has the drawback that one has to provide the full matrix rather than fast routines for matrix-vector multiplication which are available for instance in the case of partial Fourier matrices. In order to obtain the full performance of such methods one would therefore need to re-implement them, which is a daunting task because interior point methods usually require much fine tuning. On the contrary the three specialized methods described below are rather simple to implement and very efficient. Many more methods are available nowadays, including greedy methods, such as Orthogonal Matching Pursuit [78] and CoSaMP [77]. However, only the three methods below are explained in detail because they highlight the fundamental concepts which are useful to comprehend also other algorithms.

3.1 Direct and Iterative Methods

3.1.1 The Homotopy Method

The homotopy method – or modified LARS – [34,35,65,67] solves (3.1) and is a direct method, i.e., it solves the problem exactly in a finite number of steps.

One considers the ℓ_1-regularized least squares functionals

$$J_\lambda(x) = \frac{1}{2}\|Ax - y\|_2^2 + \lambda\|x\|_{\ell_1^N}, \quad x \in \mathbb{R}^N,\ \lambda > 0, \tag{3.3}$$

and their minimizers x_λ. When $\lambda = \hat{\lambda}$ is large enough then $x_{\hat{\lambda}} = 0$, and furthermore, $\lim_{\lambda \to 0} x_\lambda = x^*$, where x^* is the solution to (3.1). The idea of the homotopy method is to trace the solution x_λ from $x_{\hat{\lambda}} = 0$ to x^*. The crucial observation is that the solution path $\lambda \mapsto x_\lambda$ is piecewise linear, and it is enough to trace the endpoints of the linear pieces.

The minimizer of (3.3) can be characterized using the *subdifferential* [36], which is defined for a general convex function $F : \mathbb{R}^N \to \mathbb{R}$ at a point $x \in \mathbb{R}^N$ by

$$\partial F(x) = \{v \in \mathbb{R}^N,\ F(y) - F(x) \geq \langle v, y - x\rangle \ \text{ for all } y \in \mathbb{R}^N\}.$$

Clearly, x is a minimizer of F if and only if $0 \in \partial F(x)$. The subdifferential of J_λ is given by

$$\partial J_\lambda(x) = A^*(Ax - y) + \lambda \partial \| \cdot \|_{\ell_1^N}(x),$$

where the subdifferential of the ℓ_1-norm is given by

$$\partial \| \cdot \|_{\ell_1^N}(x) = \{v \in \mathbb{R}^N : v_\ell \in \partial | \cdot |(x_\ell), \ell = 1, \dots, N\},$$

with the subdifferential of the absolute value being

$$\partial | \cdot |(z) = \begin{cases} \{\operatorname{sgn}(z)\}, & \text{if } z \neq 0, \\ [-1, 1] & \text{if } z = 0. \end{cases}$$

See also Example 5.1 in Section 5.1.1 where we will repeat these concepts in more generality. The inclusion $0 \in \partial J_\lambda(x)$ is equivalent to

$$(A^*(Ax - y))_\ell = \lambda \operatorname{sgn}(x_\ell) \qquad \text{if } x_\ell \neq 0, \tag{3.4}$$

$$|(A^*(Ax - y))_\ell| \leq \lambda \qquad \text{if } x_\ell = 0, \tag{3.5}$$

for all $\ell = 1, \dots, N$.

As already mentioned above the homotopy method starts with $x^{(0)} = x_{\hat{\lambda}} = 0$. By conditions (3.4) and (3.5) the corresponding λ can be chosen as $\hat{\lambda} = \lambda^{(0)} = \|A^* y\|_\infty$. In the further steps $j = 1, 2, \dots$ the algorithm computes minimizers $x^{(1)}, x^{(2)}, \dots$ and maintains an active (support) set Λ_j. Denote by

$$c^{(j)} = A^*(y - Ax^{(j-1)})$$

the current residual vector. The columns of the matrix A are denoted by $a_\ell, \ell = 1, \dots, N$ and for a subset $\Lambda \subset \{1, \dots, N\}$ we let A_Λ be the submatrix of A corresponding to the columns indexed by Λ.

Step 1: Let

$$\ell^{(1)} := \arg \max_{\ell=1,\dots,N} |(A^* y)_\ell| = \arg \max_{\ell=1,\dots,N} |c_\ell^{(1)}|.$$

One assumes here and also in the further steps that the maximum is attained at only one index ℓ. The case that the maximum is attained simultaneously at two or more indices ℓ (which almost never happens) requires more complications, which we would like to avoid here. One may refer to [35] for such details.

Now set $\Lambda_1 = \{\ell^{(1)}\}$. The vector $d^{(1)} \in \mathbb{R}^N$ describing the direction of the solution (homotopy) path has components

$$d^{(1)}_{\ell^{(1)}} = \|a_{\ell^{(1)}}\|_2^{-2} \operatorname{sgn}((Ay)_{\ell^{(1)}}), \quad d^{(1)}_\ell = 0, \ \ell \neq \ell^{(1)}.$$

The first linear piece of the solution path then takes the form

$$x = x(\gamma) = x^{(0)} + \gamma d^{(1)} = \gamma d^{(1)}, \quad \gamma \in [0, \gamma^{(1)}].$$

One verifies with the definition of $d^{(1)}$ that (3.4) is always satisfied for $x = x(\gamma)$ and $\lambda = \lambda(\gamma) = \lambda^{(0)} - \gamma, \gamma \in [0, \lambda^{(0)}]$. The next breakpoint is found by determining the maximal $\gamma = \gamma^{(1)} > 0$ for which (3.5) is satisfied, which is

$$\gamma^{(1)} = \min_{\ell \neq \ell^{(1)}} \left\{ \frac{\lambda^{(0)} - c_\ell^{(1)}}{1 - (A^* A\, d^{(1)})_\ell}, \frac{\lambda^{(0)} + c_\ell^{(1)}}{1 + (A^* A\, d^{(1)})_\ell} \right\}, \tag{3.6}$$

where the minimum is taken only over positive arguments. Then $x^{(1)} = x(\gamma^{(1)}) = \gamma^{(1)} d^{(1)}$ is the next minimizer of J_λ for $\lambda = \lambda^{(1)} := \lambda^{(0)} - \gamma^{(1)}$. This $\lambda^{(1)}$ satisfies $\lambda^{(1)} = \|c^{(1)}\|_\infty$. Let $\ell^{(2)}$ be the index where the minimum in (3.6) is attained (where we again assume that the minimum is attained only at one index) and put $\Lambda_2 = \{\ell^{(1)}, \ell^{(2)}\}$.

Step j: Determine the new direction $d^{(j)}$ of the homotopy path by solving

$$A^*_{\Lambda_j} A_{\Lambda_j} d^{(j)}_{\Lambda_j} = \operatorname{sgn}(c^{(j)}_{\Lambda_j}), \tag{3.7}$$

which is a linear system of equations of size at most $|\Lambda_j| \times |\Lambda_j|$. Outside the components in Λ_j one sets $d_\ell^{(j)} = 0$, $\ell \notin \Lambda_j$. The next piece of the path is then given by

$$x(\gamma) = x^{(j-1)} + \gamma d^{(j)}, \quad \gamma \in [0, \gamma^{(j)}].$$

The maximal γ such that $x(\gamma)$ satisfies (3.5) is

$$\gamma_+^{(j)} = \min_{\ell \notin \Lambda_j} \left\{ \frac{\lambda^{(j-1)} - c_\ell^{(j)}}{1 - (A^* A d^{(j)})_\ell}, \frac{\lambda^{(j-1)} + c_\ell^{(j)}}{1 + (A^* A d^{(j)})_\ell} \right\}. \tag{3.8}$$

The maximal γ such that $x(\gamma)$ satisfies (3.4) is determined as

$$\gamma_-^{(j)} = \min_{\ell \in \Lambda_j} \{-x_\ell^{(j-1)} / d_\ell^{(j)}\}. \tag{3.9}$$

Both in (3.8) and (3.9) the minimum is taken only over positive arguments. The next breakpoint is given by $x^{(j+1)} = x(\gamma^{(j)})$ with $\gamma^{(j)} = \min\{\gamma_+^{(j)}, \gamma_-^{(j)}\}$. If $\gamma_+^{(j)}$ determines the minimum then the index $\ell_+^{(j)} \notin \Lambda_j$ providing the minimum in (3.8) is added to the active set, $\Lambda_{j+1} = \Lambda_j \cup \{\ell_+^{(j)}\}$. If $\gamma^{(j)} = \gamma_-^{(j)}$ then the index $\ell_-^{(j)} \in \Lambda_j$ is removed from the active set, $\Lambda_{j+1} = \Lambda_j \setminus \{\ell_-^{(j)}\}$. Further, one updates $\lambda^{(j)} = \lambda^{(j-1)} - \gamma^{(j)}$. By construction $\lambda^{(j)} = \|c^{(j)}\|_\infty$.

The algorithm stops when $\lambda^{(j)} = \|c^{(j)}\|_\infty = 0$, i.e., when the residual vanishes, and outputs $x^* = x^{(j)}$. Indeed, this happens after a finite number of steps. In [35] the authors proved the following result.

Theorem 3.1. *If in each step the minimum in* (3.8) *and* (3.9) *is attained in only one index* ℓ, *then the homotopy algorithm as described yields the minimizer of the* ℓ_1*-minimization problem* (3.1).

If the algorithm is stopped earlier at some iteration j then obviously it yields the minimizer of $J_\lambda = J_{\lambda^{(j)}}$. In particular, obvious stopping rules may also be used to solve the problems

$$\min \|x\|_{\ell_1^N} \quad \text{subject to } \|Ax - y\|_{\ell_2^m} \leq \epsilon \tag{3.10}$$

$$\text{and} \quad \min \|Ax - y\|_{\ell_2^m} \quad \text{subject to } \|x\|_{\ell_1^N} \leq \delta. \tag{3.11}$$

The second of these is called the *lasso* [76].

The LARS (least angle regression) algorithm is a simple modification of the homotopy method, which only adds elements to the active set in each step. So $\gamma_-^{(j)}$ in (3.9) is not considered. (Sometimes the homotopy method is therefore also called modified LARS.) Clearly, LARS is not guaranteed any more to yield the solution of (3.1). However, it is observed empirically – and can be proven rigorously in certain cases [34] – that often in sparse recovery problems, the homotopy method does never remove elements from the active set, so that in this case LARS and homotopy perform the same steps. It is a crucial point that if the solution of (3.1) is k-sparse and the homotopy method never removes elements then the solution is obtained after precisely k-steps. Furthermore, the most demanding computational part at step j is then the solution of the $j \times j$ linear system of equations (3.7). In conclusion, the homotopy and LARS methods are very efficient for sparse recovery problems.

3.1.2 Iteratively Reweighted Least Squares

In this section we want to present an iterative algorithm which, under the condition that A satisfies the NSP, is guaranteed to reconstruct vectors with the same approximation guarantees (2.4) as ℓ_1-minimization. Moreover, we will also show that such algorithm has a guaranteed (local) linear rate of convergence which, with a minimal modification, can be improved to a superlinear rate. We need to make first a brief introduction which hopefully will shed light on the basic principles of this algorithm and their interplay with sparse recovery and ℓ_1-minimization.

Denote $\mathcal{F}(y) = \{x : Ax = y\}$ and $\mathcal{N} = \ker A$. Let us start with a few non-rigorous observations; next we will be more precise. For $t \neq 0$ we simply have

$$|t| = \frac{t^2}{|t|}.$$

Hence, an ℓ_1-minimization can be recast into a weighted ℓ_2-minimization, and we may expect

$$\arg\min_{x\in\mathcal{F}(y)} \sum_{j=1}^N |x_j| \approx \arg\min_{x\in\mathcal{F}(y)} \sum_{j=1}^N x_j^2 |x_j^*|^{-1},$$

as soon as x^* is the wanted ℓ_1-norm minimizer (see the following Lemma 3.3 for a precise statement). Clearly the advantage of this approximate reformulation is that minimizing a smooth quadratic function $|t|^2$ is better than addressing the minimization of the nonsmooth function $|t|$. However, the obvious drawbacks are that neither we dispose of x^* a priori (this is the vector we are interested to compute!) nor we can expect that $x_j^* \neq 0$ for all $j = 1, \ldots, N$, since we hope for k-sparse solutions. Hence, we start assuming that we dispose of a good approximation w_j^n of $|(x_j^*)^2 + \epsilon_n^2|^{-1/2} \approx |x_j^*|^{-1}$ and we compute

$$x^{n+1} = \arg\min_{x\in\mathcal{F}(y)} \sum_{j=1}^N x_j^2 w_j^n, \tag{3.12}$$

then we up-date $\epsilon_{n+1} \leq \epsilon_n$, we define

$$w_j^{n+1} = |(x_j^{n+1})^2 + \epsilon_{n+1}^2|^{-1/2}, \tag{3.13}$$

and we iterate the process. The hope is that a proper choice of $\epsilon_n \to 0$ will allow us for the computation of an ℓ_1-minimizer, although such a limit property is far from being obvious. The next sections will help us to describe the right mathematical setting where such limit is justified.

The Relationship between ℓ_1-Minimization and Reweighted ℓ_2-Minimization Let us start with a characterization of ℓ_1-minimizers.

Lemma 3.2. *An element $x^* \in \mathcal{F}(y)$ has minimal ℓ_1-norm among all elements $z \in \mathcal{F}(y)$ if and only if*

$$\left| \sum_{x_j^*\neq 0} \operatorname{sgn}(x_j^*)\eta_j \right| \leq \sum_{x_j^*=0} |\eta_j|, \quad \textit{for all } \eta \in \mathcal{N}. \tag{3.14}$$

Moreover, x^* is unique if and only if we have the strict inequality for all $\eta \in \mathcal{N}$ which are not identically zero.

Proof. If $x^* \in \mathcal{F}(y)$ has minimum ℓ_1-norm, then we have, for any $\eta \in \mathcal{N}$ and any $t \in \mathbb{R}$,

$$\sum_{j=1}^N |x_j^* + t\eta_j| \geq \sum_{j=1}^N |x_j^*|. \tag{3.15}$$

Fix $\eta \in \mathcal{N}$. If t is sufficiently small then $x_j^* + t\eta_j$ and x_j^* will have the same sign $s_j := \operatorname{sgn}(x_j^*)$ whenever $x_j^* \neq 0$. Hence, (3.15) can be written as

$$t \sum_{x_j^* \neq 0} s_j \eta_j + \sum_{x_j^* = 0} |t\eta_j| \geq 0.$$

Choosing t of an appropriate sign, we see that (3.14) is a necessary condition.

For the opposite direction, we note that if (3.14) holds then for each $\eta \in \mathcal{N}$, we have

$$\begin{aligned}\sum_{j=1}^N |x_j^*| &= \sum_{x_j^* \neq 0} s_j x_j^* = \sum_{x_j^* \neq 0} s_j (x_j^* + \eta_j) - \sum_{x_j^* \neq 0} s_j \eta_j \\ &\leq \sum_{x_j^* \neq 0} s_j (x_j^* + \eta_j) + \sum_{x_j^* = 0} |\eta_j| \leq \sum_{j=1}^N |x_j^* + \eta_j|, \end{aligned} \tag{3.16}$$

where the first inequality uses (3.14).

If x^* is unique then we have strict inequality in (3.15) and hence subsequently in (3.14). If we have strict inequality in (3.14) then the subsequent strict inequality in (3.16) implies uniqueness. □

Next, consider the minimization in a weighted $\ell_2(w)$-norm. Suppose that the weight w is *strictly positive* which we define to mean that $w_j > 0$ for all $j \in \{1, \dots, N\}$. In this case, $\ell_2(w)$ is a Hilbert space with the inner product

$$\langle u, v \rangle_w := \sum_{j=1}^N w_j u_j v_j. \tag{3.17}$$

Define

$$x^w := \arg \min_{z \in \mathcal{F}(y)} \|z\|_{\ell_2^N(w)}. \tag{3.18}$$

Because the $\|\cdot\|_{\ell_2^N(w)}$-norm is strictly convex, the minimizer x^w is necessarily unique; we leave as an easy exercise that x^w is completely characterized by the orthogonality conditions

$$\langle x^w, \eta \rangle_w = 0, \qquad \text{for all } \eta \in \mathcal{N}. \tag{3.19}$$

A fundamental relationship between ℓ_1-minimization and weighted ℓ_2-minimization, which might seem totally unrelated at first sight, due to the different characterization of respective minimizers, is now easily shown.

Lemma 3.3. *Assume that x^* is an ℓ_1-minimizer and that x^* has no vanishing coordinates. Then the (unique) solution x^w of the weighted least squares problem*

$$x^w := \arg \min_{z \in \mathcal{F}(y)} \|z\|_{\ell_2^N(w)}, \quad w := (w_1, \dots, w_N), \quad \textit{where } w_j := |x_j^*|^{-1},$$

coincides with x^*.

Proof. Assume that x^* is not the $\ell_2^N(w)$-minimizer. Then there exists $\eta \in \mathcal{N}$ such that $0 < \langle x^*, \eta\rangle_w = \sum_{j=1}^N w_j \eta_j x_j^* = \sum_{j=1}^N \eta_j \operatorname{sgn}(x_j^*)$. However, by Lemma 3.2 and because x^* is an ℓ_1-minimizer, we have $\sum_{j=1}^N \eta_j \operatorname{sgn}(x_j^*) = 0$, a contradiction. □

An Iteratively Re-weighted Least Squares Algorithm (IRLS) Since we do not know x^*, this observation cannot be used directly. However, it leads to the following paradigm for finding x^*. We choose a starting weight w^0 and solve the weighted ℓ_2-minimization for this weight. We then use this solution to define a new weight w^1 and repeat this process. An Iteratively Re-weighted Least Squares (IRLS) algorithm of this type appeared for the first time in the approximation practice in the Ph.D. thesis of Lawson in 1961 [53], in the form of an algorithm for solving uniform approximation problems, in particular by Chebyshev polynomials, by means of limits of weighted ℓ_p–norm solutions. This iterative algorithm is now well-known in classical approximation theory as Lawson's algorithm. In [19] it is proved that this algorithm has in principle a linear convergence rate. In the 1970s extensions of Lawson's algorithm for ℓ_p-minimization, and in particular ℓ_1-minimization, were proposed. In signal analysis, IRLS was proposed as a technique to build algorithms for sparse signal reconstruction in [48]. Perhaps the most comprehensive mathematical analysis of the performance of IRLS for ℓ_p-minimization was given in the work of Osborne [66]. However, the interplay of NSP, ℓ_1-minimization, and a reweighted least squares algorithm has been clarified only recently in the work [30]. In the following we describe the essential lines of the analysis of this algorithm, by taking advantage of results and terminology already introduced in previous sections. Our analysis of the algorithm in (3.12) and (3.13) starts from the observation that

$$|t| = \min_{w>0} \frac{1}{2}(wt^2 + w^{-1}),$$

the minimum being reached for $w = \frac{1}{|t|}$. Inspired by this simple relationship, given a real number $\epsilon > 0$ and a weight vector $w \in \mathbb{R}^N$, with $w_j > 0$, $j = 1, \dots, N$, we define

$$\mathcal{J}(z, w, \epsilon) := \frac{1}{2}\left[\sum_{j=1}^N z_j^2 w_j + \sum_{j=1}^N (\epsilon^2 w_j + w_j^{-1})\right], \quad z \in \mathbb{R}^N. \tag{3.20}$$

The algorithm roughly described in (3.12) and (3.13) can be recast as an alternating method for choosing minimizers and weights based on the functional $\mathcal{J}$.

To describe this more rigorously, we define for $z \in \mathbb{R}^N$ the nonincreasing rearrangement $r(z)$ of the absolute values of the entries of z. Thus $r(z)_i$ is the i-th largest element of the set $\{|z_j|,\ j = 1, \dots, N\}$, and a vector v is k-sparse if and only if $r(v)_{k+1} = 0$.

Algorithm 1. We initialize by taking $w^0 := (1, \dots, 1)$. We also set $\epsilon_0 := 1$. We then recursively define for $n = 0, 1, \dots,$

$$x^{n+1} := \arg\min_{z \in \mathcal{F}(y)} \mathcal{J}(z, w^n, \epsilon_n) = \arg\min_{z \in \mathcal{F}(y)} \|z\|_{\ell_2(w^n)} \tag{3.21}$$

and

$$\epsilon_{n+1} := \min\left(\epsilon_n, \frac{r(x^{n+1})_{K+1}}{N}\right), \tag{3.22}$$

where K is a fixed integer that will be described more fully later. We also define

$$w^{n+1} := \arg\min_{w>0} \mathcal{J}(x^{n+1}, w, \epsilon_{n+1}). \tag{3.23}$$

We stop the algorithm if $\epsilon_n = 0$; in this case we define $x^\ell := x^n$ for $\ell > n$. However, in general, the algorithm will generate an infinite sequence $(x^n)_{n \in \mathbb{N}}$ of distinct vectors.

Each step of the algorithm requires the solution of a weighted least squares problem. In matrix form

$$x^{n+1} = D_n^{-1} A^* (A D_n^{-1} A^*)^{-1} y, \tag{3.24}$$

where D_n is the $N \times N$ diagonal matrix whose j-th diagonal entry is w_j^n and A^* denotes the transpose of the matrix A. Once x^{n+1} is found, the weight w^{n+1} is given by

$$w_j^{n+1} = [(x_j^{n+1})^2 + \epsilon_{n+1}^2]^{-1/2}, \quad j = 1, \dots, N. \tag{3.25}$$

Preliminary Results We first make some observations about the nonincreasing rearrangement $r(z)$ and the j-term approximation errors for vectors in $\mathbb{R}^N$. We have the following lemma:

Lemma 3.4. *The map $z \mapsto r(z)$ is Lipschitz continuous on $(\mathbb{R}^N, \|\cdot\|_{\ell_\infty^N})$: for any $z, z' \in \mathbb{R}^N$, we have*

$$\|r(z) - r(z')\|_{\ell_\infty^N} \le \|z - z'\|_{\ell_\infty^N}. \tag{3.26}$$

Moreover, for any j, we have

$$|\sigma_j(z)_{\ell_1^N} - \sigma_j(z')_{\ell_1^N}| \le \|z - z'\|_{\ell_1^N}, \tag{3.27}$$

and for any $J > j$, we have

$$(J - j) r(z)_J \le \|z - z'\|_{\ell_1^N} + \sigma_j(z')_{\ell_1^N}. \tag{3.28}$$

Proof. For any pair of vectors z and z', and any $j \in \{1, \dots, N\}$, let Λ be a set of $j-1$ indices corresponding to the $j-1$ largest entries in z'. Then

$$r(z)_j \le \max_{i \in \Lambda^c} |z_i| \le \max_{i \in \Lambda^c} |z_i'| + \|z - z'\|_{\ell_\infty^N} = r(z')_j + \|z - z'\|_{\ell_\infty^N}. \tag{3.29}$$

We can also reverse the roles of z and z'. Therefore, we obtain (3.26). To prove (3.27), we approximate z by a j-term best approximation $z'_{[j]} \in \Sigma_j$ of z' in ℓ_1^N. Then

$$\sigma_j(z)_{\ell_1^N} \leq \|z - z'_{[j]}\|_{\ell_1^N} \leq \|z - z'\|_{\ell_1^N} + \sigma_j(z')_{\ell_1^N},$$

and the result follows from symmetry.

To prove (3.28), it suffices to note that $(J - j)\,r(z)_J \leq \sigma_j(z)_{\ell_1^N}$. □

Our next result is an approximate reverse triangle inequality for points in $\mathcal{F}(y)$. Its importance to us lies in its implication that whenever two points $z, z' \in \mathcal{F}(y)$ have close ℓ_1-norms and one of them is close to a k-sparse vector, then they necessarily are close to each other. (Note that it also implies that the other vector must then also be close to that k-sparse vector.) This is a geometric property of the null space.

Lemma 3.5 (Inverse triangle inequality). *Assume that the NSP holds with order L and $0 < \gamma < 1$. Then, for any $z, z' \in \mathcal{F}(y)$, we have*

$$\|z' - z\|_{\ell_1^N} \leq \frac{1+\gamma}{1-\gamma}\left(\|z'\|_{\ell_1^N} - \|z\|_{\ell_1^N} + 2\sigma_L(z)_{\ell_1^N}\right). \tag{3.30}$$

Proof. Let Λ be a set of indices of the L largest entries in z. Then

$$\begin{aligned}
\|(z' - z)_{\Lambda^c}\|_{\ell_1^N} &\leq \|z'_{\Lambda^c}\|_{\ell_1^N} + \|z_{\Lambda^c}\|_{\ell_1^N} \\
&= \|z'\|_{\ell_1^N} - \|z'_\Lambda\|_{\ell_1^N} + \sigma_L(z)_{\ell_1^N} \\
&= \|z\|_{\ell_1^N} + \|z'\|_{\ell_1^N} - \|z\|_{\ell_1^N} - \|z'_\Lambda\|_{\ell_1^N} + \sigma_L(z)_{\ell_1^N} \\
&= \|z_\Lambda\|_{\ell_1^N} - \|z'_\Lambda\|_{\ell_1^N} + \|z'\|_{\ell_1^N} - \|z\|_{\ell_1^N} + 2\sigma_L(z)_{\ell_1^N} \\
&\leq \|(z' - z)_\Lambda\|_{\ell_1^N} + \|z'\|_{\ell_1^N} - \|z\|_{\ell_1^N} + 2\sigma_L(z)_{\ell_1^N}.
\end{aligned} \tag{3.31}$$

Using the NSP, this gives

$$\|(z'-z)_\Lambda\|_{\ell_1^N} \leq \gamma\|(z'-z)_{\Lambda^c}\|_{\ell_1^N} \leq \gamma\left(\|(z'-z)_\Lambda\|_{\ell_1^N} + \|z'\|_{\ell_1^N} - \|z\|_{\ell_1^N} + 2\sigma_L(z)_{\ell_1^N}\right). \tag{3.32}$$

In other words,

$$\|(z' - z)_\Lambda\|_{\ell_1^N} \leq \frac{\gamma}{1-\gamma}\left(\|z'\|_{\ell_1^N} - \|z\|_{\ell_1^N} + 2\sigma_L(z)_{\ell_1^N}\right). \tag{3.33}$$

Using this, together with (3.31), we obtain

$$\|z'-z\|_{\ell_1^N} = \|(z'-z)_{\Lambda^c}\|_{\ell_1^N} + \|(z'-z)_\Lambda\|_{\ell_1^N} \leq \frac{1+\gamma}{1-\gamma}\left(\|z'\|_{\ell_1^N} - \|z\|_{\ell_1^N} + 2\sigma_L(z)_{\ell_1^N}\right), \tag{3.34}$$

as desired. □

By using the previous lemma we obtain the following estimate.

Lemma 3.6. *Assume that the NSP holds with order L and* $0 < \gamma < 1$. *Suppose that* $\mathcal{F}(y)$ *contains an L-sparse vector. Then this vector is the unique* ℓ_1*-minimizer in* $\mathcal{F}(y)$*; denoting it by* x^**, we have moreover, for all* $v \in \mathcal{F}(y)$,

$$\|v - x^*\|_{\ell_1^N} \le 2\frac{1+\gamma}{1-\gamma}\sigma_L(v)_{\ell_1^N}. \tag{3.35}$$

Proof. We may immediately see that x^* is the unique ℓ_1-minimizer, by an application of Theorem 2.10. However, we would like to show this statement below, as consequence of the inverse triangle inequality in Lemma 3.5. For the time being, we denote the L-sparse vector in $\mathcal{F}(y)$ by x_s.

Applying (3.30) with $z' = v$ and $z = x_s$, we find

$$\|v - x_s\|_{\ell_1^N} \le \frac{1+\gamma}{1-\gamma}[\|v\|_{\ell_1^N} - \|x_s\|_{\ell_1^N}];$$

since $v \in \mathcal{F}(y)$ is arbitrary, this implies that $\|v\|_{\ell_1^N} - \|x_s\|_{\ell_1^N} \ge 0$ for all $v \in \mathcal{F}(y)$, so that x_s is an ℓ_1-norm minimizer in $\mathcal{F}(y)$.

If x' were another ℓ_1-minimizer in $\mathcal{F}(y)$, then it would follow that $\|x'\|_{\ell_1^N} = \|x_s\|_{\ell_1^N}$, and the inequality we just derived would imply $\|x' - x_s\|_{\ell_1^N} = 0$, or $x' = x_s$. It follows that x_s is the unique ℓ_1-minimizer in $\mathcal{F}(y)$, which we denote by x^*, as proposed earlier.

Finally, we apply (3.30) with $z' = x^*$ and $z = v$, and we obtain

$$\|v - x^*\| \le \frac{1+\gamma}{1-\gamma}(\|x^*\|_{\ell_1^N} - \|v\|_{\ell_1^N} + 2\sigma_L(v)_{\ell_1^N}) \le 2\frac{1+\gamma}{1-\gamma}\sigma_L(v)_{\ell_1^N},$$

where we have used the ℓ_1-minimization property of x^*. □

Our next set of remarks centers around the functional $\mathcal{J}$ defined by (3.20). Note that for each $n = 1, 2, \ldots,$ we have

$$\mathcal{J}(x^{n+1}, w^{n+1}, \epsilon_{n+1}) = \sum_{j=1}^{N}[(x_j^{n+1})^2 + \epsilon_{n+1}^2]^{1/2}. \tag{3.36}$$

We also have the following monotonicity property which holds for all $n \ge 0$:

$$\mathcal{J}(x^{n+1}, w^{n+1}, \epsilon_{n+1}) \le \mathcal{J}(x^{n+1}, w^n, \epsilon_{n+1}) \le \mathcal{J}(x^{n+1}, w^n, \epsilon_n) \le \mathcal{J}(x^n, w^n, \epsilon_n). \tag{3.37}$$

Here the first inequality follows from the minimization property that defines w^{n+1}, the second inequality from $\epsilon_{n+1} \le \epsilon_n$, and the last inequality from the minimization property that defines x^{n+1}. For each n, x^{n+1} is completely determined by w^n; for $n = 0$, in particular, x^1 is determined solely by w^0, and independent of the choice of $x^0 \in \mathcal{F}(y)$. (With the initial weight vector defined by $w^0 = (1, \dots, 1)$, x^1 is the classical minimum ℓ_2-norm element of $\mathcal{F}(y)$.) The inequality (3.37) for $n = 0$ thus holds for arbitrary $x^0 \in \mathcal{F}(y)$.

Lemma 3.7. *For each $n \geq 1$ we have*

$$\|x^n\|_{\ell_1^N} \leq \mathcal{J}(x^1, w^0, \epsilon_0) =: \mathcal{A} \tag{3.38}$$

and

$$w_j^n \geq \mathcal{A}^{-1}, \quad j = 1, \dots, N. \tag{3.39}$$

Proof. The bound (3.38) follows from (3.37) and

$$\|x^n\|_{\ell_1^N} \leq \sum_{j=1}^N [(x_j^n)^2 + \epsilon_n^2]^{1/2} = \mathcal{J}(x^n, w^n, \epsilon_n).$$

The bound (3.39) follows from

$$(w_j^n)^{-1} = [(x_j^n)^2 + \epsilon_n^2]^{1/2} \leq \mathcal{J}(x^n, w^n, \epsilon_n) \leq \mathcal{A},$$

where the last inequality uses (3.37). □

Convergence of the Algorithm In this section, we prove that the algorithm converges. Our starting point is the following lemma that establishes $(x^n - x^{n+1}) \to 0$ for $n \to \infty$.

Lemma 3.8. *Given any $y \in \mathbb{R}^m$, the x^n satisfy*

$$\sum_{n=1}^{\infty} \|x^{n+1} - x^n\|_{\ell_2^N}^2 \leq 2\mathcal{A}^2, \tag{3.40}$$

where $\mathcal{A}$ is the constant of Lemma 3.7. In particular, we have

$$\lim_{n\to\infty} (x^n - x^{n+1}) = 0. \tag{3.41}$$

Proof. For each $n = 1, 2, \dots,$ we have

$$\begin{aligned}
2[\mathcal{J}(x^n, w^n, \epsilon_n) - \mathcal{J}(x^{n+1}, w^{n+1}, \epsilon_{n+1})] &\geq 2[\mathcal{J}(x^n, w^n, \epsilon_n) - \mathcal{J}(x^{n+1}, w^n, \epsilon_n)] \\
&= \langle x^n, x^n\rangle_{w^n} - \langle x^{n+1}, x^{n+1}\rangle_{w^n} \\
&= \langle x^n + x^{n+1}, x^n - x^{n+1}\rangle_{w^n} \\
&= \langle x^n - x^{n+1}, x^n - x^{n+1}\rangle_{w^n}
\end{aligned}$$

$$= \sum_{j=1}^{N} w_j^n (x_j^n - x_j^{n+1})^2$$

$$\geq \mathcal{A}^{-1} \|x^n - x^{n+1}\|_{\ell_2^N}^2, \tag{3.42}$$

where the third equality uses the fact that $\langle x^{n+1}, x^n - x^{n+1}\rangle_{w^n} = 0$ (observe that $x^{n+1} - x^n \in \mathcal{N}$ and invoke (3.19)), and the inequality uses the bound (3.39) on the weights. If we now sum these inequalities over $n \geq 1$, we arrive at (3.40). □

From the monotonicity of ϵ_n, we know that $\epsilon := \lim_{n\to\infty} \epsilon_n$ exists and is non-negative. The following functional will play an important role in our proof of convergence:

$$f_\epsilon(z) := \sum_{j=1}^{N} (z_j^2 + \epsilon^2)^{1/2}. \tag{3.43}$$

Notice that if we knew that x^n converged to x then, in view of (3.36), $f_\epsilon(x)$ would be the limit of $\mathcal{J}(x^n, w^n, \epsilon_n)$. When $\epsilon > 0$ the functional f_ϵ is strictly convex and therefore has a unique minimizer

$$x^\varepsilon := \arg \min_{z\in\mathcal{F}(y)} f_\epsilon(z). \tag{3.44}$$

This minimizer is characterized by the following lemma:

Lemma 3.9. *Let $\epsilon > 0$ and $z \in \mathcal{F}(y)$. Then $z = x^\epsilon$ if and only if $\langle z, \eta\rangle_{\tilde{w}(z,\epsilon)} = 0$ for all $\eta \in \mathcal{N}$, where $\tilde{w}(z,\epsilon)_j = [z_j^2 + \epsilon^2]^{-1/2}$.*

Proof. For the "only if" part, let $z = x^\epsilon$ and $\eta \in \mathcal{N}$ be arbitrary. Consider the analytic function

$$G_\epsilon(t) := f_\epsilon(z + t\eta) - f_\epsilon(z).$$

We have $G_\epsilon(0) = 0$, and by the minimization property $G_\epsilon(t) \geq 0$ for all $t \in \mathbb{R}$. Hence, $G_\epsilon'(0) = 0$. A simple calculation reveals that

$$G_\epsilon'(0) = \sum_{j=1}^{N} \frac{\eta_j z_j}{[z_j^2 + \epsilon^2]^{1/2}} = \langle z, \eta\rangle_{\tilde{w}(z,\epsilon)},$$

which gives the desired result.

For the "if" part, assume that $z \in \mathcal{F}(y)$ and

$$\langle z, \eta\rangle_{\tilde{w}(z,\epsilon)} = 0 \quad \text{for all } \eta \in \mathcal{N}, \tag{3.45}$$

where $\tilde{w}(z,\epsilon)$ is defined as above. We shall show that z is a minimizer of f_ϵ on $\mathcal{F}(y)$. Indeed, consider the convex univariate function $[u^2 + \epsilon^2]^{1/2}$. For any point u_0 we have from convexity that

$$[u^2 + \epsilon^2]^{1/2} \geq [u_0^2 + \epsilon^2]^{1/2} + [u_0^2 + \epsilon^2]^{-1/2} u_0 (u - u_0), \tag{3.46}$$

because the right side is the linear function which is tangent to this function at u_0. It follows that for any point $v \in \mathcal{F}(y)$ we have

$$f_\epsilon(v) \geq f_\epsilon(z) + \sum_{j=1}^{N} [z_j^2 + \epsilon^2]^{-1/2} z_j (v_j - z_j) = f_\epsilon(z) + \langle z, v - z \rangle_{\tilde{w}(z,\epsilon)} = f_\epsilon(z), \tag{3.47}$$

where we have used the orthogonality condition (3.45) and the fact that $v - z$ is in $\mathcal{N}$. Since v is arbitrary, it follows that $z = x^\epsilon$, as claimed. □

We now prove the convergence of the algorithm.

Theorem 3.10. *Let K (the same index as used in the update rule (3.22)) be chosen so that A satisfies the Null Space Property of order K, with $\gamma < 1$. Then, for each $y \in \mathbb{R}^m$, the output of Algorithm 1 converges to a vector $\bar{x}$, with $r(\bar{x})_{K+1} = N \lim_{n\to\infty} \epsilon_n$ and the following hold:*

(i) *If $\epsilon = \lim_{n\to\infty} \epsilon_n = 0$, then $\bar{x}$ is K-sparse; in this case there is therefore a unique ℓ_1-minimizer x^*, and $\bar{x} = x^*$; moreover, we have, for $k \leq K$, and any $z \in \mathcal{F}(y)$,*

$$\|z - \bar{x}\|_{\ell_1^N} \leq c\,\sigma_k(z)_{\ell_1^N}, \quad \text{with } c := \frac{2(1+\gamma)}{1-\gamma}. \tag{3.48}$$

(ii) *If $\epsilon = \lim_{n\to\infty} \epsilon_n > 0$, then $\bar{x} = x^\epsilon$.*

(iii) *In this last case, if γ satisfies the stricter bound $\gamma < 1 - \frac{2}{K+2}$ (or, equivalently, if $\frac{2\gamma}{1-\gamma} < K$), then we have, for all $z \in \mathcal{F}(y)$ and any $k < K - \frac{2\gamma}{1-\gamma}$, that*

$$\|z - \bar{x}\|_{\ell_1^N} \leq \tilde{c}\,\sigma_k(z)_{\ell_1^N}, \quad \text{with } \tilde{c} := \frac{2(1+\gamma)}{1-\gamma}\left[\frac{K - k + \frac{3}{2}}{K - k - \frac{2\gamma}{1-\gamma}}\right]. \tag{3.49}$$

As a consequence, this case is excluded if $\mathcal{F}(y)$ contains a vector of sparsity $k < K - \frac{2\gamma}{1-\gamma}$.

Note that the approximation properties (3.48) and (3.49) are exactly of the same order as the one (2.4) provided by ℓ_1-minimization. However, in general, $\bar{x}$ is not necessarily an ℓ_1-minimizer, unless it coincides with a sparse solution.

The constant $\tilde{c}$ can be quite reasonable; for instance, if $\gamma \leq 1/2$ and $k \leq K - 3$, then we have $\tilde{c} \leq 9\frac{1+\gamma}{1-\gamma} \leq 27$.

Proof. Note that since $\epsilon_{n+1} \leq \epsilon_n$, the ϵ_n always converge. We start by considering the case $\epsilon := \lim_{n\to\infty} \epsilon_n = 0$.

Case $\epsilon = 0$: In this case, we want to prove that x^n converges , and that its limit is an ℓ_1-minimizer. Suppose that $\epsilon_{n_0} = 0$ for some n_0. Then by the definition of the algorithm, we know that the iteration is stopped at $n = n_0$, and $x^n = x^{n_0}$, $n \geq n_0$. Therefore $\bar{x} = x^{n_0}$. From the definition of ϵ_n, it then also follows that $r(x^{n_0})_{K+1} = 0$

and so $\bar{x} = x^{n_0}$ is K-sparse. As noted in Lemma 3.6, if a K-sparse solution exists when A satisfies the NSP of order K with $\gamma < 1$, then it is the unique ℓ_1-minimizer. Therefore, $\bar{x}$ equals x^*, this unique minimizer.

Suppose now that $\epsilon_n > 0$ for all n. Since $\epsilon_n \to 0$, there is an increasing sequence of indices (n_i) such that $\epsilon_{n_i} < \epsilon_{n_i-1}$ for all i. By the definition (3.22) of $(\epsilon_n)_{n\in\mathbb{N}}$, we must have $r(x^{n_i})_{K+1} < N\epsilon_{n_i-1}$ for all i. Noting that $(x^n)_{n\in\mathbb{N}}$ is a bounded sequence, there exists a subsequence $(p_j)_{j\in\mathbb{N}}$ of $(n_i)_{i\in\mathbb{N}}$ such that $(x^{p_j})_{j\in\mathbb{N}}$ converges to a point $\tilde{x} \in \mathcal{F}(y)$. By Lemma 3.4, we know that $r(x^{p_j})_{K+1}$ converges to $r(\tilde{x})_{K+1}$. Hence we get

$$r(\tilde{x})_{K+1} = \lim_{j\to\infty} r(x^{p_j})_{K+1} \leq \lim_{j\to\infty} N\epsilon_{p_j-1} = 0, \tag{3.50}$$

which means that the support-width of $\tilde{x}$ is at most K, i.e. $\tilde{x}$ is K-sparse. By the same token used above, we again have that $\tilde{x} = x^*$, the unique ℓ_1-minimizer. We must still show that $x^n \to x^*$. Since $x^{p_j} \to x^*$ and $\epsilon_{p_j} \to 0$, (3.36) implies $\mathcal{J}(x^{p_j}, w^{p_j}, \epsilon_{p_j}) \to \|x^*\|_{\ell_1^N}$. By the monotonicity property stated in (3.37), we get $\mathcal{J}(x^n, w^n, \epsilon_n) \to \|x^*\|_{\ell_1^N}$. Since (3.36) implies

$$\mathcal{J}(x^n, w^n, \epsilon_n) - N\epsilon_n \leq \|x^n\|_{\ell_1^N} \leq \mathcal{J}(x^n, w^n, \epsilon_n), \tag{3.51}$$

we obtain $\|x^n\|_{\ell_1^N} \to \|x^*\|_{\ell_1^N}$. Finally, we invoke Lemma 3.5 with $z' = x^n$, $z = x^*$, and $k = K$ to get

$$\limsup_{n\to\infty} \|x^n - x^*\|_{\ell_1^N} \leq \frac{1+\gamma}{1-\gamma}\Big(\lim_{n\to\infty} \|x^n\|_{\ell_1^N} - \|x^*\|_{\ell_1^N}\Big) = 0, \tag{3.52}$$

which completes the proof that $x^n \to x^*$ in this case.

Finally, (3.48) follows from (3.35) of Lemma 3.6 (with $L = K$), and the observation that $\sigma_n(z) \geq \sigma_{n'}(z)$ if $n \leq n'$.

Case $\epsilon > 0$: We shall first show that $x^n \to x^\epsilon$, $n \to \infty$, with x^ϵ as defined by (3.44). By Lemma 3.7, we know that $(x^n)_{n=1}^\infty$ is a bounded sequence in $\mathbb{R}^N$ and hence this sequence has accumulation points. Let (x^{n_i}) be any convergent subsequence of (x^n) and let $\tilde{x} \in \mathcal{F}(y)$ be its limit. We want to show that $\tilde{x} = x^\epsilon$.

Since $w_j^n = [(x_j^n)^2 + \epsilon_n^2]^{-1/2} \leq \epsilon^{-1}$, it follows that $\lim_{i\to\infty} w_j^{n_i} = [(\tilde{x}_j)^2 + \epsilon^2]^{-1/2} = \tilde{w}(\tilde{x}, \epsilon)_j =: \tilde{w}_j$, $j = 1, \ldots, N$. On the other hand, by invoking Lemma 3.8, we now find that $x^{n_i+1} \to \tilde{x}$, $i \to \infty$. It then follows from the orthogonality relations (3.19) that for every $\eta \in \mathcal{N}$, we have

$$\langle \tilde{x}, \eta \rangle_{\tilde{w}} = \lim_{i\to\infty} \langle x^{n_i+1}, \eta \rangle_{w^{n_i}} = 0. \tag{3.53}$$

Now the "if" part of Lemma 3.9 implies that $\tilde{x} = x^\epsilon$. Hence x^ϵ is the unique accumulation point of $(x^n)_{n\in\mathbb{N}}$ and therefore its limit. This establishes (ii).

To prove the error estimate (3.49) stated in (iii), we first note that for any $z \in \mathcal{F}(y)$, we have

$$\|x^\epsilon\|_{\ell_1^N} \leq f_\epsilon(x^\epsilon) \leq f_\epsilon(z) \leq \|z\|_{\ell_1^N} + N\epsilon, \tag{3.54}$$

where the second inequality uses the minimizing property of x^ϵ. Hence it follows that $\|x^\epsilon\|_{\ell_1^N} - \|z\|_{\ell_1^N} \leq N\epsilon$. We now invoke Lemma 3.5 to obtain

$$\|x^\epsilon - z\|_{\ell_1^N} \leq \frac{1+\gamma}{1-\gamma}[N\epsilon + 2\sigma_k(z)_{\ell_1^N}]. \tag{3.55}$$

From Lemma 3.4 and (3.22), we obtain

$$N\epsilon = \lim_{n\to\infty} N\epsilon_n \leq \lim_{n\to\infty} r(x^n)_{K+1} = r(x^\epsilon)_{K+1}. \tag{3.56}$$

It follows from (3.28) that

$$\begin{aligned}(K+1-k)N\epsilon &\leq (K+1-k)r(x^\epsilon)_{K+1}\\ &\leq \|x^\epsilon - z\|_{\ell_1^N} + \sigma_k(z)_{\ell_1^N}\\ &\leq \frac{1+\gamma}{1-\gamma}[N\epsilon + 2\sigma_k(z)_{\ell_1^N}] + \sigma_k(z)_{\ell_1^N},\end{aligned} \tag{3.57}$$

where the last inequality uses (3.55). Since by assumption on K, we have $K - k > \frac{2\gamma}{1-\gamma}$, i.e. $K + 1 - k > \frac{1+\gamma}{1-\gamma}$, we obtain

$$N\epsilon + 2\sigma_k(z)_{\ell_1^N} \leq \frac{2(K-k)+3}{(K-k) - \frac{2\gamma}{1-\gamma}}\sigma_k(z)_{\ell_1^N}.$$

Using this back in (3.55), we arrive at (3.49).

Finally, notice that if $\mathcal{F}(y)$ contains a k-sparse vector (with $k < K - \frac{2\gamma}{1-\gamma}$), then we know already that this must be the unique ℓ_1-minimizer x^*; it then follows from our arguments above that we must have $\epsilon = 0$. Indeed, if we had $\epsilon > 0$, then (3.57) would hold for $z = x^*$; since x^* is k-sparse, $\sigma_k(x^*)_{\ell_1^N} = 0$, implying $\epsilon = 0$, a contradiction with the assumption $\epsilon > 0$. This finishes the proof. □

Local Linear Rate of Convergence It is instructive to show a further very interesting result concerning the local rate of convergence of this algorithm, which makes heavy use of the NSP as well as the optimality properties we introduced above. One assumes here that $\mathcal{F}(y)$ contains the k-sparse vector x^*. The algorithm produces the sequence x^n, which converges to x^*, as established above. One denotes the (unknown) support of the k-sparse vector x^* by Λ.

We introduce an auxiliary sequence of error vectors $\eta^n \in \mathcal{N}$ via $\eta^n := x^n - x^*$ and

$$E_n := \|\eta^n\|_{\ell_1^N} = \|x^* - x^n\|_{\ell_1^N}.$$

We know that $E_n \to 0$.

The following theorem gives a bound on the rate of convergence of E_n to zero.

Theorem 3.11. *Assume A satisfies NSP of order K with constant γ such that $0 < \gamma < 1 - \frac{2}{K+2}$. Suppose that $k < K - \frac{2\gamma}{1-\gamma}$, $0 < \rho < 1$, and $0 < \gamma < 1 - \frac{2}{K+2}$ are such that*

$$\mu := \frac{\gamma(1+\gamma)}{1-\rho}\left(1 + \frac{1}{K+1-k}\right) < 1.$$

Assume that $\mathcal{F}(y)$ contains a k-sparse vector x^* and let $\Lambda = \operatorname{supp}(x^*)$. Let n_0 be such that

$$E_{n_0} \le R^* := \rho \min_{j\in\Lambda} |x_j^*|. \tag{3.58}$$

Then for all $n \ge n_0$, we have

$$E_{n+1} \le \mu\, E_n. \tag{3.59}$$

Consequently x^n converges to x^* exponentially.

Proof. We start with the relation (3.19) with $w = w^n$, $x^w = x^{n+1} = x^* + \eta^{n+1}$, and $\eta = x^{n+1} - x^* = \eta^{n+1}$, which gives

$$\sum_{j=1}^{N} (x_j^* + \eta_j^{n+1})\eta_j^{n+1} w_j^n = 0.$$

Rearranging the terms and using the fact that x^* is supported on Λ, we get

$$\sum_{j=1}^{N} |\eta_j^{n+1}|^2 w_j^n = -\sum_{j\in\Lambda} x_j^* \eta_j^{n+1} w_j^n = -\sum_{j\in\Lambda} \frac{x_j^*}{[(x_j^n)^2 + \epsilon_n^2]^{1/2}} \eta_j^{n+1}. \tag{3.60}$$

Prove of the theorem is by induction. One assumes that we have shown $E_n \le R^*$ already. We then have, for all $j \in \Lambda$,

$$|\eta_j^n| \le \|\eta^n\|_{\ell_1^N} = E_n \le \rho |x_j^*|,$$

so that

$$\frac{|x_j^*|}{[(x_j^n)^2 + \epsilon_n^2]^{1/2}} \le \frac{|x_j^*|}{|x_j^n|} = \frac{|x_j^*|}{|x_j^* + \eta_j^n|} \le \frac{1}{1-\rho}, \tag{3.61}$$

and hence (3.60) combined with (3.61) and NSP gives

$$\sum_{j=1}^{N} |\eta_j^{n+1}|^2 w_j^n \le \frac{1}{1-\rho} \|\eta_\Lambda^{n+1}\|_{\ell_1^N} \le \frac{\gamma}{1-\rho} \|\eta_{\Lambda^c}^{n+1}\|_{\ell_1^N}.$$

At the same time, the Cauchy–Schwarz inequality combined with the above estimate yields

$$\begin{aligned}
\|\eta^{n+1}_{\Lambda^c}\|^2_{\ell_1^N} &\le \Big(\sum_{j\in\Lambda^c} |\eta_j^{n+1}|^2 w_j^n\Big)\Big(\sum_{j\in\Lambda^c} [(x_j^n)^2+\epsilon_n^2]^{1/2}\Big)\\
&\le \Big(\sum_{j=1}^{N} |\eta_j^{n+1}|^2 w_j^n\Big)\Big(\sum_{j\in\Lambda^c} [(\eta_j^n)^2+\epsilon_n^2]^{1/2}\Big)\\
&\le \frac{\gamma}{1-\rho}\|\eta^{n+1}_{\Lambda^c}\|_{\ell_1^N}(\|\eta^n\|_{\ell_1^N}+N\epsilon_n).
\end{aligned} \tag{3.62}$$

If $\eta^{n+1}_{\Lambda^c}=0$, then $x^{n+1}_{\Lambda^c}=0$. In this case x^{n+1} is k-sparse and the algorithm has stopped by definition; since $x^{n+1}-x^*$ is in the null space $\mathcal{N}$, which contains no k-sparse elements other than 0, we have already obtained the solution $x^{n+1}=x^*$. If $\eta^{n+1}_{\Lambda^c}\neq 0$, then after canceling the factor $\|\eta^{n+1}_{\Lambda^c}\|_{\ell_1^N}$ in (3.62), we get

$$\|\eta^{n+1}_{\Lambda^c}\|_{\ell_1^N} \le \frac{\gamma}{1-\rho}(\|\eta^n\|_{\ell_1^N}+N\epsilon_n),$$

and thus

$$\begin{aligned}
\|\eta^{n+1}\|_{\ell_1^N} = \|\eta^{n+1}_{\Lambda}\|_{\ell_1^N} + \|\eta^{n+1}_{\Lambda^c}\|_{\ell_1^N} &\le (1+\gamma)\|\eta^{n+1}_{\Lambda^c}\|_{\ell_1^N}\\
&\le \frac{\gamma(1+\gamma)}{1-\rho}(\|\eta^n\|_{\ell_1^N}+N\epsilon_n).
\end{aligned} \tag{3.63}$$

Now, we also have by (3.22) and (3.28)

$$N\epsilon_n \le r(x^n)_{K+1} \le \frac{1}{K+1-k}(\|x^n-x^*\|_{\ell_1^N}+\sigma_k(x^*)_{\ell_1^N}) = \frac{\|\eta^n\|_{\ell_1^N}}{K+1-k}, \tag{3.64}$$

since by assumption $\sigma_k(x^*)=0$. This, together with (3.63), yields the desired bound,

$$E_{n+1} = \|\eta^{n+1}\|_{\ell_1^N} \le \frac{\gamma(1+\gamma)}{1-\rho}\Big(1+\frac{1}{K+1-k}\Big)\|\eta^n\|_{\ell_1^N} = \mu E_n.$$

In particular, since $\mu<1$, we have $E_{n+1}\le R^*$, which completes the induction step. It follows that $E_{n+1}\le \mu E_n$ for all $n\ge n_0$. □

A Surprising Superlinear Convergence Promoting ℓ_τ-Minimization for $\tau<1$
The linear rate (3.59) can be improved significantly, by a very simple modification of the rule of updating the weight:

$$w_j^{n+1} = ((x_j^{n+1})^2+\epsilon_{n+1}^2)^{-\frac{2-\tau}{2}}, \quad j=1,\dots,N, \text{ for any } 0<\tau<1.$$

This corresponds to the substitution of the function $\mathcal{J}$ with

$$\mathcal{J}_\tau(z, w, \epsilon) := \frac{\tau}{2}\Big[\sum_{j=1}^{N} z_j^2 w_j + \sum_{j=1}^{N}\Big(\epsilon^2 w_j + \frac{2-\tau}{\tau}\frac{1}{w_j^{\frac{\tau}{2-\tau}}}\Big)\Big],$$
$$z \in \mathbb{R}^N,\ w \in \mathbb{R}_+^N,\ \epsilon \in \mathbb{R}_+.$$

Surprisingly the rate of local convergence of this modified algorithm is superlinear; the rate is larger for smaller τ, increasing to approach a quadratic regime as $\tau \to 0$. More precisely the local error $E_n := \|x^n - x^*\|_{\ell_\tau^N}^\tau$ satisfies

$$E_{n+1} \leq \mu(\gamma, \tau) E_n^{2-\tau}, \tag{3.65}$$

where $\mu(\gamma, \tau) < 1$ for $\gamma > 0$ sufficiently small. The validity of (3.65) is restricted to x^n in a (small) ball centered at x^*. In particular, if x^0 is close enough to x^* then (3.65) ensures the convergence of the algorithm to the k-sparse solution x^*. We refer the reader to [30] for more details.

Some Open Problems

1. In practice this algorithm appears very robust and its convergence is either linear or even superlinear when properly tuned as previously indicated. However, such guarantees of rate of convergence are valid only in a neighborhood of a solution which is presently very difficult to estimate. This does not allow us yet to properly estimate the complexity of this method.
2. For $\tau < 1$ the algorithm seems to converge properly when τ is not too small, but when, say, $\tau < 0.5$, then the algorithm tends to fail to reach the region of guaranteed convergence. It is an open problem to very sharply characterize such phase transitions, and heuristic methods to avoid local minima are also of great interest.
3. While error guarantees of the type (2.4) are given, it is open whether (2.5) and (2.12) can hold for this algorithm. In this case one expects that the RIP plays a relevant role, instead of the NSP, as we show in Section 3.1.4 below.

3.1.3 Extensions to the Minimization of Functionals with Total Variation Terms

In concrete applications, e.g., for image processing, one might be interested in recovering at best a digital image provided only partial linear or nonlinear measurements, possibly corrupted by noise. Given the observation that natural and man-made images can be characterized by a relatively small number of edges and extensive, relatively uniform parts, one may want to help the reconstruction by imposing that the interesting solution is the one which matches the given data and also has a few discontinuities localized on sets of lower dimension.

In the context of *compressed sensing* as described in the previous sections, we have already clarified that the minimization of ℓ_1-norms occupies a fundamental role for the promotion of sparse solutions. This understanding furnishes an important interpretation of *total variation minimization*, i.e., the minimization of the L^1-norm of derivatives [72], as a regularization technique for image restoration. The problem can be modelled as follows; let $\Omega \subset \mathbb{R}^d$, for $d = 1, 2$ be a bounded open set with Lipschitz boundary, and $\mathcal{H} = L^2(\Omega)$. For $u \in L^1_{\mathrm{loc}}(\Omega)$

$$V(u, \Omega) := \sup \left\{ \int_\Omega u \operatorname{div} \varphi \, dx : \varphi \in [C_c^1(\Omega)]^d, \|\varphi\|_\infty \le 1 \right\}$$

is the variation of u, actually in the literature this is called in a popular way the *total variation* of u. Further, $u \in BV(\Omega)$, the space of bounded variation functions [1,38], if and only if $V(u, \Omega) < \infty$. In this case, we denote $|D(u)|(\Omega) = V(u, \Omega)$. If $u \in W^{1,1}(\Omega)$ (the Sobolev space of L^1-functions with L^1-distributional derivatives), then $|D(u)|(\Omega) = \int_\Omega |\nabla u| \, d\, x$. We consider as in [16,81] the minimization in $BV(\Omega)$ of the functional

$$\mathcal{J}(u) := \|Ku - g\|^2_{L^2(\Omega)} + 2\alpha |D(u)|(\Omega), \tag{3.66}$$

where $K : L^2(\Omega) \to L^2(\Omega)$ is a bounded linear operator, $g \in L^2(\Omega)$ is a datum, and $\alpha > 0$ is a fixed *regularization parameter.* Several numerical strategies to efficiently perform total variation minimization have been proposed in the literature, see for instance [15,26,49,68,83]. However, in the following we will discuss only how to adapt an iteratively reweighted least squares algorithm to this particular situation. For simplicity, we would like to work in a discrete setting [32] and we refer to [16,44] for more details in the continuous setting.

Let us fix the main notations. Since we are interested in a discrete setting we define the *discrete d-orthotope* $\Omega = \{x_1^1 < \cdots < x_{N_1}^1\} \times \cdots \times \{x_1^d < \cdots < x_{N_d}^d\} \subset \mathbb{R}^d$, $d \in \mathbb{N}$ and the considered function spaces are $\mathcal{H} = \mathbb{R}^{N_1 \times N_2 \times \cdots \times N_d}$, where $N_i \in \mathbb{N}$ for $i = 1, \ldots, d$. For $u \in \mathcal{H}$ we write $u = u(x_{\mathbf{i}})_{\mathbf{i} \in \mathcal{I}}$ with

$$\mathcal{I} := \prod_{k=1}^d \{1, \ldots, N_k\}$$

and

$$u(x_{\mathbf{i}}) = u(x_{i_1}^1, \ldots, x_{i_d}^d)$$

where $i_k \in \{1, \ldots, N_k\}$. Then we endow $\mathcal{H}$ with the Euclidean norm

$$\|u\|_{\mathcal{H}} = \|u\|_2 = \left(\sum_{\mathbf{i} \in \mathcal{I}} |u(x_{\mathbf{i}})|^2 \right)^{1/2} = \left(\sum_{x \in \Omega} |u(x)|^2 \right)^{1/2}.$$

We define the scalar product of $u, v \in \mathcal{H}$ as

$$\langle u, v\rangle_{\mathcal{H}} = \sum_{\mathbf{i}\in\mathcal{I}} u(x_{\mathbf{i}})v(x_{\mathbf{i}}),$$

and the scalar product of $p, q \in \mathcal{H}^d$ as

$$\langle p, q\rangle_{\mathcal{H}^d} = \sum_{\mathbf{i}\in\mathcal{I}} \langle p(x_{\mathbf{i}}), q(x_{\mathbf{i}})\rangle_{\mathbb{R}^d},$$

with $\langle y, z\rangle_{\mathbb{R}^d} = \sum_{j=1}^d y_j z_j$ for every $y = (y_1, \dots, y_d) \in \mathbb{R}^d$ and $z = (z_1, \dots, z_d) \in \mathbb{R}^d$. We will consider also other norms, in particular

$$\|u\|_p = \left(\sum_{\mathbf{i}\in\mathcal{I}} |u(x_{\mathbf{i}})|^p\right)^{1/p}, \quad 1 \le p < \infty,$$

and

$$\|u\|_\infty = \sup_{\mathbf{i}\in\mathcal{I}} |u(x_{\mathbf{i}})|.$$

We denote the discrete gradient ∇u by

$$(\nabla u)(x_{\mathbf{i}}) = ((\nabla u)^1(x_{\mathbf{i}}), \dots, (\nabla u)^d(x_{\mathbf{i}})),$$

with

$$(\nabla u)^j(x_{\mathbf{i}}) = \begin{cases} u(x^1_{i_1}, \dots, x^j_{i_j+1}, \dots, x^d_{i_d}) - u(x^1_{i_1}, \dots, x^j_{i_j}, \dots, x^d_{i_d}) & \text{if } i_j < N_j \\ 0 & \text{if } i_j = N_j \end{cases}$$

for all $j = 1, \dots, d$ and for all $\mathbf{i} = (i_1, \dots, i_d) \in \mathcal{I}$.

Let $\varphi : \mathbb{R} \to \mathbb{R}$, we define for $\omega \in \mathcal{H}^d$

$$\varphi(|\omega|)(\Omega) = \sum_{\mathbf{i}\in\mathcal{I}} \varphi(|\omega(x_{\mathbf{i}})|) = \sum_{x\in\Omega} \varphi(|\omega(x)|),$$

where $|y| = \sqrt{y_1^2 + \dots + y_d^2}$. In particular we define the *total variation* of u by setting $\varphi(s) = s$ and $\omega = \nabla u$, i.e.,

$$|\nabla u|(\Omega) := \sum_{\mathbf{i}\in\mathcal{I}} |\nabla u(x_{\mathbf{i}})| = \sum_{x\in\Omega} |\nabla u(x)|.$$

For an operator K we denote K^* its adjoint. Further we introduce the *discrete divergence* $\operatorname{div} : \mathcal{H}^d \to \mathcal{H}$ defined, in analogy with the continuous setting, by $\operatorname{div} = -\nabla^*$

(∇^* is the adjoint of the gradient ∇). The discrete divergence operator is explicitly given by

$$(\mathrm{div}\, p)(x_{\mathbf{i}}) = \begin{cases} p^1(x^1_{i_1},\dots,x^d_{i_d}) - p^1(x^1_{i_1-1},\dots,x^d_{i_d}) & \text{if } 1 < i_1 < N_1 \\ p^1(x^1_{i_1},\dots,x^d_{i_d}) & \text{if } i_1 = 1 \\ -p^1(x^1_{i_1-1},\dots,x^d_{i_d}) & \text{if } i_1 = N_1 \end{cases}$$
$$+ \dots + \begin{cases} p^d(x^1_{i_1},\dots,x^d_{i_d}) - p^d(x^1_{i_1},\dots,x^d_{i_d-1}) & \text{if } 1 < i_d < N_d \\ p^d(x^1_{i_1},\dots,x^d_{i_d}) & \text{if } i_d = 1 \\ -p^d(x^1_{i_1},\dots,x^d_{i_d-1}) & \text{if } i_d = N_d, \end{cases}$$

for every $p = (p^1,\dots,p^d) \in \mathcal{H}^d$ and for all $\mathbf{i} = (i_1,\dots,i_d) \in \mathcal{I}$. (Note that if we considered discrete domains Ω which are not discrete d-orthotopes, then the definitions of gradient and divergence operators should be adjusted accordingly.) We will use the symbol 1 to indicate the constant vector with entry values 1 and 1_D to indicate the characteristic function of the domain $D \subset \Omega$. We are interested in the minimization of the functional

$$\mathcal{J}(u) := \|Ku - g\|_2^2 + 2\alpha |\nabla(u)|(\Omega), \tag{3.67}$$

where $K \in \mathcal{L}(\mathcal{H})$ is a linear operator, $g \in \mathcal{H}$ is a datum, and $\alpha > 0$ is a fixed constant. In order to guarantee the existence of minimizers for (3.67) we assume that:

(C) $\mathcal{J}$ is coercive in $\mathcal{H}$, i.e., there exists a constant $C > 0$ such that $\{\mathcal{J} \le C\} := \{u \in \mathcal{H} : \mathcal{J}(u) \le C\}$ is bounded in $\mathcal{H}$.

It is well known that if $1 \notin \ker(K)$ then condition (C) is satisfied, see [81, Proposition 3.1], and we will assume this condition in the following.

Similarly to (3.20) for the minimization of the ℓ_1-norm, we consider the augmented functional

$$\mathcal{J}(u,w) := \|Ku - g\|_2^2 + \alpha\left(\sum_{x\in\Omega} w(x)|\nabla u(x)|^2 + \frac{1}{w(x)}\right). \tag{3.68}$$

We used again the notation $\mathcal{J}$ with the clear understanding that, when applied to one variable only, it refers to (3.67), otherwise to (3.68). Then, as the IRLS method for compressed sensing, we consider the following

Algorithm 2. We initialize by taking $w^0 := 1$. We also set $1 \ge \varepsilon > 0$. We then recursively define for $n = 0, 1, \dots,$

$$u^{n+1} := \arg\min_{u\in\mathcal{H}} \mathcal{J}(u, w^n) \tag{3.69}$$

and

$$w^{n+1} := \arg\min_{\varepsilon \le w_{\mathbf{i}} \le 1/\varepsilon, \mathbf{i}\in\mathcal{I}} \mathcal{J}(u^{n+1}, w). \tag{3.70}$$

Note that, by considering the Euler–Lagrange equations, (3.69) is equivalent to the solution of the following linear second order partial difference equation

$$\operatorname{div}(w^n \nabla u) - \frac{2}{\alpha} K^*(Ku - g) = 0, \tag{3.71}$$

which can be solved, e.g., by a preconditioned conjugate gradient method. Note that $\varepsilon \leq w^n(x_{\mathbf{i}}) \leq 1/\varepsilon$, $\mathbf{i} \in \mathcal{I}$ and therefore the equation can be recast into a symmetric positive definite linear system. Moreover, as perhaps already expected, the solution to (3.70) is explicitly computed by

$$w^{n+1} = \max\left(\varepsilon, \min\left(\frac{1}{|\nabla u^{n+1}|}, 1/\varepsilon\right)\right).$$

For the sake of the analysis of the convergence of this algorithm, let us introduce the following C^1 function (i.e., it is continuously differentiable):

$$\varphi_\varepsilon(z) = \begin{cases} \frac{1}{2\varepsilon} z^2 + \frac{\varepsilon}{2} & 0 \leq z \leq \varepsilon \\ z & \varepsilon \leq z \leq 1/\varepsilon \\ \frac{\varepsilon}{2} z^2 + \frac{1}{2\varepsilon} & z \geq 1/\varepsilon. \end{cases}$$

Note that

$$\varphi_\varepsilon(z) \geq |z|,$$

and

$$|z| = \lim_{\varepsilon \to 0} \varphi_\varepsilon(z), \quad \text{pointwise.}$$

We consider the following functional:

$$\mathcal{J}_\varepsilon(u) := \|Ku - g\|_2^2 + 2\alpha \varphi_\varepsilon(|\nabla(u)|)(\Omega), \tag{3.72}$$

which is clearly approximating $\mathcal{J}$ from above, i.e.,

$$\mathcal{J}_\varepsilon(u) \geq \mathcal{J}(u), \quad \text{and} \quad \lim_{\varepsilon \to 0} \mathcal{J}_\varepsilon(u) = \mathcal{J}(u), \quad \text{pointwise.} \tag{3.73}$$

Moreover, since $\mathcal{J}_\varepsilon$ is convex and smooth, by taking the Euler–Lagrange equations, we have that u_ε is a minimizer for $\mathcal{J}_\varepsilon$ if and only if

$$\operatorname{div}\left(\frac{\varphi_\varepsilon'(|\nabla u)|}{|\nabla u|} \nabla u\right) - \frac{2}{\alpha} K^*(Ku - g) = 0, \tag{3.74}$$

We have the following result of convergence of the algorithm.

Theorem 3.12. *The sequence $(u^n)_{n \in \mathbb{N}}$ has subsequences that converge to a minimizer $u_\varepsilon := u^\infty$ of $\mathcal{J}_\varepsilon$. If the minimizer were unique, then the full sequence would converge to it.*

Proof. Observe that

$$\mathcal{J}(u^n, w^n) - \mathcal{J}(u^{n+1}, w^{n+1}) = \underbrace{(\mathcal{J}(u^n, w^n) - \mathcal{J}(u^{n+1}, w^n))}_{A_n} + \underbrace{(\mathcal{J}(u^{n+1}, w^n) - \mathcal{J}(u^{n+1}, w^{n+1}))}_{B_n} \geq 0.$$

Therefore $\mathcal{J}(u^n, w^n)$ is a nonincreasing sequence and moreover it is bounded from below, since

$$\inf_{\varepsilon \leq w \leq 1/\varepsilon} \left(\sum_{x\in\Omega} w(x)|\nabla u(x)|^2 + \frac{1}{w(x)} \right) \geq 0.$$

This implies that $\mathcal{J}(u^n, w^n)$ converges. Moreover, we can write

$$B_n = \alpha \sum_{x\in\Omega} c(w^n(x), |\nabla u^{n+1}(x)|) - c(w^{n+1}(x), |\nabla u^{n+1}(x)|),$$

where $c(t, z) := tz^2 + \frac{1}{t}$. By Taylor's formula, we have

$$c(w^n, z) = c(w^{n+1}, z) + \frac{\partial c}{\partial t}(w^{n+1}, z)(w^n - w^{n+1}) + \frac{1}{2}\frac{\partial^2 c}{\partial t^2}(\xi, z)|w^n - w^{n+1}|^2,$$

for $\xi \in \mathrm{conv}(w^n, w^{n+1})$ (the segment between w^n and w^{n+1}). By definition of w^{n+1}, and taking into account that $\varepsilon \leq w^{n+1} \leq \frac{1}{\varepsilon}$, we have

$$\frac{\partial c}{\partial t}(w^{n+1}, |\nabla u^{n+1}(x)|)(w^n - w^{n+1}) \geq 0,$$

and $\frac{\partial^2 c}{\partial t^2}(t, z) = \frac{2}{t^3} \geq 2\varepsilon^3$, for any $t \leq 1/\varepsilon$. This implies that

$$\mathcal{J}(u^n, w^n) - \mathcal{J}(u^{n+1}, w^{n+1}) \geq B_n \geq \alpha\varepsilon^3 \sum_{x\in\Omega} |w^n(x) - w^{n+1}(x)|^2,$$

and since $\mathcal{J}(u^n, w^n)$ is convergent, we have

$$\|w^n - w^{n+1}\|_2 \to 0, \tag{3.75}$$

for $n \to \infty$. Since u^{n+1} is a minimizer of $\mathcal{J}(u, w^n)$ it solves the following system of variational equations

$$0 = \sum_{x\in\Omega} \left(w^n \nabla u^{n+1}(x) \cdot \nabla\varphi(x) + \frac{2}{\alpha}(Ku^{n+1} - g)(x) K\varphi(x) \right), \tag{3.76}$$

for all $\varphi \in \mathcal{H}$. Therefore we can write

$$\sum_{x\in\Omega}\left(w^{n+1}\nabla u^{n+1}(x)\cdot\nabla\varphi(x)+\frac{2}{\alpha}(Ku^{n+1}-g)(x)K\varphi(x)\right)$$
$$=\sum_{x\in\Omega}(w^{n+1}-w^n)\nabla u^{n+1}(x)\cdot\nabla\varphi(x),$$

and

$$\left|\sum_{x\in\Omega}\left(w^{n+1}\nabla u^{n+1}(x)\cdot\nabla\varphi(x)+\frac{2}{\alpha}(Ku^{n+1}-g)(x)K\varphi(x)\right)\right|$$
$$\le \|w^{n+1}-w^n\|_2\|\nabla u^{n+1}\|_2\|\nabla\varphi\|_2.$$

By monotonicity of $(\mathcal{J}(u^{n+1},w^{n+1}))_n$, and since $w^{n+1}=\frac{\varphi_\varepsilon'(|\nabla u^{n+1}|)}{|\nabla u^{n+1}|}$, we have

$$\mathcal{J}(u^1,w^0)\ge\mathcal{J}(u^{n+1},w^{n+1})=\mathcal{J}_\varepsilon(u^{n+1})\ge\mathcal{J}(u^{n+1})\ge c_1|\nabla u|(\Omega)$$
$$\ge c_2\|\nabla u^{n+1}\|_2.$$

Moreover, since $\mathcal{J}_\varepsilon(u^{n+1})\ge\mathcal{J}(u^{n+1})$ and $\mathcal{J}$ is coercive, by condition (C), we have that $\|u^{n+1}\|_2$ and $\|\nabla u^{n+1}\|_2$ are uniformly bounded with respect to n. Therefore, using (3.75), we can conclude that

$$\left|\sum_{x\in\Omega}\left(w^{n+1}\nabla u^{n+1}(x)\cdot\nabla\varphi(x)+\frac{2}{\alpha}(Ku^{n+1}-g)(x)K\varphi(x)\right)\right|$$
$$\le \|w^{n+1}-w^n\|_2\|\nabla u^{n+1}\|_2\|\nabla\varphi\|_2\to 0,$$

for $n\to\infty$, and there exists a subsequence $(u^{n_k+1})_{k\in\mathbb{N}}$ that converges in $\mathcal{H}$ to a function u^∞. Since $w^{n_k+1}=\frac{\varphi_\varepsilon'(|\nabla u^{n_k+1}|)}{|\nabla u^{n_k+1}|}$, and by taking the limit for $k\to\infty$, we obtain that in fact

$$\operatorname{div}\left(\frac{\varphi_\varepsilon'(|\nabla u^\infty|)}{|\nabla u^\infty|}\nabla u^\infty\right)-\frac{2}{\alpha}K^*(Ku^\infty-g)=0. \tag{3.77}$$

The latter are the Euler–Lagrange equations (3.74) associated to the functional $\mathcal{J}_\varepsilon$ and therefore u^∞ is a minimizer of $\mathcal{J}_\varepsilon$. □

It is left as a – not simple – exercise to prove the following result. One has to make use of the monotonicity of the approximation (3.73), of the coerciveness of $\mathcal{J}$ (property (C)), and of the continuity of $\mathcal{J}_\varepsilon$. See also [25] for more general tools from so-called Γ-*convergence* for achieving such variational limits.

Proposition 3.13. *Let us assume that $(\varepsilon_h)_h$ is a sequence of positive numbers monotonically converging to zero. The accumulation points of the sequence $(u_{\varepsilon_h})_h$ of minimizers of $\mathcal{J}_{\varepsilon_h}$ are minimizers of $\mathcal{J}$.*

Let us note a few differences between Algorithm 1 and Algorithm 2. In Algorithm 1 we have been able to establish a rule for up-dating the parameter ϵ according to the iterations. This was not done for Algorithm 2, where we considered the limit for $\varepsilon \to 0$ only at the end. It is an interesting open question how can we simultaneously address a choice of a decreasing sequence $(\varepsilon_n)_n$ during the iterations and show directly the convergence of the resulting sequence to a minimizer of $\mathcal{J}$.

3.1.4 Iterative Hard Thresholding

Let us now return to the sparse recovery problem (2.6) and address a new iterative algorithm which, under the RIP for A, has stability properties as in (2.12), which are reached in a finite number of iterations. In this section we address the following

Algorithm 3. We initialize by taking $x^0 = 0$. We iterate

$$x^{n+1} = \mathbb{H}_k(x^n + A^*(y - Ax^n)), \tag{3.78}$$

where

$$\mathbb{H}_k(x) = x_{[k]}, \tag{3.79}$$

is the operator which returns the best k-term approximation to x, see (2.1).

Note that if x^* is k-sparse and $Ax^* = y$, then x^* is a fixed point of

$$x^* = \mathbb{H}_k(x^* + A^*(y - Ax^*)).$$

This algorithm can be seen as a minimizing method for the functional

$$\mathcal{J}(x) = \|y - Ax\|_{\ell_2^N}^2 + 2\alpha\|x\|_{\ell_0^N}, \tag{3.80}$$

for a suitable $\alpha = \alpha(k) > 0$ or equivalently for the solution of the optimization problem

$$\min_x \|y - Ax\|_{\ell_2^N}^2 \quad \text{subject to } \|x\|_{\ell_0^N} \leq k.$$

Actually, it was shown in [6] that if $\|A\| < 1$ then this algorithm converges to a local minimizer of (3.80). We would like to analyze this algorithm following [7] in the case A satisfies the RIP. We start with a few technical lemmas which shed light on fundamental properties of RIP matrices and sparse approximations, as established in [77].

Lemma 3.14. *For all index sets $\Lambda \subset \{1, \dots, N\}$ and all A for which the RIP holds with order $k = |\Lambda|$, we have*

$$\|A_\Lambda^* y\|_{\ell_2^N} \leq (1 + \delta_k)\|y\|_{\ell_2^N}, \tag{3.81}$$

$$(1 - \delta_k)^2\|x_\Lambda\|_{\ell_2^N} \leq \|A_\Lambda^* A_\Lambda x_\Lambda\|_{\ell_2^N} \leq (1 + \delta_k)^2\|x_\Lambda\|_{\ell_2^N}, \tag{3.82}$$

and

$$\|(I - A_\Lambda^* A_\Lambda)x_\Lambda\|_{\ell_2^N} \le \delta_k^2 \|x_\Lambda\|_{\ell_2^N}. \tag{3.83}$$

Furthermore, for two disjoint sets Λ_1 and Λ_2 and all A for which the RIP holds with order $k = |\Lambda|$, $\Lambda = \Lambda_1 \cup \Lambda_2$,

$$\|A_{\Lambda_1}^* A_{\Lambda_2} x_{\Lambda_2}\|_{\ell_2^N} \le \delta_k^2 \|x_\Lambda\|_{\ell_2^N}. \tag{3.84}$$

Proof. The proof of (3.81)–(3.83) is a simple exercise, and it is left to the reader. For (3.84), just note that $A_{\Lambda_1}^* A_{\Lambda_2}$ is a submatrix of $A_{\Lambda_1\cup\Lambda_2}^* A_{\Lambda_1\cup\Lambda_2} - I$, and therefore $\|A_{\Lambda_1}^* A_{\Lambda_2}\| \le \|I - A_{\Lambda_1\cup\Lambda_2}^* A_{\Lambda_1\cup\Lambda_2}\|$. One concludes by (3.83). □

Lemma 3.15. *Suppose the matrix A satisfies the RIP of order k with constant $\delta_k > 0$. Then for all vectors x, the following bound holds*

$$\|Ax\|_{\ell_2^N} \le (1 + \delta_k)\left(\|x\|_{\ell_2^N} + \frac{\|x\|_{\ell_1^N}}{k^{1/2}} \right). \tag{3.85}$$

Proof. In this proof we consider $\mathbb{R}^N$ as a Banach space endowed with several different norms. In particular, the statement of the lemma can be regarded as a result about the operator norm of A as a map between two Banach spaces. For a set $\Lambda \subset \{1, 2, \dots, N\}$, we consider $B_{\ell_2^\Lambda}$ the ℓ_2-norm unit ball of vectors supported in Λ and we define the convex set

$$S = \operatorname{conv}\left\{ \bigcup_{|\Lambda| \le k} B_{\ell_2^\Lambda} \right\} \subset \mathbb{R}^N.$$

The set S can be considered the unit ball of a norm $\|\cdot\|_S$ on $\mathbb{R}^N$, and the upper bound of the RIP property a statement about the norm of A between $S := (\mathbb{R}^N, \|\cdot\|_S)$ and $\ell_2^N := (\ell_2^N, \|\cdot\|_{\ell_2^N})$, i.e., (with a slight abuse of notation)

$$\|A\|_{S \to \ell_2^N} = \max_{x \in S} \|Ax\|_{\ell_2^N} \le (1 + \delta_k).$$

Let us define a second convex body,

$$K = \left\{ x : \|x\|_{\ell_2^N} + \frac{\|x\|_{\ell_1^N}}{k^{1/2}} \le 1 \right\} \subset \mathbb{R}^N,$$

and we consider, analogously (and with the same abuse of notation), the operator norm

$$\|A\|_{K \to \ell_2^N} = \max_{x \in K} \|Ax\|_{\ell_2^N}.$$

The content of the lemma is in fact the claim that

$$\|A\|_{K \to \ell_2^N} \le \|A\|_{S \to \ell_2^N}.$$

To establish this point it is sufficient to check that $K \subset S$. To do that we prove the reverse inclusion of the polar sets, i.e.,

$$S^\circ \subset K^\circ.$$

We recall here that the polar set of $\Omega \subset \mathbb{R}^N$ is

$$\Omega^\circ := \left\{ y : \sup_{x \in \Omega} \langle x, y \rangle \leq 1 \right\}.$$

If Ω is convex than $\Omega^{\circ\circ} = \Omega$. Moreover, the norm associated to a convex body Ω can also be expressed by

$$\|x\|_\Omega = \sup_{y \in \Omega^\circ} \langle x, y \rangle.$$

In particular, the norm with unit ball S° is easily calculated as

$$\|x\|_{S^\circ} = \max_{|\Lambda| \leq k} \|x_\Lambda\|_2.$$

Now, consider a vector x in the unit ball S° and let Λ be the support of the k-best approximation of x. We must have

$$\|x_{\Lambda^c}\|_\infty \leq \frac{1}{\sqrt{k}},$$

otherwise $|x_j| > \frac{1}{\sqrt{k}}$ for all $j \in \Lambda$, but then $\|x\|_{S^\circ} \geq \|x_\Lambda\|_2 > 1$, a contradiction. Therefore, we can write

$$x = x_\Lambda + x_{\Lambda^c} \in B_{\ell_2^N} + \frac{1}{\sqrt{k}} B_{\ell_\infty^N}.$$

But the set on the right-hand side is precisely K° since

$$\begin{aligned} \sup_{y \in K^\circ} \langle x, y \rangle = \|x\|_K &= \|x\|_{\ell_2^N} + \frac{\|x\|_{\ell_1^N}}{k^{1/2}} \\ &= \sup_{y \in B_{\ell_2^N}} \langle x, y \rangle + \sup_{z \in \frac{1}{k^{1/2}} B_{\ell_\infty^N}} \langle x, z \rangle = \sup_{y \in B_{\ell_2^N} + \frac{1}{k^{1/2}} B_{\ell_\infty^N}} \langle x, y \rangle. \end{aligned}$$

In summary $S^\circ \subset K^\circ$. □

Lemma 3.16. *For any x we denote $x^{[k]} = x - x_{[k]}$, where $x_{[k]}$ is the best k-term approximation to x. Let*

$$y = Ax + e = Ax_{[k]} + Ax^{[k]} + e = Ax_{[k]} + \tilde{e}.$$

If A has the RIP of order k, then the norm of the error $\tilde{e}$ can be bounded by

$$\|\tilde{e}\|_{\ell_2^N} \leq (1 + \delta_k) \left(\sigma_k(x)_{\ell_2^N} + \frac{\sigma_k(x)_{\ell_1^N}}{\sqrt{k}} \right) + \|e\|_{\ell_2^N}. \tag{3.86}$$

Proof. Decompose $x = x_{[k]} + x^{[k]}$ and $\tilde{e} = Ax^{[k]} + e$. To compute the norm of the error term, we simply apply the triangle inequality and Lemma 3.15. □

After this collection of technical results, we are able to establish a first convergence result.

Theorem 3.17. *Let us assume that $y = Ax + e$ is a noisy encoding of x via A, where x is k-sparse. If A has the RIP of order $3k$ and constant $\delta_{3k}^2 < \frac{1}{\sqrt{32}}$, then, at iteration n, Algorithm 3 will recover an approximation x^n satisfying*

$$\|x - x^n\|_{\ell_2^N} \leq 2^{-n}\|x\|_{\ell_2^N} + 5\|e\|_{\ell_2^N}. \tag{3.87}$$

Furthermore, after at most

$$n^* = \left\lceil \log_2 \left(\frac{\|x\|_{\ell_2^N}}{\|e\|_{\ell_2^N}} \right) \right\rceil \tag{3.88}$$

iterations, the algorithm estimates x with accuracy

$$\|x - x^{n^*}\|_{\ell_2^N} \leq 6\|e\|_{\ell_2^N}. \tag{3.89}$$

Proof. Let us denote $z^n := x^n + A^*(y - Ax^n)$, $r^n = x - x^n$, and $\Lambda^n := \operatorname{supp}(r^n)$. By triangle inequality we can write

$$\begin{aligned}\|x - x^{n+1}\|_{\ell_2^N} &= \|(x - x^{n+1})_{\Lambda^{(n+1)}}\|_{\ell_2^N} \\ &\leq \|x_{\Lambda^{n+1}} - z^n_{\Lambda^{n+1}}\|_{\ell_2^N} + \|x^{n+1}_{\Lambda^{n+1}} - z^n_{\Lambda^{n+1}}\|_{\ell_2^N}.\end{aligned}$$

By definition $x^{n+1} = \mathbb{H}_k(z^n) = z^{(n)}_{[k]}$. This implies

$$\|x^{n+1}_{\Lambda^{n+1}} - z^n_{\Lambda^{n+1}}\|_{\ell_2^N} \leq \|x_{\Lambda^{n+1}} - z^n_{\Lambda^{n+1}}\|_{\ell_2^N},$$

and

$$\|x - x^{n+1}\|_{\ell_2^N} \leq 2\|x_{\Lambda^{n+1}} - z^n_{\Lambda^{n+1}}\|_{\ell_2^N}.$$

We can also write

$$z^n_{\Lambda^{n+1}} = x^n_{\Lambda^{n+1}} + A^*_{\Lambda^{n+1}} Ar^n + A^*_{\Lambda^{n+1}} e.$$

We then have

$$\begin{aligned}\|x - x^{n+1}\|_{\ell_2^N} &\leq 2\|x_{\Lambda^{n+1}} - x^n_{\Lambda^{n+1}} - A^*_{\Lambda^{n+1}} Ar^n - A^*_{\Lambda^{n+1}} e\|_{\ell_2^N} \\ &\leq 2\|(I - A^*_{\Lambda^{n+1}} A_{\Lambda^{n+1}})r^n\|_{\ell_2^N} + 2\|A^*_{\Lambda^{n+1}} A_{\Lambda^n \setminus \Lambda^{n+1}} r^n\|_{\ell_2^N} \\ &\quad + 2\|A^*_{\Lambda^{n+1}} e\|_{\ell_2^N}.\end{aligned}$$

Note that $|\Lambda^n| \leq 2k$ and that $|\Lambda^{n+1} \cup \Lambda^n| \leq 3k$. By an application of the bounds in Lemma 3.14, and by using the fact that $\delta_{2k} \leq \delta_{3k}$ (note that a $2k$-sparse vector is also $3k$-sparse)

$$\|r^{n+1}\|_{\ell_2^N} \leq 2\delta_{2k}^2 \|r^n_{\Lambda^{n+1}}\|_{\ell_2^N} + 2\delta_{3k}^2 \|r^n_{\Lambda^n \setminus \Lambda^{n+1}}\|_{\ell_2^N} + 2(1+\delta_{2k})\|e\|_{\ell_2^N}.$$

Moreover $\|r^n_{\Lambda^{n+1}}\|_{\ell_2^N} + \|r^n_{\Lambda^n \setminus \Lambda^{n+1}}\|_{\ell_2^N} \leq \sqrt{2}\|r^n\|_{\ell_2^N}$. Therefore we have the bound

$$\|r^{n+1}\|_{\ell_2^N} \leq 2\sqrt{2}\delta_{3k}^2 \|r^n\|_{\ell_2^N} + 2(1+\delta_{3k})\|e\|_{\ell_2^N}.$$

By assumption $\delta_{3k}^2 < \frac{1}{\sqrt{32}}$ and $2\sqrt{2}\delta_{3k}^2 < \frac{1}{2}$. (Note that here we could simply choose any value $\delta_{3k}^2 < \frac{1}{\sqrt{8}}$ and obtain a slightly different estimate!) Then we get the recursion

$$\|r^{n+1}\|_{\ell_2^N} \leq 2^{-1}\|r^n\|_{\ell_2^N} + 2.17\|e\|_{\ell_2^N},$$

which iterated (note that $x^0 = 0$ and $2.17\sum_{n=0}^{\infty} 2^{-n} \leq 4.34$) gives

$$\|r^{n+1}\|_{\ell_2^N} \leq 2^{-n}\|x\|_{\ell_2^N} + 4.34\|e\|_{\ell_2^N}.$$

This is precisely the bound we were looking for. The rest of the statements of the theorem is left as an exercise. □

We have also the following result.

Corollary 3.18. *Let us assume that $y = Ax + e$ is a noisy encoding of x via A, where x is* an arbitrary vector. *If A has the RIP of order* $3k$ *and constant $\delta_{3k}^2 < \frac{1}{\sqrt{32}}$, then, at iteration n, Algorithm* 3 *will recover an approximation x^n satisfying*

$$\|x - x^n\|_{\ell_2^N} \leq 2^{-n}\|x\|_{\ell_2^N} + 6\left(\sigma_k(x)_{\ell_2^N} + \frac{\sigma_k(x)_{\ell_1^N}}{\sqrt{k}} + \|e\|_{\ell_2^N}\right). \tag{3.90}$$

Furthermore, after at most

$$n^* = \left\lceil \log_2\left(\frac{\|x\|_{\ell_2^N}}{\|e\|_{\ell_2^N}}\right)\right\rceil \tag{3.91}$$

iterations, the algorithm estimates x with accuracy

$$\|x - x^{n^*}\|_{\ell_2^N} \leq 7\left(\sigma_k(x)_{\ell_2^N} + \frac{\sigma_k(x)_{\ell_1^N}}{\sqrt{k}} + \|e\|_{\ell_2^N}\right). \tag{3.92}$$

Proof. We first note

$$\|x - x^n\|_{\ell_2^N} \leq \sigma_k(x)_{\ell_2^N} + \|x_{[k]} - x^n\|_{\ell_2^N}.$$

The proof now follows by bounding $\|x_{[k]} - x^n\|_{\ell_2^N}$. For this we simply apply Theorem 3.17 to $x_{[k]}$ with $\tilde{e}$ instead of e, and use Lemma 3.16 to bound $\|\tilde{e}\|_{\ell_2^N}$. The rest is left as an exercise. □

A Brief Discussion This algorithm has a guaranteed error reduction from the very beginning of the iteration, and it is robust to noise, i.e., an estimate of the type (2.12) holds. Moreover, each iteration costs mainly as much as an application of A^*A. At first glance this algorithm is greatly superior with respect to IRLS; however, we have to stress that IRLS can converge superlinearly and a fine analysis of its complexity is widely open.

4 Numerical Methods for Sparse Recovery

In the previous chapters we put most of the emphasis on finite dimensional linear problems (also of relatively small size) where the model matrix A has the RIP or the NSP. This setting is suitable for applications in coding/decoding or compressed acquisition problems, hence from human-made problems coming from technology; however it does not fit many possible applications where we are interested in recovering quantities from partial real-life measurements. In this case we may need to work with large dimensional problems (even infinite dimensional) where the model linear (or nonlinear) operator which defines the measurements has not such nice properties as the RIP and NSP.

Here and later we are concerned with the more general setting and the efficient minimization of functionals of the type:

$$\mathcal{J}(u) := \|Ku - y\|_Y^2 + 2\|(\langle u, \tilde{\psi}_\lambda \rangle)_{\lambda \in \mathcal{I}}\|_{\ell_{1,\alpha}(\mathcal{I})}, \tag{4.1}$$

where $K : X \to Y$ is a bounded linear operator acting between two separable Hilbert spaces X and Y, $y \in Y$ is a given measurement, and $\Psi := (\psi_\lambda)_{\lambda \in \mathcal{I}}$ is a prescribed countable basis for X with associated dual $\tilde{\Psi} := (\tilde{\psi}_\lambda)_{\lambda \in \mathcal{I}}$. For $1 \leq p < \infty$, the sequence norm $\|\mathbf{u}\|_{\ell_{p,\alpha}(\mathcal{I})} := (\sum_{\lambda \in \mathcal{I}} |u_\lambda|^p \alpha_\lambda)^{1/p}$ is the usual norm for weighted p-summable sequences, with weight $\alpha = (\alpha_\lambda)_{\lambda \in \mathcal{I}} \in \mathbb{R}_+^{\mathcal{I}}$, such that $\alpha_\lambda \geq \bar{\alpha} > 0$. Associated to the basis, we are given the synthesis map $F : \ell_2(\mathcal{I}) \to X$ defined by

$$F\mathbf{u} := \sum_{\lambda \in \mathcal{I}} u_\lambda \psi_\lambda, \quad \mathbf{u} \in \ell_2(\mathcal{I}). \tag{4.2}$$

We can re-formulate equivalently the functional in terms of sequences in $\ell_2(\mathcal{I})$ as follows:

$$\mathcal{J}(\mathbf{u}) := \mathcal{J}_\alpha(\mathbf{u}) = \|(K \circ F)\mathbf{u} - y\|_Y^2 + 2\|\mathbf{u}\|_{\ell_{1,\alpha}(\mathcal{I})}. \tag{4.3}$$

For ease of notation let us write $A := K \circ F$. Such functional turns out to be very useful in many practical problems, where one cannot observe directly the quantities of most interest; instead their values have to be inferred from their effect on observable quantities. When this relationship between the observable y and the interesting quantity u is (approximately) linear the situation can be modeled mathematically by the equation

$$y = Ku. \tag{4.4}$$

If K is a "nice" (e.g., well-conditioned), easily invertible operator, and if the data y are free of noise, then this is a well-known task which can be addressed with standard numerical analysis methods. Often, however, the mapping K is not invertible or ill-conditioned. Moreover, typically (4.4) is only an idealized version in which noise has been neglected; a more accurate model is

$$y = Ku + e, \tag{4.5}$$

in which the data are corrupted by an (unknown) noise e. In order to deal with this type of reconstruction problem a *regularization* mechanism is required [37]. Regularization techniques try, as much as possible, to take advantage of (often vague) prior knowledge one may have about the nature of u, which is embedded into the model. The approach modelled by the functional $\mathcal{J}$ in (4.1) is indeed tailored to the case when u can be represented by a *sparse* expansion, i.e., when u can be represented by a series expansion (4.2) with respect to an orthonormal basis (or a frame [27]) that has only a small number of large coefficients. The previous chapters should convince the reader that imposing an additional ℓ_1-norm term as in (4.2) has indeed the effect of sparsifying possible solutions. Hence, we model the sparsity constraint by a regularizing ℓ_1-term in the functional to be minimized; of course, we could consider also a minimization of the type (3.80), but that has the disadvantage of being nonconvex and not being necessarily robust to noise, when no RIP conditions are imposed on the model operator A.

In the following we will not use anymore the bold form $\mathbf{u}$ for a sequence in $\ell_2(\mathcal{I})$, since here and later we will exclusively work with the space $\ell_2(\mathcal{I})$.

4.1 Iterative Soft-Thresholding in Hilbert Spaces

Several authors have proposed independently an iterative soft-thresholding algorithm to approximate minimizers $u^* := u^*_\alpha$ of the functional in (4.2), see [35, 39, 74, 75]. More precisely, u^* is the limit of sequences $u^{(n)}$ defined recursively by

$$u^{(n+1)} = \mathbb{S}_\alpha[u^{(n)} + A^*y - A^*Au^{(n)}], \tag{4.6}$$

starting from an arbitrary $u^{(0)}$, where $\mathbb{S}_\alpha$ is the soft-thresholding operation defined by $\mathbb{S}_\alpha(u)_\lambda = S_{\alpha_\lambda}(u_\lambda)$ with

$$S_\tau(x) = \begin{cases} x - \tau & x > \tau \\ 0 & |x| \leq \tau \\ x + \tau & x < -\tau. \end{cases} \tag{4.7}$$

This is our starting point and the reference iteration on which we want to work out several innovations. Strong convergence of this algorithm was proved in [28], under the assumption that $\|A\| < 1$ (actually, convergence can be shown also for $\|A\| < \sqrt{2}$ [21]; nevertheless, the condition $\|A\| < 1$ is by no means a restriction, since it can always be met by a suitable rescaling of the functional $\mathcal{J}$, in particular of K, y, and α). Soft-thresholding plays a role in this problem because it leads to the unique minimizer of a functional combining ℓ_2 and ℓ_1-norms, i.e., (see Lemma 4.1)

$$\mathbb{S}_\alpha(a) = \arg \min_{u \in \ell_2(\mathcal{I})} (\|u - a\|^2_{\ell_2(\mathcal{I})} + 2\|u\|_{\ell_{1,\alpha}(\mathcal{I})}). \tag{4.8}$$

We will call the iteration (4.6) the *iterative soft-thresholding algorithm* or the *thresholded Landweber iteration* (ISTA).

In this section we would like to provide the analysis of the convergence of this algorithm. Due to the lack of assumptions such as the RIP or the NSP, the methods we use come exclusively from convex analysis and we cannot take advantage of relatively simple estimates as we did for the convergence analysis of Algorithm 1 and Algorithm 3.

4.1.1 The Surrogate Functional

The first relevant observation is that the algorithm can be recast into an iterated minimization of a properly augmented functional, which we call the *surrogate functional* of $\mathcal{J}$, and which is defined by

$$\mathcal{J}^S(u, a) := \|Au - y\|^2_Y + 2\|u\|_{\ell_{1,\alpha}(\mathcal{I})} + \|u - a\|^2_{\ell_2(\mathcal{I})} - \|Au - Aa\|^2_Y. \tag{4.9}$$

Assume here and later that $\|A\| < 1$. Observe that, in this case, we have

$$\|u - a\|^2_{\ell_2(\mathcal{I})} - \|Au - Aa\|^2_Y \geq C\|u - a\|^2_{\ell_2(\mathcal{I})}, \tag{4.10}$$

for $C = (1 - \|A\|^2) > 0$. Hence

$$\mathcal{J}(u) = \mathcal{J}^S(u, u) \leq \mathcal{J}^S(u, a), \tag{4.11}$$

and

$$\mathcal{J}^S(u, a) - \mathcal{J}^S(u, u) \geq C\|u - a\|^2_{\ell_2(\mathcal{I})}. \tag{4.12}$$

In particular, $\mathcal{J}^S$ is strictly convex with respect to u and it has a unique minimizer with respect to u once a is fixed. We have the following technical lemmas.

Lemma 4.1. *The soft-thresholding operator is the solution of the following optimization problem:*

$$\mathbb{S}_\alpha(a) = \arg \min_{u \in \ell_2(\mathcal{I})} (\|u - a\|^2_{\ell_2(\mathcal{I})} + 2\|u\|_{\ell_{1,\alpha}(\mathcal{I})}).$$

Proof. By componentwise optimization, we can reduce the problem to a scalar problem, i.e., we need to prove that

$$S_{\alpha_\lambda}(a_\lambda) = \arg \min_x (x - a_\lambda)^2 + 2\alpha_\lambda |x|,$$

which is shown by a simple direct computation. Let x^* be the minimizer. It is clear that $\operatorname{sgn}(x^*)\operatorname{sgn}(a_\lambda) \geq 0$ otherwise the function is increased. Hence we need to optimize $(x-a_\lambda)^2 + 2\alpha_\lambda \operatorname{sgn}(a_\lambda)x$ which has minimum at $\bar{x} = (a_\lambda - \operatorname{sgn}(a_\lambda)\alpha_\lambda)$. If $|a_\lambda| > \alpha_\lambda$ than $x^* = \bar{x}$. Otherwise $\operatorname{sgn}(\bar{x})\operatorname{sgn}(a_\lambda) < 0$ and $\bar{x}$ cannot be the minimizer, and we have to choose $x^* = 0$. □

Lemma 4.2. *We can express the optimization of $\mathcal{J}^S(u,a)$ with respect to u explicitly by*

$$\mathbb{S}_\alpha(a + A^*(y - Aa)) = \arg\min_{u \in \ell_2(\mathcal{I})} \mathcal{J}^S(u,a).$$

Proof. By developing the norm squares in (4.9) it is a straightforward computation to show

$$\mathcal{J}^S(u,a) = \|u - (a + A^*(y - Aa))\|^2_{\ell_2(\mathcal{I})} + 2\|u\|_{\ell_{1,\alpha}(\mathcal{I})} + \Phi(a, A, y),$$

where $\Phi(a, A, y)$ is a function which does not depend on u. The statement follows now from an application of Lemma 4.1 and by the observation that the addition of constants to a functional does not modify its minimizer. □

4.1.2 The Algorithm and Preliminary Convergence Properties

By Lemma 4.2 we achieve the following formulation of the algorithm;

Algorithm 4. We initialize by taking any $u^{(0)} \in \ell_2(\mathcal{I})$, for instance $u^{(0)} = 0$. We iterate

$$\begin{aligned} u^{(n+1)} &= \mathbb{S}_\alpha[u^{(n)} + A^*y - A^*Au^{(n)}] \\ &= \arg\min_{u \in \ell_2(\mathcal{I})} \mathcal{J}^S(u, u^{(n)}). \end{aligned}$$

Lemma 4.3. *The sequence $(\mathcal{J}(u^{(n)}))_{n\in\mathbb{N}}$ is nonincreasing. Moreover $(u^{(n)})_{n\in\mathbb{N}}$ is bounded in $\ell_2(\mathcal{I})$ and*

$$\lim_{n\to\infty} \|u^{(n+1)} - u^{(n)}\|^2_{\ell_2(\mathcal{I})} = 0. \tag{4.13}$$

Proof. Let us consider the estimates, which follow from the optimality of $u^{(n+1)}$, (4.11), and (4.12),

$$\begin{aligned} \mathcal{J}(u^{(n)}) &= \mathcal{J}^S(u^{(n)}, u^{(n)}) \\ &\geq \mathcal{J}^S(u^{(n+1)}, u^{(n)}) \\ &\geq \mathcal{J}^S(u^{(n+1)}, u^{(n+1)}) = \mathcal{J}(u^{(n+1)}). \end{aligned}$$

Hence, the sequence $\mathcal{J}(u^{(n)})$ is nonincreasing, and

$$\mathcal{J}(u^{(0)}) \geq \mathcal{J}(u^{(n)}) \geq 2\bar{\alpha}\|u^{(n)}\|_{\ell_1(\mathcal{I})} \geq 2\bar{\alpha}\|u^{(n)}\|_{\ell_2(\mathcal{I})}.$$

Therefore, $(u^{(n)})_{n\in\mathbb{N}}$ is bounded in $\ell_2(\mathcal{I})$. By (4.12), we have

$$\mathcal{J}(u^{(n)}) - \mathcal{J}(u^{(n+1)}) \geq C\|u^{(n)} - u^{(n+1)}\|^2_{\ell_2(\mathcal{I})}.$$

Since $\mathcal{J}(u^{(n)}) \geq 0$ is a nonincreasing sequence and is bounded below, it also converges, and

$$\lim_{n\to\infty} \|u^{(n+1)} - u^{(n)}\|^2_{\ell_2(\mathcal{I})} = 0. \qquad \square$$

This lemma already gives strong hints that the algorithm converges. In particular, two successive iterations become closer and closer (4.13), and by the uniform boundedness of $(u^{(n)})_{n\in\mathbb{N}}$, we know already that there are weakly converging subsequences. However, in order to conclude the convergence of the full sequence to a minimizer of $\mathcal{J}$ we need more technical work.

4.1.3 Weak Convergence of the Algorithm

As as simple exercise we state the following

Lemma 4.4. *The operator $\mathbb{S}_\alpha$ is nonexpansive, i.e.,*

$$\|\mathbb{S}_\alpha(u) - \mathbb{S}_\alpha(a)\|_{\ell_2(\mathcal{I})} \leq \|u - a\|_{\ell_2(\mathcal{I})}, \tag{4.14}$$

for all $u, a \in \ell_2(\mathcal{I})$.

Proof. Sketch: reason again componentwise and distinguish cases whether u_λ and/or a_λ are smaller or larger than the threshold $\pm\alpha_\lambda$. $\square$

Moreover, we can characterize minimizers of $\mathcal{J}$ in the following way.

Proposition 4.5. *Define*

$$\Gamma(u) = \mathbb{S}_\alpha[u + A^*y - A^*Au].$$

Then the set of minimizers of $\mathcal{J}$ coincides with the set Fix(Γ) *of fixed points of Γ. In particular, since $\mathcal{J}$ is a coercive functional (i.e., $\{u : \mathcal{J}(u) \leq C\}$ is weakly compact for all $C > 0$), it has minimizers, and therefore Γ has fixed points.*

Proof. Assume that u is the minimizer of $\mathcal{J}^S(\cdot, a)$, for a fixed. Let us now observe, first of all, that

$$\begin{aligned}\mathcal{J}^S(u+h, a) = \mathcal{J}^S(u, a) &+ 2\langle h, u - a - A^*(y - Aa)\rangle \\ &+ \sum_{\lambda\in\mathcal{I}} 2\alpha_\lambda(|u_\lambda + h_\lambda| - |u_\lambda|) + \|h\|^2_{\ell_2(\mathcal{I})}.\end{aligned}$$

We define now $\mathcal{I}_0 = \{\lambda : u_\lambda = 0\}$ and $\mathcal{I}_1 = \mathcal{I} \setminus \mathcal{I}_0$. Since by Lemma 4.2 we have $u = \mathbb{S}_\alpha(a + A^*(y - Aa))$, substituting it for u, we then have

$$\mathcal{J}^S(u+h,a)-\mathcal{J}^S(u,a) = \|h\|^2_{\ell_2(\mathcal{I})} + \sum_{\lambda\in\mathcal{I}_0}[2\alpha_\lambda|h_\lambda|-2h_\lambda(a-A^*(y-Aa))_\lambda]$$
$$+\sum_{\lambda\in\mathcal{I}_1}[2\alpha_\lambda|u_\lambda+h_\lambda|-2\alpha_\lambda|u_\lambda|+h_\lambda(-2\alpha_\lambda\,\mathrm{sgn}(u_\lambda))].$$

If $\lambda \in \mathcal{I}_0$ then $|(a-A^*(y-Aa))_\lambda| \le \alpha_\lambda$, so that $2\alpha_\lambda|h_\lambda|-2h_\lambda(a-A^*(y-Aa))_\lambda \ge 0$. If $\lambda \in \mathcal{I}_1$, we distinguish two cases: if $u_\lambda > 0$, then

$$2\alpha_\lambda|u_\lambda + h_\lambda| - 2\alpha_\lambda|u_\lambda| + h_\lambda(-2\alpha_\lambda\,\mathrm{sgn}(u_\lambda)) = 2\alpha_\lambda[|u_\lambda + h_\lambda| - (u_\lambda + h_\lambda)] \ge 0.$$

If $u_\lambda < 0$, then

$$2\alpha_\lambda|u_\lambda + h_\lambda| - 2\alpha_\lambda|u_\lambda| + h_\lambda(-2\alpha_\lambda\,\mathrm{sgn}(u_\lambda)) = 2\alpha_\lambda[|u_\lambda + h_\lambda| + (u_\lambda + h_\lambda)] \ge 0.$$

It follows

$$\mathcal{J}^S(u + h, a) - \mathcal{J}^S(u, a) \ge \|h\|^2_{\ell_2(\mathcal{I})}. \tag{4.15}$$

Let us assume now that

$$u = \mathbb{S}_\alpha[u + A^*y - A^*Au].$$

Then u is the minimizer of $\mathcal{J}^S(\cdot, u)$, and therefore

$$\mathcal{J}^S(u + h, u) \ge \mathcal{J}^S(u, u) + \|h\|^2_{\ell_2(\mathcal{I})}.$$

Observing now that $\mathcal{J}(u) = \mathcal{J}^S(u, u)$ and that $\mathcal{J}^S(u+h, u) = \mathcal{J}(u+h) + \|h\|^2_{\ell_2(\mathcal{I})} - \|Ah\|^2_Y$, we conclude that $\mathcal{J}(u + h) \ge \mathcal{J}(u) + \|Ah\|^2_Y$ for every h. Hence u is a minimizer of $\mathcal{J}$. Vice versa, if u is a minimizer of $\mathcal{J}$, then it is a minimizer of $\mathcal{J}^S(\cdot, u)$, and hence a fixed point of Γ. □

We need now to recall an important and well-known result related to iterations of nonexpansive maps [64]. We report it without proof; a simplified version of it can be also found in the Appendix B of [28].

Theorem 4.6 (Opial's theorem). *Let a mapping Γ from $\ell_2(\mathcal{I})$ to itself satisfy the following conditions:*

(i) *Γ is nonexpansive, i.e., $\|\Gamma(u) - \Gamma(a)\|_{\ell_2(\mathcal{I})} \le \|u - a\|_{\ell_2(\mathcal{I})}$, for all $u, a \in \ell_2(\mathcal{I})$;*

(ii) *Γ is asymptotically regular, i.e., $\|\Gamma^{n+1}(u) - \Gamma^n(u)\|_{\ell_2(\mathcal{I})} \to 0$ for $n \to \infty$;*

(iii) *the set* Fix(Γ) *of its fixed points is not empty.*

Then, for all u, the sequence $(\Gamma^n(u))_{n\in\mathbb{N}}$ converges weakly to a fixed point in Fix(Γ).

Eventually we have the weak convergence of the algorithm.

Theorem 4.7. *For any initial choice $u^{(0)} \in \ell_2(\mathcal{I})$, Algorithm 4 produces a sequence $(u^{(n)})_{n\in\mathbb{N}}$ which converges weakly to a minimizer of $\mathcal{J}$.*

Proof. It is sufficient to observe that, due to our previous results, Lemma 4.4, Lemma 4.3, and Proposition 4.5, and the assumption $\|A\| < 1$, the map

$$\Gamma(u) = \mathbb{S}_\alpha[u + A^*y - A^*Au]$$

fulfills the requirements of Opial's theorem. □

4.1.4 Strong Convergence of the Algorithm

In this section we shall prove the convergence of the successive iterates $u^{(n)}$ not only in the weak topology, but also in norm. Let us start by introducing some useful notations:

$$u^* = w - \lim_n u^{(n)}, \quad \xi^{(n)} = u^{(n)} - u^*, \quad h = u^* + A^*(y - Au^*),$$

where "$w - \lim$" denotes the symbol for the weak limit. We again split the proof into several intermediate lemmas.

Lemma 4.8. *We have*

$$\|A\xi^{(n)}\|_Y^2 \to 0,$$

for $n \to \infty$.

Proof. Since

$$\xi^{(n+1)} - \xi^{(n)} = \mathbb{S}_\alpha(h + (I - A^*A)\xi^{(n)}) - \mathbb{S}_\alpha(h) - \xi^{(n)},$$

and for (4.13) $\|\xi^{(n+1)} - \xi^{(n)}\|_{\ell_2(\mathcal{I})} = \|u^{(n+1)} - u^{(n)}\|_{\ell_2(\mathcal{I})} \to 0$ for $n \to \infty$, we have

$$\|\mathbb{S}_\alpha(h + (I - A^*A)\xi^{(n)}) - \mathbb{S}_\alpha(h) - \xi^{(n)}\|_{\ell_2(\mathcal{I})} \to 0, \tag{4.16}$$

for $n \to \infty$, and hence, also

$$\max(0, \|\xi^n\|_{\ell_2(\mathcal{I})} - \|\mathbb{S}_\alpha(h + (I - A^*A)\xi^{(n)}) - \mathbb{S}_\alpha(h)\|_{\ell_2(\mathcal{I})}) \to 0, \tag{4.17}$$

for $n \to \infty$. Since $\mathbb{S}_\alpha$ is nonexpansive we have

$$-\|\mathbb{S}_\alpha(h + (I - A^*A)\xi^{(n)}) - \mathbb{S}_\alpha(h)\|_{\ell_2(\mathcal{I})} \geq -\|(I - A^*A)\xi^{(n)}\|_{\ell_2(\mathcal{I})} \geq -\|\xi^{(n)}\|_{\ell_2(\mathcal{I})};$$

therefore the "max" in (4.17) can be dropped, and it follows that

$$0 \leq \|\xi^n\|_{\ell_2(\mathcal{I})} - \|(I - A^*A)\xi^{(n)}\|_{\ell_2(\mathcal{I})} \to 0, \tag{4.18}$$

for $n \to \infty$. Because

$$\|\xi^n\|_{\ell_2(\mathcal{I})} + \|(I - A^*A)\xi^{(n)}\|_{\ell_2(\mathcal{I})} \leq 2\|\xi^n\|_{\ell_2(\mathcal{I})} = 2\|u^{(n)} - u^*\|_{\ell_2(\mathcal{I})} \leq C,$$

(Remember that $u^{(n)}$ is uniformly bounded by Lemma 4.3.), we obtain

$$\|\xi^n\|^2_{\ell_2(\mathcal{J})} - \|(I - A^*A)\xi^{(n)}\|^2_{\ell_2(\mathcal{J})} \to 0,$$

for $n \to \infty$ by (4.18). The inequality

$$\|\xi^n\|^2_{\ell_2(\mathcal{J})} - \|(I - A^*A)\xi^{(n)}\|^2_{\ell_2(\mathcal{J})} = 2\|A\xi^{(n)}\|^2_Y - \|A^*A\xi^{(n)}\|^2_{\ell_2(\mathcal{J})} \geq \|A\xi^{(n)}\|^2_Y,$$

then implies the statement. □

The previous lemma allows us to derive the following fundamental property.

Lemma 4.9. *For h given as above, $\|\mathbb{S}_\alpha(h + \xi^{(n)}) - \mathbb{S}_\alpha(h) - \xi^{(n)}\|_{\ell_2(\mathcal{J})} \to 0$, for $n \to \infty$.*

Proof. We have

$$\begin{aligned}
&\|\mathbb{S}_\alpha(h + \xi^{(n)}) - \mathbb{S}_\alpha(h) - \xi^{(n)}\|_{\ell_2(\mathcal{J})} \\
&\quad \leq \|\mathbb{S}_\alpha(h + (I - A^*A)\xi^{(n)}) - \mathbb{S}_\alpha(h) - \xi^{(n)}\|_{\ell_2(\mathcal{J})} \\
&\qquad + \|\mathbb{S}_\alpha(h + \xi^{(n)}) - \mathbb{S}_\alpha(h + (I - A^*A)\xi^{(n)})\|_{\ell_2(\mathcal{J})} \\
&\quad \leq \|\mathbb{S}_\alpha(h + (I - A^*A)\xi^{(n)}) - \mathbb{S}_\alpha(h) - \xi^{(n)}\|_{\ell_2(\mathcal{J})} + \|A^*A\xi^{(n)}\|_{\ell_2(\mathcal{J})}.
\end{aligned}$$

Both terms tend to 0, the first because of (4.16) and the second because of Lemma 4.8. □

Lemma 4.10. *If for some $a \in \ell_2(\mathcal{J})$ and some sequence $(v^n)_{n\in\mathbb{N}}$, $w - \lim_n v^n = 0$, and $\lim_n \|\mathbb{S}_\alpha(a + v^n) - \mathbb{S}_\alpha(a) - v^n\|_{\ell_2(\mathcal{J})} = 0$, then $\|v^n\|_{\ell_2(\mathcal{J})} \to 0$, for $n \to \infty$.*

Proof. Let us define a finite set $\mathcal{J}_0 \subset \mathcal{J}$ such that $\sum_{\lambda\in\mathcal{J}\setminus\mathcal{J}_0} |a_\lambda|^2 \leq (\frac{\bar{\alpha}}{4})^2$, where $\bar{\alpha} = \inf_\lambda \alpha_\lambda$. Because this is a finite set $\sum_{\lambda\in\mathcal{J}_0} |v^n_\lambda|^2 \to 0$ for $n \to \infty$, and hence we can concentrate on $\sum_{\lambda\in\mathcal{J}\setminus\mathcal{J}_0} |v^n_\lambda|^2$ only. For each n, we split $\mathcal{J}_1 = \mathcal{J} \setminus \mathcal{J}_0$ into two subsets: $\mathcal{J}_{1,n} = \{\lambda \in \mathcal{J}_1 : |v^n_\lambda + a_\lambda| < \alpha_\lambda\}$ and $\tilde{\mathcal{J}}_{1,n} = \mathcal{J}_1 \setminus \mathcal{J}_{1,n}$. If $\lambda \in \mathcal{J}_{1,n}$ then $S_{\alpha_\lambda}(a_\lambda + v^n_\lambda) = S_{\alpha_\lambda}(a_\lambda) = 0$ (since $|a_\lambda| \leq \frac{\bar{\alpha}}{4} \leq \alpha_\lambda$), so that $|S_{\alpha_\lambda}(a_\lambda + v^n_\lambda) - S_{\alpha_\lambda}(a_\lambda) - v^n_\lambda| = |v^n_\lambda|$. It follows

$$\sum_{\lambda\in\mathcal{J}_{1,n}} |v^n_\lambda|^2 \leq \sum_{\lambda\in\mathcal{J}} |S_{\alpha_\lambda}(a_\lambda + v^n_\lambda) - S_{\alpha_\lambda}(a_\lambda) - v^n_\lambda|^2 \to 0,$$

for $n \to \infty$. It remains to prove that $\sum_{\lambda\in\tilde{\mathcal{J}}_{1,n}} |v^n_\lambda|^2 \to 0$ as $n \to \infty$. If $\lambda \in \mathcal{J}_1$ and $|a_\lambda + v^n_\lambda| \geq \alpha_\lambda$, then $|v^n_\lambda| \geq |a_\lambda + v^n_\lambda| - |a_\lambda| \geq \alpha_\lambda - \frac{\bar{\alpha}}{4} > \frac{\bar{\alpha}}{4} \geq |a_\lambda|$, so that $a_\lambda + v^n_\lambda$

and v_λ^n have the same sign. In particular, $\alpha_\lambda - \frac{\bar{\alpha}}{4} > |a_\lambda|$ implies $\alpha_\lambda - |a_\lambda| \geq \frac{\bar{\alpha}}{4}$. It follows that

$$\begin{aligned}|v_\lambda^n - S_{\alpha_\lambda}(a_\lambda + v_\lambda^n) + S_{\alpha_\lambda}(a_\lambda)| &= |v_\lambda^n - S_{\alpha_\lambda}(a_\lambda + v_\lambda^n)| \\ &= |v_\lambda^n - (a_\lambda + v_\lambda^n) + \alpha_\lambda \operatorname{sgn}(v_\lambda^n)| \\ &\geq \alpha_\lambda - |a_\lambda| \geq \frac{\bar{\alpha}}{4}.\end{aligned}$$

This implies that

$$\sum_{\lambda \in \tilde{\mathcal{J}}_{1,n}} |S_{\alpha_\lambda}(a_\lambda + v_\lambda^n) - S_{\alpha_\lambda}(a_\lambda) - v_\lambda^n|^2 \geq \left(\frac{\bar{\alpha}}{4}\right)^2 |\tilde{\mathcal{J}}_{1,n}|.$$

But, $\sum_{\lambda \in \tilde{\mathcal{J}}_{1,n}} |S_{\alpha_\lambda}(a_\lambda + v_\lambda^n) - S_{\alpha_\lambda}(a_\lambda) - v_\lambda^n|^2 \to 0$ for $n \to \infty$ and therefore $\tilde{\mathcal{J}}_{1,n}$ must be empty for n large enough. □

The combination of Lemma 4.8 and Lemma 4.9, together with the weak convergence Theorem 4.7 allows us to have norm convergence.

Theorem 4.11. *For any initial choice $u^{(0)} \in \ell_2(\mathcal{J})$, Algorithm 4 produces a sequence $(u^{(n)})_{n \in \mathbb{N}}$ which converges strongly to a minimizer u^* of $\mathcal{J}$.*

4.2 Principles of Acceleration

Recently, also the qualitative convergence properties of iterative soft-thresholding have been investigated. Note first that the aforementioned condition $\|A\| < 1$ (or even $\|A\| < \sqrt{2}$) does not guarantee contractivity of the iteration operator $I - A^*A$, since A^*A may not be boundedly invertible. The insertion of $\mathbb{S}_\alpha$ does not improve the situation since $\mathbb{S}_\alpha$ is nonexpansive, but also noncontractive. Hence, for any minimizer u^* (which is also a fixed point of (4.6)), the estimate

$$\|u^* - u^{(n+1)}\|_{\ell_2(\mathcal{J})} \leq \left\|(I - A^*A)(u^* - u^{(n)})\right\|_{\ell_2(\mathcal{J})} \leq \|I - A^*A\| \|u^* - u^{(n)}\|_{\ell_2(\mathcal{J})} \tag{4.19}$$

does not give rise to a linear error reduction. However, under additional assumptions on the operator A or on minimizers u^*, linear convergence of (4.6) can be easily ensured. In particular, if A fulfills the so-called *finite basis injectivity* (FBI) condition (see [10] where this terminology is introduced), i.e., for any finite set $\Lambda \subset \mathcal{J}$, the restriction A_Λ is injective, then (4.6) converges linearly to a minimizer u^* of $\mathcal{J}$. The following simple argument shows indeed that the FBI condition implies linear error reduction as soon as $\|A\| < 1$. In that case, we have strong convergence (Theorem 4.11) of the $u^{(n)}$ to a finitely supported limit sequence u^*. We can therefore find a finite index set $\Lambda \subset \mathcal{J}$ such that all iterates $u^{(n)}$ and u^* are supported in Λ for n large enough.

By the FBI condition, A_Λ is injective and hence $(A^*A)_{\Lambda\times\Lambda}$ is boundedly invertible, so that $I - A^*_\Lambda A_\Lambda$ is a contraction on $\ell_2(\Lambda)$. Using

$$u_\Lambda^{(n+1)} = \mathbb{S}_\alpha(u_\Lambda^{(n)} + A^*_\Lambda(y - A_\Lambda u_\Lambda^{(n)}))$$

and an analogous argument as in (4.19), it follows that $\|u^* - u^{(n+1)}\|_{\ell_2(\mathcal{I})} \leq \gamma \|u^* - u^{(n)}\|_{\ell_2(\mathcal{I})}$, where $\gamma = \max\{|1 - \|(A^*A_{\Lambda\times\Lambda})^{-1}\|^{-1}|, |\|A^*A_{\Lambda\times\Lambda}\| - 1|\} \in (0,1)$. Typical examples where $A = K \circ F$ fulfills the FBI condition arise when K is injective and Ψ is either a Riesz basis for X or a so-called FBI frame, i.e., each finite subsystem of Ψ is linearly independent. However, depending on Λ, the matrix $A^*A_{\Lambda\times\Lambda}$ can be arbitrarily badly conditioned, resulting in a constant error reduction γ, arbitrarily close to 1.

However, it is possible to show that for several FBI operators K and for certain choices of Ψ, the matrix A^*A can be preconditioned by a matrix $D^{-1/2}$, resulting in the matrix $D^{-1/2}A^*AD^{-1/2}$, in such a way that any restriction $(D^{-1/2}A^*AD^{-1/2})_{\Lambda\times\Lambda}$ turns out to be well-conditioned as soon as $\Lambda \subset \mathcal{I}$ is a small set, but independently of its "selection" within $\mathcal{I}$. Let us remark that, in particular, we do not claim to be able to have entirely well-conditioned matrices (as it happens in well-posed problems [23, 24] by simple diagonal preconditioning), but that only small arbitrary finite dimensional submatrices are indeed well-conditioned. Let us say that one can promote a "local" well-conditioning of the matrices.

Typically one considers injective (non local) compact operators K with Schwartz kernel having certain polynomial decay properties of the derivatives, i.e.,

$$Ku(x) = \int_\Omega \Phi(x,\xi)u(\xi)d\xi, \quad x \in \tilde{\Omega},$$

for $\tilde{\Omega}, \Omega \subset \mathbb{R}^d$, $u \in X := H^t(\Omega)$, and

$$|\partial_x^\alpha \partial_\xi^\beta \Phi(x,\xi)| \leq c_{\alpha,\beta}|x-\xi|^{-(d+2t+|\alpha|+|\beta|)}, \quad t \in \mathbb{R}, \text{ and multi-indexes } \alpha, \beta \in \mathbb{N}^d.$$

Moreover, for the proper choice of the discrete matrix $A^*A := F^*K^*KF$, one uses multiscale bases Ψ, such as wavelets, which do make a good job in this situation. We refer the reader to [22] for more details.

4.2.1 From the Projected Gradient Method to Iterative Soft-Thresholding with Decreasing Thresholding Parameter

With such "local" well-conditioning, it should also be clear that iterating on small sets Λ will also improve the convergence rate. Unfortunately, iterative soft-thresholding does not act initially on small sets (see also Figure 7), but it rather starts iterating on relatively large sets, slowly shrinking to the size of the support of the limit u^*.

Let us take a closer look at the characteristic dynamics of Algorithm 4 in Figure 4. Let us assume for simplicity here that $\alpha_\lambda = \bar{\alpha} > 0$ for all $\lambda \in \mathcal{I}$. As this plot of the

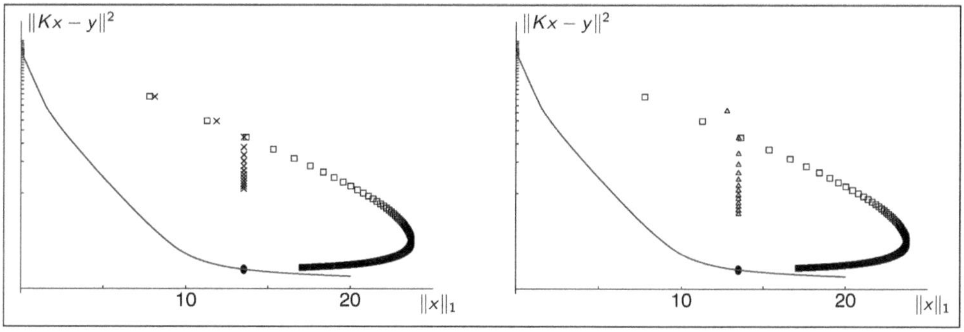

(a) (b)

Figure 4. The path, in the $\|x\|_1$ vs. $\|Kx-y\|^2$ plane, followed by the iterates $u^{(n)}$ of three different iterative algorithms. The operator K and the data y are taken from a seismic tomography problem [55]. The boxes (in both (a) and (b)) correspond to the thresholded Landweber algorithm. In this example, iterative thresholded Landweber (4.6) first overshoots the ℓ_1-norm of the limit (represented by the fat dot), and then requires a large number of iterations to reduce $\|u^{(n)}\|_1$ again (500 are shown in this figure). In (a) the crosses correspond to the path followed by the iterates of the projected Landweber iteration (which is given as in (4.20) for $\beta^{(n)} = 1$); in (b) the triangles correspond to the projected steepest descent iteration (4.20); in both cases, only 15 iterates are shown. The discrepancy decreases more quickly for projected steepest descent than for the projected Landweber algorithm. The solid line corresponds to the limit *trade-off curve*, generated by $u^*(\bar{\alpha})$ for decreasing values of $\bar{\alpha} > 0$. The vertical axes uses a logarithmic scale for clarity.

discrepancy $\mathcal{D}(u^{(n)}) = \|Ku^{(n)} - y\|_Y^2 = \|Au^{(n)} - y\|_Y^2$ versus $\|u^{(n)}\|_{\ell_1(\mathcal{I})}$ shows, the algorithm converges initially relatively fast, then it overshoots the value $\|u^*\|_{\ell_1(\mathcal{I})}$ and it takes very long to re-correct back. In other words, starting from $u^{(0)} = 0$, the algorithm generates a path $\{u^{(n)}; n \in \mathbb{N}\}$ that is initially fully contained in the ℓ_1-ball $B_R := B_{\ell_1(\mathcal{I})}(R) := \{u \in \ell_2(\Lambda); \|u\|_{\ell_1(\mathcal{I})} \leq R\}$, with $R := \|u^*\|_{\ell_1(\mathcal{I})}$. Then it gets out of the ball to slowly inch back to it in the limit.

The way to avoid this long "external" detour was proposed in [29] by forcing the successive iterates to remain within the ball B_R. One method to achieve this is to substitute for the thresholding operations the projection $\mathbb{P}_{B_R}$, where, for any closed convex set C, and any u, we define $\mathbb{P}_C(u)$ to be the unique point in C for which the ℓ_2-distance to u is minimal. With a slight abuse of notation, we shall denote $\mathbb{P}_{B_R}$ by $\mathbb{P}_R$; this will not cause confusion, because it will be clear from the context whether the subscript of $\mathbb{P}$ is a set or a positive number.

Furthermore, modifying the iterations by introducing an adaptive "descent parameter" $\beta^{(n)} > 0$ in each iteration, defining $u^{(n+1)}$ by

$$u^{(n+1)} = \mathbb{P}_R[u^{(n)} + \beta^{(n)} A^*(y - Au^{(n)})], \tag{4.20}$$

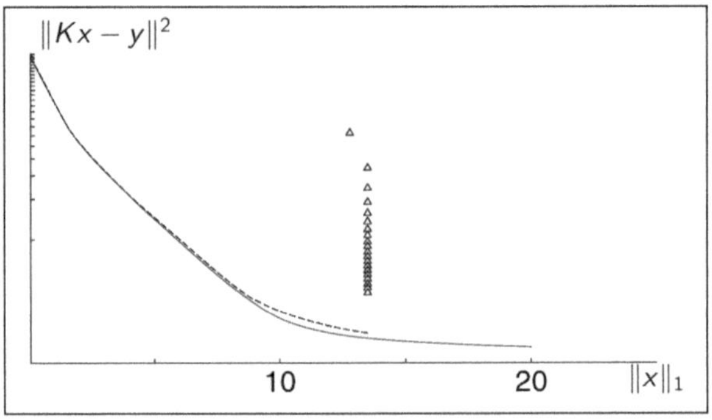

Figure 5. Trade-off curve and its approximation with algorithm (4.21) in 200 steps.

does lead, in numerical simulations, to much faster convergence. The typical dynamics of this modified algorithm are illustrated in Figure 4(b), which clearly shows the larger steps and faster convergence (when compared with the *projected Landweber iteration* in Figure 4(a) which is the iteration (4.20) for $\beta^{(n)} = 1$ for all n). We shall refer to this modified algorithm as the *projected gradient iteration* or the *projected steepest descent* (PSD). The motivation of the faster convergence behavior is the fact that we never leave the target ℓ_1-ball, and we tend not to iterate on large index sets. On the basis of this intuition we find even more promising results for an 'interior' algorithm in which we still project on ℓ_1-balls, but now with a slowly increasing radius, i.e.

$$u^{(n+1)} = \mathbb{P}_{R^{(n)}}(u^{(n)} + \beta^{(n)} A^*(y - Au^{(n)})) \quad \text{and} \quad R^{(n+1)} = (n+1)R/N \quad (4.21)$$

where N is the prescribed maximum number of iterations (the origin is chosen as the starting point of this iteration). The better performance of this algorithm can be explained by the fact that the projection $\mathbb{P}_R(u)$ onto an ℓ_1-ball of radius R coincides with a thresholding $\mathbb{P}_R(u) = \mathbb{S}_{\mu(u;R)}(u)$ for a suitable thresholding parameter $\mu = \mu(u; R)$ depending on u and R, which is larger for smaller R.

Lemma 4.12. *For any fixed $a \in \ell_2(\mathcal{I})$ and for $\tau > 0$, $\|\mathbb{S}_\tau(a)\|_{\ell_1(\mathcal{I})}$ is a piecewise linear, continuous, decreasing function of τ; moreover, if $a \in \ell_1(\mathcal{I})$ then $\|\mathbb{S}_0(a)\|_{\ell_1(\mathcal{I})} = \|a\|_{\ell_1(\mathcal{I})}$ and $\|\mathbb{S}_\tau(a)\|_{\ell_1(\mathcal{I})} = 0$ for $\tau \geq \max_\lambda |a_\lambda|$.*

Proof. $\|\mathbb{S}_\tau(a)\|_{\ell_1(\mathcal{I})} = \sum_\lambda |S_\tau(a_\lambda)| = \sum_\lambda S_\tau(|a_\lambda|) = \sum_{|a_\lambda|>\tau}(|a_\lambda| - \tau)$; the sum in the right hand side is finite for $\tau > 0$. □

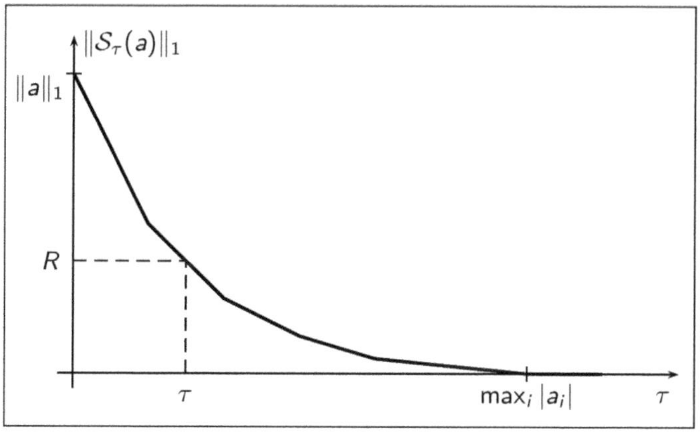

Figure 6. For a given vector $a \in \ell_2$, $\|\mathbb{S}_\tau(a)\|_{\ell_1^m}$ is a piecewise linear continuous and decreasing function of τ (strictly decreasing for $\tau < \max_i |a_i|$). The knots are located at $\{|a_i|, i = 1, \dots, m\}$ and 0. Finding τ such that $\|\mathbb{S}_\tau(a)\|_{\ell_1^m} = R$ ultimately comes down to a linear interpolation. The figure is made for the finite dimensional case.

A schematic illustration is given in Figure 6.

Lemma 4.13. *If $\|a\|_{\ell_1(\mathcal{I})} > R$, then the $\ell_2(\mathcal{I})$ projection of a on the ℓ_1-ball with radius R is given by $\mathbb{P}_R(a) = \mathbb{S}_\mu(a)$ where μ (depending on a and R) is chosen such that $\|\mathbb{S}_\mu(a)\|_{\ell_1(\mathcal{I})} = R$. If $\|a\|_{\ell_1(\mathcal{I})} \le R$ then $\mathbb{P}_R(a) = \mathbb{S}_0(a) = a$.*

Proof. Suppose $\|a\|_{\ell_1(\mathcal{I})} > R$. Because, by Lemma 4.12, $\|\mathbb{S}_\mu(a)\|_{\ell_1(\mathcal{I})}$ is continuous in μ and $\|\mathbb{S}_\mu(a)\|_{\ell_1(\mathcal{I})} = 0$ for sufficiently large μ, we can choose μ such that $\|\mathbb{S}_\mu(a)\|_{\ell_1(\mathcal{I})} = R$. (See Figure 6.) On the other hand, $u^* = \mathbb{S}_\mu(a)$ is the unique minimizer of $\|u - a\|^2_{\ell_2(\mathcal{I})} + 2\mu\|u\|_{\ell_1(\mathcal{I})}$ (see Lemma 4.1), i.e.,

$$\|u^* - a\|^2_{\ell_2(\mathcal{I})} + 2\mu\|u^*\|_{\ell_1(\mathcal{I})} < \|u - a\|^2_{\ell_2(\mathcal{I})} + 2\mu\|u\|_{\ell_1(\mathcal{I})},$$

for all $u \neq u^*$. Since $\|u^*\|_{\ell_1(\mathcal{I})} = R$, it follows that

$$\text{for all } u \in B_R,\ u \neq u^* : \quad \|u^* - a\|^2 < \|u - a\|^2.$$

Hence u^* is closer to a than any other u in B_R. In other words, $\mathbb{P}_R(a) = u^* = \mathbb{S}_\mu(a)$. □

This in particular implies that the algorithm (4.21) iterates initially on very small sets which inflate by growing during the process and approach the size of the support of the target minimizer u^*. Unlike the thresholded Landweber iteration and the projected steepest descent [28, 29], unfortunately there is no proof yet of convergence of this 'interior' algorithm, this being a very interesting open problem.

However, we can provide an algorithm which mimics the behavior of (4.21), i.e., it starts with large thresholding parameters $\alpha^{(n)}$ and geometrically reduces them during the iterations to a target limit $\alpha > 0$, for which the convergence is guaranteed:

$$u^{(n+1)} = \mathbb{S}_{\alpha^{(n)}}[u^{(n)} + A^* y - A^* A u^{(n)}]. \tag{4.22}$$

For matrices A for which the restrictions $(A^* A)_{\Lambda \times \Lambda}$ are uniformly well-conditioned with respect to Λ of small size, our analysis provides also a prescribed linear rate of convergence of the iteration (4.22).

4.2.2 Sample of Analysis of Acceleration Methods

Technical Lemmas We are particularly interested in computing approximations with the smallest possible number of nonzero entries. As a benchmark, we recall that the most economical approximations of a given vector $v \in \ell_2(\mathcal{J})$ are provided again by the *best k-term approximations* $v_{[k]}$, defined by discarding in v all but the $k \in \mathbb{N}_0$ largest coefficients in absolute value. The error of best k-term approximation is defined as

$$\sigma_k(v)_{\ell_2} := \|v - v_{[k]}\|_{\ell_2(\mathcal{J})}. \tag{4.23}$$

The subspace of all ℓ_2 vectors with best k-term approximation rate $s > 0$, i.e., $\sigma_k(v)_{\ell_2} \lesssim k^{-s}$ for some decay rate $s > 0$, is commonly referred to as the *weak ℓ_τ space $\ell_\tau^w(\mathcal{J})$*, for $\tau = (s + \frac{1}{2})^{-1}$, which, endowed with

$$|v|_{\ell_\tau^w(\mathcal{J})} := \sup_{k \in \mathbb{N}_0} (k+1)^s \sigma_k(v)_{\ell_2}, \tag{4.24}$$

becomes the quasi-Banach space $(\ell_\tau^w(\mathcal{J}), |\cdot|_{\ell_\tau^w(\mathcal{J})})$. Moreover, for any $0 < \epsilon \leq 2 - \tau$, we have the continuous embedding $\ell_\tau(\mathcal{J}) \hookrightarrow \ell_\tau^w(\mathcal{J}) \hookrightarrow \ell_{\tau+\epsilon}(\mathcal{J})$, justifying why $\ell_\tau^w(\mathcal{J})$ is called weak $\ell_\tau(\mathcal{J})$.

When it comes to the concrete computations of good approximations with a small number of active coefficients, one frequently utilizes certain thresholding procedures. Here small entries of a given vector are simply discarded, whereas the large entries may be slightly modified. As we have discussed so far, we shall make use of *soft-thresholding* that we have already introduced in (4.7). We recall from Lemma 4.4 that $\mathbb{S}_\alpha$ is non-expansive for any $\alpha \in \mathbb{R}_+^{\mathcal{J}}$,

$$\|\mathbb{S}_\alpha(v) - \mathbb{S}_\alpha(w)\|_{\ell_2(\mathcal{J})} \leq \|v - w\|_{\ell_2(\mathcal{J})}, \quad \text{for all } v, w \in \ell_2(\mathcal{J}). \tag{4.25}$$

Moreover, for any fixed $x \in \mathbb{R}$, the mapping $\tau \mapsto S_\tau(x)$ is Lipschitz continuous and

$$|S_\tau(x) - S_{\tau'}(x)| \leq |\tau - \tau'|, \quad \text{for all } \tau, \tau' \geq 0. \tag{4.26}$$

We readily infer the following technical estimate.

Lemma 4.14. *Assume* $v \in \ell_2(\mathcal{J})$, $\alpha, \beta \in \mathbb{R}_+^{\mathcal{J}}$ *such that* $\bar{\alpha} = \inf_\lambda \alpha_\lambda = \inf_\lambda \beta_\lambda = \bar{\beta} > 0$, *and define* $\Lambda_{\bar{\alpha}}(v) := \{\lambda \in \mathcal{J} : |v_\lambda| > \bar{\alpha}\}$. *Then*

$$\|\mathbb{S}_\alpha(v) - \mathbb{S}_\beta(v)\|_{\ell_2(\mathcal{J})} \leq (\#\Lambda_{\bar{\alpha}}(v))^{1/2} \max_{\lambda \in \Lambda_{\bar{\alpha}}(v)} |\alpha_\lambda - \beta_\lambda|. \tag{4.27}$$

Proof. It is left as a simple exercise. □

Let $v \in \ell_\tau^w(\mathcal{J})$, it is well-known [31, §7]

$$\#\Lambda_{\bar{\alpha}}(v) \leq C|v|_{\ell_\tau^w(\mathcal{J})}^\tau \bar{\alpha}^{-\tau}, \tag{4.28}$$

and, for $\alpha_\lambda = \bar{\alpha}$ for all $\lambda \in \mathcal{J}$, we have

$$\|v - \mathbb{S}_\alpha(v)\|_{\ell_2(\mathcal{J})} \leq C|v|_{\ell_\tau^w(\mathcal{J})}^{\tau/2} \bar{\alpha}^{1-\tau/2}, \tag{4.29}$$

where the constants are given by $C = C(\tau) > 0$.

In the sequel, we shall also use the following support size estimate.

Lemma 4.15. *Let* $v \in \ell_\tau^w(\mathcal{J})$ *and* $w \in \ell_2(\mathcal{J})$ *with* $\|v - w\|_{\ell_2(\mathcal{J})} \leq \epsilon$. *Assume* $\alpha = (\alpha_\lambda)_{\lambda \in \mathcal{J}} \in \mathbb{R}_+^{\mathcal{J}}$ *and* $\inf_\lambda \alpha_\lambda = \bar{\alpha} > 0$. *Then it holds*

$$\#\operatorname{supp} \mathbb{S}_\alpha(w) \leq \#\Lambda_{\bar{\alpha}}(w) \leq \frac{4\epsilon^2}{\bar{\alpha}^2} + 4C|v|_{\ell_\tau^w(\mathcal{J})}^\tau \bar{\alpha}^{-\tau}, \tag{4.30}$$

where $C = C(\tau) > 0$.

Proof. We consider two sets $\mathcal{J}_1 = \{\lambda \in \mathcal{J} : |w_\lambda| \geq \bar{\alpha}, \text{ and } |v_\lambda| > \bar{\alpha}/2\}$, and $\mathcal{J}_2 = \{\lambda \in \mathcal{J} : |w_\lambda| \geq \bar{\alpha}, \text{ and } |v_\lambda| \leq \bar{\alpha}/2\}$. Then from (4.28)

$$\#\mathcal{J}_1 \leq \#\{\lambda \in \mathcal{J} : |v_\lambda| > \bar{\alpha}/2\} \leq 2^\tau C|v|_{\ell_\tau^w(\mathcal{J})}^\tau \bar{\alpha}^{-\tau} \leq 4C|v|_{\ell_\tau^w(\mathcal{J})}^\tau \bar{\alpha}^{-\tau},$$

and

$$(\bar{\alpha}/2)^2(\#\mathcal{J}_2) \leq \sum_{\lambda \in \mathcal{J}_2} |v_\lambda - w_\lambda|^2 \leq \varepsilon^2.$$

These estimates imply (4.30). □

Decreasing Iterative Soft-thresholding For threshold parameters $\alpha, \alpha^{(n)} \in \mathbb{R}_+^{\mathcal{J}}$, where $\alpha^{(n)} \geq \alpha$, i.e., $\alpha_\lambda^{(n)} \geq \alpha_\lambda$ for all $\lambda \in \Lambda$, and $\bar{\alpha} = \inf_{\lambda \in \mathcal{J}} \alpha_\lambda > 0$, we consider the iteration

Algorithm 5.

$$u^{(0)} = 0, \quad u^{(n+1)} = \mathbb{S}_{\alpha^{(n)}}(u^{(n)} + A^*(y - Au^{(n)})), \quad n = 0, 1, \ldots \tag{4.31}$$

which, for $\alpha_\lambda^{(n)} \geq \alpha_\lambda^{(n+1)}$, we call the *decreasing iterative soft-thresholding algorithm* (D-ISTA).

Theorem 4.16. *Let* $\|A\| < \sqrt{2}$ *and let* $\bar{u} := (I - A^*A)u^* + A^*y \in \ell_\tau^w(\mathcal{I})$ *for some* $0 < \tau < 2$. *Moreover, let* $L = L(\alpha) := \frac{4\|u^*\|^2_{\ell_2(\mathcal{I})}}{\bar{\alpha}^2} + 4C\|\bar{u}\|^\tau_{\ell_\tau^w(\mathcal{I})}\bar{\alpha}^{-\tau}$, *and assume that for* $S^* := \operatorname{supp} u^*$ *and all finite subsets* $\Lambda \subset \mathcal{I}$ *with at most* $\#\Lambda \leq 2L$ *elements, the operator* $(I - A^*A)_{(S^*\cup\Lambda)\times(S^*\cup\Lambda)}$ *is contractive on* $\ell_2(S^* \cup \Lambda)$, *i.e.,* $\|(I - A^*A)_{S^*\cup\Lambda\times S^*\cup\Lambda} w\|_{\ell_2(S^*\cup\Lambda)} \leq \gamma_0\|w\|_{\ell_2(S^*\cup\Lambda)}$, *for all* $w \in \ell_2(S^* \cup \Lambda)$, *or*

$$\|(I - A^*A)_{S^*\cup\Lambda\times S^*\cup\Lambda}\| \leq \gamma_0, \tag{4.32}$$

where $0 < \gamma_0 < 1$. *Then, for any* $\gamma_0 < \gamma < 1$, *the iterates* $u^{(n)}$ *from* (4.31) *fulfill* $\#\operatorname{supp} u^{(n)} \leq L$ *and they converge to* u^* *at a linear rate*

$$\|u^* - u^{(n)}\|_{\ell_2(\mathcal{I})} \leq \gamma^n \|u^*\|_{\ell_2(\mathcal{I})} =: \epsilon_n \tag{4.33}$$

whenever the $\alpha^{(n)}$ *are chosen according to*

$$\alpha_\lambda \leq \alpha_\lambda^{(n)} \leq \alpha_\lambda + (\gamma - \gamma_0)L^{-1/2}\epsilon_n, \quad \text{for all } \lambda \in \Lambda. \tag{4.34}$$

Proof. We develop the proof by induction. For the initial iterate, we have $u^{(0)} = 0$, so that $\#\operatorname{supp} u^{(0)} \leq L$ and (4.33) is trivially true. Assume as an induction hypothesis that $S^{(n)} := \operatorname{supp}(u^{(n)})$ is such that $\#S^{(n)} \leq L$, and $\|u^* - u^{(n)}\|_{\ell_2(\mathcal{I})} \leq \epsilon_n$. Abbreviating $w^{(n)} := u^{(n)} + A^*(y - Au^{(n)})$, by $\|A^*A\| \leq 2$ and the induction hypothesis, it follows that

$$\|\bar{u} - w^{(n)}\|_{\ell_2(\mathcal{I})} = \|(I - A^*A)(u^* - u^{(n)})\|_{\ell_2(\mathcal{I})} \leq \|u^* - u^{(n)}\|_{\ell_2(\mathcal{I})} \leq \epsilon_n. \tag{4.35}$$

Hence, using (4.30), we obtain the estimate

$$\#S^{(n+1)} = \#\operatorname{supp} \mathbb{S}_{\alpha^{(n)}}(w^{(n)}) \leq \Lambda_{\bar{\alpha}}(w^{(n)}) \leq \frac{4\epsilon_n^2}{\bar{\alpha}^2} + 4C|\bar{u}|^\tau_{\ell_\tau^w(\mathcal{I})}\bar{\alpha}^{-\tau} \leq L. \tag{4.36}$$

Since also $\#S^{(n)} \leq L$ by induction hypothesis, the set $\Lambda^{(n)} := S^{(n)} \cup S^{(n+1)}$ has at most $2L$ elements, so that, by assumption, $(I - A^*A)_{S\cup\Lambda^{(n)}\times S\cup\Lambda^{(n)}}$ is contractive with contraction constant γ_0. Using the identities

$$\begin{aligned} u^*_{S\cup\Lambda^{(n)}} &= \mathbb{S}_\alpha(\bar{u}_{S\cup\Lambda^{(n)}}) \\ &= \mathbb{S}_\alpha(u^*_{S\cup\Lambda^{(n)}} + A^*_{S\cup\Lambda^{(n)}}(y - A_{S\cup\Lambda^{(n)}}u^*_{S\cup\Lambda^{(n)}})), \end{aligned}$$

and

$$\begin{aligned} u^{(n+1)}_{S\cup\Lambda^{(n)}} &= \mathbb{S}_{\alpha^{(n)}}(w^{(n)}_{S\cup\Lambda^{(n)}}) \\ &= \mathbb{S}_{\alpha^{(n)}}(u^{(n)}_{S\cup\Lambda^{(n)}} + A^*_{S\cup\Lambda^{(n)}}(y - A_{S\cup\Lambda^{(n)}}u^{(n)}_{S\cup\Lambda^{(n)}})), \end{aligned}$$

it follows from (4.25), (4.27), (4.19), and $\alpha^{(n)} \geq \alpha$ that

$$\begin{aligned} &\|u^* - u^{(n+1)}\|_{\ell_2(\mathcal{I})} \\ &\quad = \|(u^* - u^{(n+1)})_{S\cup\Lambda^{(n)}}\|_{\ell_2(S\cup\Lambda^{(n)})} \\ &\quad = \|\mathbb{S}_{\alpha}(\bar{u}_{S\cup\Lambda^{(n)}}) - \mathbb{S}_{\alpha^{(n)}}(w^{(n)}_{S\cup\Lambda^{(n)}})\|_{\ell_2(S\cup\Lambda^{(n)})} \\ &\quad \leq \|\mathbb{S}_{\alpha}(\bar{u}_{S\cup\Lambda^{(n)}}) - \mathbb{S}_{\alpha}(w^{(n)}_{S\cup\Lambda^{(n)}})\|_{\ell_2(S\cup\Lambda^{(n)})} \\ &\qquad + \|\mathbb{S}_{\alpha}(w^{(n)}_{S\cup\Lambda^{(n)}}) - \mathbb{S}_{\alpha^{(n)}}(w^{(n)}_{S\cup\Lambda^{(n)}})\|_{\ell_2(S\cup\Lambda^{(n)})} \\ &\quad \leq \|(I - A^*A)_{S\cup\Lambda^{(n)}\times S\cup\Lambda^{(n)}}(u^* - u^{(n)})_{S\cup\Lambda^{(n)}}\|_{\ell_2(S\cup\Lambda^{(n)})} \\ &\qquad + (\#\Lambda_{\bar{\alpha}}(w^{(n)}))^{1/2}\Big(\max_{\lambda\in\Lambda_{\bar{\alpha}}(w^{(n)})} |\alpha_\lambda - \alpha^{(n)}_\lambda|\Big) \\ &\quad \leq \gamma_0\epsilon_n + (\#\Lambda_{\bar{\alpha}}(w^{(n)}))^{1/2}\Big(\max_{\lambda\in\Lambda_{\bar{\alpha}}(w^{(n)})} |\alpha_\lambda - \alpha^{(n)}_\lambda|\Big). \end{aligned}$$

Using (4.36) we obtain $\|u - u^{(n+1)}\|_{\ell_2(\mathcal{I})} \leq \gamma_0\epsilon_n + \sqrt{L}(\max_{\lambda\in\Lambda_{\bar{\alpha}}(w^{(n)})} |\alpha^{(n)}_\lambda - \alpha_\lambda|)$, and, since the $\alpha^{(n)}$ are chosen according to (4.34), the claim follows. □

Note that assumption (4.32) in finite dimension essentially coincides with the request that the matrix A satisfies the RIP (see Lemma 3.14). With these results at hand and those related to RIP matrices in finite dimension, we are in the situation of estimating the relevant parameters in order to apply Theorem 4.16 when we are dealing with a compressed sensing problem. We proceed to a numerical comparison of the algorithm D-ISTA in (4.31) and the iterative soft-thresholding ISTA. In Figure 7 we show the behavior of the algorithms in the computation of a sparse minimizer u^* for A being a 500×2500 matrix with i.i.d. Gaussian entries, $\alpha = 10^{-4}$, $\gamma_0 = 0.01$ and $\gamma = 0.998$. In this case, related to a small value of α (we reiterate that a small range of α is the most crucial situation for the efficiency of iterative algorithms, see the subsection below), ISTA tends to iterate initially on vectors with a large number of nonzero entries, while D-ISTA slowly inflates the support size of the iterations to eventually converge to the right support of u^*. The iteration on an inflating support allows D-ISTA to take advantage of the local well-conditioning of the matrix A from the very beginning of the iterations. This effect results in a *controlled* linear rate of convergence which is much steeper than the one of ISTA. In particular in Figure 7 after 1500 iterations D-ISTA has correctly detected the support of the minimizer u^* and reached already an accuracy of $10^{-0.5}$, whereas it is clear that the convergence of ISTA is simply dramatically slow.

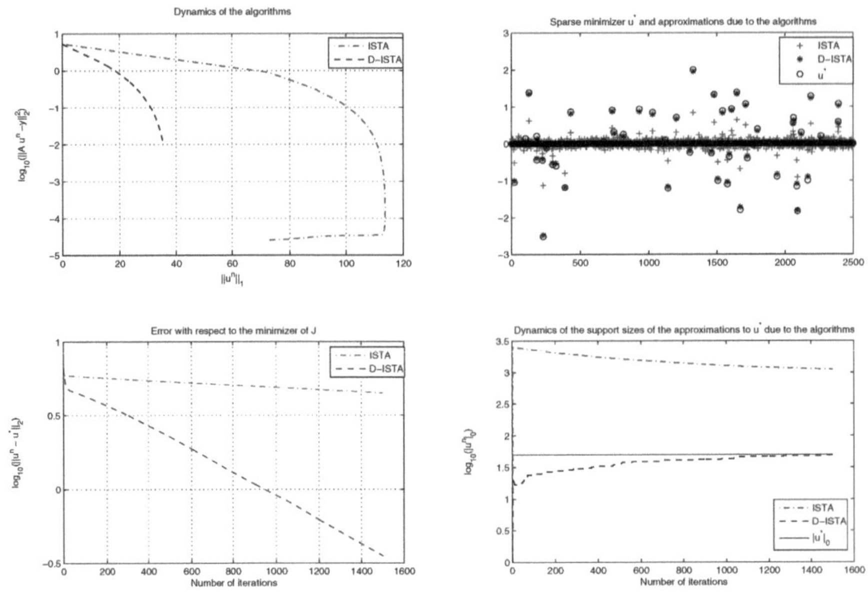

Figure 7. We show the behavior of the algorithms ISTA and D-ISTA in the computation of a sparse minimizer u^* for A being a 500×2500 matrix with i.i.d. Gaussian entries, $\alpha = 10^{-4}$, $\gamma_0 = 0.01$ and $\gamma = 0.998$. In the top left figure we present the dynamics of the algorithms in the plane $\|u\|_{\ell_1} - \log(\|Au - y\|_2^2)$. On the bottom left, we show the absolute error to the precomputed minimizer u^* with respect to the number of iterations. On the bottom right we show how the size of the supports of the iterations grow with the number of iterations. The figure on the top right shows the vector u^*, and the approximations due to the algorithms. In this case D-ISTA detects the right support of u^* after 1500 iterations, whereas ISTA keeps dramatically far behind.

Related Algorithms There exist by now several iterative methods that can be used for the minimization problem (4.3) in *finite dimensions*. We shall account a few of the most recently analyzed and discussed:

(a) the *GPSR-algorithm* (gradient projection for sparse reconstruction), another iterative projection method, in the auxiliary variables $x, z \geq 0$ with $u = x - z$, [40];

(b) the $\ell_1 - \ell_s$ *algorithm*, an interior point method using preconditioned conjugate gradient substeps (this method solves a linear system in each outer iteration step), [52];

(c) *FISTA* (fast iterative soft-thresholding algorithm) is a variation of the iterative soft-thresholding [5]. Define the operator $\Gamma(u) = \mathbb{S}_\alpha(u + A^*(y - Au))$. The

FISTA is defined as the iteration, starting for $u^{(0)} = 0$,

$$u^{(n+1)} = \Gamma\left(u^{(n)} + \frac{t^{(n)} - 1}{t^{(n+1)}}(u^{(n)} - u^{(n-1)})\right),$$

where $t^{(n+1)} = \frac{1+\sqrt{1+4(t^{(n)})^2}}{2}$ and $t^{(0)} = 1$.

As is addressed in the recent paper [56] which accounts a very detailed comparison of these different algorithms, they do perform quite well when the regularization parameter α is sufficiently large, with a small advantage for GPSR. When α gets quite small all the algorithms, except for FISTA, deteriorate significantly their performances. Moreover, local conditioning properties of the linear operator A seem particularly affecting the performances of iterative algorithms.

While these methods are particularly suited for finite dimensional problems, it would be interesting to produce an effective strategy, for any range of the parameter α, for a large class of infinite dimensional problems. In the recent paper [22] the following ingredients are combined for this scope:

- *multiscale preconditioning* allows for local well-conditioning of the matrix A and therefore reproduces at infinite dimension the conditions of best performances for iterative algorithms;
- *adaptivity* combined with a *decreasing thresholding strategy* allow for a *controlled* inflation of the support size of the iterations, promoting the minimal computational cost in terms of number of algebraic equations, as well as from the very beginning of the iteration the exploitation of the local well-conditioning of the matrix A.

In [69] the authors propose also an adaptive method similar to [22] where, instead of the soft-thresholding, a *coarsening function*, i.e., a compressed hard-thresholding procedure, is implemented. The emphasis in the latter contribution is on the regularization properties of such an adaptive method which does not dispose of a reference energy functional (4.1).

5 Large Scale Computing

5.1 Domain Decomposition Methods for ℓ_1-Minimization

Besides the elegant mathematics needed for the convergence proof, one of the major features of Algorithm 4 is its simplicity, also in terms of implementation. Indeed thresholding methods combined with wavelets have been often presented, e.g., in image processing, as a possible good alternative to total variation minimization which requires instead, as we have already discussed in the previous sections, the solution of a degenerate partial differential equation. However, as pointed out in the previous sections, in general, iterative soft-thresholding can converge very slowly.

In particular, it is practically not possible to use such algorithm when the dimension of the problem is really large, unless we provide all the modifications we accounted above (i.e., preconditioning, decreasing thresholding strategy, adaptivity etc.). And still for certain very large scale problems, this might not be enough. For that we need to consider further dimensionality reduction techniques. In this section we introduce a sequential domain decomposition method for the linear inverse problem with sparsity constraints modelled by (4.3). The goal is to join the simplicity of Algorithm 4 with a dimension reduction technique provided by a decomposition which will improve the convergence and the complexity of the algorithm without increasing the sophistication of the algorithm.

For simplicity, we start by decomposing the "domain" of the sequences $\mathcal{I}$ into two disjoint sets $\mathcal{I}_1, \mathcal{I}_2$ so that $\mathcal{I} = \mathcal{I}_1 \cup \mathcal{I}_2$. The extension to decompositions into multiple subsets $\mathcal{I} = \mathcal{I}_1 \cup \cdots \cup \mathcal{I}_{\mathcal{N}}$ follows from an analysis similar to the basic case $\mathcal{N} = 2$. Associated to a decomposition $\mathcal{C} = \{\mathcal{I}_1, \mathcal{I}_2\}$ we define the *extension operators* $E_i : \ell_2(\mathcal{I}_i) \to \ell_2(\mathcal{I})$, $(E_i v)_\lambda = v_\lambda$, if $\lambda \in \mathcal{I}_i$, $(E_i v)_\lambda = 0$, otherwise, $i = 1, 2$. The adjoint operator, which we call the *restriction operator*, is denoted by $R_i := E_i^*$. With these operators we may define the functional, $\mathcal{J} : \ell_2(\mathcal{I}_1) \times \ell_2(\mathcal{I}_2) \to \mathbb{R}$, given by

$$\mathcal{J}(u_1, u_2) := \mathcal{J}(E_1 u_1 + E_2 u_2).$$

For the sequence u_i we use the notation $u_{\lambda,i}$ in order to denote its components. We want to formulate and analyze the following algorithm: Pick an initial $E_1 u_1^{(0)} + E_2 u_2^{(0)} := u^{(0)} \in \ell_1(\mathcal{I})$, for example $u^{(0)} = 0$ and iterate

$$\begin{cases} u_1^{(n+1)} = \arg\min_{v_1 \in \ell_2(\mathcal{I}_1)} \mathcal{J}(v_1, u_2^{(n)}), \\ u_2^{(n+1)} = \arg\min_{v_2 \in \ell_2(\mathcal{I}_2)} \mathcal{J}(u_1^{(n+1)}, v_2), \\ u^{(n+1)} := E_1 u_1^{(n+1)} + E_2 u_2^{(n+1)}. \end{cases} \tag{5.1}$$

Let us observe that $\|E_1 u_1 + E_2 u_2\|_{\ell_1(\mathcal{I})} = \|u_1\|_{\ell_1(\mathcal{I}_1)} + \|u_2\|_{\ell_1(\mathcal{I}_2)}$, hence

$$\arg \min_{v_1 \in \ell_2(\mathcal{I}_1)} \mathcal{J}(v_1, u_2^{(n)}) = \arg \min_{v_1 \in \ell_2(\mathcal{I}_1)} \|(y - A E_2 u_2^{(n)}) - A E_1 v_1\|_Y^2 + \tau \|v_1\|_{\ell_1(\mathcal{I})}.$$

A similar formulation holds for $\arg\min_{v_2 \in \ell_2(\mathcal{I}_2)} \mathcal{J}(u_1^{(n+1)}, v_2)$. This means that the solution of the local problems on $\mathcal{I}_i$ is of the *same* kind as the original problem $\arg\min_{u \in \ell_2(\mathcal{I})} \mathcal{J}(u)$, but the dimension for each has been reduced. Unfortunately the functionals $\mathcal{J}(u, u_2^{(n)})$ and $\mathcal{J}(u_1^{(n+1)}, v)$ do not need to have a unique minimizer. Therefore the formulation as in (5.1) is not in principle well defined. In the following we will consider a particular choice of the minimizers and in particular we will implement Algorithm 4 in order to solve each local problem. This choice leads to the following algorithm.

Algorithm 6. Pick an initial $E_1 u_1^{(0)} + E_2 u_2^{(0)} := u^{(0)} \in \ell_1(\mathcal{I})$, for example $u^{(0)} = 0$, fix $L, M \in \mathbb{N}$, and iterate

$$\begin{cases} \begin{cases} u_1^{(n+1,0)} = u_1^{(n,L)} \\ u_1^{(n+1,\ell+1)} = \mathbb{S}_\alpha(u_1^{(n+1,\ell)} + R_1 A^*((y - AE_2 u_2^{(n,M)}) - AE_1 u_1^{(n+1,\ell)})) \\ \ell = 0, \ldots, L-1 \end{cases} \\ \begin{cases} u_2^{(n+1,0)} = u_2^{(n,M)} \\ u_2^{(n+1,\ell+1)} = \mathbb{S}_\alpha(u_2^{(n+1,\ell)} + R_2 A^*((y - AE_1 u_1^{(n+1,L)}) - AE_2 u_2^{(n+1,\ell)})) \\ \ell = 0, \ldots, M-1 \end{cases} \\ u^{(n+1)} := E_1 u_1^{(n+1,L)} + E_2 u_2^{(n+1,M)}. \end{cases} \tag{5.2}$$

Of course, for $L = M = \infty$ the previous algorithm realizes a particular instance of (5.1). However, in practice we will never execute an infinite number of inner iterations and therefore it is important to analyze the convergence of the algorithm when $L, M \in \mathbb{N}$ are finite.

At this point the question is whether algorithm (5.2) really converges to a minimizer of the original functional $\mathcal{J}$. This is the scope of the following sections. Only for ease of notation, we assume now that the thresholding parameter $\alpha > 0$ is a scalar, hence $\mathbb{S}_\alpha(u)$ acts on u with the same thresholding $S_\alpha(u_\lambda)$ for each vector component u_λ.

5.1.1 Weak Convergence of the Sequential Algorithm

A useful tool in the analysis of non-smooth functionals and their minima is the concept of the subdifferential. We shortly introduced it already in the presentation of the homotopy method in Section 3.1.1. Recall that for a convex functional F on some Banach space V its subdifferential $\partial F(x)$ at a point $x \in V$ with $F(x) < \infty$ is defined as the set

$$\partial F(x) = \{v \in V^*, F(y) - F(x) \geq v(y - x) \text{ for all } y \in V\},$$

where V^* denotes the dual space of V. It is obvious from this definition that $0 \in \partial F(x)$ if and only if x is a minimizer of F.

Example 5.1. Let $V = \ell_1(\mathcal{I})$ and $F(x) := \|x\|_{\ell_1(\mathcal{I})}$ is the ℓ_1-norm. We have

$$\partial\|\cdot\|_{\ell_1(\mathcal{I})}(x) = \{\xi \in \ell_\infty(\mathcal{I}) : \xi_\lambda \in \partial|\cdot|(x_\lambda), \lambda \in \mathcal{I}\} \tag{5.3}$$

where $\partial|\cdot|(z) = \{\operatorname{sgn}(z)\}$ if $z \neq 0$ and $\partial|\cdot|(0) = [-1, 1]$.

By observing that $\partial(\|A\cdot - y\|_Y^2)(u) = \{2A^*(Au - y)\}$ and by an application of [36, Proposition 5.2] combined with the example above, we obtain the following characterizations of the subdifferentials of $\mathcal{J}$ and $\mathcal{J}^S$ (we recall that $\mathcal{J}^S$ is the surrogate functional defined in (4.9)).

Lemma 5.2. i) *The subdifferential of $\mathcal{J}$ at u is given by*

$$\begin{aligned}\partial\mathcal{J}(u) &= 2A^*(Au-y) + 2\alpha\partial\|\cdot\|_{\ell_1(\mathcal{I})}(u)\\ &= \{\xi \in \ell_\infty(\mathcal{I}) : \xi_\lambda \in [2A^*(Au-y)]_\lambda + 2\alpha\partial|\cdot|(u_\lambda)\}.\end{aligned}$$

ii) *The subdifferential of $\mathcal{J}^S$ with respect to the sole component u is given by*

$$\begin{aligned}\partial_u\mathcal{J}^S(u,a) &= -2(a + A^*(y-Aa)) + 2u + 2\alpha\partial\|\cdot\|_{\ell_1(\mathcal{I})}(u)\\ &= \{\xi \in \ell_\infty(\mathcal{I}) : \xi_\lambda \in [-2(a + A^*(y-Aa))]_\lambda + 2u_\lambda\\ &\quad + 2\alpha\partial|\cdot|(u_\lambda)\}.\end{aligned}$$

In light of Lemma 4.2 we can reformulate Algorithm 6 by

$$\begin{cases}\begin{cases}u_1^{(n+1,0)} = u_1^{(n,L)}\\ u_1^{(n+1,\ell+1)} = \arg\min_{u_1\in\ell_2(\mathcal{I}_1)} \mathcal{J}^S(E_1u_1+E_2u_2^{(n,M)}, E_1u_1^{(n+1,\ell)}+E_2u_2^{(n,M)})\\ \ell = 0,\dots,L-1\end{cases}\\ \begin{cases}u_2^{(n+1,0)} = u_2^{(n,M)}\\ u_2^{(n+1,\ell+1)} = \arg\min_{u_2\in\ell_2(\mathcal{I}_2)} \mathcal{J}^S(E_1u_1^{(n+1,L)}+E_2u_2, E_1u_1^{(n+1,L)}+E_2u_2^{(n+1,\ell)})\\ \ell = 0,\dots,M-1\end{cases}\\ u^{(n+1)} := E_1u_1^{(n+1,L)} + E_2u_2^{(n+1,M)}.\end{cases} \tag{5.4}$$

Before we actually start proving the weak convergence of the algorithm in (5.4) we recall the following definition [71].

Definition 5.3. Let V be a topological space and $\mathcal{A} = (A_n)_{n\in\mathbb{N}}$ a sequence of subsets of V. The subset $A \subseteq V$ is called the *limit of the sequence* $\mathcal{A}$, and we write $A = \lim_n A_n$, if

$$A = \{a \in V : \exists a_n \in A_n, a = \lim_n a_n\}.$$

The following observation will be useful to us, see, e.g., [71, Proposition 8.7].

Lemma 5.4. *Assume that Γ is a convex function on $\mathbb{R}^M$ and $(x_n)_{n\in\mathbb{N}} \subset \mathbb{R}^M$ a convergent sequence with limit x such that $\Gamma(x_n), \Gamma(x) < \infty$. Then the subdifferentials satisfy*

$$\lim_{n\to\infty} \partial\Gamma(x_n) \subseteq \partial\Gamma(x).$$

In other words, the subdifferential $\partial\Gamma$ of a convex function is an outer semicontinuous set-valued function.

Theorem 5.5 (Weak convergence). *The algorithm in* (5.4) *produces a sequence* $(u^{(n)})_{n\in\mathbb{N}}$ *in* $\ell_2(\mathcal{I})$ *whose weak accumulation points are minimizers of the functional* $\mathcal{J}$. *In particular, the set of the weak accumulation points is non-empty and if* $u^{(\infty)}$ *is a weak accumulation point then*

$$u^{(\infty)} = \mathbb{S}_\alpha(u^{(\infty)} + A^*(g - Au^{(\infty)})).$$

Proof. The proof is partially inspired by the one of Theorem 3.12, where also alternating minimizations were considered. Let us first observe that by (4.11)

$$\begin{aligned}\mathcal{J}(u^{(n)}) = \mathcal{J}^S(u^{(n)}, u^{(n)}) &= \mathcal{J}^S(E_1u_1^{(n,L)} + E_2u_2^{(n,M)}, E_1u_1^{(n,L)} + E_2u_2^{(n,M)}) \\ &= \mathcal{J}^S(E_1u_1^{(n,L)} + E_2u_2^{(n,M)}, E_1u_1^{(n+1,0)} + E_2u_2^{(n,M)}).\end{aligned}$$

By definition of $u_1^{(n+1,1)}$ and its minimal properties in (5.4) we have

$$\begin{aligned}&\mathcal{J}^S(E_1u_1^{(n,L)} + E_2u_2^{(n,M)}, E_1u_1^{(n+1,0)} + E_2u_2^{(n,M)}) \\ &\quad\ge \mathcal{J}^S(E_1u_1^{(n+1,1)} + E_2u_2^{(n,M)}, E_1u_1^{(n+1,0)} + E_2u_2^{(n,M)}).\end{aligned}$$

Again, an application of (4.11) gives

$$\begin{aligned}&\mathcal{J}^S(E_1u_1^{(n+1,1)} + E_2u_2^{(n,M)}, E_1u_1^{(n+1,0)} + E_2u_2^{(n,M)}) \\ &\quad\ge \mathcal{J}^S(E_1u_1^{(n+1,1)} + E_2u_2^{(n,M)}, E_1u_1^{(n+1,1)} + E_2u_2^{(n,M)}).\end{aligned}$$

Concatenating these inequalities we obtain

$$\mathcal{J}(u^{(n)}) \ge \mathcal{J}^S(E_1u_1^{(n+1,1)} + E_2u_2^{(n,M)}, E_1u_1^{(n+1,1)} + E_2u_2^{(n,M)}).$$

In particular, from (4.12) we have

$$\begin{aligned}&\mathcal{J}(u^{(n)}) - \mathcal{J}^S(E_1u_1^{(n+1,1)} + E_2u_2^{(n,M)}, E_1u_1^{(n+1,1)} + E_2u_2^{(n,M)}) \\ &\quad\ge C\|u_1^{(n+1,1)} - u_1^{(n+1,0)}\|^2_{\ell_2(\mathcal{I}_1)}.\end{aligned}$$

By induction we obtain

$$\begin{aligned}\mathcal{J}(u^{(n)}) &\ge \mathcal{J}^S(E_1u_1^{(n+1,1)} + E_2u_2^{(n,M)}, E_1u_1^{(n+1,1)} + E_2u_2^{(n,M)}) \ge \cdots \\ &\ge \mathcal{J}^S(E_1u_1^{(n+1,L)} + E_2u_2^{(n,M)}, E_1u_1^{(n+1,L)} + E_2u_2^{(n,M)}) \\ &= \mathcal{J}(E_1u_1^{(n+1,L)} + E_2u_2^{(n,M)}),\end{aligned}$$

and

$$\mathcal{J}(u^{(n)}) - \mathcal{J}(E_1u_1^{(n+1,L)} + E_2u_2^{(n,M)}) \ge C\sum_{\ell=0}^{L-1}\|u_1^{(n+1,\ell+1)} - u_1^{(n+1,\ell)}\|^2_{\ell_2(\mathcal{I}_1)}.$$

By definition of $u_2^{(n+1,1)}$ and its minimal properties we have

$$\begin{aligned}&\mathcal{J}^S(E_1u_1^{(n+1,L)} + E_2u_2^{(n,M)}, E_1u_1^{(n+1,L)} + E_2u_2^{(n,M)})\\ &\quad\geq \mathcal{J}^S(E_1u_1^{(n+1,L)} + E_2u_2^{(n+1,1)}, E_1u_1^{(n+1,L)} + E_2u_2^{(n+1,0)}).\end{aligned}$$

By similar arguments as above we find

$$\mathcal{J}(u^{(n)}) \geq \mathcal{J}^S(E_1u_1^{(n+1,L)} + E_2u_2^{(n+1,M)}, E_1u_1^{(n+1,L)} + E_2u_2^{(n+1,M)}) = \mathcal{J}(u^{(n+1)}), \tag{5.5}$$

and

$$\begin{aligned}&\mathcal{J}(u^{(n)}) - \mathcal{J}(u^{(n+1)})\\ &\geq C\Big(\sum_{\ell=0}^{L-1} \|u_1^{(n+1,\ell+1)} - u_1^{(n+1,\ell)}\|^2_{\ell_2(\mathcal{J}_1)} + \sum_{m=0}^{M-1} \|u_2^{(n+1,m+1)} - u_2^{(n+1,m)}\|^2_{\ell_2(\mathcal{J}_2)}\Big).\end{aligned} \tag{5.6}$$

From (5.5) we have $\mathcal{J}(u^{(0)}) \geq \mathcal{J}(u^{(n)}) \geq 2\alpha\|u^{(n)}\|_{\ell_1(\mathcal{J})} \geq 2\alpha\|u^{(n)}\|_{\ell_2(\mathcal{J})}$. This means that $(u^{(n)})_{n\in\mathbb{N}}$ is uniformly bounded in $\ell_2(\mathcal{J})$, hence there exists a weakly convergent subsequence $(u^{(n_j)})_{j\in\mathbb{N}}$. Let us denote $u^{(\infty)}$ the weak limit of the subsequence. For simplicity, we rename such subsequence by $(u^{(n)})_{n\in\mathbb{N}}$. Moreover, since the sequence $(\mathcal{J}(u^{(n)}))_{n\in\mathbb{N}}$ is monotonically decreasing and bounded from below by 0, it is also convergent. From (5.6) and the latter convergence we deduce

$$\Big(\sum_{\ell=0}^{L-1} \|u_1^{(n+1,\ell+1)} - u_1^{(n+1,\ell)}\|^2_{\ell_2(\mathcal{J}_1)} + \sum_{m=0}^{M-1} \|u_2^{(n+1,m+1)} - u_2^{(n+1,m)}\|^2_{\ell_2(\mathcal{J}_2)}\Big) \to 0, \tag{5.7}$$

for $n \to \infty$. In particular, by the standard inequality $(a^2 + b^2) \geq \frac{1}{2}(a + b)^2$ for $a, b > 0$ and the triangle inequality, we have also

$$\|u^{(n)} - u^{(n+1)}\|_{\ell_2(\mathcal{J})} \to 0, \quad n \to \infty. \tag{5.8}$$

We would like to show now that

$$0 \in \lim_{n\to\infty} \partial\mathcal{J}(u^{(n)}) \subset \partial\mathcal{J}(u^{(\infty)}).$$

To this end, and in light of Lemma 5.2, we reason componentwise. By definition of $u_1^{(n+1,L)}$ we have

$$\begin{aligned}0 \in [&-2(u_1^{(n+1,L-1)} + R_1A^*((y - AE_2u_2^{(n,M)}) - AE_1u_1^{(n+1,L-1)}))]_\lambda\\ &+ 2u_{\lambda,1}^{(n+1,L)} + 2\alpha\partial|\cdot|(u_{\lambda,1}^{(n+1,L)}),\end{aligned} \tag{5.9}$$

for $\lambda \in \mathcal{I}_1$, and by definition of $u_2^{(n+1,M)}$ we have

$$0 \in [-2(u_2^{(n+1,M-1)} + R_2 A^*((y - AE_1 u_1^{(n+1,L)}) - AE_2 u_2^{(n+1,M-1)}))]_\lambda \\ + 2u_{\lambda,2}^{(n+1,M)} + 2\alpha\partial|\cdot|(u_{\lambda,2}^{(n+1,M)}), \tag{5.10}$$

for $\lambda \in \mathcal{I}_2$. Let us compute $\partial\mathcal{J}(u^{(n+1)})_\lambda$,

$$\partial\mathcal{J}(u^{(n+1)})_\lambda = [-2A^*(y - AE_1 u_1^{(n+1,L)} - AE_2 u_2^{(n+1,M)})]_\lambda + 2\alpha\partial|\cdot|(u_{\lambda,i}^{(n+1,K)}), \tag{5.11}$$

where $\lambda \in \mathcal{I}_i$ and $K = L, M$ for $i = 1, 2$ respectively. We would like to find a $\xi_\lambda^{(n+1)} \in \partial\mathcal{J}(u^{(n+1)})_\lambda$ such that $\xi_\lambda^{(n+1)} \to 0$ for $n \to \infty$. By (5.9) we have that for $\lambda \in \mathcal{I}_1$

$$0 = [-2(u_1^{(n+1,L-1)} + R_1 A^*((y - AE_2 u_2^{(n,M)}) - AE_1 u_1^{(n+1,L-1)}))]_\lambda \\ + 2u_{\lambda,1}^{(n+1,L)} + 2\alpha\xi_{\lambda,1}^{(n+1)},$$

for a $\xi_{\lambda,1}^{(n+1)} \in \partial|\cdot|(u_{\lambda,1}^{(n+1,L)})$, and, by (5.10), for $\lambda \in \mathcal{I}_2$

$$0 = [-2(u_2^{(n+1,M-1)} + R_2 A^*((y - AE_1 u_1^{(n+1,L)}) - AE_2 u_2^{(n+1,M-1)}))]_\lambda \\ + 2u_{\lambda,2}^{(n+1,M)} + 2\alpha\xi_{\lambda,2}^{(n+1)},$$

for a $\xi_{\lambda,2}^{(n+1)} \in \partial|\cdot|(u_{\lambda,2}^{(n+1,M)})$. Thus by adding zero to (5.11) as represented by the previous two formulas, we can choose

$$\xi_\lambda^{(n+1)} = 2(u_{\lambda,1}^{(n+1,L)} - u_{\lambda,1}^{(n+1,L-1)}) + [R_1 A^* AE_1(u_1^{(n+1,L)} - u_1^{(n+1,L-1)})]_\lambda \\ + [R_1 A^* AE_2(u_2^{(n+1,M)} - u_1^{(n,M)})]_\lambda,$$

if $\lambda \in \mathcal{I}_1$ and

$$\xi_\lambda^{(n+1)} = 2(u_{\lambda,2}^{(n+1,M)} - u_{\lambda,2}^{(n+1,M-1)}) + [R_2 A^* AE_1(u_2^{(n+1,M)} - u_1^{(n+1,M-1)})]_\lambda,$$

if $\lambda \in \mathcal{I}_2$. For both these choices, from (5.7) and (5.8), and by continuity of A, we have $\xi_\lambda^{(n+1)} \to 0$ for $n \to \infty$. Again by continuity of A, weak convergence of $u^{(n)}$ (which implies componentwise convergence), and Lemma 5.4 we obtain

$$0 \in \lim_{n\to\infty} \partial\mathcal{J}(u^{(n)})_\lambda \subset \partial\mathcal{J}(u^{(\infty)})_\lambda, \quad \text{for all } \lambda \in \mathcal{I}.$$

It follows from Lemma 5.2 that $0 \in \partial\mathcal{J}(u^{(\infty)})$. By the properties of the subdifferential we have that $u^{(\infty)}$ is a minimizer of $\mathcal{J}$. Of course, the reasoning above holds for any weakly convergent subsequence and therefore all weak accumulation points of the original sequence $(u^{(n)})_n$ are minimizers of $\mathcal{J}$.

Similarly, by taking now the limit for $n \to \infty$ in (5.9) and (5.10), and by using (5.7) we obtain

$$0 \in [-2(R_1 u^{(\infty)} + R_1 A^*((y - AE_2 R_2 u^{(\infty)}) - AE_1 R_1 u^{(\infty)}))]_\lambda + 2u_\lambda^{(\infty)} \\ + 2\alpha \partial |\cdot|(u_\lambda^{(\infty)}),$$

for $\lambda \in \mathcal{J}_1$ and

$$0 \in [-2(R_2 u^{(\infty)} + R_2 A^*((y - AE_1 R_1 u^{(\infty)}) - AE_2 R_2 u^{(\infty)})]_\lambda + 2u_\lambda^{(\infty)} \\ + 2\alpha \partial |\cdot|(u_\lambda^{(\infty)}).$$

for $\lambda \in \mathcal{J}_2$. In other words, we have

$$0 \in \partial_u \mathcal{J}^S(u^{(\infty)}, u^{(\infty)}).$$

An application of Lemma 5.2 and Proposition 4.5 imply

$$u^{(\infty)} = \mathbb{S}_\alpha(u^{(\infty)} + A^*(y - Au^{(\infty)})). \qquad \square$$

Remark 5.6. 1. Because $u^{(\infty)} = \mathbb{S}_\alpha(u^{(\infty)} + A^*(y - Au^{(\infty)}))$, we could infer the minimality of $u^{(\infty)}$ by invoking Proposition 4.5. In the previous proof we wanted to present an alternative argument based on differential inclusions.

2. Since $(u^{(n)})_{n \in \mathbb{N}}$ is bounded and (5.7) holds, also $(u_i^{n,\ell})_{n,\ell}$ are bounded for $i = 1, 2$.

5.1.2 Strong Convergence of the Sequential Algorithm

In this section we want to show that the convergence of a subsequence $(u^{n_j})_j$ to any accumulation point $u^{(\infty)}$ holds not only in the weak topology, but also in the Hilbert space $\ell_2(\mathcal{J})$ norm. Let us define

$$\eta^{(n+1)} := u_1^{(n+1,L)} - u_1^{(\infty)}, \qquad \eta^{(n+1/2)} := u_1^{(n+1,L-1)} - u_1^{(\infty)},$$
$$\mu^{(n+1)} := u_2^{(n+1,M)} - u_2^{(\infty)}, \qquad \mu^{(n+1/2)} := u_2^{(n+1,M-1)} - u_2^{(\infty)},$$

where $u_i^{(\infty)} := R_i u^{(\infty)}$. From Theorem 5.5 we also have

$$u_i^{(\infty)} = \mathbb{S}_\alpha(\underbrace{u_i^{(\infty)} + R_i A^*(y - AE_1 u_1^{(\infty)} - AE_2 u_2^{(\infty)})}_{:=h_i}), \quad i = 1, 2.$$

Let us also denote $h := E_1 h_1 + E_2 h_2$ and $\xi^{(n)} := E_1 \eta^{(n+1/2)} + E_2 \mu^{(n+1/2)}$.

For the proof of strong convergence we need the following technical lemmas. The strategy of their proofs is similar to that of Lemma 4.8 and Lemma 4.9.

Lemma 5.7. $\|A\xi^{(n)}\|_Y^2 \to 0$ *for* $n \to \infty$.

Proof. Since

$$\begin{aligned}
\eta^{(n+1)} - \eta^{(n+1/2)} &= \mathbb{S}_\alpha(h_1 + (I - R_1 A^* A E_1)\eta^{(n+1/2)} \\
&\qquad - R_1 A^* A E_2 \mu^{(n)}) - \mathbb{S}_\alpha(h_1) - \eta^{(n+1/2)}, \\
\mu^{(n+1)} - \mu^{(n+1/2)} &= \mathbb{S}_\alpha(h_2 + (I - R_2 A^* A E_2)\mu^{(n+1/2)} \\
&\qquad - R_2 A^* A E_1 \eta^{(n+1)}) - \mathbb{S}_\alpha(h_2) - \mu^{(n+1/2)},
\end{aligned}$$

and $\|\eta^{(n+1)} - \eta^{(n+1/2)}\|_{\ell_2(\mathcal{I}_1)} = \|u_1^{(n+1,L)} - u_1^{(n+1,L-1)}\|_{\ell_2(\mathcal{I}_1)} \to 0$, $\|\mu^{(n+1)} - \mu^{(n+1/2)}\|_{\ell_2(\mathcal{I}_1)} = \|u_2^{(n+1,M)} - u_2^{(n+1,M-1)}\|_{\ell_2(\mathcal{I}_2)} \to 0$ by (5.7), we have

$$\begin{aligned}
&\|\mathbb{S}_\alpha(h_1 + (I - R_1 A^* A E_1)\eta^{(n+1/2)} - R_1 A^* A E_2 \mu^{(n)}) - \mathbb{S}_\alpha(h_1) - \eta^{(n+1/2)}\|_{\ell_2(\mathcal{I}_1)} \\
&\quad \geq \big| \|\mathbb{S}_\alpha(h_1 + (I - R_1 A^* A E_1)\eta^{(n+1/2)} - R_1 A^* A E_2 \mu^{(n)}) \\
&\quad - \mathbb{S}_\alpha(h_1)\|_{\ell_2(\mathcal{I}_1)} - \|\eta^{(n+1/2)}\|_{\ell_2(\mathcal{I}_1)} \big| \to 0, \qquad (5.12)
\end{aligned}$$

and

$$\begin{aligned}
&\|\mathbb{S}_\alpha(h_2 + (I - R_2 A^* A E_2)\mu^{(n+1/2)} - R_2 A^* A E_1 \eta^{(n+1)}) - \mathbb{S}_\alpha(h_2) - \mu^{(n+1/2)}\|_{\ell_2(\mathcal{I}_2)} \\
&\quad \geq \big| \|\mathbb{S}_\alpha(h_2 + (I - R_2 A^* A E_2)\mu^{(n+1/2)} - R_2 A^* A E_1 \eta^{(n+1)}) \\
&\quad - \mathbb{S}_\alpha(h_2)\|_{\ell_2(\mathcal{I}_2)} - \|\mu^{(n+1/2)}\|_{\ell_2(\mathcal{I}_2)} \big| \to 0. \qquad (5.13)
\end{aligned}$$

By nonexpansiveness of $\mathbb{S}_\alpha$ we have the estimates

$$\begin{aligned}
&\|\mathbb{S}_\alpha(h_2 + (I - R_2 A^* A E_2)\mu^{(n+1/2)} - R_2 A^* A E_1 \eta^{(n+1)}) - \mathbb{S}_\alpha(h_2)\|_{\ell_2(\mathcal{I}_2)} \\
&\quad \leq \|(I - R_2 A^* A E_2)\mu^{(n+1/2)} - R_2 A^* A E_1 \eta^{(n+1)}\|_{\ell_2(\mathcal{I}_2)} \\
&\quad \leq \|(I - R_2 A^* A E_2)\mu^{(n+1/2)} - R_2 A^* A E_1 \eta^{(n+1/2)}\|_{\ell_2(\mathcal{I}_2)} \\
&\qquad + \underbrace{\|R_2 A^* A E_1 (\eta^{(n+1/2)}) - \eta^{(n+1)})\|_{\ell_2(\mathcal{I}_2)}}_{:=\varepsilon^{(n)}}.
\end{aligned}$$

Similarly, we have

$$\begin{aligned}
&\|\mathbb{S}_\alpha(h_1 + (I - R_1 A^* A E_1)\eta^{(n+1/2)} - R_1 A^* A E_2 \mu^{(n)}) - \mathbb{S}_\alpha(h_1)\|_{\ell_2(\mathcal{I}_1)} \\
&\quad \leq \|(I - R_1 A^* A E_1)\eta^{(n+1/2)} - R_1 A^* A E_2 \mu^{(n)}\|_{\ell_2(\mathcal{I}_1)} \\
&\quad \leq \|(I - R_1 A^* A E_1)\eta^{(n+1/2)} - R_1 A^* A E_2 \mu^{(n+1/2)}\|_{\ell_2(\mathcal{I}_1)} \\
&\qquad + \underbrace{\|R_1 A^* A E_2 (\mu^{(n+1/2)} - \mu^{(n)})\|_{\ell_2(\mathcal{I}_1)}}_{\delta^{(n)}}.
\end{aligned}$$

Combining the previous inequalities, we obtain the estimates

$$\begin{aligned}
&\|\mathbb{S}_\alpha(h_1 + (I - R_1 A^* A E_1)\eta^{(n+1/2)} - R_1 A^* A E_2 \mu^{(n)} - \mathbb{S}_\alpha(h_1)\|^2_{\ell_2(\mathcal{J}_1)} \\
&\qquad + \|\mathbb{S}_\alpha(h_2 + (I - R_2 A^* A E_2)\mu^{(n+1/2)} - R_2 A^* A E_1 \eta^{(n+1)} - \mathbb{S}_\alpha(h_2)\|^2_{\ell_2(\mathcal{J}_2)} \\
&\quad \le \|(I - R_1 A^* A E_1)\eta^{(n+1/2)} - R_1 A^* A E_2 \mu^{(n)}\|^2_{\ell_2(\mathcal{J}_1)} \\
&\qquad + \|(I - R_2 A^* A E_2)\mu^{(n+1/2)} - R_2 A^* A E_1 \eta^{(n+1)}\|^2_{\ell_2(\mathcal{J}_2)} \\
&\quad = (\|(I - R_1 A^* A E_1)\eta^{(n+1/2)} - R_1 A^* A E_2 \mu^{(n+1/2)}\|_{\ell_2(\mathcal{J}_1)} \\
&\qquad + \|R_1 A^* A E_2(\mu^{(n+1/2)} - \mu^{(n)})\|_{\ell_2(\mathcal{J}_1)})^2 \\
&\qquad + (\|(I - R_2 A^* A E_2)\mu^{(n+1/2)} - R_2 A^* A E_1 \eta^{(n+1/2)}\|_{\ell_2(\mathcal{J}_2)} \\
&\qquad + \|R_2 A^* A E_1(\eta^{(n+1/2)}) - \eta^{(n+1)}\|_{\ell_2(\mathcal{J}_2)})^2 \\
&\quad \le \|(I - A^* A)\xi^{(n)}\|^2_{\ell_2(\mathcal{J})} + ((\varepsilon^{(n)})^2 + (\delta^{(n)})^2 + C'(\varepsilon^{(n)} + \delta^{(n)})) \\
&\quad \le \|\xi^{(n)}\|^2_{\ell_2(\mathcal{J})} + ((\varepsilon^{(n)})^2 + (\delta^{(n)})^2 + C'(\varepsilon^{(n)} + \delta^{(n)})).
\end{aligned}$$

The constant $C' > 0$ is due to the boundedness of $u^{(n,\ell)}$. Certainly, by (5.7), for every $\varepsilon > 0$ there exists n_0 such that for $n > n_0$ we have $(\varepsilon^{(n)})^2 + (\delta^{(n)})^2 + C'(\varepsilon^{(n)} + \delta^{(n)})) \le \varepsilon$. Therefore, if

$$\begin{aligned}
&\|\mathbb{S}_\alpha(h_1 + (I - R_1 A^* A E_1)\eta^{(n+1/2)} - R_1 A^* A E_2 \mu^{(n)} - \mathbb{S}_\alpha(h_1)\|^2_{\ell_2(\mathcal{J}_1)} \\
&\qquad + \|\mathbb{S}_\alpha(h_2 + (I - R_2 A^* A E_2)\mu^{(n+1/2)} - R_2 A^* A E_1 \eta^{(n+1)} - \mathbb{S}_\alpha(h_2)\|^2_{\ell_2(\mathcal{J}_2)} \\
&\quad \ge \|\xi^{(n)}\|^2_{\ell_2(\mathcal{J})},
\end{aligned}$$

then

$$\begin{aligned}
0 &\le \|(I - R_1 A^* A E_1)\mu^{(n+1/2)} - R_1 A^* A E_2 \mu^{(n)}\|^2_{\ell_2(\mathcal{J}_1)} \\
&\quad + \|(I - R_2 A^* A E_2)\mu^{(n+1/2)} - R_2 A^* A E_1 \eta^{(n+1)}\|^2_{\ell_2(\mathcal{J})} - \|\xi^{(n)}\|^2_{\ell_2(\mathcal{J})} \\
&\le (\varepsilon^{(n)})^2 + (\delta^{(n)})^2 + C'(\varepsilon^{(n)} + \delta^{(n)})) \le \varepsilon.
\end{aligned}$$

If, instead, we have

$$\begin{aligned}
&\|\mathbb{S}_\alpha(h_1 + (I - R_1 A^* A E_1)\eta^{(n+1/2)} - R_1 A^* A E_2 \mu^{(n)}) - \mathbb{S}_\alpha(h_1)\|^2_{\ell_2(\mathcal{J}_1)} \\
&\qquad + \|\mathbb{S}_\alpha(h_2 + (I - R_2 A^* A E_2)\mu^{(n+1/2)} - R_2 A^* A E_1 \eta^{(n+1)}) - \mathbb{S}_\alpha(h_2)\|^2_{\ell_2(\mathcal{J}_2)} \\
&\quad < \|\xi^{(n)}\|^2_{\ell_2(\mathcal{J})},
\end{aligned}$$

then by (5.12) and (5.13)

$$\begin{aligned}
&\|\xi^{(n)}\|^2_{\ell_2(\mathcal{I})} - (\|(I - R_1 A^* A E_1)\mu^{(n+1/2)} - R_1 A^* A E_2 \mu^{(n)})\|^2_{\ell_2(\mathcal{I}_1)} \\
&\qquad + \|(I - R_2 A^* A E_2)\mu^{(n+1/2)} - R_2 A^* A E_1 \eta^{(n+1)})\|^2_{\ell_2(\mathcal{I}_2)}) \\
&\le \|\xi^{(n)}\|^2_{\ell_2(\mathcal{I})} \\
&\quad - \|\mathbb{S}_\alpha(h_1 + (I - R_1 A^* A E_1)\eta^{(n+1/2)} - R_1 A^* A E_2 \mu^{(n)}) - \mathbb{S}_\alpha(h_1)\|^2_{\ell_2(\mathcal{I}_1)} \\
&\quad - \|\mathbb{S}_\alpha(h_2 + (I - R_2 A^* A E_2)\mu^{(n+1/2)} - R_2 A^* A E_1 \eta^{(n+1)}) - \mathbb{S}_\alpha(h_2)\|^2_{\ell_2(\mathcal{I}_2)} \\
&= |\|\xi^{(n)}\|^2_{\ell_2(\mathcal{I})} \\
&\quad - \|\mathbb{S}_\alpha(h_1 + (I - R_1 A^* A E_1)\eta^{(n+1/2)} - R_1 A^* A E_2 \mu^{(n)}) - \mathbb{S}_\alpha(h_1)\|^2_{\ell_2(\mathcal{I}_1)} \\
&\quad - \|\mathbb{S}_\alpha(h_2 + (I - R_2 A^* A E_2)\mu^{(n+1/2)} - R_2 A^* A E_1 \eta^{(n+1)}) \\
&\qquad - \mathbb{S}_\alpha(h_2)\|^2_{\ell_2(\mathcal{I}_2)}| \\
&\le |\|\eta^{(n+1/2)}\|^2_{\ell_2(\mathcal{I}_1)} \\
&\quad - \|\mathbb{S}_\alpha(h_1 + (I - R_1 A^* A E_1)\eta^{(n+1/2)} - R_1 A^* A E_2 \mu^{(n)}) - \mathbb{S}_\alpha(h_1)\|^2_{\ell_2(\mathcal{I}_1)}| \\
&\quad + |\|\mu^{(n+1/2)}\|^2_{\ell_2(\mathcal{I}_2)} - \|\mathbb{S}_\alpha(h_2 + (I - R_2 A^* A E_2)\mu^{(n+1/2)} \\
&\qquad\qquad - R_2 A^* A E_1 \eta^{(n+1)}) - \mathbb{S}_\alpha(h_2)\|^2_{\ell_2(\mathcal{I}_2)}| \le \varepsilon
\end{aligned}$$

for n large enough. This implies

$$\begin{aligned}
\lim_{n\to\infty} [&\|\xi^{(n)}\|^2_{\ell_2(\mathcal{I})} - (\|(I - R_1 A^* A E_1)\mu^{(n+1/2)} - R_1 A^* A E_2 \mu^{(n)}\|^2_{\ell_2(\mathcal{I}_1)} \\
&+ \|(I - R_2 A^* A E_2)\mu^{(n+1/2)} - R_2 A^* A E_1 \eta^{(n+1)}\|^2_{\ell_2(\mathcal{I}_2)})] = 0.
\end{aligned}$$

Observe now that

$$\begin{aligned}
&\|(I - R_1 A^* A E_1)\mu^{(n+1/2)} - R_1 A^* A E_2 \mu^{(n)}\|^2_{\ell_2(\mathcal{I}_1)} \\
&\qquad + \|(I - R_2 A^* A E_2)\mu^{(n+1/2)} - R_2 A^* A E_1 \eta^{(n+1)}\|^2_{\ell_2(\mathcal{I}_2)} \\
&\quad \le (\|(I - R_1 A^* A E_1)\mu^{(n+1/2)} - R_1 A^* A E_2 \mu^{(n+1/2)}\|_{\ell_2(\mathcal{I}_1)} + \delta^{(n)})^2 \\
&\qquad + (\|(I - R_2 A^* A E_2)\mu^{(n+1/2)} - R_2 A^* A E_1 \eta^{(n+1/2)}\|_{\ell_2(\mathcal{I}_2)} + \varepsilon^{(n)})^2 \\
&\quad \le \|(I - A^* A)\xi^{(n)}\|^2_{\ell_2(\mathcal{I})} + ((\varepsilon^{(n)})^2 + (\delta^{(n)})^2 + 2C'(\varepsilon^{(n)} + \delta^{(n)})),
\end{aligned}$$

for a suitable constant $C' > 0$ as above. Therefore we have

$$\begin{aligned}
&\|\xi^{(n)}\|^2_{\ell_2(\mathcal{I})} - (\|(I - R_1 A^* A E_1)\mu^{(n+1/2)} - R_1 A^* A E_2 \mu^{(n)}\|^2_{\ell_2(\mathcal{I}_1)} \\
&\qquad + \|(I - R_2 A^* A E_2)\mu^{(n+1/2)} - R_2 A^* A E_1 \eta^{(n+1)}\|^2_{\ell_2(\mathcal{I}_2)}) \\
&\geq \|\xi^{(n)}\|^2_{\ell_2(\mathcal{I})} - \|(I - A^* A)\xi^{(n)}\|^2_{\ell_2(\mathcal{I})} - ((\varepsilon^{(n)})^2 + (\delta^{(n)})^2 + 2C'(\varepsilon^{(n)} + \delta^{(n)})) \\
&= 2\|A\xi^{(n)}\|^2_Y - \|A^* A \xi^{(n)}\|^2_{\ell_2(\mathcal{I})} - ((\varepsilon^{(n)})^2 + (\delta^{(n)})^2 + 2C'(\varepsilon^{(n)} + \delta^{(n)})) \\
&\geq \|A\xi^{(n)}\|^2_Y - ((\varepsilon^{(n)})^2 + (\delta^{(n)})^2 + 2C'(\varepsilon^{(n)} + \delta^{(n)})).
\end{aligned}$$

This implies $\|A\xi^{(n)}\|^2_Y \to 0$ for $n \to \infty$. □

Lemma 5.8. *For* $h = E_1 h_1 + E_2 h_2$, $\|\mathbb{S}_\alpha(h + \xi^{(n)}) - \mathbb{S}_\alpha(h) - \xi^{(n)}\|_{\ell_2(\mathcal{I})} \to 0$, *for* $n \to \infty$.

Proof. We have

$$\begin{aligned}
&\mathbb{S}_\alpha(h + \xi^{(n)} - A^* A \xi^{(n)}) \\
&\quad = E_1(\mathbb{S}_\alpha(h_1 + (I - R_1 A^* A E_1)\eta^{(n+1/2)} - R_1 A^* A E_2 \mu^{(n+1/2)})) \\
&\qquad + E_2(\mathbb{S}_\alpha(h_2 + (I - R_2 A^* A E_2)\mu^{(n+1/2)} - R_2 A^* A E_1 \eta^{(n+1/2)})).
\end{aligned}$$

Therefore, we can write

$$\begin{aligned}
&\mathbb{S}_\alpha(h + \xi^{(n)} - A^* A \xi^{(n)}) \\
&\quad = E_1[\mathbb{S}_\alpha(h_1 + (I - R_1 A^* A E_1)\eta^{(n+1/2)} - R_1 A^* A E_2 \mu^{(n)}) \\
&\qquad\quad + \mathbb{S}_\alpha(h_1 + (I - R_1 A^* A E_1)\eta^{(n+1/2)} - R_1 A^* A E_2 \mu^{(n+1/2)}) \\
&\qquad\quad - \mathbb{S}_\alpha(h_1 + (I - R_1 A^* A E_1)\eta^{(n+1/2)} - R_1 A^* A E_2 \mu^{(n)})] \\
&\qquad + E_2[\mathbb{S}_\alpha(h_2 + (I - R_2 A^* A E_2)\mu^{(n+1/2)} - R_2 A^* A E_1 \eta^{(n+1)}) \\
&\qquad\quad + \mathbb{S}_\alpha(h_2 + (I - R_2 A^* A E_2)\mu^{(n+1/2)} - R_2 A^* A E_1 \eta^{(n+1/2)}) \\
&\qquad\quad - \mathbb{S}_\alpha(h_2 + (I - R_2 A^* A E_2)\mu^{(n+1/2)} - R_2 A^* A E_1 \eta^{(n+1)})].
\end{aligned}$$

By using the nonexpansiveness of $\mathbb{S}_\alpha$, the boundedness of the operators E_i, R_i, A^*A, and the triangle inequality we obtain,

$$
\begin{aligned}
&\|\mathbb{S}_\alpha(h+\xi^{(n)}) - \mathbb{S}_\alpha(h) - \xi^{(n)}\|_{\ell_2(\mathcal{I})} \\
&\quad \le \|\mathbb{S}_\alpha(h+\xi^{(n)} - A^*A\xi^{(n)}) - \mathbb{S}_\alpha(h) - \xi^{(n)}\|_{\ell_2(\mathcal{I})} \\
&\qquad + \|\mathbb{S}_\alpha(h+\xi^{(n)}) - \mathbb{S}_\alpha(h+\xi^{(n)} - A^*A\xi^{(n)})\|_{\ell_2(\mathcal{I})} \\
&\quad \le \Big(\underbrace{\|\mathbb{S}_\alpha(h_1 + (I - R_1A^*AE_1)\eta^{(n+1/2)} - R_1A^*AE_2\mu^{(n)}) - \mathbb{S}_\alpha(h_1) - \eta^{(n+1/2)}\|_{\ell_2(\mathcal{I}_1)}}_{:=A^{(n)}} \\
&\qquad + \underbrace{\|\mathbb{S}_\alpha(h_2 + (I - R_2A^*AE_2)\mu^{(n+1/2)} - R_2A^*AE_1\eta^{(n+1)}) - \mathbb{S}_\alpha(h_2) - \mu^{(n+1/2)}\|_{\ell_2(\mathcal{I}_2)}}_{:=B^{(n)}} \\
&\qquad + \underbrace{\|\mu^{(n+1/2)} - \mu^{(n)}\|_{\ell_2(\mathcal{I}_2)} + \|\eta^{(n+1)} - \eta^{(n+1/2)}\|_{\ell_2(\mathcal{I}_1)}}_{:=C^{(n)}} \\
&\qquad + \underbrace{\|A^*A\xi^{(n)}\|_{\ell_2(\mathcal{I})}}_{:=D^{(n)}} \Big).
\end{aligned}
$$

The quantities $A^{(n)}$, $B^{(n)}$ vanish for $n \to \infty$ because of (5.12) and (5.13). The quantity $C^{(n)}$ vanishes for $n \to \infty$ because of (5.7), and $D^{(n)}$ vanishes for $n \to \infty$ thanks to Lemma 5.7. □

By combining the previous technical achievements, we can now state the strong convergence.

Theorem 5.9 (Strong convergence). *Algorithm 6 produces a sequence $(u^{(n)})_{n\in\mathbb{N}}$ in $\ell_2(\mathcal{I})$ whose strong accumulation points are minimizers of the functional $\mathcal{J}$. In particular, the set of strong accumulation points is non-empty.*

Proof. Let $u^{(\infty)}$ be a weak accumulation point and let $(u^{(n_j)})_{j\in\mathbb{N}}$ be a subsequence weakly convergent to $u^{(\infty)}$. Let us again denote the latter sequence $(u^{(n)})_{n\in\mathbb{N}}$. With the notation used in this section, by Theorem 5.5 and (5.7) we have that $\xi^{(n)} = E_1\eta^{(n+1/2)} + E_2\mu^{(n+1/2)}$ weakly converges to zero. By Lemma 5.8 we have

$$\lim_{n\to\infty} \|\mathbb{S}_\alpha(h+\xi^{(n)}) - \mathbb{S}_\alpha(h) - \xi^{(n)}\|_{\ell_2(\mathcal{I})} = 0.$$

From Lemma 4.10 we conclude that $\xi^{(n)} = E_1\eta^{(n+1/2)} + E_2\mu^{(n+1/2)}$ converges to zero strongly. Again by (5.7) we have that $(u^{(n)})_{n\in\mathbb{N}}$ converges to $u^{(\infty)}$ strongly. □

5.1.3 A Parallel Domain Decomposition Method

The most natural modification to (5.2) in order to obtain a parallelizable algorithm is to substitute the term $u^{(n+1,L)}$ with $R_1 u^{(n)}$ in the second inner iterations. This makes the inner iterations on $\mathcal{I}_1$ and $\mathcal{I}_2$ mutually independent, hence executable by two processors at the same time. We obtain the following algorithm: Pick an initial $u^{(0)} \in \ell_1(\mathcal{I})$, for example $u^{(0)} = 0$, and iterate

$$\begin{cases} \begin{cases} u_1^{(n+1,0)} = R_1 u^{(n)} \\ u_1^{(n+1,\ell+1)} = \mathbb{S}_\alpha(u_1^{(n+1,\ell)} + R_1 A^*((y - AE_2R_2u^{(n)}) - AE_1u_1^{(n+1,\ell)})) \\ \ell = 0,\dots,L-1 \end{cases} \\ \begin{cases} u_2^{(n+1,0)} = R_2 u^{(n)} \\ u_2^{(n+1,\ell+1)} = \mathbb{S}_\alpha(u_2^{(n+1,\ell)} + R_2 A^*((y - AE_1R_1u^{(n)}) - AE_2u_2^{(n+1,\ell)})) \\ \ell = 0,\dots,M-1 \end{cases} \\ u^{(n+1)} := E_1 u_1^{(n+1,L)} + E_2 u_2^{(n+1,M)}. \end{cases} \tag{5.14}$$

The behavior of this algorithm is somehow bizarre. Indeed, the algorithm usually alternates between the two subsequences given by $u^{(2n)}$ and its consecutive iteration $u^{(2n+1)}$. These two sequences are complementary, in the sense that they encode independent patterns of the solution. In particular, for $u^{(\infty)} = u' + u''$, $u^{(2n)} \approx u'$ and $u^{(2n+1)} \approx u''$ for n not too large. During the iterations and for n large the two subsequences slowly approach each other, merging the complementary features and shaping the final limit which usually coincides with the wanted minimal solution, see Figure 8. Unfortunately, this "oscillatory behavior" makes it impossible to prove monotonicity of the sequence $(\mathcal{J}(u^{(n)}))_{n\in\mathbb{N}}$ and we have no proof of convergence. However,

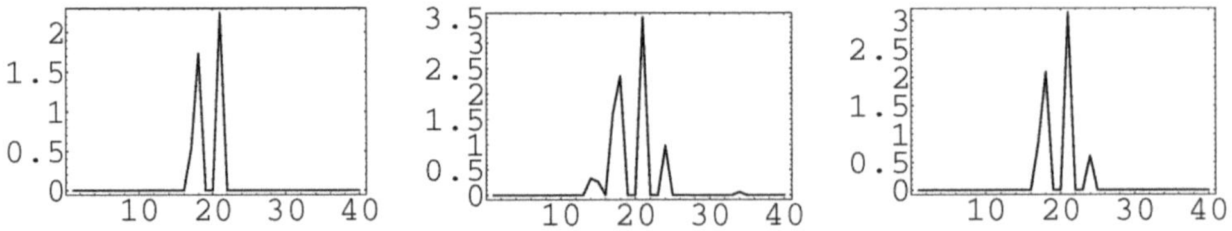

Figure 8. On the left we show $u^{(2n)}$, in the center $u^{(2n+1)}$, and on the right $u^{(\infty)}$. The two consecutive iterations contain different features which will appear in the solution.

since the subsequences are early indicating different features of the final limit, we may modify the algorithm by substituting $u^{(n+1)} := E_1 u_1^{(n+1,L)} + E_2 u_2^{(n+1,M)}$ with $u^{(n+1)} := \frac{(E_1 u_1^{(n+1,L)} + E_2 u_2^{(n+1,M)}) + u^{(n)}}{2}$ which is the average of the current iteration and the previous one. This enforces an early merging of complementary features and leads to the following algorithm:

Algorithm 7. Pick an initial $u^{(0)} \in \ell_1(\mathcal{I})$, for example $u^{(0)} = 0$, and iterate

$$\begin{cases} \begin{cases} u_1^{(n+1,0)} = R_1 u^{(n)} \\ u_1^{(n+1,\ell+1)} = \mathbb{S}_\alpha(u_1^{(n+1,\ell)} + R_1 A^*((y - AE_2R_2u^{(n)}) - AE_1 u_1^{(n+1,\ell)})) \\ \ell = 0, \dots, L-1 \end{cases} \\ \begin{cases} u_2^{(n+1,0)} = R_2 u^{(n)} \\ u_2^{(n+1,\ell+1)} = \mathbb{S}_\alpha(u_2^{(n+1,\ell)} + R_2 A^*((y - AE_1R_1u^{(n)}) - AE_2 u_2^{(n+1,\ell)})) \\ \ell = 0, \dots, M-1 \end{cases} \\ u^{(n+1)} := \frac{(E_1 u_1^{(n+1,L)} + E_2 u_2^{(n+1,M)}) + u^{(n)}}{2}. \end{cases} \tag{5.15}$$

The proof of strong convergence of this algorithm is very similar to the one of Algorithm 6. For the details, we refer the reader to [42].

5.2 Domain Decomposition Methods for Total Variation Minimization

We would like to continue here our parallel discussion of ℓ_1-minimization and total variation minimization as we did in Section 3.1.3. In particular, we would like to show that also for total variation minimization it is possible to formulate domain decomposition methods. Several numerical strategies to efficiently perform total variation minimization have been proposed in the literature as well [2,14,15,17,21,26,68,81,83]. In particular in Section 3.1.3 we mentioned specifically the linearization approach by iteratively least squares presented in [16,32]. These approaches differ significantly, and they provide a convincing view of the interest this problem has been able to generate and of its applicative impact. However, because of their iterative-sequential formulation, none of the mentioned methods is able to address extremely large problems, such as 4D imaging (spatial plus temporal dimensions) from functional magnetic-resonance in nuclear medical imaging, astronomical imaging or global terrestrial seismic tomography, in real-time, or at least in an acceptable computational time. For such large scale simulations we need to address methods which allow us to reduce the problem to a finite sequence of sub-problems of a more manageable size, perhaps computable by one of the methods listed above. We address the interested reader to the broad literature included in [45] for an introduction to domain decomposition methods both for PDEs and convex minimization. This section will make use of more advanced concepts and results in convex analysis, which we will not introduce, and we refer the interested reader to the books [36,50].

Difficulty of the Problem Due to the nonsmoothness and nonadditivity of the total variation with respect to a nonoverlapping domain decomposition (note that the total variation of a function on the whole domain equals the sum of the total variations on the subdomains plus the size of the jumps at the interfaces [45, formula (3.4)]; this is one of the main differences to the situation we already encountered with ℓ_1-minimization), one encounters additional difficulties in showing convergence of such decomposition strategies to global minimizers. In particular, we stress very clearly that well-known approaches as in [13, 18, 79, 80] are not directly applicable to this problem, because either they address additive problems (as the one of ℓ_1-minimization) or smooth convex minimizations, which is *not* the case of total variation minimization. We emphasize that the successful convergence of such alternating algorithms is far from being obvious for nonsmooth and nonadditive problems, as many counterexamples can be constructed, see for instance [82].

The Approach, Results, and Technical Issues In this section we show how to adapt Algorithm 6 and Algorithm 7 to the case of an *overlapping* domain decomposition for total variation minimization. The setting of an overlapping domain decomposition eventually provides us with a framework in which one can successfully prove its convergence to minimizers of $\mathcal{J}$ in (3.67), both in its sequential and parallel forms. Let us stress that to our knowledge this is the first method which addresses a domain decomposition strategy for total variation minimization with a formal, theoretical justification of convergence [45, 60, 84].

The analysis below is performed again for the discrete approximation of the continuous functional (3.66), for ease again denoted $\mathcal{J}$ in (3.67). Essentially we approximate functions u by their sampling on a regular grid and their gradient Du by finite differences ∇u. For ease of presentation, and in order to avoid unnecessary technicalities, we limit our analysis to splitting the problem into two subdomains Ω_1 and Ω_2. This is by no means a restriction. The generalization to multiple domains comes quite natural in our specific setting, see also [45, Remark 5.3]. When dealing with discrete subdomains Ω_i, for technical reasons, we will require a certain splitting property for the total variation, i.e.,

$$\begin{aligned} |\nabla u|(\Omega) &= |\nabla u|_{\Omega_1}|(\Omega_1) + c_1(u|_{(\Omega_2 \setminus \Omega_1) \cup \Gamma_1}), \\ |\nabla u|(\Omega) &= |\nabla u|_{\Omega_2}|(\Omega_2) + c_2(u|_{(\Omega_1 \setminus \Omega_2) \cup \Gamma_2}), \end{aligned} \tag{5.16}$$

where c_1 and c_2 are suitable functions which depend only on the restrictions $u|_{(\Omega_2 \setminus \Omega_1) \cup \Gamma_1}$ and $u|_{(\Omega_1 \setminus \Omega_2) \cup \Gamma_2}$ respectively, see (5.23) (symbols and notations are those introduced in Section 3.1.3). Note that this formula is the discrete analogue of [45, formula (3.4)] in the continuous setting. The simplest examples of discrete domains with such a property are discrete d-dimensional rectangles (d*-orthotopes*). For instance, with our notations, it is easy to check that for $d = 1$ and for Ω being a discrete interval, one computes $c_1(u|_{(\Omega_2 \setminus \Omega_1) \cup \Gamma_1}) = |\nabla u|_{(\Omega_2 \setminus \Omega_1) \cup \Gamma_1}|((\Omega_2 \setminus \Omega_1) \cup \Gamma_1)$, $c_2(u|_{(\Omega_1 \setminus \Omega_2) \cup \Gamma_2}) = |\nabla u|_{(\Omega_1 \setminus \Omega_2) \cup \Gamma_2}|((\Omega_1 \setminus \Omega_2) \cup \Gamma_2))$; it is straightforward to generalize the computation to $d > 1$. Hence, for ease of presentation, we will assume to work with d-orthotope domains, also noting that such decompositions are already sufficient for any practical use in image processing, and stressing that the results can be generalized also to subdomains with different shapes as long as (5.16) is satisfied.

Additional Notations Additionally to the notations already introduced in Section 3.1.3 for the total variation minimization setting, we consider also the closed convex set

$$\mathcal{K} := \{\operatorname{div} p : p \in \mathcal{H}^d, |p(x)|_\infty \leq 1 \text{ for all } x \in \Omega\},$$

where $|p(x)|_\infty = \max\{|p^1(x)|, \dots, |p^d(x)|\}$, and denote $P_{\mathcal{K}}(u) = \arg\min_{v \in \mathcal{K}} \|u - v\|_2$ the *orthogonal projection onto* $\mathcal{K}$.

5.2.1 The Overlapping Domain Decomposition Algorithm

As before we are interested in the minimization of the functional

$$\mathcal{J}(u) := \|Ku - g\|_2^2 + 2\alpha|\nabla(u)|(\Omega), \tag{5.17}$$

where $K \in \mathcal{L}(\mathcal{H})$ is a linear operator, $g \in \mathcal{H}$ is a datum, and $\alpha > 0$ is a fixed constant. We assume that $1 \notin \ker(K)$.

Now, instead of minimizing (5.17) on the whole domain we decompose Ω into two overlapping subdomains Ω_1 and Ω_2 such that $\Omega = \Omega_1 \cup \Omega_2$, $\Omega_1 \cap \Omega_2 \neq \emptyset$, and (5.16) is fulfilled. For consistency of the definitions of gradient and divergence, we assume that also the subdomains Ω_i are discrete d-orthotopes as well as Ω, stressing that this is by no means a restriction, but only for ease of presentation. Due to this domain decomposition $\mathcal{H}$ is split into two closed subspaces $V_j = \{u \in \mathcal{H} : \operatorname{supp}(u) \subset \Omega_j\}$, for $j = 1, 2$. Note that $\mathcal{H} = V_1 + V_2$ is not a direct sum of V_1 and V_2, but just a linear sum of subspaces. Thus any $u \in \mathcal{H}$ has a nonunique representation

$$u(x) = \begin{cases} u_1(x) & x \in \Omega_1 \setminus \Omega_2 \\ u_1(x) + u_2(x) & x \in \Omega_1 \cap \Omega_2 \\ u_2(x) & x \in \Omega_2 \setminus \Omega_1 \end{cases}, \quad u_i \in V_i,\ i = 1, 2. \tag{5.18}$$

We denote by Γ_1 the interface between Ω_1 and $\Omega_2 \setminus \Omega_1$ and by Γ_2 the interface between Ω_2 and $\Omega_1 \setminus \Omega_2$ (the interfaces are naturally defined in the discrete setting).

We introduce the trace operator of the restriction to a boundary Γ_i

$$\operatorname{Tr}|_{\Gamma_i} : V_i \to \mathbb{R}^{\Gamma_i}, \quad i = 1, 2,$$

with $\operatorname{Tr}|_{\Gamma_i} v_i = v_i|_{\Gamma_i}$, the restriction of v_i on Γ_i. Note that $\mathbb{R}^{\Gamma_i}$ is as usual the set of maps from Γ_i to $\mathbb{R}$. The trace operator is clearly a linear and continuous operator. We additionally fix a *bounded uniform partition of unity* (BUPU) $\{\chi_1, \chi_2\} \subset \mathcal{H}$ such that

(a) $\operatorname{Tr}|_{\Gamma_i} \chi_i = 0$ for $i = 1, 2$,

(b) $\chi_1 + \chi_2 = 1$,

(c) $\operatorname{supp} \chi_i \subset \Omega_i$ for $i = 1, 2$,

(d) $\max\{\|\chi_1\|_\infty, \|\chi_2\|_\infty\} = \kappa < \infty$.

We would like to solve

$$\arg\min_{u \in \mathcal{H}} \mathcal{J}(u),$$

by picking an initial $V_1 + V_2 \ni \tilde{u}_1^{(0)} + \tilde{u}_2^{(0)} := u^{(0)} \in \mathcal{H}$, e.g., $\tilde{u}_i^{(0)} = 0, i = 1, 2$, and iterate

$$\begin{cases} u_1^{(n+1)} \approx \arg\min_{\substack{v_1 \in V_1 \\ \operatorname{Tr}|_{\Gamma_1} v_1 = 0}} \mathcal{J}(v_1 + \tilde{u}_2^{(n)}) \\ u_2^{(n+1)} \approx \arg\min_{\substack{v_2 \in V_2 \\ \operatorname{Tr}|_{\Gamma_2} v_2 = 0}} \mathcal{J}(u_1^{(n+1)} + v_2) \\ u^{(n+1)} := u_1^{(n+1)} + u_2^{(n+1)} \\ \tilde{u}_1^{(n+1)} := \chi_1 \cdot u^{(n+1)} \\ \tilde{u}_2^{(n+1)} := \chi_2 \cdot u^{(n+1)}. \end{cases} \tag{5.19}$$

Note that we are minimizing over functions $v_i \in V_i$ for $i = 1, 2$ which vanish on the interior boundaries, i.e., $\operatorname{Tr}|_{\Gamma_i} v_i = 0$. Moreover $u^{(n)}$ is the sum of the local minimizers $u_1^{(n)}$ and $u_2^{(n)}$, which are not uniquely determined on the overlapping part. Therefore we introduced a suitable correction by χ_1 and χ_2 in order to force the subminimizing sequences $(u_1^{(n)})_{n\in\mathbb{N}}$ and $(u_2^{(n)})_{n\in\mathbb{N}}$ to keep uniformly bounded. This issue will be explained in detail below, see Lemma 5.20. From the definition of χ_i, $i = 1, 2$, it is clear that

$$u_1^{(n+1)} + u_2^{(n+1)} = u^{(n+1)} = (\chi_1 + \chi_2) u^{(n+1)} = \tilde{u}_1^{(n+1)} + \tilde{u}_2^{(n+1)}.$$

Note that in general $u_1^{(n)} \neq \tilde{u}_1^{(n)}$ and $u_2^{(n)} \neq \tilde{u}_2^{(n)}$. In (5.19) we use "$\approx$" (the approximation symbol) because in practice we never perform the exact minimization. In the following section we discuss how to realize the approximation to the individual subspace minimizations.

5.2.2 Local Minimization by Lagrange Multipliers

Let us consider, for example, the subspace minimization on Ω_1

$$\begin{aligned} &\arg \min_{\substack{v_1 \in V_1 \\ \operatorname{Tr}|_{\Gamma_1} v_1 = 0}} \mathcal{J}(v_1 + u_2) \\ &= \arg \min_{\substack{v_1 \in V_1 \\ \operatorname{Tr}|_{\Gamma_1} v_1 = 0}} \|K v_1 - (g - K u_2)\|_2^2 + 2\alpha |\nabla(v_1 + u_2)|(\Omega). \end{aligned} \tag{5.20}$$

First of all, observe that $\{u \in \mathcal{H} : \operatorname{Tr}|_{\Gamma_1} u = \operatorname{Tr}|_{\Gamma_1} u_2, \ \mathcal{J}(u) \leq C\} \subset \{\mathcal{J} \leq C\}$, hence the former set is also bounded by assumption (C) and the minimization problem (5.20) has solutions.

It is useful to us to consider again a surrogate functional $\mathcal{J}_1^s$ of $\mathcal{J}$: Assume $a, u_1 \in V_1, u_2 \in V_2$, and define

$$\mathcal{J}_1^s(u_1 + u_2, a) := \mathcal{J}(u_1 + u_2) + \|u_1 - a\|_2^2 - \|K(u_1 - a)\|_2^2. \tag{5.21}$$

A straightforward computation shows that

$$\begin{aligned}\mathcal{J}_1^s(u_1 + u_2, a) &= \|u_1 - (a + (K^*(g - Ku_2 - Ka))|_{\Omega_1})\|_2^2 + 2\alpha|\nabla(u_1 + u_2)|(\Omega) \\ &\quad + \Phi(a, g, u_2),\end{aligned}$$

where Φ is a function of a, g, u_2 only. Note that now the variable u_1 is not anymore affected by the action of K. Consequently, we want to realize an approximate solution to (5.20) by using the following algorithm: For $u_1^{(0)} = \tilde{u}_1^{(0)} \in V_1$,

$$u_1^{(\ell+1)} = \arg \min_{\substack{u_1 \in V_1 \\ \mathrm{Tr}|_{\Gamma_1} u_1 = 0}} \mathcal{J}_1^s(u_1 + u_2, u_1^{(\ell)}), \quad \ell \geq 0. \tag{5.22}$$

Additionally in (5.22) we can restrict the total variation on Ω_1 only, since we have

$$|\nabla(u_1 + u_2)|(\Omega) = |\nabla(u_1 + u_2)|_{\Omega_1}|(\Omega_1) + c_1(u_2|_{(\Omega_2 \setminus \Omega_1) \cup \Gamma_1}), \tag{5.23}$$

where we used (5.16) and the assumption that u_1 vanishes on the interior boundary Γ_1. Hence (5.22) is equivalent to

$$\begin{aligned}&\arg \min_{\substack{u_1 \in V_1 \\ \mathrm{Tr}|_{\Gamma_1} u_1 = 0}} \mathcal{J}_1^s(u_1 + u_2, u_1^{(\ell)}) \\ &\quad = \arg \min_{\substack{u_1 \in V_1 \\ \mathrm{Tr}|_{\Gamma_1} u_1 = 0}} \|u_1 - z_1\|_2^2 + 2\alpha|\nabla(u_1 + u_2)|_{\Omega_1}|(\Omega_1),\end{aligned}$$

where $z_1 = u_1^{(\ell)} + (K^*(g - Ku_2 - Ku_1^{(\ell)}))|_{\Omega_1}$.

Before proving the convergence of this algorithm, we first need to clarify how to practically compute $u_1^{(\ell+1)}$ for $u_1^{(\ell)}$ given. To this end we need to introduce further notions and to recall some useful results.

Generalized Lagrange Multipliers for Nonsmooth Objective Functions We consider the following problem

$$\arg \min_{x \in V} \{F(x) : Gx = b\}, \tag{5.24}$$

where $G : V \to V$ is a linear operator on V. We have the following useful result, which is simply a suitable generalization of the well-known Lagrange multiplier theorem.

Theorem 5.10 ([50, Theorem 2.1.4, p. 305]). *Let* $N = \{G^*\lambda : \lambda \in V\} = \text{Range}(G^*)$. *Then,* $x_0 \in \{x \in V : Gx = b\}$ *solves the constrained minimization problem* (5.24) *if and only if*

$$0 \in \partial F(x_0) + N.$$

We want to exploit Theorem 5.10 in order to produce an algorithmic solution to each iteration step (5.22), which practically stems from the solution of a problem of this type

$$\arg \min_{\substack{u_1 \in V_1 \\ \text{Tr}|_{\Gamma_1} u_1 = 0}} \|u_1 - z_1\|_2^2 + 2\alpha |\nabla(u_1 + u_2|_{\Omega_1})|(\Omega_1).$$

It is well known how to solve this problem if $u_2 \equiv 0$ in Ω_1 and the trace condition is not imposed. For the general case we propose the following solution strategy. In what follows all the involved quantities are restricted to Ω_1, e.g., $z_1 = z_1|_{\Omega_1}, u_2 = u_2|_{\Omega_1}$.

Theorem 5.11 (Oblique thresholding). *For* $u_2 \in V_2$ *and for* $z_1 \in V_1$ *the following statements are equivalent:*

(i) $u_1^* = \arg\min_{\substack{u_1 \in V_1 \\ \text{Tr}|_{\Gamma_1} u_1 = 0}} \|u_1 - z_1\|_2^2 + 2\alpha |\nabla(u_1 + u_2)|(\Omega_1)$;

(ii) *there exists* $\eta \in \text{Range}(\text{Tr}|_{\Gamma_1})^* = \{\eta \in V_1 \text{ with } \text{supp}(\eta) = \Gamma_1\}$ *such that* $0 \in u_1^* - (z_1 - \eta) + \alpha \partial_{V_1} |\nabla(\cdot + u_2)|(\Omega_1)(u_1^*)$;

(iii) *there exists* $\eta \in V_1$ *with* $\text{supp}(\eta) = \Gamma_1$ *such that* $u_1^* = (I - P_{\alpha\mathcal{K}})(z_1 + u_2 - \eta) - u_2 \in V_1$ *and* $\text{Tr}|_{\Gamma_1} u_1^* = 0$;

(iv) *there exists* $\eta \in V_1$ *with* $\text{supp}(\eta) = \Gamma_1$ *such that* $\text{Tr}|_{\Gamma_1} \eta = \text{Tr}|_{\Gamma_1} z_1 + \text{Tr}|_{\Gamma_1} P_{\alpha\mathcal{K}}(\eta - (z_1 + u_2))$ *or equivalently* $\eta = (\text{Tr}|_{\Gamma_1})^* \text{Tr}|_{\Gamma_1}(z_1 + P_{\alpha\mathcal{K}}(\eta - (z_1 + u_2)))$.

We call the solution operation provided by this theorem an *oblique thresholding*, in analogy to the terminology for ℓ_1-minimization (see Lemma 4.1), because it performs a thresholding of the derivatives, i.e., it sets to zero most of the derivatives of $u = u_1 + u_2 \approx z_1$ on Ω_1, provided u_2, which is a fixed vector in V_2.

Proof. Let us show the equivalence between (i) and (ii). The problem in (i) can be reformulated as

$$u_1^* = \arg\min_{u_1 \in V_1} \{F(u_1) := \|u_1 - z_1\|_2^2 + 2\alpha |\nabla(u_1 + u_2)|(\Omega_1), \text{Tr}|_{\Gamma_1} u_1 = 0\}. \quad (5.25)$$

Recall that $\text{Tr}|_{\Gamma_1} : V_1 \to \mathbb{R}^{\Gamma_1}$ is a surjective map with closed range. This means that $(\text{Tr}|_{\Gamma_1})^*$ is injective and that $\text{Range}(\text{Tr}|_{\Gamma_1})^* = \{\eta \in V_1 \text{ with } \text{supp}(\eta) = \Gamma_1\}$ is closed. Using Theorem 5.10 the optimality of u_1^* is equivalent to the existence of $\eta \in \text{Range}(\text{Tr}|_{\Gamma_1})^*$ such that

$$0 \in \partial_{V_1} F(u_1^*) + 2\eta. \quad (5.26)$$

Due to the continuity of $\|u_1 - z_1\|_2^2$ in V_1, we have, by [36, Proposition 5.6], that

$$\partial_{V_1} F(u_1^*) = 2(u_1^* - z_1) + 2\alpha \partial_{V_1} |\nabla(\cdot + u_2)|(\Omega_1)(u_1^*). \tag{5.27}$$

Thus, the optimality of u_1^* is equivalent to

$$0 \in u_1^* - z_1 + \eta + \alpha \partial_{V_1} |\nabla(\cdot + u_2)|(\Omega_1)(u_1^*). \tag{5.28}$$

This concludes the equivalence of (i) and (ii). Let us show now that (iii) is equivalent to (ii). The condition in (iii) can be rewritten as

$$\xi^* = (I - P_{\alpha\mathcal{K}})(z_1 + u_2 - \eta), \quad \xi^* = u_1^* + u_2.$$

Since $|\nabla(\cdot)|(\Omega_1) \geq 0$ is 1-homogeneous and lower-semicontinuous, by [45, Example 4.2.2], the latter is equivalent to

$$0 \in \xi^* - (z_1 + u_2 - \eta) + \alpha \partial_{V_1} |\nabla(\cdot)|(\Omega_1)(\xi^*),$$

and equivalent to (ii). Note that in particular we have

$$\partial_{V_1} |\nabla(\cdot)|(\Omega_1)(\xi^*) = \partial_{V_1} |\nabla(\cdot + u_2)|(\Omega_1)(u_1^*),$$

which is easily shown by a direct computation from the definition of subdifferential. We prove now the equivalence between (iii) and (iv). We have

$$\begin{aligned} u_1^* &= (I - P_{\alpha\mathcal{K}})(z_1 + u_2 - \eta) - u_2 \in V_1, \\ &\qquad \text{(for some } \eta \in V_1 \text{ with } \operatorname{supp}(\eta) = \Gamma_1, \operatorname{Tr}|_{\Gamma_1} u_1^* = 0), \\ &= z_1 - \eta - P_{\alpha\mathcal{K}}(z_1 + u_2 - \eta). \end{aligned}$$

By applying $\operatorname{Tr}|_{\Gamma_1}$ to both sides of the latter equality we get

$$0 = \operatorname{Tr}|_{\Gamma_1} z_1 - \operatorname{Tr}|_{\Gamma_1} \eta - \operatorname{Tr}|_{\Gamma_1} P_{\alpha\mathcal{K}}(z_1 + u_2 - \eta).$$

By observing that $-\operatorname{Tr}|_{\Gamma_1} P_{\alpha\mathcal{K}}(\xi) = \operatorname{Tr}|_{\Gamma_1} P_{\alpha\mathcal{K}}(-\xi)$, we obtain the fixed point equation

$$\operatorname{Tr}|_{\Gamma_1} \eta = \operatorname{Tr}|_{\Gamma_1} z_1 + \operatorname{Tr}|_{\Gamma_1} P_{\alpha\mathcal{K}}(\eta - (z_1 + u_2)). \tag{5.29}$$

Conversely, since all the considered quantities in

$$(I - P_{\alpha\mathcal{K}})(z_1 + u_2 - \eta) - u_2$$

are in V_1, the whole expression is an element in V_1 and hence u_1^* as defined in (iii) is an element in V_1 and $\operatorname{Tr}|_{\Gamma_1} u_1^* = 0$. This shows the equivalence between (iii) and (iv) and therewith finishes the proof. □

We wonder now whether any of the conditions in Theorem 5.11 is indeed practically satisfied. In particular, we want to show that $\eta \in V_1$ as in (iii) or (iv) of the previous theorem is provided as the limit of the following iterative algorithm: for $m \geq 0$

$$\eta^{(0)} \in V_1, \ \mathrm{supp}\,\eta^{(0)} = \Gamma_1 \quad \eta^{(m+1)} = (\mathrm{Tr}\,|_{\Gamma_1})^* \,\mathrm{Tr}\,|_{\Gamma_1}(z_1 + P_{\alpha\mathcal{K}}(\eta^{(m)} - (z_1 + u_2))). \tag{5.30}$$

Proposition 5.12. *The following statements are equivalent:*

(i) *there exists* $\eta \in V_1$ *such that* $\eta = (\mathrm{Tr}\,|_{\Gamma_1})^* \,\mathrm{Tr}\,|_{\Gamma_1}(z_1 + P_{\alpha\mathcal{K}}(\eta - (z_1 + u_2)))$ *(which is in turn the condition* (iv) *of Theorem* 5.11*)*

(ii) *the iteration* (5.30) *converges to any* $\eta \in V_1$ *that satisfies* (5.29).

For the proof of this Proposition we need to recall some well-known notions and results.

Definition 5.13. A nonexpansive map $\mathcal{T} : \mathcal{H} \to \mathcal{H}$ is strongly nonexpansive if for $(u_n - v_n)_{n\in\mathbb{N}}$ bounded and $\|\mathcal{T}(u_n) - \mathcal{T}(v_n)\|_2 - \|u_n - v_n\|_2 \to 0$ we have

$$u_n - v_n - (\mathcal{T}(u_n) - \mathcal{T}(v_n)) \to 0, \quad n \to \infty.$$

Proposition 5.14 (Corollaries 1.3, 1.4, and 1.5 [11]). *Let* $\mathcal{T} : \mathcal{H} \to \mathcal{H}$ *be a strongly nonexpansive map. Then* $\mathrm{Fix}\,\mathcal{T} = \{u \in \mathcal{H} : \mathcal{T}(u) = u\} \neq \emptyset$ *if and only if* $(\mathcal{T}^n u)_{n\in\mathbb{N}}$ *converges to a fixed point* $u_0 \in \mathrm{Fix}\,\mathcal{T}$ *for any choice of* $u \in \mathcal{H}$.

Proof of Proposition 5.12. Projections onto convex sets are strongly nonexpansive [4, Corollary 4.2.3]. Moreover, the composition of strongly nonexpansive maps is strongly nonexpansive [11, Lemma 2.1]. By an application of Proposition 5.14 we immediately have the result, since any map of the type $\mathcal{T}(\xi) = Q(\xi) + \xi_0$ is strongly nonexpansive whenever Q is (this is a simple observation from the definition of strongly nonexpansive maps). Indeed, we are looking for fixed points η satisfying $\eta = (\mathrm{Tr}\,|_{\Gamma_1})^* \,\mathrm{Tr}\,|_{\Gamma_1}(z_1 + P_{\alpha\mathcal{K}}(\eta - (z_1 + u_2)))$ or, equivalently, ξ satisfying

$$\xi = \underbrace{(\mathrm{Tr}\,|_{\Gamma_1})^* \,\mathrm{Tr}\,|_{\Gamma_1} P_{\alpha\mathcal{K}}(\xi)}_{:=Q} - \underbrace{((\mathrm{Tr}\,|_{\Gamma_1})^* \,\mathrm{Tr}\,|_{\Gamma_1} u_2)}_{:=\xi_0},$$

where $\xi = (\mathrm{Tr}\,|_{\Gamma_1})^* \,\mathrm{Tr}\,|_{\Gamma_1}(\eta - (z_1 + u_2))$. □

Convergence of the Subspace Minimization From the results of the previous section it follows that the iteration (5.22) can be explicitly computed by

$$u_1^{(\ell+1)} = S_\alpha(u_1^{(\ell)} + K^*(g - Ku_2 - Ku_1^{(\ell)}) + u_2 - \eta^{(\ell)}) - u_2, \tag{5.31}$$

where, in analogy with the soft-thresholding, we denote $S_\alpha := I - P_{\alpha\mathcal{K}}$, and $\eta^{(\ell)} \in V_1$ is any solution of the fixed point equation

$$\eta = (\operatorname{Tr}|_{\Gamma_1})^* \operatorname{Tr}|_{\Gamma_1} ((u_1^{(\ell)} + K^*(g - Ku_2 - Ku_1^{(\ell)})) \\ - P_{\alpha\mathcal{K}}(u_1^{(\ell)} + K^*(g - Ku_2 - Ku_1^{(\ell)}) + u_2 - \eta)).$$

The computation of $\eta^{(\ell)}$ can be implemented by the algorithm (5.30).

Proposition 5.15. *Assume $u_2 \in V_2$ and $\|K\| < 1$. Then the iteration* (5.31) *converges to a solution $u_1^* \in V_1$ of* (5.20) *for any initial choice of $u_1^{(0)} \in V_1$.*

The proof of this proposition is similar to the one of Theorem 4.7 and it is omitted.

Let us conclude this section mentioning that all the results presented here hold symmetrically for the minimization on V_2, and that the notations should be just adjusted accordingly.

5.2.3 Convergence of the Sequential Alternating Subspace Minimization

In this section we want to prove the convergence of the algorithm (5.19) to minimizers of $\mathcal{J}$. In order to do that, we need a characterization of solutions of the minimization problem (3.67) as the one provided in [81, Proposition 4.1] for the continuous setting. We specify the arguments in [81, Proposition 4.1] for our discrete setting and we highlight the significant differences with respect to the continuous one.

Characterization of Solutions We make the following assumptions:

(A_φ) $\varphi : \mathbb{R} \to \mathbb{R}$ is a convex function, nondecreasing in $\mathbb{R}^+$ such that

(i) $\varphi(0) = 0$.

(ii) There exist $c > 0$ and $b \geq 0$ such that $cz - b \leq \varphi(z) \leq cz + b$, for all $z \in \mathbb{R}^+$.

The particular example we have in mind is simply $\varphi(s) = s$, but we keep a more general notation for uniformity with respect to the continuous version in [81, Proposition 4.1]. In this section we are concerned with the following more general minimization problem

$$\arg\min_{u \in \mathcal{H}} \{\mathcal{J}_\varphi(u) := \|Ku - g\|_2^2 + 2\alpha\varphi(|\nabla u|)(\Omega)\} \tag{5.32}$$

where $g \in \mathcal{H}$ is a datum, $\alpha > 0$ is a fixed constant (in particular for $\varphi(s) = s$).

To characterize the solution of the minimization problem (5.32) we use duality results from [36]. Therefore we recall the definition of the *conjugate (or Legendre transform)* of a function (for example see [36, Definition 4.1, p. 17]):

Definition 5.16. Let V and V^* be two vector spaces placed in the duality by a bilinear pairing denoted by $\langle \cdot, \cdot \rangle$ and $\phi : V \to \mathbb{R}$ be a convex function. The *conjugate function (or Legendre transform)* $\phi^* : V^* \to \mathbb{R}$ is defined by

$$\phi^*(u^*) = \sup_{u \in V} \{\langle u, u^* \rangle - \phi(u)\}.$$

Proposition 5.17. *Let $\zeta, u \in \mathcal{H}$. If the assumption (A_φ) is fulfilled, then $\zeta \in \partial \mathcal{J}_\varphi(u)$ if and only if there exists $M = (M_0, \bar{M}) \in \mathcal{H} \times \mathcal{H}^d$, $\frac{|\bar{M}(x)|}{2\alpha} \le c_1 \in [0, +\infty)$ for all $x \in \Omega$ such that*

$$\langle \bar{M}(x), (\nabla u)(x) \rangle_{\mathbb{R}^d} + 2\alpha\varphi(|(\nabla u)(x)|) + 2\alpha\varphi_1^*\left(\frac{|\bar{M}(x)|}{2\alpha}\right) = 0 \quad \textit{for all } x \in \Omega, \tag{5.33}$$

$$K^* M_0 - \operatorname{div} \bar{M} + \zeta = 0, \tag{5.34}$$

$$-M_0 = 2(Ku - g), \tag{5.35}$$

where φ_1^* is the conjugate function of φ_1 defined by $\varphi_1(s) = \varphi(|s|)$, for $s \in \mathbb{R}$.

If additionally φ is differentiable and $|(\nabla u)(x)| \neq 0$ for $x \in \Omega$, then we can compute $\bar{M}$ as

$$\bar{M}(x) = -2\alpha \frac{\varphi'(|(\nabla u)(x)|)}{|(\nabla u)(x)|} (\nabla u)(x). \tag{5.36}$$

Remark 5.18. (i) For $\varphi(s) = s$ the function φ_1 from Proposition 5.17 turns out to be $\varphi_1(s) = |s|$. Its conjugate function φ_1^* is then given by

$$\varphi_1^*(s^*) = \sup_{s \in \mathbb{R}} \{\langle s^*, s \rangle - |s|\} = \begin{cases} 0 & \text{for } |s^*| \le 1 \\ \infty & \text{else.} \end{cases}$$

Hence condition (5.33) specifies as follows

$$\langle \bar{M}(x), (\nabla u)(x) \rangle_{\mathbb{R}^d} + 2\alpha |(\nabla u)(x)| = 0,$$

and, directly from the proof of Proposition 5.17 in the Appendix, $|\bar{M}(x)| \le 2\alpha$ for all $x \in \Omega$.

(ii) We want to highlight a few important differences with respect to the continuous case. Due to our definition of the gradient and its relationship with the divergence operator $-\operatorname{div} = \nabla^*$ no boundary conditions are needed. Therefore condition (10) of [81, Proposition 4.1] has no discrete analogue in our setting. The continuous total variation of a function can be decomposed into an absolutely continuous part with respect to the Lebesgue measure and a singular part, whereas no singular part appears in the discrete setting. Therefore condition (6) and (7) of [81, Proposition 4.1] have not a discrete analogue either.

(iii) An interesting consequence of Proposition 5.17 is that the map $S_\alpha = (I - P_{\alpha\mathcal{K}})$ is bounded, i.e., $\|S_\alpha(z^k)\|_2 \to \infty$ if and only if $\|z^k\|_2 \to \infty$, for $k \to \infty$. In fact, since

$$S_\alpha(z) = \arg\min_{u\in\mathcal{H}} \|u - z\|_2^2 + 2\alpha|\nabla u|(\Omega),$$

from (5.34) and (5.35), we immediately obtain

$$S_\alpha(z) = z - \frac{1}{2}\operatorname{div}\bar{M},$$

and $\bar{M}$ is uniformly bounded.

Proof of Proposition 5.17. It is clear that $\zeta \in \partial\mathcal{J}_\varphi(u)$ if and only if

$$u = \arg\min_{v\in\mathcal{H}}\{\mathcal{J}_\varphi(v) - \langle\zeta, v\rangle_{\mathcal{H}}\}.$$

Let us consider the following variational problem:

$$\inf_{v\in\mathcal{H}}\{\mathcal{J}_\varphi(v) - \langle\zeta, v\rangle_{\mathcal{H}}\} = \inf_{v\in\mathcal{H}}\{\|Kv - g\|_2^2 + 2\alpha\varphi(|\nabla v|)(\Omega) - \langle\zeta, v\rangle_{\mathcal{H}}\}, \qquad (\mathcal{P}) \tag{5.37}$$

We denote such an infimum by $\inf(\mathcal{P})$. Now we compute $(\mathcal{P}^*)$ the dual of (5.37). Let $\mathcal{F} : \mathcal{H} \to \mathbb{R}$, $\mathcal{G} : \mathcal{H} \times \mathcal{H}^d \to \mathbb{R}$, $\mathcal{G}_1 : \mathcal{H} \to \mathbb{R}$, $\mathcal{G}_2 : \mathcal{H}^d \to \mathbb{R}$, such that

$$\begin{aligned}
\mathcal{F}(v) &= -\langle\zeta, v\rangle_{\mathcal{H}},\\
\mathcal{G}_1(w_0) &= \|w_0 - g\|_2^2,\\
\mathcal{G}_2(\bar{w}) &= 2\alpha\varphi(|\bar{w}|)(\Omega),\\
\mathcal{G}(w) &= \mathcal{G}_1(w_0) + \mathcal{G}_2(\bar{w})
\end{aligned}$$

with $w = (w_0, \bar{w}) \in \mathcal{H} \times \mathcal{H}^d$. Then the dual problem of (5.37) is given by (cf. [36, p. 60])

$$\sup_{p^*\in\mathcal{H}\times\mathcal{H}^d}\{-\mathcal{F}^*(\mathcal{M}^*p^*) - \mathcal{G}^*(-p^*)\}, \qquad (\mathcal{P}^*) \tag{5.38}$$

where $\mathcal{M} : \mathcal{H} \to \mathcal{H} \times \mathcal{H}^d$ is defined by

$$\mathcal{M}v = (Kv, (\nabla v)^1, \ldots, (\nabla v)^d)$$

and $\mathcal{M}^*$ is its adjoint. We denote the supremum in (5.38) by $\sup(\mathcal{P}^*)$. Using the definition of the conjugate function we compute $\mathcal{F}^*$ and $\mathcal{G}^*$. In particular

$$\begin{aligned}
\mathcal{F}^*(\mathcal{M}^*p^*) &= \sup_{v\in\mathcal{H}}\{\langle\mathcal{M}^*p^*, v\rangle_{\mathcal{H}} - \mathcal{F}(v)\} = \sup_{v\in\mathcal{H}}\langle\mathcal{M}^*p^* + \zeta, v\rangle_{\mathcal{H}}\\
&= \begin{cases} 0 & \mathcal{M}^*p^* + \zeta = 0\\ \infty & \text{otherwise}\end{cases}
\end{aligned}$$

where $p^* = (p_0^*, \bar{p}^*)$ and

$$
\begin{aligned}
\mathcal{G}^*(p^*) &= \sup_{w \in \mathcal{H} \times \mathcal{H}^d} \{\langle p^*, w\rangle_{\mathcal{H}\times\mathcal{H}^d} - \mathcal{G}(w)\} \\
&= \sup_{w=(w_0,\bar{w}) \in \mathcal{H} \times \mathcal{H}^d} \{\langle p_0^*, w_0\rangle_{\mathcal{H}} + \langle \bar{p}^*, \bar{w}\rangle_{\mathcal{H}^d} - \mathcal{G}_1(w_0) - \mathcal{G}_2(\bar{w})\} \\
&= \sup_{w_0 \in \mathcal{H}} \{\langle p_0^*, w_0\rangle_{\mathcal{H}} - \mathcal{G}_1(w_0)\} + \sup_{\bar{w} \in \mathcal{H}^d} \{\langle \bar{p}^*, \bar{w}\rangle_{\mathcal{H}^d} - \mathcal{G}_2(\bar{w})\} \\
&= \mathcal{G}_1^*(p_0^*) + \mathcal{G}_2^*(\bar{p}^*).
\end{aligned}
$$

We have that

$$
\mathcal{G}_1^*(p_0^*) = \left\langle \frac{p_0^*}{4} + g, p_0^* \right\rangle_{\mathcal{H}},
$$

and (see [36])

$$
\mathcal{G}_2^*(\bar{p}^*) = 2\alpha\varphi_1^*\left(\frac{|\bar{p}^*|}{2\alpha}\right)(\Omega),
$$

if $\frac{|\bar{p}^*(x)|}{2\alpha} \in \operatorname{Dom} \varphi_1^*$, where φ_1^* is the conjugate function of φ_1 defined by

$$
\varphi_1(s) := \varphi(|s|), \quad s \in \mathbb{R}.
$$

For simplicity we include in the following subsection all the explicit computation of these conjugate functions. We can write (5.38) in the following way

$$
\sup_{p^* \in \mathcal{K}} \left\{ -\left\langle \frac{-p_0^*}{4} + g, -p_0^* \right\rangle_{\mathcal{H}} - 2\alpha\varphi_1^*\left(\frac{|\bar{p}^*|}{2\alpha}\right)(\Omega) \right\}, \tag{5.39}
$$

where

$$
\mathcal{K} = \left\{ p^* \in \mathcal{H} \times \mathcal{H}^d : \frac{|\bar{p}^*(x)|}{2\alpha} \in \operatorname{Dom} \varphi_1^* \text{ for all } x \in \Omega, \mathcal{M}^* p^* + \zeta = 0 \right\}.
$$

The function φ_1 also fulfills assumption (A_φ)(ii) (i.e., there exists $c_1 > 0, b \geq 0$ such that $c_1 z - b \leq \varphi_1(z) \leq c_1 z + b$, for all $z \in \mathbb{R}^+$). The conjugate function of φ_1 is given by $\varphi_1^*(s) = \sup_{z\in\mathbb{R}}\{\langle s, z\rangle - \varphi_1(z)\}$. Using the previous inequalities and that φ_1 is even (i.e., $\varphi_1(z) = \varphi_1(-z)$ for all $z \in \mathbb{R}$) we have

$$
\begin{aligned}
\sup_{z\in\mathbb{R}}\{\langle s, z\rangle - c_1|z| + b\} \geq \sup_{z\in\mathbb{R}}\{\langle s, z\rangle - \varphi_1(z)\} &\geq \sup_{z\in\mathbb{R}}\{\langle s, z\rangle - c_1|z| - b\} \\
&= \begin{cases} -b & \text{if } |s| \leq c_1 \\ \infty & \text{else.} \end{cases}
\end{aligned}
$$

In particular, one can see that $s \in \operatorname{Dom} \varphi_1^*$ if and only if $|s| \leq c_1$.

From $\mathscr{M}^* p^* + \zeta = 0$ we obtain

$$\begin{aligned}\langle \mathscr{M}^* p^*, \omega\rangle_{\mathcal{H}} + \langle \zeta, \omega\rangle_{\mathcal{H}} &= \langle p^*, \mathscr{M}\omega\rangle_{\mathcal{H}^{d+1}} + \langle \zeta, \omega\rangle_{\mathcal{H}} \\ &= \langle p_0^*, K\omega\rangle_{\mathcal{H}} + \langle \bar{p}^*, \nabla\omega\rangle_{\mathcal{H}^d} + \langle \zeta, \omega\rangle_{\mathcal{H}} = 0,\end{aligned}$$

for all $\omega \in \mathcal{H}$. Then, since $\langle \bar{p}^*, \nabla\omega\rangle_{\mathcal{H}^d} = \langle -\operatorname{div} \bar{p}^*, \omega\rangle_{\mathcal{H}}$ (see Section 3.1.3), we have

$$K^* p_0^* - \operatorname{div} \bar{p}^* + \zeta = 0.$$

Hence we can write $\mathscr{K}$ in the following way

$$\mathscr{K} = \left\{ p^* = (p_0^*, \bar{p}^*) \in \mathcal{H} \times \mathcal{H}^d : \frac{|\bar{p}^*(x)|}{2\alpha} \le c_1 \text{ for all } x \in \Omega,\ K^* p_0^* - \operatorname{div} \bar{p}^* + \zeta = 0 \right\}.$$

We now apply the duality results from [36, Theorem III.4.1], since the functional in (5.37) is convex, continuous with respect to $\mathscr{M}v$ in $\mathcal{H} \times \mathcal{H}^d$, and $\inf(\mathscr{P})$ is finite. Then $\inf(\mathscr{P}) = \sup(\mathscr{P}^*) \in \mathbb{R}$ and (5.38) has a solution $M = (M_0, \bar{M}) \in \mathscr{K}$.

Let us assume that u is a solution of (5.37) and M is a solution of (5.38). From $\inf(\mathscr{P}) = \sup(\mathscr{P}^*)$ we get

$$\|Ku - g\|_2^2 + 2\alpha\varphi(|\nabla u|)(\Omega) - \langle \zeta, u\rangle_{\mathcal{H}} = -\left\langle \frac{-M_0}{4} + g, -M_0 \right\rangle_{\mathcal{H}} - 2\alpha\varphi_1^*\left(\frac{|\bar{M}|}{2\alpha}\right)(\Omega) \tag{5.40}$$

where $M = (M_0, \bar{M}) \in \mathcal{H} \times \mathcal{H}^d$, $\frac{|\bar{M}(x)|}{2\alpha} \le c_1$ and $K^* M_0 - \operatorname{div} \bar{M} + \zeta = 0$, which verifies the direct implication of (5.34). In particular

$$-\langle \zeta, u\rangle_{\mathcal{H}} = \langle K^* M_0, u\rangle_{\mathcal{H}} - \langle \operatorname{div} \bar{M}, u\rangle_{\mathcal{H}} = \langle M_0, Ku\rangle_{\mathcal{H}} + \langle \bar{M}, \nabla u\rangle_{\mathcal{H}^d},$$

and

$$\begin{aligned}&\|Ku - g\|_2^2 + \langle M_0, Ku\rangle_{\mathcal{H}} + \langle \bar{M}, \nabla u\rangle_{\mathcal{H}^d} \\ &\quad + 2\alpha\varphi(|\nabla u|)(\Omega) + \left\langle \frac{-M_0}{4} + g, -M_0 \right\rangle_{\mathcal{H}} \\ &\quad + 2\alpha\varphi_1^*\left(\frac{|\bar{M}|}{2\alpha}\right)(\Omega) = 0.\end{aligned} \tag{5.41}$$

Let us write (5.41) again in the following form

$$\begin{aligned}&\sum_{x\in\Omega} |(Ku - g)(x)|^2 + \sum_{x\in\Omega} M_0(x)(Ku)(x) + \sum_{x\in\Omega}\sum_{j=1}^{d} \bar{M}^j(x)(\nabla u)^j(x) \\ &\quad + \sum_{x\in\Omega} 2\alpha\varphi(|(\nabla u)(x)|) \\ &\quad + \sum_{x\in\Omega} \left(\frac{-M_0(x)}{4} + g(x)\right)(-M_0(x)) + \sum_{x\in\Omega} 2\alpha\varphi_1^*\left(\frac{|\bar{M}(x)|}{2\alpha}\right) = 0.\end{aligned} \tag{5.42}$$

Now we have

(i) $2\alpha\varphi(|(\nabla u)(x)|)+\sum_{j=1}^{d} \bar{M}^j(x)(\nabla u)^j(x)+2\alpha\varphi_1^*(\frac{|\bar{M}(x)|}{2\alpha}) \geq 2\alpha\varphi(|(\nabla u)(x)|)-\sum_{j=1}^{d} |\bar{M}^j(x)||(\nabla u)^j(x)| + 2\alpha\varphi_1^*(\frac{|\bar{M}(x)|}{2\alpha}) \geq 0$ by the definition of φ_1^*, since

$$\begin{aligned}2\alpha\varphi_1^*\left(\frac{|\bar{M}(x)|}{2\alpha}\right) &= \sup_{S\in\mathbb{R}^d} \{\langle \bar{M}(x), S\rangle_{\mathbb{R}^d} - 2\alpha\varphi(|S|)\} \\ &= \sup_{S=(S^1,\dots,S^d)\in\mathbb{R}^d} \left\{\sum_{j=1}^{d} |\bar{M}^j(x)||S^j| - 2\alpha\varphi(|S|)\right\}.\end{aligned}$$

(ii) $|(Ku-g)(x)|^2 + M_0(x)(Ku)(x) + (\frac{-M_0(x)}{4} + g(x))(-M_0(x)) = (((Ku)(x)-g(x)))^2 + M_0(x)((Ku)(x)-g(x)) + (\frac{M_0(x)}{2})^2 = (((Ku)(x)-g(x)) + \frac{M_0(x)}{2})^2 \geq 0.$

Hence condition (5.41) reduces to

$$2\alpha\varphi(|(\nabla u)(x)|) + \sum_{j=1}^{d} \bar{M}^j(x)(\nabla u)^j(x) + 2\alpha\varphi_1^*\left(\frac{|\bar{M}(x)|}{2\alpha}\right) = 0, \tag{5.43}$$

$$-M_0(x) = 2((Ku)(x) - g(x)), \tag{5.44}$$

for all $x \in \Omega$. Conversely, if such an $M = (M_0, \bar{M}) \in \mathcal{H} \times \mathcal{H}^d$ with $\frac{|\bar{M}(x)|}{2\alpha} \leq c_1$ exists, which fulfills conditions (5.33)–(5.35), it is clear from previous considerations that equation (5.40) holds. Let us denote the functional on the left-hand side of (5.40) by

$$P(u) := \|Ku - g\|_2^2 + 2\alpha\varphi(|\nabla u|)(\Omega) - \langle \zeta, u\rangle_{\mathcal{H}},$$

and the functional on the right-hand side of (5.40) by

$$P^*(M) := -\left\langle \frac{-M_0}{4} + g, -M_0 \right\rangle_{\mathcal{H}} - 2\alpha\varphi_1^*\left(\frac{|\bar{M}|}{2\alpha}\right)(\Omega).$$

We know that the functional P is the functional of (5.37) and P^* is the functional of (5.38). Hence $\inf P = \inf(\mathcal{P})$ and $\sup P^* = \sup(\mathcal{P}^*)$. Since P is convex, continuous with respect to $\mathcal{M}u$ in $\mathcal{H} \times \mathcal{H}^d$, and $\inf(\mathcal{P})$ is finite we know again from [36, Theorem III.4.1] that $\inf(\mathcal{P}) = \sup(\mathcal{P}^*) \in \mathbb{R}$. We assume that M is no solution of (5.38), i.e., $P^*(M) < \sup(\mathcal{P}^*)$, and u is no solution of (5.37), i.e, $P(u) > \inf(\mathcal{P})$. Then we have that

$$P(u) > \inf(\mathcal{P}) = \sup(\mathcal{P}^*) > P^*(M).$$

Thus (5.40) is valid if and only if M is a solution of (5.38) and u is a solution of (5.37) which amounts to saying that $\zeta \in \partial\mathcal{J}_\varphi(u)$.

If additionally φ is differentiable and $|(\nabla u)(x)| \neq 0$ for $x \in \Omega$, we show that we can compute $\bar{M}(x)$ explicitly. From equation (5.33) (resp. (5.43)) we have

$$2\alpha\varphi_1^*\left(\frac{|-\bar{M}(x)|}{2\alpha}\right) = -\langle \bar{M}(x), (\nabla u)(x)\rangle_{\mathbb{R}^d} - 2\alpha\varphi(|(\nabla u)(x)|). \tag{5.45}$$

From the definition of conjugate function we have

$$\begin{aligned} 2\alpha\varphi_1^*\left(\frac{|-\bar{M}(x)|}{2\alpha}\right) &= 2\alpha \sup_{t\in\mathbb{R}} \left\{\left\langle\frac{|-\bar{M}(x)|}{2\alpha}, t\right\rangle - \varphi_1(t)\right\} \\ &= 2\alpha \sup_{t\geq 0} \left\{\left\langle\frac{|-\bar{M}(x)|}{2\alpha}, t\right\rangle - \varphi_1(t)\right\} \\ &= 2\alpha \sup_{t\geq 0} \sup_{\substack{S\in\mathbb{R}^d \\ |S|=t}} \left\{\left\langle\frac{-\bar{M}(x)}{2\alpha}, S\right\rangle_{\mathbb{R}^d} - \varphi_1(|S|)\right\} \\ &= \sup_{S\in\mathbb{R}^d} \{\langle -\bar{M}(x), S\rangle_{\mathbb{R}^d} - 2\alpha\varphi(|S|)\}. \end{aligned} \tag{5.46}$$

Now, if $|(\nabla u)(x)| \neq 0$ for $x \in \Omega$, then it follows from (5.45) that the supremum is obtained at $S = |(\nabla u)(x)|$ and we have

$$\nabla_S(-\langle \bar{M}(x), S\rangle_{\mathbb{R}^d} - 2\alpha\varphi(|S|)) = 0,$$

which implies

$$\bar{M}^j(x) = -2\alpha\frac{\varphi'(|(\nabla u)(x)|)}{|(\nabla u)(x)|}(\nabla u)^j(x) \quad j = 1, \ldots, d,$$

and verifies (5.36). This finishes the proof. □

Computation of Conjugate Functions Let us compute the conjugate function of the convex function $\mathcal{G}_1(w_0) = \|w_0 - g\|_2^2$. From Definition 5.16 we have

$$\mathcal{G}_1^*(p_0^*) = \sup_{w_0\in\mathcal{H}} \{\langle w_0, p_0^*\rangle_{\mathcal{H}} - \mathcal{G}_1(w_0)\} = \sup_{w_0\in\mathcal{H}} \{\langle w_0, p_0^*\rangle_{\mathcal{H}} - \langle w_0 - g, w_0 - g\rangle_{\mathcal{H}}\}.$$

We set $H(w_0) := \langle w_0, p_0^*\rangle_{\mathcal{H}} - \langle w_0 - g, w_0 - g\rangle_{\mathcal{H}}$. To get the maximum of H we compute the Gâteaux-differential at w_0 of H,

$$H'(w_0) = p_0^* - 2(w_0 - g) = 0$$

and we set it to zero $H'(w_0) = 0$, since $H''(w_0) < 0$, and we get $w_0 = \frac{p_0^*}{2} + g$. Thus we have that

$$\sup_{w_0\in\mathcal{H}} H(w_0) = \left\langle\frac{p_0^*}{4} + g, p_0^*\right\rangle_{\mathcal{H}} = \mathcal{G}_1^*(p_0^*).$$

Now we are going to compute the conjugate function of $\mathcal{G}_2(\bar{w}) = 2\alpha\varphi(|\bar{w}|)(\Omega)$. Associated to our notations we define the space $\mathcal{H}_0^+ = \mathbb{R}_0^{+N_1\times\cdots\times N_d}$. From Definition 5.16 we have

$$\begin{aligned}
\mathcal{G}_2^*(\bar{p}^*) &= \sup_{\bar{w}\in\mathcal{H}^d} \{\langle \bar{w}, \bar{p}^*\rangle_{\mathcal{H}^d} - 2\alpha\varphi(|\bar{w}|)(\Omega)\} \\
&= \sup_{t\in\mathcal{H}_0^+} \sup_{\substack{\bar{w}\in\mathcal{H}^d \\ |\bar{w}(x)|=t(x)}} \{\langle \bar{w}, \bar{p}^*\rangle_{\mathcal{H}^d} - 2\alpha\varphi(|\bar{w}|)(\Omega)\} \\
&= \sup_{t\in\mathcal{H}_0^+} \{\langle t, |\bar{p}^*|\rangle_{\mathcal{H}} - 2\alpha\varphi(t)(\Omega)\}.
\end{aligned}$$

If φ were an even function then

$$\begin{aligned}
\sup_{t\in\mathcal{H}_0^+} \{\langle t, |\bar{p}^*|\rangle_{\mathcal{H}} - 2\alpha\varphi(t)(\Omega)\} &= \sup_{t\in\mathcal{H}} \{\langle t, |\bar{p}^*|\rangle_{\mathcal{H}} - 2\alpha\varphi(t)(\Omega)\} \\
&= 2\alpha \sup_{t\in\mathcal{H}} \left\{\left\langle t, \frac{|\bar{p}^*|}{2\alpha}\right\rangle_{\mathcal{H}} - \varphi(t)(\Omega)\right\} \\
&= 2\alpha\varphi^*\left(\frac{|\bar{p}^*|}{2\alpha}\right)(\Omega),
\end{aligned}$$

where φ^* is the conjugate function of φ.

Unfortunately φ is not even in general. To overcome this difficulty we have to choose a function which is equal to $\varphi(s)$ for $s \geq 0$ and does not change the supremum for $s < 0$. For instance, one can choose $\varphi_1(s) = \varphi(|s|)$ for $s \in \mathbb{R}$. Then we have

$$\begin{aligned}
\sup_{t\in\mathcal{H}_0^+} \{\langle t, |\bar{p}^*|\rangle_{\mathcal{H}} - 2\alpha\varphi(t)(\Omega)\} &= \sup_{t\in\mathcal{H}} \{\langle t, |\bar{p}^*|\rangle_{\mathcal{H}} - 2\alpha\varphi_1(t)(\Omega)\} \\
&= 2\alpha \sup_{t\in\mathcal{H}} \left\{\left\langle t, \frac{|\bar{p}^*|}{2\alpha}\right\rangle_{\mathcal{H}} - \varphi_1(t)(\Omega)\right\} \\
&= 2\alpha\varphi_1^*\left(\frac{|\bar{p}^*|}{2\alpha}\right)(\Omega),
\end{aligned}$$

where φ_1^* is the conjugate function of φ_1. Note that one can also choose $\varphi_1(s) = \varphi(s)$ for $s \geq 0$ and $\varphi_1(s) = \infty$ for $s < 0$.

Convergence Properties We return to the sequential algorithm (5.19). Let us explicitly express the algorithm as follows:

Algorithm 8. Pick an initial $V_1 + V_2 \ni \tilde{u}_1^{(0)} + \tilde{u}_2^{(0)} := u^{(0)} \in \mathcal{H}$, for example, $\tilde{u}_i^{(0)} = 0, i = 1, 2$, and iterate

$$\begin{cases} \begin{cases} u_1^{(n+1,0)} = \tilde{u}_1^{(n)} \\ u_1^{(n+1,\ell+1)} = \arg\min_{\substack{u_1 \in V_1 \\ \mathrm{Tr}|_{\Gamma_1} u_1 = 0}} \mathcal{J}_1^s(u_1 + \tilde{u}_2^{(n)}, u_1^{(n+1,\ell)}) \\ \ell = 0, \dots, L-1 \end{cases} \\ \begin{cases} u_2^{(n+1,0)} = \tilde{u}_2^{(n)} \\ u_2^{(n+1,m+1)} = \arg\min_{\substack{u_2 \in V_2 \\ \mathrm{Tr}|_{\Gamma_2} u_2 = 0}} \mathcal{J}_2^s(u_1^{(n+1,L)} + u_2, u_2^{(n+1,m)}) \\ m = 0, \dots, M-1 \end{cases} \\ u^{(n+1)} := u_1^{(n+1,L)} + u_2^{(n+1,M)} \\ \tilde{u}_1^{(n+1)} := \chi_1 \cdot u^{(n+1)} \\ \tilde{u}_2^{(n+1)} := \chi_2 \cdot u^{(n+1)}. \end{cases} \tag{5.47}$$

Note that we do prescribe a finite number L and M of inner iterations for each subspace respectively and that $u^{(n+1)} = \tilde{u}_1^{(n+1)} + \tilde{u}_2^{(n+1)}$, with $u_i^{(n+1)} \neq \tilde{u}_i^{(n+1)}$, $i = 1, 2$, in general. In this section we want to prove its convergence for any choice of L and M.

Observe that, for $a \in V_i$ and $\|K\| < 1$,

$$\|u_i - a\|_2^2 - \|Ku_i - Ka\|_2^2 \geq C\|u_i - a\|_2^2, \tag{5.48}$$

for $C = (1 - \|K\|^2) > 0$. Hence

$$\mathcal{J}(u) = \mathcal{J}_i^s(u, u_i) \leq \mathcal{J}_i^s(u, a), \tag{5.49}$$

and

$$\mathcal{J}_i^s(u, a) - \mathcal{J}_i^s(u, u_i) \geq C\|u_i - a\|_2^2. \tag{5.50}$$

Proposition 5.19 (Convergence properties). *Let us assume that $\|K\| < 1$. The algorithm in (5.47) produces a sequence $(u^{(n)})_{n\in\mathbb{N}}$ in $\mathcal{H}$ with the following properties:*

(i) $\mathcal{J}(u^{(n)}) > \mathcal{J}(u^{(n+1)})$ *for all $n \in \mathbb{N}$ (unless $u^{(n)} = u^{(n+1)}$);*

(ii) $\lim_{n\to\infty} \|u^{(n+1)} - u^{(n)}\|_2 = 0$;

(iii) *the sequence $(u^{(n)})_{n\in\mathbb{N}}$ has subsequences which converge in $\mathcal{H}$.*

Proof. The proof of this proposition follows the lines of the one of Theorem 5.5, with minor differences, and it is omitted. In particular one can show

$$\left(\sum_{\ell=0}^{L-1} \|u_1^{(n+1,\ell+1)} - u_1^{(n+1,\ell)}\|_2^2 + \sum_{m=0}^{M-1} \|u_2^{(n+1,m+1)} - u_2^{(n+1,m)}\|_2^2\right) \to 0, \quad n \to \infty. \tag{5.51}$$

This limit property will be crucial in the following. □

The use of the partition of unity $\{\chi_1, \chi_2\}$ allows us not only to guarantee the boundedness of $(u^{(n)})_{n\in\mathbb{N}}$, but also of the sequences $(\tilde{u}_1^{(n)})_{n\in\mathbb{N}}$ and $(\tilde{u}_2^{(n)})_{n\in\mathbb{N}}$.

Lemma 5.20. *The sequences $(\tilde{u}_1^{(n)})_{n\in\mathbb{N}}$ and $(\tilde{u}_2^{(n)})_{n\in\mathbb{N}}$ produced by the algorithm (5.47) are bounded, i.e., there exists a constant $\tilde{C} > 0$ such that $\|\tilde{u}_i^{(n)}\|_2 \leq \tilde{C}$ for $i = 1, 2$.*

Proof. From the boundedness of $(u^{(n)})_{n\in\mathbb{N}}$ we have

$$\|\tilde{u}_i^{(n)}\|_2 = \|\chi_i u^{(n)}\|_2 \leq \kappa \|u^{(n)}\|_2 \leq \tilde{C} \quad \text{for } i = 1, 2. \qquad \square$$

From Remark 5.18(iii) we can also show the following auxiliary lemma.

Lemma 5.21. *The sequences $(\eta_1^{(n,L)})_{n\in\mathbb{N}}$ and $(\eta_2^{(n,M)})_{n\in\mathbb{N}}$ are bounded.*

Proof. From previous considerations we know that

$$\begin{aligned} u_1^{(n,L)} &= S_\alpha(z_1^{(n,L-1)} + \tilde{u}_2^{(n-1)} - \eta_1^{(n,L)}) - \tilde{u}_2^{(n-1)}, \\ u_2^{(n,M)} &= S_\alpha(z_2^{(n,M-1)} + u_1^{(n,L)} - \eta_2^{(n,M)}) - u_1^{(n,L)}. \end{aligned}$$

Assume $(\eta_1^{(n,L)})_{n\in\mathbb{N}}$ were unbounded, then by Remark 5.18(iii), also $S_\alpha(z_1^{(n,L-1)} + \tilde{u}_2^{(n-1)} - \eta_1^{(n,L)})$ would be unbounded. Since $(\tilde{u}_2^{(n)})_{n\in\mathbb{N}}$ and $(u_1^{(n,L)})_{n\in\mathbb{N}}$ are bounded by Lemma 5.20 and formula (5.51), we have a contradiction. Thus $(\eta_1^{(n,L)})_{n\in\mathbb{N}}$ has to be bounded. With the same argument we can show that $(\eta_2^{(n,M)})_{n\in\mathbb{N}}$ is bounded. $\square$

We can eventually show the convergence of the algorithm to minimizers of $\mathcal{J}$.

Theorem 5.22 (Convergence to minimizers). *Assume $\|K\| < 1$. Then accumulation points of the sequence $(u^{(n)})_{n\in\mathbb{N}}$ produced by algorithm (5.47) are minimizers of $\mathcal{J}$. If $\mathcal{J}$ has a unique minimizer then the sequence $(u^{(n)})_{n\in\mathbb{N}}$ converges to it.*

Proof. Let us denote $u^{(\infty)}$ the limit of a subsequence. For simplicity, we rename such a subsequence by $(u^{(n)})_{n\in\mathbb{N}}$. From Lemma 5.20 we know that $(\tilde{u}_1^{(n)})_{n\in\mathbb{N}}$, $(\tilde{u}_2^{(n)})_{n\in\mathbb{N}}$ and consequently $(u_1^{(n,L)})_{n\in\mathbb{N}}$, $(u_2^{(n,M)})_{n\in\mathbb{N}}$ are bounded. So the limit $u^{(\infty)}$ can be written as

$$u^{(\infty)} = u_1^{(\infty)} + u_2^{(\infty)} = \tilde{u}_1^{(\infty)} + \tilde{u}_2^{(\infty)} \tag{5.52}$$

where $u_1^{(\infty)}$ is the limit of $(u_1^{(n,L)})_{n\in\mathbb{N}}$, $u_2^{(\infty)}$ is the limit of $(u_2^{(n,M)})_{n\in\mathbb{N}}$, and $\tilde{u}_i^{(\infty)}$ is the limit of $(\tilde{u}_i^{(n)})_{n\in\mathbb{N}}$ for $i = 1, 2$. Now we show that $\tilde{u}_2^{(\infty)} = u_2^{(\infty)}$. By using the triangle inequality, from (5.51) it directly follows that

$$\|u_2^{(n+1,M)} - \tilde{u}_2^{(n)}\|_2 \to 0, \quad n \to \infty. \tag{5.53}$$

Moreover, since $\chi_2 \in V_2$ is a fixed vector which is independent of n, we obtain from Proposition 5.19(ii) that

$$\|\chi_2(u^{(n)} - u^{(n+1)})\|_2 \to 0, \quad n \to \infty,$$

and hence

$$\|\tilde{u}_2^{(n)} - \tilde{u}_2^{(n+1)}\|_2 \to 0, \quad n \to \infty. \tag{5.54}$$

Putting (5.53) and (5.54) together and noting that

$$\|u_2^{(n+1,M)} - \tilde{u}_2^{(n)}\|_2 + \|\tilde{u}_2^{(n)} - \tilde{u}_2^{(n+1)}\|_2 \geq \|u_2^{(n+1,M)} - \tilde{u}_2^{(n+1)}\|_2,$$

we have

$$\|u_2^{(n+1,M)} - \tilde{u}_2^{(n+1)}\|_2 \to 0, \quad n \to \infty, \tag{5.55}$$

which means that the sequences $(u_2^{(n,M)})_{n\in\mathbb{N}}$ and $(\tilde{u}_2^{(n)})_{n\in\mathbb{N}}$ have the same limit, i.e., $\tilde{u}_2^{(\infty)} = u_2^{(\infty)}$, which we denote by $u_2^{(\infty)}$. Then from (5.55) and (5.52) it directly follows that $\tilde{u}_1^{(\infty)} = u_1^{(\infty)}$.

As in the proof of the oblique thresholding theorem we set

$$F_1(u_1^{(n+1,L)}) := \|u_1^{(n+1,L)} - z_1^{(n+1,L)}\|_2^2 + 2\alpha|\nabla(u_1^{(n+1,L)} + \tilde{u}_2^{(n)}|_{\Omega_1})|(\Omega_1),$$

where

$$z_1^{(n+1,L)} := u_1^{(n+1,L-1)} + K^*(g - K\tilde{u}_2^{(n)} - Ku_1^{(n+1,L-1)})|_{\Omega_1}.$$

The optimality condition for $u_1^{(n+1,L)}$ is

$$0 \in \partial_{V_1} F_1(u_1^{(n+1,L)}) + 2\eta_1^{(n+1,L)},$$

where

$$\eta_1^{(n+1,L)} = (\operatorname{Tr}|_{\Gamma_1})^* \operatorname{Tr}|_{\Gamma_1}(z_1^{(n+1,L)} + P_{\alpha\mathcal{K}}(\eta_1^{(n+1,L)} - z_1^{(n+1,L)} - \tilde{u}_2^{(n)})).$$

In order to use the characterization of elements in the subdifferential of $|\nabla u|(\Omega)$, i.e., Proposition 5.17, we have to rewrite the minimization problem for F_1. More precisely, we define

$$\hat{F}_1(\xi_1^{(n+1,L)}) := \|\xi_1^{(n+1,L)} - \tilde{u}_2^{(n)}|_{\Omega_1} - z_1^{(n+1,L)}\|_2^2 + 2\alpha|\nabla(\xi_1^{(n+1,L)})|(\Omega_1),$$

for $\xi_1^{(n+1,L)} \in V_1$ with $\operatorname{Tr}|_{\Gamma_1}\xi_1^{(n+1,L)} = \tilde{u}_2^{(n)}$. Then the optimality condition for $\xi_1^{(n+1,L)}$ is

$$0 \in \partial\hat{F}_1(\xi_1^{(n+1,L)}) + 2\eta_1^{(n+1,L)}. \tag{5.56}$$

Note that indeed $\xi_1^{(n+1,L)}$ is optimal if and only if $u_1^{(n+1,L)} = \xi_1^{(n+1,L)} - \tilde{u}_2^{(n)}|_{\Omega_1}$ is optimal.

Analogously we define

$$\hat{F}_2(\xi_2^{(n+1,M)}) := \|\xi_2^{(n+1,M)} - u_1^{(n+1,L)}|_{\Omega_2} - z_2^{(n+1,M)}\|_2^2 + 2\alpha|\nabla(\xi_2^{(n+1,M)})|(\Omega_2),$$

for $\xi_2^{(n+1,M)} \in V_2$ with $\operatorname{Tr}|_{\Gamma_2}\xi_2^{(n+1,M)} = u_1^{(n+1,L)}$, and the optimality condition for $\xi_2^{(n+1,M)}$ is

$$0 \in \partial\hat{F}_2(\xi_2^{(n+1,M)}) + 2\eta_2^{(n+1,M)}, \tag{5.57}$$

where

$$\eta_2^{(n+1,M)} = (\operatorname{Tr}|_{\Gamma_2})^* \operatorname{Tr}|_{\Gamma_2}(z_2^{(n+1,M)} + P_{\alpha\mathcal{K}}(\eta_2^{(n+1,M)} - z_2^{(n+1,M)} - u_1^{(n+1,L)})).$$

Let us recall that now we are considering functionals as in Proposition 5.17 with $\varphi(s) = s$, $K = I$, and $\Omega = \Omega_i$, $i = 1, 2$. From Proposition 5.17 and Remark 5.18 we get that $\xi_1^{(n+1,L)}$, and consequently $u_1^{(n+1,L)}$ is optimal, i.e., $-2\eta_1^{(n+1,L)} \in \partial\hat{F}_1(\xi_1^{(n+1,L)})$, if and only if there exists an $M_1^{(n+1)} = (M_{0,1}^{(n+1)}, \bar{M}_1^{(n+1)}) \in V_1 \times V_1^d$ with $|\bar{M}_1^{(n+1)}(x)| \le 2\alpha$ for all $x \in \Omega_1$ such that

$$\langle \bar{M}_1^{(n+1)}(x), (\nabla(u_1^{(n+1,L)} + \tilde{u}_2^{(n)}))(x)\rangle_{\mathbb{R}^d} + 2\alpha\varphi(|(\nabla(u_1^{(n+1,L)} + \tilde{u}_2^{(n)}))(x)|) = 0, \tag{5.58}$$

$$-2(u_1^{(n+1,L)}(x) - z_1^{(n+1,L)}(x)) - \operatorname{div}\bar{M}_1^{(n+1)}(x) - 2\eta_1^{(n+1,L)}(x) = 0, \tag{5.59}$$

for all $x \in \Omega_1$. Analogously we get that $\xi_2^{(n+1,M)}$, and consequently $u_2^{(n+1,M)}$ is optimal, i.e., $-2\eta_2^{(n+1,M)} \in \partial\hat{F}_2(\xi_2^{(n+1,M)})$, if and only if there exists an $M_2^{(n+1)} = (M_{0,2}^{(n+1)}, \bar{M}_2^{(n+1)}) \in V_2 \times V_2^d$ with $|\bar{M}_2^{(n+1)}(x)| \le 2\alpha$ for all $x \in \Omega_2$ such that

$$\begin{aligned} &\langle \bar{M}_2^{(n+1)}(x), (\nabla(u_1^{(n+1,L)} + u_2^{(n+1,M)}))(x)\rangle_{\mathbb{R}^d} \\ &\qquad + 2\alpha\varphi(|(\nabla(u_1^{(n+1,L)} + \tilde{u}_2^{(n+1,M)}))(x)|) = 0, \end{aligned} \tag{5.60}$$

$$\begin{aligned} -2(u_2^{(n+1,M)}(x) - z_2^{(n+1,M)}(x)) - \operatorname{div}\bar{M}_2^{(n+1)}(x)& \\ -2\eta_2^{(n+1,M)}(x) = 0&, \end{aligned} \tag{5.61}$$

for all $x \in \Omega_2$. Since $(\bar{M}_1^{(n)}(x))_{n\in\mathbb{N}}$ is bounded for all $x \in \Omega_1$ and $(\bar{M}_2^{(n)}(x))_{n\in\mathbb{N}}$ is bounded for all $x \in \Omega_2$, there exist convergent subsequences $(\bar{M}_1^{(n_k)}(x))_{k\in\mathbb{N}}$ and $(\bar{M}_2^{(n_k)}(x))_{k\in\mathbb{N}}$. Let us denote $\bar{M}_1^{(\infty)}(x)$ and $\bar{M}_2^{(\infty)}(x)$ the respective limits of the sequences. For simplicity we rename such sequences by

$$(\bar{M}_1^{(n)}(x))_{n\in\mathbb{N}} \quad \text{and} \quad (\bar{M}_2^{(n)}(x))_{n\in\mathbb{N}}.$$

Note that, by Lemma 5.21 (or simply from (5.59) and (5.61)) the sequences $(\eta_1^{(n,L)})_{n\in\mathbb{N}}$ and $(\eta_2^{(n,M)})_{n\in\mathbb{N}}$ are also bounded. Hence there exist convergent subsequences which we denote, for simplicity, again by $(\eta_1^{(n,L)})_{n\in\mathbb{N}}$ and $(\eta_2^{(n,M)})_{n\in\mathbb{N}}$ with limits $\eta_i^{(\infty)}$, $i = 1, 2$. By taking the limits for $n \to \infty$ in (5.58)–(5.61) we obtain for all $x \in \Omega_1$

$$\begin{aligned}\langle \bar{M}_1^{(\infty)}(x), \nabla(u_1^{(\infty)} + u_2^{(\infty)})(x)\rangle_{\mathbb{R}^d} + 2\alpha\varphi(|\nabla(u_1^{(\infty)} + u_2^{(\infty)})(x)|) = 0,\\ -2(u_1^{(\infty)}(x) - z_1^{(\infty)}(x)) - \operatorname{div} \bar{M}_1^{(\infty)}(x) - 2\eta_1^{(\infty)}(x) = 0,\end{aligned}$$

and for all $x \in \Omega_2$

$$\begin{aligned}\langle \bar{M}_2^{(\infty)}(x), \nabla(u_1^{(\infty)} + u_2^{(\infty)})(x)\rangle_{\mathbb{R}^d} + 2\alpha\varphi(|\nabla(u_1^{(\infty)} + u_2^{(\infty)})(x)|) = 0,\\ -2(u_2^{(\infty)}(x) - z_2^{(\infty)}(x)) - \operatorname{div} \bar{M}_2^{(\infty)}(x) - 2\eta_2^{(\infty)}(x) = 0.\end{aligned}$$

Since $\operatorname{supp} \eta_1^{(\infty)} = \Gamma_1$ and $\operatorname{supp} \eta_2^{(\infty)} = \Gamma_2$ we have

$$\begin{aligned}\langle \bar{M}_1^{(\infty)}(x), \nabla u^{(\infty)}(x)\rangle_{\mathbb{R}^d} + 2\alpha\varphi(|\nabla u^{(\infty)}(x)|) = 0 \quad &\text{for all } x \in \Omega_1,\\ -2K^*((Ku^{(\infty)})(x) - g(x)) - \operatorname{div} \bar{M}_1^{(\infty)}(x) = 0 \quad &\text{for all } x \in \Omega_1 \setminus \Gamma_1,\end{aligned} \tag{5.62}$$

$$\begin{aligned}\langle \bar{M}_2^{(\infty)}(x), \nabla u^{(\infty)}(x)\rangle_{\mathbb{R}^d} + 2\alpha\varphi(|\nabla u^{(\infty)}(x)|) = 0 \quad &\text{for all } x \in \Omega_2,\\ -2K^*((Ku^{(\infty)})(x) - g^{(\infty)}(x)) - \operatorname{div} \bar{M}_2^{(\infty)}(x) = 0 \quad &\text{for all } x \in \Omega_2 \setminus \Gamma_2.\end{aligned} \tag{5.63}$$

Observe now that from Proposition 5.17 we also have that $0 \in \mathcal{J}(u^{(\infty)})$ if and only if there exists $M^{(\infty)} = (M_0^{(\infty)}, \bar{M}^{(\infty)})$ with $|\bar{M}_0^{(\infty)}(x)| \le 2\alpha$ for all $x \in \Omega$ such that

$$\begin{aligned}\langle \bar{M}^{(\infty)}(x), (\nabla(u^{(\infty)})(x)\rangle_{\mathbb{R}^d} + 2\alpha\varphi(|(\nabla(u^{(\infty)})(x)|) = 0 \quad &\text{for all } x \in \Omega,\\ -2K^*((Ku^{(\infty)})(x) - g^{(\infty)}(x)) - \operatorname{div} \bar{M}^{(\infty)}(x) = 0 \quad &\text{for all } x \in \Omega.\end{aligned} \tag{5.64}$$

Note that $\bar{M}_j^{(\infty)}(x)$, $j = 1, 2$, for $x \in \Omega_1 \cap \Omega_2$ satisfies both (5.62) and (5.63). Hence let us choose

$$M^{(\infty)}(x) = \begin{cases} M_1^{(\infty)}(x) & \text{if } x \in \Omega_1 \setminus \Gamma_1\\ M_2^{(\infty)}(x) & \text{if } x \in (\Omega_2 \setminus \Omega_1) \cup \Gamma_1.\end{cases}$$

With this choice of $M^{(\infty)}$ equations (5.62)–(5.64) are valid and hence $u^{(\infty)}$ is optimal in Ω. □

Remark 5.23. (i) If $\nabla u^{(\infty)}(x) \neq 0$ for $x \in \Omega_j$, $j = 1, 2$, then $\bar{M}_j^{(\infty)}$ is given as in equation (5.36) by $\bar{M}_j^{(\infty)}(x) = -2\alpha \frac{(\nabla u^{(\infty)}|_{\Omega_j})(x)}{|(\nabla u^{(\infty)}|_{\Omega_j})(x)|}$.

(ii) The boundedness of the sequences $(\tilde{u}_1^{(n)})_{n\in\mathbb{N}}$ and $(\tilde{u}_2^{(n)})_{n\in\mathbb{N}}$ has been technically used for showing the existence of an optimal decomposition $u^{(\infty)} = u_1^{(\infty)} + u_2^{(\infty)}$ in the proof of Theorem 5.22. Their boundedness is guaranteed as in Lemma 5.20 by the use of the partition of unity $\{\chi_1, \chi_2\}$. Let us emphasize that there is no way of obtaining the boundedness of the local sequences $(u_1^{(n,L)})_{n\in\mathbb{N}}$ and $(u_2^{(n,M)})_{n\in\mathbb{N}}$ otherwise. In Figure 11 we show that the local sequences can become unbounded in case we do not modify them by means of the partition of unity.

(iii) Note that for deriving the optimality condition (5.64) for $u^{(\infty)}$ we combined the respective conditions (5.62) and (5.63) for $u_1^{(\infty)}$ and $u_2^{(\infty)}$. In doing that, we strongly took advantage of the overlapping property of the subdomains, hence avoiding a fine analysis of $\eta_1^{(\infty)}$ and $\eta_2^{(\infty)}$ on the interfaces Γ_1 and Γ_2.

5.2.4 A Parallel Algorithm

The parallel version of the previous algorithm (5.47) reads as follows:

Algorithm 9. Pick an initial $V_1 + V_2 \ni \tilde{u}_1^{(0)} + \tilde{u}_2^{(0)} := u^{(0)} \in \mathcal{H}$, for example, $\tilde{u}_i^{(0)} = 0$, $i = 1, 2$, and iterate

$$\begin{cases} \begin{cases} u_1^{(n+1,0)} = \tilde{u}_1^{(n)} \\ u_1^{(n+1,\ell+1)} = \arg\min_{\substack{u_1\in V_1 \\ \mathrm{Tr}|_{\Gamma_1} u_1 = 0}} \mathcal{J}_1^s(u_1 + \tilde{u}_2^{(n)}, u_1^{(n+1,\ell)}) \\ \ell = 0, \ldots, L-1 \end{cases} \\ \begin{cases} u_2^{(n+1,0)} = \tilde{u}_2^{(n)} \\ u_2^{(n+1,m+1)} = \arg\min_{\substack{u_2\in V_2 \\ \mathrm{Tr}|_{\Gamma_2} u_2 = 0}} \mathcal{J}_2^s(\tilde{u}_1^{(n)} + u_2, u_2^{(n+1,m)}) \\ m = 0, \ldots, M-1 \end{cases} \\ u^{(n+1)} := \frac{(u_1^{(n+1,L)} + u_2^{(n+1,M)}) + u^{(n)}}{2} \\ \tilde{u}_1^{(n+1)} := \chi_1 \cdot u^{(n+1)} \\ \tilde{u}_2^{(n+1)} := \chi_2 \cdot u^{(n+1)}. \end{cases} \tag{5.65}$$

As for ℓ_1-minimization also for this version the parallel algorithm is shown to converge in a similar way as its sequential counterpart.

5.2.5 Applications and Numerics

In this section we shall present the application of the sequential algorithm (5.19) for the minimization of $\mathcal{J}$ in one and two dimensions. In particular, we show how to implement the dual method of Chambolle [14] in order to compute the orthogonal projection $P_{\alpha\mathcal{K}}(g)$ in the oblique thresholding. Furthermore we present numerical examples for image *inpainting*, i.e., the recovery of missing parts of images by minimal total variation interpolation, and compressed sensing, i.e., the nonadaptive compressed acquisition of images for a classical toy problem inspired by magnetic resonance imaging (MRI) [57]. The numerical examples of this section and respective Matlab codes can be found at [85].

Computation of $P_{\alpha\mathcal{K}}(g)$ To solve the subiterations in (5.19), we compute the minimizer by means of oblique thresholding. More precisely, let us denote $u_2 = \tilde{u}_2^{(n)}$, $u_1 = u_1^{(n+1,\ell+1)}$, and $z_1 = u_1^{(n+1,\ell)} + K^*(g - Ku_2 - Ku_1^{(n+1,\ell)})$. We shall compute the minimizer u_1 of the first subminimization problem by

$$u_1 = (I - P_{\alpha\mathcal{K}})(z_1 + u_2 - \eta) - u_2 \in V_1,$$

for an $\eta \in V_1$ with $\operatorname{supp} \eta = \Gamma_1$ which fulfills

$$\operatorname{Tr}|_{\Gamma_1}(\eta) = \operatorname{Tr}|_{\Gamma_1}(z_1 + P_{\alpha\mathcal{K}}(\eta - z_1 - u_2)).$$

Hence the element $\eta \in V_1$ is a limit of the corresponding fixed point iteration for $m \geq 0$,

$$\eta^{(0)} \in V_1,\ \operatorname{supp} \eta^{(0)} = \Gamma_1, \quad \eta^{(m+1)} = (\operatorname{Tr}|_{\Gamma_1})^* \operatorname{Tr}|_{\Gamma_1}(z_1 + P_{\alpha\mathcal{K}}(\eta^{(m)} - z_1 - u_2)). \tag{5.66}$$

Here $\mathcal{K}$ is defined as in Section 5.2, i.e.,

$$\mathcal{K} = \{\operatorname{div} p : p \in \mathcal{H}^d, |p(x)|_\infty \leq 1 \text{ for all } x \in \Omega\}.$$

To compute the projection onto $\alpha\mathcal{K}$ in the oblique thresholding we use an algorithm proposed by Chambolle in [14]. His algorithm is based on considerations of the convex conjugate of the total variation and on exploiting the corresponding optimality condition. It amounts to compute $P_{\alpha\mathcal{K}}(g)$ approximately by $\alpha \operatorname{div} p^{(n)}$, where $p^{(n)}$ is the nth iterate of the following semi-implicit gradient descent algorithm:

Choose $\tau > 0$, let $p^{(0)} = 0$ and, for any $n \geq 0$, iterate

$$p^{(n+1)}(x) = \frac{p^{(n)}(x) + \tau(\nabla(\operatorname{div} p^{(n)} - g/\alpha))(x)}{1 + \tau|(\nabla(\operatorname{div} p^{(n)} - g/\alpha))(x)|}.$$

For $\tau > 0$ sufficiently small, i.e., $\tau < 1/8$, the iteration $\alpha \operatorname{div} p^{(n)}$ was shown to converge to $P_{\alpha\mathcal{K}}(g)$ as $n \to \infty$ (compare [14, Theorem 3.1]). Let us stress that we propose this algorithm here just for the ease of its presentation; its choice for the approximation of projections is of course by no means a restriction and one may want to implement other recent, and perhaps faster strategies, e.g., [15, 26, 49, 68, 83].

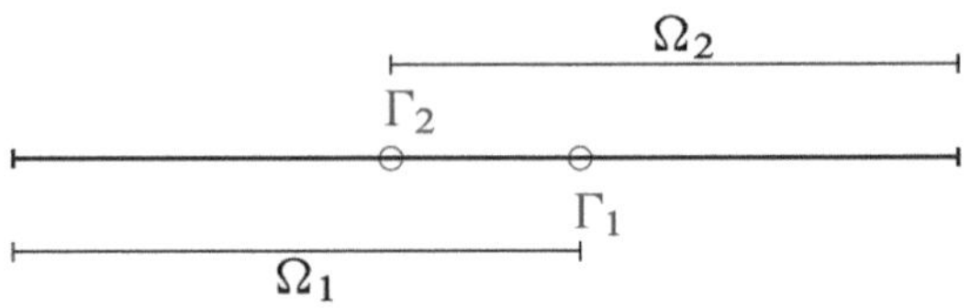

Figure 9. Overlapping domain decomposition in 1D.

Domain Decompositions In one dimension the domain Ω is a set of N equidistant points on an interval $[a, b]$, i.e., $\Omega = \{a = x_1, \dots, x_N = b\}$ and is split into two overlapping intervals Ω_1 and Ω_2. Let $|\Omega_1 \cap \Omega_2| =: G$ be the size of the overlap of Ω_1 and Ω_2. Then we set $\Omega_1 = \{a = x_1, \dots, x_{n_1}\}$ and $\Omega_2 = \{x_{n_1-G+1}, \dots, x_N = b\}$ with $|\Omega_1| := n_1 = \lceil \frac{N+G}{2} \rceil$. The interfaces Γ_1 and Γ_2 are located in $i = n_1 + 1$ and $n_1 - G$ respectively (cf. Figure 9). The auxiliary functions χ_1 and χ_2 can be chosen in the following way (cf. Figure 10):

$$\chi_1(x_i) = \begin{cases} 1 & x_i \in \Omega_1 \setminus \Omega_2, \\ 1 - \frac{1}{G}(i - (n_1 - G + 1)) & x_i \in \Omega_1 \cap \Omega_2, \end{cases}$$

$$\chi_2(x_i) = \begin{cases} 1 & x_i \in \Omega_2 \setminus \Omega_1, \\ \frac{1}{G}(i - (n_1 - G + 1)) & x_i \in \Omega_1 \cap \Omega_2. \end{cases}$$

Note that $\chi_1(x_i) + \chi_2(x_i) = 1$ for all $x_i \in \Omega$ (i.e for all $i = 1, \dots, N$).

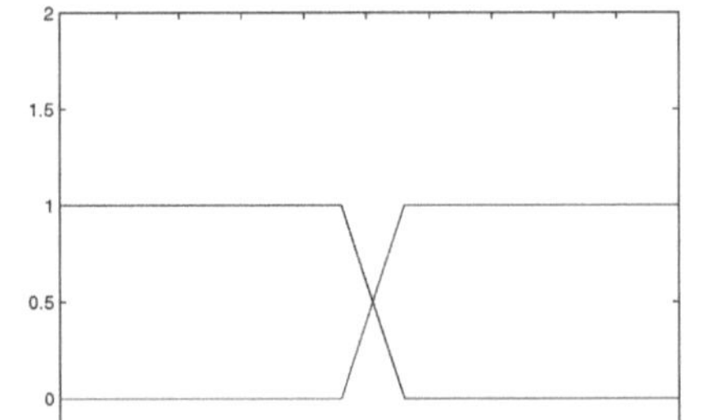

Figure 10. Auxiliary functions χ_1 and χ_2 for an overlapping domain decomposition with two subdomains.

In two dimensions the domain Ω, i.e., the set of $N_1 \times N_2$ equidistant points on the 2-dimensional rectangle $[a, b] \times [c, d]$, is split in an analogous way with respect to its rows. In particular we have Ω_1 and Ω_2 consist of equidistant points on $[a, x_{n_1}] \times [c, d]$ and $[x_{n_1-G+1}, b] \times [c, d]$ respectively. The splitting in more than two domains is done similarly. The auxiliary functions χ_i can be chosen in an analogous way as in the one dimensional case.

Numerical Experiments In the following we present numerical examples for the sequential algorithm (5.47) in two particular applications: signal interpolation/image inpainting, and compressed sensing [85].

In Figure 11 we show a partially corrupted 1D signal on an interval Ω of 100 sampling points, with a loss of information on an interval $D \subset \Omega$. The domain D of the missing signal points is marked in green. These signal points are reconstructed by total variation interpolation, i.e., minimizing the functional $\mathcal{J}$ in (3.67) with $\alpha = 1$ and $Ku = 1_{\Omega \setminus D} \cdot u$, where $1_{\Omega \setminus D}$ is the indicator function of $\Omega \setminus D$. In Figure 11 we also illustrate the effect of implementing the BUPU within the domain decomposition algorithm.

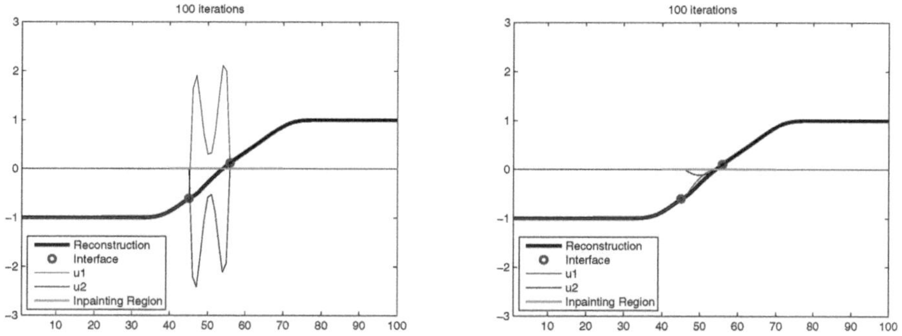

(a) (b)

Figure 11. Here we present two numerical experiments related to the interpolation of a 1D signal by total variation minimization. The original signal is only provided outside of the green subinterval. On the left we show an application of algorithm (5.47) when no correction with the partition of unity is provided. In this case, the sequence of the local iterations $u_1^{(n)}, u_2^{(n)}$ is unbounded. On the right we show an application of algorithm (5.47) with the use of the partition of unity which enforces the uniform boundedness of the local iterations $u_1^{(n)}, u_2^{(n)}$.

Figure 12 shows an example of the domain decomposition algorithm (5.47) for total variation inpainting. As for the 1D example in Figure 11 the operator K is a multiplier, i.e., $Ku = 1_{\Omega \setminus D} \cdot u$, where Ω denotes the rectangular image domain and $D \subset \Omega$ the missing domain in which the original image content got lost. The regularization parameter α is fixed at the value 10^{-2}. In Figure 12 the missing domain D is the black writing which covers parts of the image. Here, the image domain of size 449×570 pixels is split into five overlapping subdomains with an overlap size $G = 28 \times 570$. Finally, in Figure 13 we illustrate the successful application of our domain decomposition algorithm (5.47) for a compressed sensing problem. Here, we consider a medical-type image (the so-called *Logan–Shepp phantom*) and its reconstruction from only partial Fourier data. In this case the linear operator $K = S \circ \mathcal{F}$, where $\mathcal{F}$ denotes the $2D$ Fourier matrix and S is a *downsampling operator* which

Figure 12. This figure shows an application of algorithm (5.47) for image inpainting. In this simulation the problem was split into five subproblems on overlapping subdomains.

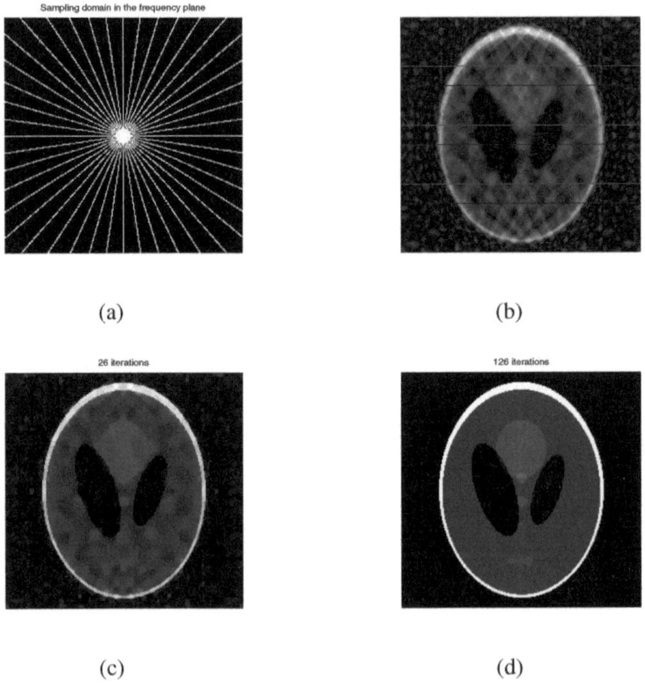

Figure 13. We show an application of algorithm (5.47) in a classical compressed sensing problem for recovering piecewise constant medical-type images from given partial Fourier data. In this simulation the problem was split via decomposition into four overlapping subdomains. On the top-left figure, we show the sampling data of the image in the Fourier domain. On the top-right the back-projection provided by the sampled frequency data together with the highlighted partition of the physical domain into four subdomains is shown. The bottom figures present intermediate iterations of the algorithm, i.e., $u^{(26)}$ and $u^{(125)}$.

selects only a few frequencies as output. We minimize $\mathcal{J}$ with α set at 0.4×10^{-2}. In the application of algorithm (5.47) the image domain of size 256×256 pixels is split into four overlapping subdomains with an overlap size $G = 20 \times 256$.

Acknowledgments. I'm very grateful to the colleagues Antonin Chambolle, Ronny Ramlau, Holger Rauhut, Jared Tanner, Gerd Teschke, and Mariya Zhariy for accepting the invitation to prepare and present a course at the Johann Radon Institute for Computational and Applied Mathematics (RICAM) of the Austrian Academy of Sciences during the Summer School "Theoretical Foundations and Numerical Methods for Sparse Recovery". Also these lecture notes were prepared on this occasion. Furthermore I would like to thank especially Wolfgang Forsthuber, Madgalena Fuchs, Andreas Langer, and Annette Weihs for the enormous help they provided us for the organization of the Summer School. I acknowledge the financial support of RICAM, of the Doctoral Program in Computational Mathematics of the Johannes Kepler University of Linz, and the support by the project Y 432-N15 START-Preis "Sparse Approximation and Optimization in High Dimensions".

Bibliography

[1] L. Ambrosio, N. Fusco and D. Pallara, *Functions of Bounded Variation and Free Discontinuity Problems*, Oxford Mathematical Monographs, Oxford University Press, 2000.

[2] G. Aubert and P. Kornprobst, *Mathematical Problems in Image Processing: Partial Differential Equations and the Calculus of Variation*, Springer, 2002.

[3] R. G. Baraniuk, M. Davenport, R. A. DeVore and M. Wakin, A simple proof of the restricted isometry property for random matrices, *Constr. Approx.* **28** (2008), 253–263.

[4] H. H. Bauschke, J. M. Borwein and A. S. Lewis, *The Method of Cyclic Projections for Closed Convex Sets in Hilbert Space. Recent Developments in Optimization Theory and Nonlinear Analysis*, (Jerusalem, 1995), 1–38, Contemp. Math. 204, Amer. Math. Soc., Providence, RI, 1997

[5] A. Beck and M. Teboulle, A fast iterative shrinkage-thresholding algorithm for linear inverse problems, *SIAM J. Imaging Sci.* **2**:1 (2009), 183–202.

[6] T. Blumensath and M. E. Davies, Iterative thresholding for sparse approximations, *J. Fourier Anal. Appl.* **14** (2008), 629–654.

[7] ———, Iterative hard thresholding for compressed sensing, *Appl. Comput. Harmon. Anal.* **27**:3 (2009), 265–274.

[8] T. Bonesky, S. Dahlke, P. Maass and T. Raasch, Adaptive wavelet methods and sparsity reconstruction for inverse heat conduction problems, Preprint series DFG-SPP 1324, Philipps University of Marburg, 2009.

[9] A. Braides, *Γ-Convergence for Beginners*, Oxford Lecture Series in Mathematics and Its Applications 22, Oxford University Press, 2002.

[10] K. Bredies and D. Lorenz, Linear convergence of iterative soft-thresholding, *J. Fourier Anal. Appl.* **14**:5–6 (2008), 813–837.

[11] R. E. Bruck and S. Reich, Nonexpansive projections and resolvents of accretive operators in Banach spaces *Houston J. Math.* **3**:4 (1977), 459–470.

[12] E. J. Candès, J. Romberg and T. Tao, Stable signal recovery from incomplete and inaccurate measurements, *Comm. Pure Appl. Math.* **59** (2006), 1207–1223.

[13] C. Carstensen, Domain decomposition for a non-smooth convex minimalization problem and its application to plasticity, *Numerical Linear Algebra with Applications* **4**:3 (1998), 177–190.

[14] A. Chambolle, An algorithm for total variation minimization and applications, *J. Math. Imaging Vision* **20**:1–2 (2004), 89–97.

[15] A. Chambolle and J. Darbon, On total variation minimization and surface evolution using parametric maximum flows, *International Journal of Computer Vision* **84**:3 (2009), 288–307.

[16] A. Chambolle and P.-L. Lions, Image recovery via total variation minimization and related problems, *Numer. Math.* **76** (1997), 167–188.

[17] T. F. Chan, G. H. Golub and P. Mulet, A nonlinear primal-dual method for total variation-based image restoration, *SIAM J. Sci. Comput.* **20**:6 (1999), 1964–1977.

[18] T. F. Chan and T. P. Mathew, Domain decomposition algorithms, *Acta Numerica* **3** (1994), 61–143.

[19] A. K. Cline, Rate of convergence of Lawson's algorithm, *Math. Comp.* **26** (1972), 167–176.

[20] A. Cohen, W. Dahmen and R. A. DeVore, Compressed sensing and best k-term approximation, *J. Amer. Math. Soc.* **22** (2009), 211–231.

[21] P. L. Combettes and V. R. Wajs, Signal recovery by proximal forward-backward splitting, *Multiscale Model. Simul.* **4** (2005), 1168–1200.

[22] S. Dahlke, M. Fornasier and T. Raasch, Multilevel preconditioning for adaptive sparse optimization, Preprint series DFG-SPP 1324, Philipps University of Marburg, 2009.

[23] W. Dahmen, Wavelet and multiscale methods for operator equations, *Acta Numerica* **6** (1997), 55–228.

[24] W. Dahmen and A. Kunoth, Multilevel preconditioning, *Numer. Math.* **63** (1992), 315–344.

[25] G. Dal Maso, *An Introduction to Γ-Convergence*, Birkhäuser, Boston, 1993.

[26] J. Darbon and M. Sigelle, A fast and exact algorithm for total variation minimization, in: *LNCS Springer series, 2nd Iberian Conference on Pattern Recognition and Image Analysis (IbPria)* 3522(1), pp. 351–359, 2005.

[27] I. Daubechies, *Ten Lectures on Wavelets*, SIAM, 1992.

[28] I. Daubechies, M. Defrise and C. De Mol, An iterative thresholding algorithm for linear inverse problems with a sparsity constraint, *Comm. Pure Appl. Math.* **57** (2004), 1413–1457.

[29] I. Daubechies, M. Fornasier and I. Loris, Accelerated projected gradient methods for linear inverse problems with sparsity constraints, *J. Fourier Anal. Appl.* **14**:5–6 (2008), 764–792.

[30] I. Daubechies, R. A. DeVore, M. Fornasier and C. S. Güntürk, Iteratively re-weighted least squares minimization for sparse recovery, *Comm. Pure Appl. Math.* **63**:1 (2010), 1–38.

[31] R. A. DeVore, Nonlinear approximation, *Acta Numerica* **7** (1998), 51–150.

[32] D. Dobson and C. R. Vogel, Convergence of an iterative method for total variation denoising, *SIAM J. Numer. Anal.* **34**:5 (1997), 1779–1791.

[33] D. L. Donoho, Compressed sensing, *IEEE Trans. Inform. Theory* **52** (2006), 1289–1306.

[34] D. L. Donoho and Y. Tsaig, Fast solution of l1-norm minimization problems when the solution may be sparse, *IEEE Trans. Inform. Theory* **54** (2008), 4789–4812.

[35] B. Efron, T. Hastie, I. Johnstone and R. Tibshirani, Least angle regression, *Ann. Statist.* **32** (2004), 407–499.

[36] I. Ekeland and R. Témam, *Convex Analysis and Variational Problems*, SIAM, 1999.

[37] H. W. Engl, M. Hanke and A. Neubauer, *Regularization of Inverse Problems*, Springer, 1996.

[38] L. C. Evans and R. F. Gariepy, *Measure Theory and Fine Properties of Functions*, CRC Press, 1992.

[39] M. Figueiredo and R. D. Nowak, An EM algorithm for wavelet-based image restoration., *IEEE Trans. Image Proc.* **12** (2003), 906–916.

[40] M. Figueiredo, R. Nowak and S. J. Wright, Gradient projection for sparse reconstruction: Application to compressed sensing and other inverse problems, *IEEE J. Selected Topics in Signal Process.* **4**:1 (2007), 586–597.

[41] ———, Sparse reconstruction by separable approximation, *IEEE Trans. Signal Proc.* **57**:7 (2009), 2479–2493.

[42] M. Fornasier, Domain decomposition methods for linear inverse problems with sparsity constraints, *Inverse Problems* **23** (2007), 2505–2526.

[43] M. Fornasier, A. Langer and C.-B. Schönlieb, A convergent overlapping domain decomposition method for total variation minimization, preprint (2009).

[44] M. Fornasier and R. March, Restoration of color images by vector valued BV functions and variational calculus, *SIAM J. Appl. Math.* **68** (2007), 437–460.

[45] M. Fornasier and C.-B. Schönlieb, Subspace correction methods for total variation and ℓ_1-minimization, *SIAM J. Numer. Anal.* **47**:5 (2009), 3397–3428.

[46] A. Y. Garnaev and E. D. Gluskin, On widths of the Euclidean ball, *Sov. Math., Dokl.* **30** (1984), 200–204.

[47] E. D. Gluskin, Norms of random matrices and widths of finite-dimensional sets, *Math. USSR-Sb* **48** (1984), 173–182.

[48] I. F. Gorodnitsky and B. D. Rao, Sparse signal reconstruction from limited data using FOCUSS: a recursive weighted norm minimization algorithm, *IEEE Transactions on Signal Processing* **45** (1997), 600–616.

[49] T. Goldstein and S. Osher, The split Bregman method for L^1 regularized problems, *SIAM Journal on Imaging Sciences* **2**:2 (2009), 323–343.

[50] J.-B. Hiriart-Urruty and C. Lemaréchal, *Convex Analysis and Minimization Algorithms I*, Grundlehren der mathematischen Wissenschaften 305, Springer, Berlin, 1996

[51] B. S. Kashin, Diameters of some finite-dimensional sets and classes of smooth functions, *Math. USSR, Izv.* **11** (1977), 317–333.

[52] S.-J. Kim, K. Koh, M. Lustig, S. Boyd and D. Gorinevsky, A method for large-scale ℓ_1-regularized least squares problems with applications in signal processing and statistics, *IEEE Journal Sel. Top. Signal Process.* **1** (2007), 606–617.

[53] C. L. Lawson, *Contributions to the Theory of Linear Least Maximum Approximation*, 1961, Ph.D. thesis, University of California, Los Angeles.

[54] G. G. Lorentz, M. v. Golitschek and Y. Makovoz. *Constructive Approximation: Advanced Problems*, Springer, Berlin, 1996.

[55] I. Loris, G. Nolet, I. Daubechies and F. A. Dahlen, Tomographic inversion using ℓ_1-norm regularization of wavelet coefficients, *Geophysical Journal International* **170**:1 (2007), 359–370.

[56] I. Loris, On the performance of algorithms for the minimization of 1-penalized functionals, *Inverse Problems* **25** (2009), 035008.

[57] M. Lustig, D. Donoho and J. M. Pauly, Sparse MRI: The application of compressed sensing for rapid MR imaging, *Magnetic Resonance in Medicine* **58**:6 (2007), 1182–1195.

[58] S. Mallat, *A Wavelet Tour of Signal Processing: The Sparse Way*, Academic Press, 2009.

[59] S. G. Mallat and Z. Zhang, Matching pursuits with time-frequency dictionaries, *IEEE Trans. Signal Process.* **41** (1993), 3397–3415.

[60] J. Müller, *Parallel Methods for Nonlinear Imaging Techniques*, Master thesis, University of Münster, 2008.

[61] B. K. Natarajan, Sparse approximate solutions to linear systems, *SIAM J. Comput.* **24** (1995), 227–234.

[62] Y. Nesterov, Smooth minimization of non-smooth functions, *Mathematic Programming, Ser. A* **103** (2005), 127–152.

[63] Y. Nesterov and A. Nemirovskii, *Interior-point Polynomial Algorithms in Convex Programming*, SIAM Studies in Applied Mathematics 13, Society for Industrial and Applied Mathematics (SIAM), Philadelphia, PA, 1994.

[64] Z. Opial, Weak convergence of the sequence of successive approximations for nonexpansive mappings, *Bull. Amer. Math. Soc.* **73** (1967), 591–597.

[65] M. Osborne, B. Presnell and B. Turlach, A new approach to variable selection in least squares problems, *IMA J. Numer. Anal.* **20** (2000), 389–403.

[66] M. R. Osborne, *Finite Algorithms in Optimization and Data Analysis*, Wiley Series in Probability and Mathematical Statistics: Applied Probability and Statistics, John Wiley & Sons Ltd., Chichester, 1985.

[67] M. Osborne, B. Presnell and B. Turlach, On the LASSO and its dual, *J. Comput. Graph. Statist.* **9** (2000), 319–337.

[68] S. Osher, M. Burger, D. Goldfarb, J. Xu and W. Yin, An iterative regularization method for total variation-based image restoration, *Multiscale Model. Simul.* **4**:2 (2005) 460-489.

[69] R. Ramlau, G. Teschke and M. Zhariy, A compressive Landweber iteration for solving ill-posed inverse problems, *Inverse Problems* **24**:6 (2008), 065013.

[70] H. Rauhut, Compressive sensing and structured random matrices, in: *Theoretical Foundations and Numerical Methods for Sparse Recovery*, Radon Series Comp. Appl. Math. deGruyter, 2010.

[71] R. T. Rockafellar and R. J. B. Wets, *Variational Analysis*, Grundlehren der Mathematischen Wissenschaften 317, Springer, Berlin, 1998.

[72] L. Rudin, S. Osher and E. Fatemi, Nonlinear total variation based noise removal algorithms, *Physica D* **60** (1992), 259–268.

[73] C.-B. Schönlieb, *Total Variation Minimization with an H^{-1} Constraint*, CRM Series 9, Singularities in Nonlinear Evolution Phenomena and Applications proceedings, Scuola Normale Superiore Pisa 2009, 201–232.

[74] J.-L. Starck, E. J. Candès and D. L. Donoho, Astronomical image representation by curvelet transform, *Astronomy and Astrophysics* **298** (2003), 785–800.

[75] J.-L. Starck, M. K. Nguyen and F. Murtagh, Wavelets and curvelets for image deconvolution: a combined approach, *Signal Proc.* **83** (2003), 2279–2283.

[76] R. Tibshirani, Regression shrinkage and selection via the lasso, *J. Roy. Statist. Soc. Ser. B* **58** (1996), 267–288.

[77] J. A. Tropp and D. Needell, CoSaMP: Iterative signal recovery from incomplete and inaccurate samples, *Appl. Comput. Harmon. Anal.* **26**:3 (2008), 301–321.

[78] J. A. Tropp, Greed is good: Algorithmic results for sparse approximation, *IEEE Trans. Inform. Theory* **50** (2004), 2231–2242.

[79] X.-C. Tai and P. Tseng, Convergence rate analysis of an asynchronous space decomposition method for convex minimization, *Math. Comp.* **71**:239 (2001), 1105–1135.

[80] X.-C. Tai and J. Xu, Global convergence of subspace correction methods for convex optimization problems, *Math. Comp.* **71**:237 (2002), 105–124.

[81] L. Vese, A study in the BV space of a denoising-deblurring variational problem, *Appl. Math. Optim.* **44** (2001), 131–161.

[82] J. Warga, Minimizing certain convex functions, *J. Soc. Indust. Appl. Math.* **11** (1963), 588–593.

[83] P. Weiss, L. Blanc-Féraud and G. Aubert, Efficient schemes for total variation minimization under constraints in image processing, *SIAM J. Sci. Comput.* **31**:3 (2009), 2047–2080.

[84] J. Xu, X.-C. Tai and L.-L. Wang, *A Two-level Domain Decomposition Method for Image Restoration*, UCLA Computational and Applied Mathematics Report 09-92, November 2009.

[85] Matlab code and numerical experiments for the domain decomposition methods presented in this chapter can be downloaded at `http://homepage.univie.ac.at/carola.schoenlieb/webpage_tvdode/tv_dode_numerics.htm`

Author information

Massimo Fornasier, RICAM, Austrian Academy of Sciences, Linz, Austria.
E-mail: `massimo.fornasier@oeaw.ac.at`

Sparse Recovery in Inverse Problems

Ronny Ramlau and Gerd Teschke

Abstract. Within this chapter we present recent results on sparse recovery algorithms for inverse and ill-posed problems, i.e. we focus on those inverse problems in which we can assume that the solution has a sparse series expansion with respect to a preassigned basis or frame. The presented approaches to approximate solutions of inverse problems are limited to iterative strategies that essentially rely on the minimization of Tikhonov-like variational problems, where the sparsity constraint is integrated through ℓ_p norms. In addition to algorithmic and computational aspects, we also discuss in greater detail regularization properties that are required for cases in which the operator is ill-posed and no exact data are given. Such scenarios reflect realistic situations and manifest therefore its great suitability for "real-life" applications.

Keywords. Inverse and ill-posed problems, regularization theory, convergence rates, sparse recovery, iterated soft-shrinkage, accelerated steepest descent, nonlinear approximation.

2010 Mathematics Subject Classification. 41A46, 45Q05, 47A52, 49M05, 52A07, 65B99, 65K10.

Table of Contents.

1 Introduction

The aim of this chapter is to present technologies for the recovery of sparse signals in situations in which the given data are linked to the signal to be reconstructed through an ill-posed measurement model. In such scenarios one is typically faced with regularization issues and the construction of suitable methods that allow a stable reconstruction of sparse signals.

The first author was supported by the Austrian Science Fund, Grants P20237-N14, P19496-N18 and DK W1214-N15. The second author was partially supported by Deutsche Forschungsgemeinschaft Grants TE 354/3-1, TE 354/4-1, TE 354/5-1, TE 354/8-1.

1.1 Road Map of the Chapter

Nowadays there exist a great variety of schemes realizing sparse reconstructions. Most of them are well-suited for finite or infinite dimensional problems but where the underlying physical model is well-posed. More delicate are those cases in which ill-posed operators are involved. So far, for linear ill-posed problems, there are numerous schemes available that perform quite well sparse reconstructions, e.g. [2, 14, 18, 19, 24, 25, 52]. The majority of these approaches rely on iterative concepts in which adequate sparsity constraints are involved. Within this chapter we do not discuss the pros and contras of all these methods. In the context of linear problems we just concentrate on one new approach that involves a complete different but very powerful technology – that is adaptive approximation. The second focus of this chapter is on the generalization of conventional iterative strategies to nonlinear ill-posed problems.

Therefore the road map for this chapter is as follows: In Section 2 we collect the basics on inverse problems. To elaborate the differences between well- and ill-posedness and the concepts of regularization theory as simple as possible we limit ourselves in this introductory section to linear problems. After this preliminary part we continue in Section 3 with linear problems and present a sparse recovery principle that essentially relies on the theory of adaptive approximation. The main ingredient that ensures stable recovery are sophisticated refinement strategies. In Section 4 we turn then to nonlinear ill-posed problems and discuss in greater detail Tikhonov regularization with sparsity constraints. The established regularization properties include convergence results and convergence rates for a-priori as well as for a-posteriori parameter rules. After the general discussion on Tikhonov regularization we focus within the following Sections 5 and 6 on the development of implementable algorithms to numerically realize sparse recovery. The first method presented in Section 5 relies on the surrogate functional technology. This approach results in a Landweber-type iteration where a shrinkage operation is applied in each iteration step. This method can be generalized to general sparsity constraints, but fails to be numerically efficient. To overcome this deficiency, we introduce in Section 6 a slightly modified concept leading to a very similar iteration, but where in each iteration a projection on a preassigned ℓ_1 ball is applied. Moreover, this new iteration is designed with an adaptive step length control resulting in a numerically very efficient method.

1.2 Remarks on Sparse Recovery Algorithms

As mentioned above, we discuss in this chapter two different species of sparse recovery algorithms. The first species developed for linear inverse problems relies on nonlinear approximation, the second species designed for nonlinear inverse problems relies on linear approximation.

In principle, when it comes to numerical realizations, we are faced with the problem that we can only treat finite index sets. Therefore one has to answer the question which coefficients should be involved in the reconstruction process and which can be neglected. Linear approximation simply suggests a truncation of the infinite index set. In a wavelet framework this would mean to limit the number resolution scales. For many problems in which the solution is supposed to have a certain Sobolev smoothness, this proceeding might yield reasonable results. Nevertheless, there are still cases in which linear approximation fails to yield optimal results. Then often nonlinear approximation concepts are much better suited. The reason why nonlinear strategies perform better than standard linear methods is due to the properties of the solution and the operator. To clarify this statement, we introduce by x_N the best N-term approximation of the solution x. Considering bases or frames of sufficiently smooth wavelet type (e.g. wavelets of order d), it is known that if both

$$0 < s < \frac{d-t}{n},$$

where n is the space dimension and t denoting the smoothness of the Sobolev space, and x is in the Besov space $B_\tau^{sn+t}(L_\tau(\Omega))$ with $\tau = (1/2 + s)^{-1}$, then

$$\sup_{N \in \mathbb{N}} N^s \|x - x_N\| < \infty.$$

The condition $x \in B_\tau^{sn+t}(L_\tau(\Omega))$ is much milder than requiring $x \in H^{sn+t}(\Omega)$ that would be needed to guarantee the same rate of convergence with linear approximation. However, for inverse problems it is in general not always possible to estimate the regularity of the solution from the regularity of the right hand side due to the presence of the noise. Therefore, special a-priori information about x and/or the operator is required. In certain cases, e.g. the tomographic reconstruction problem analyzed in [35], this information can be derived. A suitable model class for the tomographic reconstruction problem are piecewise constant functions with jumps along smooth manifolds. It is shown that such functions belong to the Sobolev space $H^{sd}(\Omega)$ with $sd < 1/2$. An adaptive approximation of such functions (when carried out in $L_2(\Omega)$) pays off if the Besov regularity in the scale $B_\tau^{sd}(L_\tau(\Omega))$, $\tau = (s + 1/2)^{-1}$ is significantly higher. This issue is discussed in [50, Rem. 4.3] and indeed such functions belong to $B_\tau^{sd}(L_\tau(\Omega))$ with $sd < 1/\tau = s + 1/2$. For the two-dimensional case, which is the case of this application, we therefore have that the solution x belongs to $H^{sd}(\Omega)$ for $s < 1/4$ and to $B_\tau^{sd}(L_\tau(\Omega))$ for s < 1/2. Consequently, the Besov regularity is indeed higher than the Sobolev regularity and nonlinear approximation pays off. How nonlinear approximation strategies can be realized for linear inverse and ill-posed problems shall be discussed in great detail in Section 3. As Besov regularity directly translates into sparsity, the elaborated adaptive Landweber-type scheme performs a sparse recovery for x.

Sparse recovery algorithms for nonlinear inverse problems rely so far on linear approximation concepts that originate from the minimization of Tikhonov-like functionals. These concepts were originally developed for linear inverse problems within the last decade, see e.g. [2, 14, 18, 19, 24, 25, 52], and have led to many breakthroughs in a broad field of applications. They are due to its simple nature very easy to use and can be applied in various reformulations. The generalization of these methods to nonlinear problems has permitted an algorithmic realization of sparse recovery for problems that were by then not feasible. The main ingredients are a proper variational formulation of the data misfit term and an adequate involvement of the sparsity constraint either through an extra penalty term or an restriction of the possible solution set. These two concepts shall be elaborated in Sections 5 and 6 which are furnished with associated numerical experiments.

2 Classical Inverse Problems

In many applications in the natural sciences, medicine or imaging one has to determine the cause x of a measured effect y. A classical example is Computerized Tomography (CT), a medical application, where a patient is screened using x-rays. The observed damping of the rays is then used to reconstruct the density distribution of the body. In order to achieve such a reconstruction, the measured data and the searched for quantity have to be linked by a mathematical model, which we will denote by F (or A, if the model is linear). In an abstract setting, the determination of the cause x can be stated as follows: Solve an operator equation

$$F(x) = y, \tag{2.1}$$

$F : X \to Y$, where X, Y are Banach (Hilbert) spaces. For the CT problem, the operator describing the connection between the measurements and the density distribution (in 2 dimensions) is given by the Radon transform,

$$y(s,\omega) = (Ax)(s,\omega) = \int_{\mathbb{R}} x(s\omega + t\omega^{\perp})\,dt, \quad s \in \mathbb{R},\ \omega \in \mathcal{S}^1.$$

As in practice the observed data stems from measurements, one never has the exact data y available, but rather a noisy variant y^δ. In the following we might assume that at least a bound δ for the noise is available (e.g. if the accuracy of the measurement device is known):

$$\|y - y^\delta\| \le \delta.$$

In connection with Inverse Problems, the following questions arise:

(i) Does there exist a solution of equation (2.1) for given exact y?

(ii) Is the solution unique?

(iii) If the solution is determined from noisy data, how accurate is it?

(iv) How to solve (2.1)?

2.1 Preliminaries

In order to give a first idea on the problems that may be encountered for ill-posed problems, we will now consider a linear operator equation in finite dimensions. Assume $A \in \mathbb{R}^{n \times n}$, and we want to solve the linear system $Ax = y$ from noisy data y^δ. If we assume that A is invertible on range(A) and also $y^\delta \in$ range(A) (which is already a severe restriction), then we can define

$$\begin{aligned} x^\dagger &:= A^{-1} y, \\ x^\delta &:= A^{-1} y^\delta. \end{aligned}$$

With $x^\dagger - x^\delta = A^{-1}(y - y^\delta)$ the distance between x^δ and $x^\dagger$ can be estimated as follows,

$$\begin{aligned} \|x^\dagger - x^\delta\| &\le \|A^{-1}\| \, \|y - y^\delta\| \\ &\le \|A^{-1}\| \, \delta. \end{aligned} \tag{2.2}$$

If we additionally assume that A is symmetric and positive definite with $\|A\| \le 1$, then A has an eigensystem (λ_i, x_i) with eigenvalues $0 < \lambda_i \le 1$ and associated eigenvectors x_i. Moreover we have

$$\|A^{-1}\| = \frac{1}{\lambda_{\min}} \Rightarrow \|x^\dagger - x^\delta\| \le \frac{\delta}{\lambda_{\min}}.$$

Therefore, the reconstruction quality is of the same order $\mathcal{O}(\delta)$ as the data error, magnified only by the norm of the inverse operator. However, it turns out that $\mathcal{O}(\delta)$ estimates are only possible in a finite dimensional setting: Indeed, if we define the operator

$$Ax = \sum_{i=1}^{\infty} \lambda_i \langle x, x_i \rangle x_i$$

with orthonormal basis x_i and $\lambda_i \to 0$, then it is easily to see that the right hand side of estimate (2.2) explodes. In fact, for inverse problems with dim range$(A) = \infty$ and, e.g., compact operator, it is in general impossible to obtain convergence rates. Under additional assumptions on the solution of the equation $Ax = y$, the best possible convergence rate is given by

$$\|x - x^\delta\| = \mathcal{O}(\delta^s), \quad s < 1. \tag{2.3}$$

The above considerations were based on the assumption $y^\delta \in$ range(A). As we will see in the following example, this is a severe restriction that will not hold in practice:

Let us consider the integral equation

$$y(s) = \int_0^s x(t)dt \quad 0 \le s \le 1.$$

If $x \in C^0[0,1]$, then it immediately follows that $y \in C^1[0,1]$ and

$$x(s) = y'(s), \quad y(0) = 0.$$

For noisy measurements this condition will not hold, as the noise will not only alter the initial value but also the smoothness of y^δ, as the data noise is usually not differentiable. The same also holds for Computerized Tomography: It can be shown [35] that the exact CT data belongs to the Sobolev space $H^{1/2}(\mathbb{R} \times \mathcal{S}^1)$, but for the noisy data we only have $y^\delta \in L_2$.

Now let us define *well-posed* and *ill-posed problems*.

Definition 2.1. Let $A : X \to Y$ linear operator and X, Y be topological spaces. Then the problem (A, X, Y) is well-posed if condition (i)–(iii) are fulfilled at the same time,

(i) $Ax = y$ has a solution for each $y \in Y$

(ii) the solution is unique

(iii) the solution depends continuously on the data, i.e.

$$y_n \to y, \quad y_n = Ax_n, \quad \Longrightarrow \quad x_n \to x \quad \text{and} \quad Ax = y.$$

If one of the conditions is violated, then the problem is ill posed.

Roughly speaking, well-posed problems allow for an error estimate as in (2.2), whereas the best possible rate for ill posed problems is as in (2.3).

Let us denote by $L(X, Y)$ the set of all linear and continuous operators $A : X \to Y$. An important class of operators that lead to ill-posed problems are *compact* operators.

Definition 2.2. An operator $A \in L(X, Y)$ is compact, if it maps bounded sets to relative compact sets. Or equivalently, for any bounded sequence $(x_n)_{n\in\mathbb{N}}$, the sequence $y_n = Ax_n$ has a convergent subsequence.

Integral operators are an important class of examples for compact operators.

Definition 2.3. Let $G \in \mathbb{R}^n$ be a bounded set and $k : G \times G \to G$. We define the integral operator K by

$$(Kx)(s) = \int_G k(s,t)x(t)dt.$$

Proposition 2.4. *Let $k \in C(G, G)$ and K be an integral operator considered between any of the spaces $L_2(G)$ and $C(G)$. Then K is compact. If $k \in L_2(G, G)$, then the integral operator $K : L_2(G) \to L_2(G)$ is compact.*

Another example for compact operators are Sobolev embedding operators. For bounded G and a real number $s > 0$, let us consider the map

$$i_s : H^s(G) \to L_2(G), \quad \text{which is defined by } i_s x = x.$$

Here H^s denotes the standard Sobolev space. Then we have

Proposition 2.5. *The Sobolev embedding operator i_s is compact.*

Proposition 2.6. *Compact operators with* $\dim \operatorname{range}(K) = \infty$ *are not continuously invertible, i.e. they are ill-posed.*

Now let us assume that a given operator $A : H^s \to H^{s+t}$, $s \geq 0, t > 0$, is continuously invertible. As pointed out above, the measured data will not belong to H^{s+t} but rather to L_2. Therefore, we have to consider the operator equation between H^s and L_2, i.e. the equation $y = i_{s+t}(Ax)$. As a combination of a continuous and a compact operator, $i_s \circ A$ is also compact and therefore not continuously invertible – regardless of the invertibility of A.

A key ingredient for the stable inversion of compact operators is the spectral decomposition:

Proposition 2.7. *Let $K : X \to X$, X be a Hilbert space and assume that K is compact and self-adjoint (i.e, $\langle Kx, y\rangle = \langle x, Ky\rangle \; \forall x, y \in X$). By (λ_j, u_j) denote the set of eigenvalues λ_j and associated eigenvectors u_j with $Ku_j = \lambda_j u_j$. Then $\lambda_j \to 0$ (if* $\dim \operatorname{range}(K) = \infty$*) and the functions u_j form an orthonormal basis of* $\overline{\operatorname{range}(K)}$ *with*

$$Kx = \sum_{i=1}^{\infty} \lambda_i \langle x, u_i\rangle u_i.$$

The eigenvalue decomposition can be generalized to compact operators that are not self-adjoint. Let $K : X \to Y$ be given. The adjoint operator $K^* : Y \to X$ is formally defined by the equation

$$\langle Kx, y\rangle = \langle x, K^* y\rangle \quad \forall x, y.$$

We can then define the operator $K^*K : X \to X$ and find

$$\langle K^*Kx, y\rangle = \langle Kx, Ky\rangle = \langle x, K^*Ky\rangle,$$
$$\langle K^*Kx, x\rangle = \langle Kx, Kx\rangle = \|Kx\|^2,$$

i.e., K^*K is selfadjoint and positive semi-definite, which also guarantees that all eigenvalues λ_i of K^*K are nonnegative. Therefore we have

$$K^*Kx = \sum_i \lambda_i \langle x, u_i\rangle u_i.$$

Defining

$$\sigma_i = +\sqrt{\lambda_i},$$
$$Ku_i = \sigma_i v_i,$$

we find that the functions v_i also form an orthonormal system for Y:

$$\begin{aligned}\langle v_i, v_j\rangle &= \frac{1}{\sigma_i\sigma_j}\langle Ku_i, Ku_j\rangle\\ &= \frac{1}{\sigma_i\sigma_j}\langle K^*Ku_i, u_j\rangle\\ &= \frac{\sigma_i}{\sigma_j}\langle u_i, u_j\rangle = \delta_{ij} = \begin{cases}1 & i = j\\ 0 & i \neq j,\end{cases}\end{aligned}$$

and get

$$\begin{aligned}Kx &= K\Big(\sum_i \langle x, u_i\rangle u_i\Big) = \sum_i \langle x, u_i\rangle Ku_i = \sum_i \sigma_i \langle x, u_i\rangle v_i,\\ K^*y &= \sum_i \sigma_i \langle y, v_i\rangle u_i.\end{aligned}$$

The above decomposition of K is called the *singular value decomposition* and $\{\sigma_i, u_i, v_i\}$ is the singular system of K. The *generalized inverse* of K is defined as follows:

Definition 2.8. The generalized inverse $K^\dagger$ of K is defined as

$$\begin{aligned}\operatorname{dom}(K^\dagger) &= \operatorname{range}(K) \oplus \operatorname{range}(K)^\perp,\\ K^\dagger y &:= x^\dagger,\\ x^\dagger &= \arg\min_x \|y - Kx\|.\end{aligned}$$

If the minimizer $x^\dagger$ of the functional $\|y - Kx\|^2$ is not unique then the one with minimal norm is taken.

Proposition 2.9. *The generalized solution $x^\dagger$ has the following properties*

(i) *$x^\dagger$ is the unique solution of $K^*Kx = K^*y$,*

(ii) *$Kx^\dagger = P_{R(K)}y$, where $P_{R(K)}$ denotes the orthogonal projection on the range of K,*

(iii) *$x^\dagger$ can be decomposed w.r.t. the singular system as*

$$x^\dagger = \sum_i \frac{1}{\sigma_i}\langle y, v_i\rangle u_i, \tag{2.4}$$

(iv) *the generalized inverse is continuous if and only if* $\operatorname{range}(K)$ *is closed.*

A direct consequence of the above given representation of $x^\dagger$ is the so-called Picard condition:

$$y \in \operatorname{range}(K) \Leftrightarrow \sum_i \frac{|\langle y, v_i\rangle|^2}{\sigma_i^2} < \infty.$$

The condition states that the moments of the right hand side y (w.r.t. to the system $\{v_i\}$) have to tend to zero fast enough in order to compensate the growth of $1/\sigma_i$.

What happens if we apply noisy data to formula (2.4)? Assume $y \in \operatorname{range}(K)$, $y = Kx^\dagger$, and $y_l^\delta = y + \delta v_l$. Then for all l

$$\|y - y_l^\delta\| \leq \delta,$$

but with

$$x^\delta = \sum_i \frac{1}{\sigma_i}\langle y_l^\delta, v_i\rangle u_i$$

we obtain

$$\|x^\dagger - x^\delta\|^2 = \sum_i \frac{\delta^2}{\sigma_i^2}|\langle v_l, v_i\rangle|^2 = \frac{\delta^2}{\sigma_l^2} \to \infty \quad \text{as } l \to \infty,$$

which shows that the reconstruction error can be arbitrarily large even if the noisy data is close to the true data.

2.2 Regularization Theory

In order to get a reasonable reconstruction, we have to introduce different methods that ensure a good and stable reconstructions. These methods are often defined via functions of operators.

Definition 2.10. Let $f : \mathbb{R}^+ \to \mathbb{R}$. For compact operators, we define

$$f(K)x := \sum_i f(\sigma_i)\langle x, u_i\rangle v_i.$$

Of course, this definition is only well-defined for functions f for which the sum converges. We can now define *regularization methods*.

Definition 2.11. A regularization of an operator $K^\dagger$ is a family of operators $(R_\alpha)_{\alpha>0}$,

$$R_\alpha : Y \to X$$

with the following properties: there exists a map $\alpha = \alpha(\delta, y^\delta)$ such that for all $y \in \operatorname{dom}(K^\dagger)$ and all $y^\delta \in Y$ with $\|y - y^\delta\| \leq \delta$,

$$\lim_{\delta \to 0} R_{\alpha(\delta, y^\delta)} y^\delta = x^\dagger$$

and

$$\lim_{\delta\to 0} \alpha(\delta, y^\delta) = 0.$$

The parameter α is called regularization parameter.

In the classical setting, regularizing operators R_α are defined via filter functions F_α:

$$R_\alpha y^\delta := \sum_{i\in\mathbb{N}} \sigma_i^{-1} F_\alpha(\sigma_i)\langle y^\delta, v_i\rangle u_i .$$

The requirements of Definition 2.11 have some immediate consequences on the admissible filter functions. In particular, $\mathrm{dom}(R_\alpha) = Y$ enforces $|\sigma_i^{-1} F_\alpha(\sigma_i)| \le C$ for all i, and the pointwise convergence of R_α to $K^\dagger$ requires $\lim_{\alpha\to 0} F_\alpha(t) = 1$. Well-known regularization methods are:

(i) Truncated singular value decomposition:

$$R_\alpha y^\delta := \sum_{i}^{N} \sigma_i^{-1} \langle y^\delta, v_i\rangle u_i .$$

In this case, the filter function is given by

$$F_\alpha(\sigma) := \begin{cases} 1 & \text{if } \sigma \ge \alpha \\ 0 & \text{if } \sigma < \alpha. \end{cases}$$

(ii) Truncated Landweber iteration: For $\beta \in (0, \frac{2}{\|K\|^2})$ and $m \in \mathbb{N}$, set

$$F_{1/m}(\lambda) = 1 - (1 - \beta\lambda^2)^m .$$

Here, the regularization parameter $\alpha = 1/m$ only admits discrete values.

(iii) Tikhonov regularization: Here, the filter function is given by

$$F_\alpha(\sigma) = \frac{\sigma^2}{\sigma^2 + \alpha}.$$

The regularized solutions of Landweber's and Tikhonov's method can also be characterized as follows:

Proposition 2.12. *The regularized solution due to Landweber, $x^\delta_{1/m}$, is also given by the m-th iterate of the Landweber iteration given by*

$$x^{n+1} = x^n + K^*(y^\delta - Kx^n), \quad \text{with } x^0 = 0.$$

The regularization parameter is the reciprocal of the stopping index of the iteration.

Proposition 2.13. *The regularized solution due to Tikhonov,*

$$x_\alpha^\delta := \sum_i \frac{\sigma_i^2}{\sigma_i^2 + \alpha} \cdot \sigma_i^{-1} \langle y^\delta, v_i \rangle u_i,$$

is also the unique minimizer of the Tikhonov functional

$$J_\alpha(x) = \|y^\delta - Kx\|^2 + \alpha \|x\|^2, \tag{2.5}$$

which is minimized by the unique solution of the equation

$$(K^*K + \alpha I)x = K^* y^\delta.$$

Tikhonov's variational formulation (2.5) is important as it allows generalizations towards nonlinear operators as well as to sparse reconstructions. As mentioned above, regularization methods also require proper parameter choice rules.

Proposition 2.14. *The Tikhonov regularization combined with one of the parameter choice rules*

a) $\alpha(\delta, y^\delta) \to 0$ *and* $\frac{\delta^2}{\alpha(\delta, y^\delta)} \to 0$

b) $\alpha_*(\delta, y^\delta)$ *s.t.* $\|y^\delta - Kx_{\alpha_*}^\delta\| = \tau\delta$ *for fixed* $\tau > 1$ *(discrepancy principle)*

is a regularization method.

Proposition 2.15. *Let* $\tau > 1$. *If the Landweber iteration is stopped after* m_* *iterations, where* m_* *is the first index with*

$$\|y^\delta - Kx^{m_*}\| \le \tau\delta < \|y^\delta - Kx^{m_*-1}\| \quad \textit{(discrepancy principle)},$$

then the iteration is a regularization method with $R_{\frac{1}{m_*}} y^\delta = x^{m_*}$.

The last two propositions show that the regularized solutions for Tikhonov's or Landweber's method converge towards the true solution provided a proper parameter choice rule was applied. However, no result on the speed of convergence is provided. Due to Bakhushinsky one rather has

Proposition 2.16. *Let* $x_\alpha^\delta = R_\alpha y^\delta$, R_α *be a regularization method. Then the convergence of* $x_\alpha^\delta \to x^\dagger$ *can be arbitrary slow.*

To overcome this drawback, we have to assume a certain regularity of the solution. Indeed, convergence rates can be achieved provided the solution fulfills a so-called source-conditions. Here we limit ourselves to the Hölder-type source conditions,

$$x^\dagger = (K^*K)^{\frac{\nu}{2}} w, \quad \text{i.e. } x^\dagger \in \operatorname{range}(K^*K)^{\frac{\nu}{2}} \subseteq \operatorname{dom}(K) = X, \quad \nu > 0.$$

Definition 2.17. A regularization method is called order optimal if for a given parameter choice rule the estimate

$$\|x^\dagger - x^\delta_{\alpha(\delta,y^\delta)}\| = \mathcal{O}(\delta^{\frac{\nu}{\nu+1}}) \tag{2.6}$$

holds for all $x^\dagger = (K^*K)^{\frac{\nu}{2}} w$ and $\|y^\delta - y\| \leq \delta$.

It turns out that for $x^\dagger = (K^*K)^{\frac{\nu}{2}} w$ this is actually the best possible convergence rate, no method can do better. Also, we have $\delta^{\frac{\nu}{\nu+1}} > \delta$ for $\delta < 1$, i.e., we always loose some information in the reconstruction procedure.

Proposition 2.18. *Tikhonov regularization and Landweber iteration together with the discrepancy principle are order optimal.*

3 Nonlinear Approximation for Linear Ill-Posed Problems

Within this section we consider linear inverse problems and construct for them a Landweber-like algorithm for the sparse recovery of the solution x borrowing "leafs" from nonlinear approximation. The classical Landweber iteration provides in combination with suitable regularization parameter rules an order optimal regularization scheme (for the definition, see eq. (2.6)). However, for many applications the implementation of Landweber's method is numerically very intensive. Therefore we propose an adaptive variant of Landweber's iteration that significantly may reduce the computational expense, i.e. leading to a *compressed* version of Landweber's iteration. We lend the concept of adaptivity that was primarily developed for well-posed operator equations (in particular, for elliptic PDE's) essentially exploiting the concept of wavelets (frames), Besov regularity, best N-term approximation and combine it with classical iterative regularization schemes. As the main result we define an adaptive variant of Landweber's iteration from which we show regularization properties for exact and noisy data that hold in combination with an adequate refinement/stopping rule (a-priori as well as a-posteriori principles). The results presented in this Section where first published in [47].

3.1 Landweber Iteration and Its Discretization

The Landweber iteration is a gradient method for the minimization of $\|y^\delta - Ax\|^2$ and is therefore given through

$$x^{n+1} = x^n + \gamma A^*(y^\delta - Ax^n). \tag{3.1}$$

As it can be retrieved, e.g. in [28], iteration (3.1) is for $0 < \gamma < 2/\|A\|^2$ a linear regularization method as long as the iteration is truncated at some finite index n_*. In order to identify the optimal truncation index n_*, one may apply either an a-priori or an a-posteriori parameter rule. The Landweber method (3.1) is an order optimal linear regularization method, see [28], if the iteration is truncated at the a-priori chosen iteration index

$$n_* = \left\lfloor \gamma \left(2\frac{\gamma}{\nu} e\right)^{\nu/(\nu+1)} \left(\frac{\rho}{\delta}\right)^{2/(\nu+1)} \right\rfloor, \tag{3.2}$$

where the common notation $\lfloor p \rfloor$ denotes the smallest integer less or equal p. Here, we have assumed that the solution $x^\dagger$ of our linear equation admits the smoothness condition

$$x^\dagger = (A^*A)^{\nu/(\nu+1)}v, \quad \|v\| \leq \rho.$$

If n_* is chosen as suggested in (3.2), then optimal convergence order with respect to $x^\dagger$ can be achieved. This proceeding, however, needs exact knowledge of the parameters $\nu,\ \rho$ in the source condition. This shortfall can be avoided when applying Morozov's discrepancy principle. This principle performs the iteration as long as

$$\|Ax^n - y^\delta\| > \tau\delta \tag{3.3}$$

holds with $\tau > 1$, and truncates the iteration once

$$\|Ax^{n_*} - y^\delta\| \leq \tau\delta \tag{3.4}$$

is fulfilled for the first time. The regularization properties of this principle were investigated in [20]. The authors have shown that, as long as (3.3) holds, the next iterate will be closer to the generalized solution than the previous iterate. This property turned out to be very fruitful for the investigation of discretized variants of (3.1). This can be retracted in details in [38] where a discretization of the form

$$x^{n+1} = x^n + \gamma A^*_{r^\delta(n)}(y^\delta - A_{r^\delta(n)}x^n) \tag{3.5}$$

was suggested. The basic idea in [38] is the introduction of approximations $A_{r^\delta(m)}$ to the operator A that are updated/refined in dependence on a specific discrepancy principle.

Iteration (3.5) acts in the infinite dimensional Hilbert space X. To treat (3.5) numerically, we have to discretize the inverse problem which means that we have to find a discretized variant of (3.1) through the discretization of the corresponding normal equation of $\|y^\delta - Ax\|^2$. To this end, we assume that we have for the underlaying space X a preassigned countable system of functions $\{\phi_\lambda : \lambda \in \Lambda\} \subset X$ at our disposal for which there exist constants C_1, C_2 with $0 < C_1 \leq C_2 < \infty$ such that for all $x \in X$,

$$C_1\|x\|_X^2 \leq \sum_{\lambda\in\Lambda} |\langle x, \phi_\lambda\rangle|^2 \leq C_2\|x\|_X^2. \tag{3.6}$$

For such a system, which is often referred to as a frame for X, see [6] for further details, we may consider the operator $\mathcal{F} : X \to \ell_2$ via $x \mapsto c = \{\langle x, \phi_\lambda\rangle\}_{\lambda\in\Lambda}$ with adjoint $\mathcal{F}^* : \ell_2 \to X$ via $c \mapsto \sum_{\lambda\in\Lambda} c_\lambda\phi_\lambda$. The operator $\mathcal{F}$ is often referred in the literature to as the analysis operator, whereas $\mathcal{F}^*$ is referred to as the synthesis operator. The composition of both, $\mathcal{F}^*\mathcal{F}$, is called the frame operator which is by condition (3.6) an invertible map; guaranteeing that each $x \in X$ can be reconstructed from its moments $\langle x, \phi_\lambda \rangle$. Moreover, there is for every $x \in X$ at least one sequence c such that $x = \mathcal{F}^* c$. Consequently, we can define

$$S = \mathcal{F} A^* A \mathcal{F}^*, \quad x = \mathcal{F}^* c \quad \text{and} \quad g^\delta = \mathcal{F} A^* y^\delta$$

leading to the discretized normal equation

$$Sc = g^\delta. \tag{3.7}$$

An approximate solution for (3.7) can then be derived by the corresponding sequence space Landweber iteration,

$$c^{n+1} = c^n + \gamma(g^\delta - Sc^n). \tag{3.8}$$

Note that the operator $S : \ell_2(\Lambda) \to \ell_2(\Lambda)$ is symmetric but through the ill-posedness of A not boundedly invertible on $\ell_2(\Lambda)$ (even on the subspace Ran F). This is one major difference to [50] in which the invertibility of S on Ran F was substantially used to ensure the convergence of the Landweber iteration. Since we can neither handle the infinite dimensional vectors c^n and g^δ nor apply the infinite dimensional matrix S, iteration (3.8) is not a practical algorithm. To this end, we need to study the convergence and regularization properties of the iteration in which c^n, g^δ and S are approximated by finite length objects. Proceeding as suggested [50], we assume that we have the following three routines at our disposal:

- **RHS**$_\varepsilon[y] \to g_\varepsilon$. This routine determines a finitely supported $g_\varepsilon \in \ell_2(\Lambda)$ satisfying

$$\|g_\varepsilon - \mathcal{F} A^* y\| \le \varepsilon.$$

- **APPLY**$_\varepsilon[c] \to w_\varepsilon$. This routine determines, for a finitely supported $c \in \ell_2(\Lambda)$ and an infinite matrix S, a finitely supported w_ε satisfying

$$\|w_\varepsilon - Sc\| \le \varepsilon.$$

- **COARSE**$_\varepsilon[c] \to c_\varepsilon$. This routine creates, for a finitely supported with $c \in \ell_2(\Lambda)$, a vector c_ε by replacing all but N coefficients of c by zeros such that

$$\|c_\varepsilon - c\| \le \varepsilon,$$

  whereas N is at most a constant multiple of the minimal value N for which the latter inequality holds true.

For the detailed functionality of these routines we refer the interested reader to [10, 50]. For the sake of more flexibility in our proposed approach, we allow (in contrast to classical setup suggested in [50]) ε to be different within each iteration step and sometimes different for each of the three routines. Consequently, we set $\varepsilon = \varepsilon_n^R$ for the routine $\mathbf{RHS}_\varepsilon[\cdot]$, $\varepsilon = \varepsilon_n^A$ for $\mathbf{APPLY}_\varepsilon[\cdot]$ and, finally, $\varepsilon = \varepsilon_n^C$ for $\mathbf{COARSE}_\varepsilon[\cdot]$. The subscript n of the created error tolerance or so-called refinement sequences $\{\varepsilon_n^C\}_{n\in\mathbb{N}}$, $\{\varepsilon_n^A\}_{n\in\mathbb{N}}$ and $\{\varepsilon_n^R\}_{n\in\mathbb{N}}$ will be related to the iteration index by specific refinement strategies of the form

$$r^\delta : \mathbb{N} \to \mathbb{N}.$$

In principle, the refinement sequences are converging to zero and have to be selected in advance; the map r^δ represents a specific integer to integer map (constructed below) that allows an adjustment of the reconstruction accuracy within each iteration step m. As a simple example consider the refinement rule $r^\delta(n) = n$ that chooses for each iteration n the preselected error tolerances ε_n^C, ε_n^A and ε_n^R. Choosing proper refinement strategies $r^\delta(n)$ enables us to establish convergence results and, thanks to the introduced subtleness, several desired regularization results.

For ease of notation we write, if not otherwise stated, instead of $\varepsilon_{r^\delta(n)}^{\{C,A,R\}}$ just the index $r^\delta(n)$, i.e. we abbreviate

$$\mathbf{COARSE}_{\varepsilon^C_{r^\delta(n)}}[\cdot], \quad \mathbf{APPLY}_{\varepsilon^A_{r^\delta(n)}}[\cdot], \quad \text{and} \quad \mathbf{RHS}_{\varepsilon^R_{r^\delta(n)}}[\cdot]$$

by

$$\mathbf{COARSE}_{r^\delta(n)}[\cdot], \quad \mathbf{APPLY}_{r^\delta(n)}[\cdot], \quad \text{and} \quad \mathbf{RHS}_{r^\delta(n)}[\cdot].$$

Note, this does not mean the same accuracy for all three routines, it just means the same index for the accuracy/refinement sequences.

Summarizing the last steps results in the following inexact/approximative variant of (3.8)

$$\tilde{c}^{\,n+1} = \mathbf{COARSE}_{r^\delta(n)}[\tilde{c}^{\,n} - \gamma\mathbf{APPLY}_{r^\delta(n)}[\tilde{c}^{\,n}] + \gamma\mathbf{RHS}_{r^\delta(n)}[y^\delta]]. \tag{3.9}$$

3.2 Regularization Theory for A-Priori Parameter Rules

As mentioned above, the a-priori parameter rule (3.2) for the exact Landweber iteration (3.1) yields an order optimal regularization scheme. The natural question is whether the same holds true for the inexact (nonlinear and adaptive) Landweber iteration (3.9). A positive answer of the latter question essentially relies on the construction of a suitable refinement strategy r^δ.

In order to achieve an optimal convergence rate, we have to establish some preliminary results describing the difference between the exact iteration (3.1) and the inexact iteration (3.9).

Lemma 3.1. *Assume,* $c^0 = \tilde{c}^{\,0}$. *Then, for all* $n \geq 0$,

$$\|c^{n+1} - \tilde{c}^{\,n+1}\| \leq \gamma \sum_{i=0}^{n} (1 + \gamma \|S\|)^i (\varepsilon^C_{r^\delta(n-i)}/\gamma + \varepsilon^A_{r^\delta(n-i)} + \varepsilon^R_{r^\delta(n-i)}). \tag{3.10}$$

The latter lemma allows now to prove that the truncated inexact Landweber iteration (3.9) is an order optimal regularization method. The regularization method R_α can be described with the help of an adequate refinement map r^δ and the a-priori parameter rule (3.2).

Definition 3.2 (Regularization method with a-priori parameter rule).

i) Given sequences of error tolerances $\{\varepsilon_n^{\{C,A,R\}}\}_{n\in\mathbb{N}}$ and routines **COARSE**, **APPLY** and **RHS** defined as above,

ii) for $\delta > 0$ with $\|y^\delta - y\| \leq \delta$ derive the truncation index $n_*(\delta, \rho)$ as in (3.2),

iii) define the quantities

$$C_{n,r^\delta} := \sum_{i=0}^{n} (1 + \gamma \|S\|)^i (\varepsilon^C_{r^\delta(n-i)} + \beta(\varepsilon^A_{r^\delta(n-i)} + \varepsilon^R_{r^\delta(n-i)})),$$

iv) choose the map r^δ such that C_{n_*-1,r^δ} satisfies

$$C_{n_*-1,r^\delta} \leq \delta^{\nu/(\nu+1)} \rho^{1/(\nu+1)},$$

v) define the regularization

$$R_\alpha g^\delta := \mathcal{F}^* \tilde{c}^{\,\delta}_{n_*}$$

with regularization parameter $\alpha = 1/n_*(\delta, \rho)$.

Theorem 3.3 (Regularization result). *Let the truncation index* $n_* = n_*(\delta, \rho)$ *be as in* (3.2). *Then, the inexact Landweber iteration* (3.9) *truncated at index* n_* *and updated with the refinement strategy* r^δ *(satisfying iv) in* Definition 3.2*) yields for* $\alpha(\delta, \rho) = 1/n_*(\delta, \rho)$ *a regularization method* R_α, *which is for all* $\nu > 0$ *and* $0 < \gamma < 2/\|S\|^2$ *order optimal.*

3.3 Regularization Theory by A-Posteriori Parameter Rules

The exact Landweber iteration (3.1) combined with the discrepancy principle (3.3) and (3.4) yields a regularization method. In what follows we show how this result carries over to (3.9).

The application of the discrepancy principle (3.3) and (3.4) requires a frequent evaluation of the residual discrepancy $\|Ax^n - y^\delta\|$. Therefore, we have to propose a function that is numerical implementable and approximates the residual, preferably by means of **APPLY** and **RHS**.

Definition 3.4. For some $y \in Y$, $c \in \ell_2(\Lambda)$ and some $n \geq 0$ the approximate discrepancy is defined by

$$(\mathbf{RES}_n[c, y])^2 := \langle \mathbf{APPLY}_n[c], c\rangle - 2\langle \mathbf{RHS}_n[y], c\rangle + \|y\|^2. \tag{3.11}$$

The following lemma gives a result on the distance between the exact function space residual discrepancy $\|Ax - y\|$ and its inexact version $\mathbf{RES}_n[c, y]$.

Lemma 3.5. *For $c \in \ell_2(\Lambda)$ with $Fc = x$, $y \in Y$ and some integer $n \geq 0$ it holds*

$$|\,\|Ax - y\|^2 - (\mathbf{RES}_n[c, y])^2| \leq (\varepsilon_n^A + 2\varepsilon_n^R)\,\|c\|\,. \tag{3.12}$$

To achieve convergence of (3.9), we have to elaborate under which conditions a decay of the approximation errors $\|\tilde{c}_n - c^\dagger\|$ can be ensured.

Lemma 3.6. *Let $\delta > 0$, $0 < c < 1$, $0 < \gamma < 2/(3\|S\|)$ and $n_0 \geq 1$. If there exists for $0 \leq n \leq n_0$ a refinement strategy $r^\delta(n)$ such that $\mathbf{RES}_{r^\delta(n)}[\tilde{c}^{\,n}, y^\delta]$ fulfills*

$$c(\mathbf{RES}_{r^\delta(n)}[\tilde{c}^{\,n}, y^\delta])^2 > \frac{\delta^2 + C_{r^\delta(n)}(\tilde{c}^{\,n})}{1 - \frac{3}{2}\gamma\,\|S\|}, \tag{3.13}$$

then, for $0 \leq n \leq n_0$, the approximation errors $\|\tilde{c}^{\,n} - c^\dagger\|$ decrease monotonically.

The above Lemma 3.6 holds in particular for exact data, i.e. $\delta = 0$. In this case, condition (3.13) simplifies to

$$c(\mathbf{RES}_{r(m)}[\tilde{c}_{\,m}, y])^2 \geq \frac{C_{r(n)}(\tilde{c}^{\,n})}{1 - \frac{3}{2}\gamma\,\|S\|}. \tag{3.14}$$

To prove convergence, we follow the suggested proceeding in [38] and introduce an updating rule (U) for the refinement strategy r:

U(i) Let $r(0)$ be the smallest integer ≥ 0 with

$$c(\mathbf{RES}_{r(0)}[\tilde{c}_0, y])^2 \geq \frac{C_{r(0)}(\tilde{c}_0)}{1 - \frac{3}{2}\gamma \|S\|}, \tag{3.15}$$

if $r(0)$ with (3.15) does not exist, stop the iteration, set $n_0 = 0$.

U(ii) If for $n \geq 1$

$$c(\mathbf{RES}_{r(n-1)}[\tilde{c}^n, y])^2 \geq \frac{C_{r(n-1)}(\tilde{c}^n)}{1 - \frac{3}{2}\gamma \|S\|}, \tag{3.16}$$

set $r(n) = r(n-1)$

U(iii) if

$$c(\mathbf{RES}_{r(n-1)}[\tilde{c}^n, y])^2 < \frac{C_{r(n-1)}(\tilde{c}^n)}{1 - \frac{3}{2}\gamma \|S\|}, \tag{3.17}$$

set $r(n) = r(n-1) + j$, where j is the smallest integer with

$$c(\mathbf{RES}_{r(n-1)+j}[\tilde{c}^n, y])^2 \geq \frac{C_{r(n-1)+j}(\tilde{c}^n)}{1 - \frac{3}{2}\gamma \|S\|}, \tag{3.18}$$

U(iv) if there is no integer j with (3.18), then stop the iteration, set $n_0 = n$.

Lemma 3.7. *Let $\delta = 0$ and $\{\tilde{c}^n\}_{n\in\mathbb{N}}$ be the sequence of iterates* (3.9). *Assume the updating rule* (U) *for r was applied. Then, if the iteration never stops,*

$$\sum_{n=0}^{\infty} (\mathbf{RES}_{r(n)}[\tilde{c}^n, y])^2 \leq \frac{1}{\beta(1-c)(1-\frac{3}{2}\gamma\|S\|)} \|\tilde{c}^n - c^\dagger\|^2. \tag{3.19}$$

If the iteration stops after n_0 steps,

$$\sum_{n=0}^{n_0-1} (\mathbf{RES}_{r(n)}[\tilde{c}^n, y])^2 \leq \frac{1}{\beta(1-c)(1-\frac{3}{2}\gamma\|S\|)} \|\tilde{c}^0 - c^\dagger\|^2. \tag{3.20}$$

Combining the monotone decay of the approximation errors and the uniform boundedness of the accumulated discrepancies enables strong convergence of iteration (3.9) towards a solution of the inverse problem for exact data $y^\delta = y$.

Theorem 3.8. *Let $x^\dagger$ denote the generalized solution of the given inverse problem. Suppose $\tilde{c}^n$ is computed* (3.9) *with exact data y in combination with updating rule* (U) *for the refinement strategy r. Then, for arbitrarily chosen $\tilde{c}^0$ the sequence $\tilde{c}^n$*

converges in norm, i.e.

$$\lim_{n\to\infty} \tilde{c}^{\,n} = c^\dagger$$

with

$$x^\dagger = \mathcal{F}^* c^\dagger.$$

The convergence of the inexact Landweber iteration for noisy data relies on a comparison between the individual noise free and noisy iterations. For a comparison it is essential to analyze the δ-dependence of **COARSE**, **APPLY** and **RHS**; in particular for $\delta \to 0$. For a given error level $\|v^\delta - v\| \leq \delta$ the routines (here just exemplarily stated for **COARSE**, but must hold for all three routines) should fulfill for any fixed $\varepsilon > 0$

$$\|\mathbf{COARSE}_\varepsilon(v^\delta) - \mathbf{COARSE}_\varepsilon(v)\| \to 0 \quad \text{as } \delta \to 0. \tag{3.21}$$

To ensure (3.21), the three routines as proposed in [50] must be slightly adjusted, which is demonstrated in great details for the **COARSE** routine only but must be done for **APPLY** and **RHS** accordingly.

The definition of **COARSE** as proposed in [50] (with a slight modified ordering of the output entries) is as follows

$\mathbf{COARSE}_\varepsilon[v] \to v_\varepsilon$

i) Let V be the set of non-zero coefficients of v, ordered by their original indexing in v. Define $q := \lceil \log(\frac{(\#V)^{1/2}\|v\|}{\varepsilon}) \rceil$.

ii) Divide the elements of V into bins $V_0, \ldots, V_q$, where for $0 \leq k < q$

$$V_k := \{v_i \in V : 2^{-k-1}\|v\| < |v_i| \leq 2^{-k}\|v\|\}, \tag{3.22}$$

and possible remaining elements are put into V_q. Let the elements of a single V_k be also ordered by their original indexing in v. Denote the vector obtained by subsequently extracting the elements of $V_0, \ldots, V_q$ by $\gamma(v)$.

iii) Create v_ε by extracting elements from $\gamma(v)$ and putting them at the original indices, until the smallest l is found with

$$\|v - v_\varepsilon\|^2 = \sum_{i>l} |\gamma_i(v)|^2 < \varepsilon^2. \tag{3.23}$$

The integer q in i) is chosen such that $\sum_{v_i \in V_q} |v_i|^2 < \varepsilon^2$, i.e. the elements of V_q are not used to build v_ε in iii).

Keeping the original order of the coefficients of v in V_k, the output vector of **COARSE** becomes unique. This "natural" ordering does not cause any extra computational cost.

The goal is to construct a noise dependent output vector that converges to the noise free output vector as $\delta \to 0$. To achieve this, the uniqueness of **COARSE** must be ensured. The non-uniqueness of **COARSE** is through the bin sorting procedure which is naturally non-unique as long as the input vectors are noisy (i.e. different noisy versions of the same vector result in significantly different output vectors). This leads to the problem that the index in $\gamma(v^\delta)$ of some noisy element v_i^δ can differ to the index in $\gamma(v)$ of its noise free version v_i. To overcome this drawback for (at least) sufficiently small δ, we define a noise dependent version $\mathbf{COARSE}^\delta$.

$\mathbf{COARSE}_\varepsilon^\delta[v^\delta] \to v_\varepsilon^\delta$

i) Let V^δ be the set of non-zero coefficients of v^δ ordered by their indexing in v^δ. Define $q^\delta := \lceil \log(\frac{(\#V^\delta)^{1/2}(\|v^\delta\|+\delta)}{\varepsilon}) \rceil$.

ii) Divide the elements of V^δ into bins $V_0^\delta, \ldots, V_{q^\delta}^\delta$, where for $0 \le k < q^\delta$

$$V_k^\delta := \{v_i^\delta \in V^\delta : 2^{-k-1}(\|v^\delta\| + \delta) + \delta < |v_i^\delta| \le 2^{-k}(\|v^\delta\| + \delta) + \delta\}, \tag{3.24}$$

and possible remaining elements are put into $V_{q^\delta}^\delta$. Again, let the elements of a single V_k^δ be ordered by their indexing in v^δ. Denote the vector obtained by the bin sorting process by $\gamma^\delta(v^\delta)$.

iii) Create v_ε^δ by extracting elements from $\gamma^\delta(v^\delta)$ and putting them on the original places, until the first index l^δ is found with

$$\|v^\delta - v_\varepsilon^\delta\|^2 = \|v^\delta\|^2 - \sum_{1 \le i \le l^\delta} |\gamma_i^\delta(v^\delta)|^2 < \varepsilon^2 - (l^\delta + 1)\delta(2\|v^\delta\| + \delta). \tag{3.25}$$

The latter definition of $\mathbf{COARSE}^\delta$ enables us to achieve the desired property (3.21).

Lemma 3.9. *Given $\varepsilon > 0$ and $\delta > 0$. For arbitrary finite length vectors v, $v^\delta \in \ell_2$ with $\|v^\delta - v\| \le \delta$, the routine* $\mathbf{COARSE}^\delta$ *is convergent in the sense that*

$$\|\mathbf{COARSE}_\varepsilon^\delta[v^\delta] - \mathbf{COARSE}_\varepsilon[v]\| \to 0 \quad \text{as } \delta \to 0. \tag{3.26}$$

Achieving convergence of the inexact iteration, we introduce as for the noise free situation an updating rule which we denote (D). The updating rule (D) is based on the refinement strategy $r^\delta(n)$.

D(i) Let $r^\delta(0)$ be the smallest integer ≥ 0 with

$$c(\mathbf{RES}_{r^\delta(0)}[\tilde{c}\,^\delta_0, y^\delta])^2 \geq \frac{\delta^2 + C_{r^\delta(0)}(\tilde{c}\,^\delta_0)}{1 - \frac{3}{2}\beta\,\|S\|}, \tag{3.27}$$

if $r^\delta(0)$ does not exist, stop the iteration, set $n_* = 0$.

D(ii) If for $n \geq 1$

$$c(\mathbf{RES}_{r^\delta(n-1)}[\tilde{c}\,^\delta_n, y^\delta])^2 \geq \frac{\delta^2 + C_{r^\delta(n-1)}(\tilde{c}\,^\delta_n)}{1 - \frac{3}{2}\beta\,\|S\|}, \tag{3.28}$$

set $r^\delta(n) = r^\delta(n-1)$

D(iii) if

$$c(\mathbf{RES}_{r^\delta(n-1)}[\tilde{c}\,^\delta_n, y^\delta])^2 < \frac{\delta^2 + C_{r^\delta(n-1)}(\tilde{c}\,^\delta_n)}{1 - \frac{3}{2}\beta\,\|S\|}, \tag{3.29}$$

set $r^\delta(n) = r^\delta(n-1) + j$, where j is the smallest integer with

$$c(\mathbf{RES}_{r^\delta(n-1)+j}[\tilde{c}\,^\delta_n, y^\delta])^2 \geq \frac{\delta^2 + C_{r^\delta(n-1)+j}(\tilde{c}\,^\delta_n)}{1 - \frac{3}{2}\beta\,\|S\|} \tag{3.30}$$

and

$$C_{r^\delta(m-1)+j}(\tilde{c}\,^\delta_m) > c_1\delta^2. \tag{3.31}$$

D(iv) If (3.29) holds and no j with (3.30),(3.31) exists, then stop the iteration, set $n^\delta_* = n$.

Theorem 3.10. *Let $x^\dagger$ be the solution of the inverse problem for exact data $y \in \operatorname{Ran} A$. Suppose that for any $\delta > 0$ and y^δ with $\|y^\delta - y\| \leq \delta$ the adaptive approximation $\tilde{c}\,^\delta_n$ is derived by the inexact Landweber iteration* (3.9) *in combination with rule* (D) *for r^δ and stopping index n^δ_*. Then, the family of R_α defined through*

$$R_\alpha y^\delta := \mathcal{F}^*\tilde{c}\,^\delta_{n^\delta_*} \quad \text{with } \alpha = \alpha(\delta, y^\delta) = \frac{1}{n^\delta_*}$$

yields a regularization of the ill-posed operator A, i.e. $\|R_\alpha y^\delta - x^\dagger\|_X \to 0$ as $\delta \to 0$.

4 Tikhonov Regularization with Sparsity Constraints

In this section we turn now to nonlinear inverse and ill-posed problems. The focus is on the generalization of Tikhonov regularization as it was introduced in Section 2 (see formula (2.5)) to nonlinear problems. In particular, we consider those operator equations in which the solution x has a *sparse* series expansion $x = \sum_{\lambda\in\Lambda} c_\lambda \phi_\lambda$ with respect to a preassigned basis or frame, i.e. the series expansion of x has only a very small number of non-vanishing coefficients c_λ, or that x is compressible (meaning that x can be well-approximated by a sparse series expansion).

4.1 Regularization Result for A-Priori Parameter Rules

We consider the operator equation $F(x) = y$ and assume F is possibly ill-posed and maps between Hilbert spaces X and Y and we suppose there are only noisy data y^δ with $\|y^\delta - y\| \le \delta$ available. The natural generalization of Tikhonov's variational formulation is then given by

$$J_\alpha(x) = \|F(x) - y^\delta\|^2 + \alpha\|x\|^2. \tag{4.1}$$

The second term determines the properties of the solution. In the given setting the penalty term is a quadratic Hilbert space norm ensuring finite energy of the solution. The minimizer is due to the convexity and differentiability of $\|\cdot\|^2$ also very easy to compute. However, for certain classes of inverse problems, e.g. in medical or astrophysical imaging or signal peak analysis, such Hilbert space constraints seem not to be best suited, because they lead to over-smoothed solutions implying that jumps and edges cannot be nicely reconstructed. Therefore, alternatives are required that can perform much better. An alternative that may circumvent the mentioned drawbacks are so-called sparsity measures. Prominent examples of sparsity measures are ℓ_p-norms, $0 < p < 2$, on the coefficients of the series expansions of the solution to be reconstructed. But also much more general constraints such as the wide class of convex, one-homogeneous and weakly lower semi-continuous constraints are possible, see e.g. [3, 33, 34, 36, 49] or [9, 11].

In what follows we restrict ourselves to ℓ_p-norm constraints. Once a frame is preassigned, we know that for every $x \in X$ there is a sequence c such that $x = \mathcal{F}^*c$, and therefore the given operator equation can be expressed as $F(\mathcal{F}^*c) = y$. Consequently, we can define, for a given a-priori guess $\bar{c} \in \ell_2(\Lambda)$, an adequate Tikhonov functional by

$$J_\alpha(c) = \|F(\mathcal{F}^*c) - y^\delta\|^2 + \alpha\Psi(c, \bar{c}) \tag{4.2}$$

with minimizer

$$c_\alpha^\delta := \arg\min_{c\in\ell_2(\Lambda)} J_\alpha(c).$$

To specify Ψ, we define

$$\Psi_{p,w}(c) := \Big(\sum_{\lambda\in\Lambda} w_\lambda |c_\lambda|^p\Big)^{1/p},$$

where $w = \{w_\lambda\}_{\lambda\in\Lambda}$ is a sequence of weights with $0 < C < w_\lambda$. A popular choice for the ansatz system $\{\phi_\lambda : \lambda \in \Lambda\}$ are wavelet bases or frames. In particular, for orthonormal wavelet bases and for properly chosen weights w one has $\Psi_{p,w}(c) \sim \|c\|_{B^s_{p,p}}$, where $B^s_{p,p}$ denotes a standard Besov space. In this section we restrict the analysis to either $\Psi(c,\bar{c}) = \Psi_{p,w}(c-\bar{c})$ or $\Psi(c,\bar{c}) = \Psi^p_{p,w}(c-\bar{c})$.

For c^δ_α as a minimizer of (4.2) we can achieve for any of the introduced sparsity measures regularization properties if the following assumptions hold true:

(i) F is strongly continuous, i.e. $c^n \overset{w}{\to} c \Rightarrow F(c^n) \to F(c)$,

(ii) a-priori parameter available with $\alpha(\delta) \to 0$ and $\delta^2/\alpha(\delta) \to 0$ as $\delta \to 0$,

(iii) $x^\dagger$ as well $\bar{c}$ have finite value of $\Psi_{p,\omega}$.

Here, we denote by the symbol $\overset{w}{\to}$ the weak convergence.

Theorem 4.1. *Suppose* (i) *and* (iii) *hold and that we are given a sequences* $\delta_k \to 0$ *and* $\alpha(\delta_k)$ *with* (ii). *Then the sequence of minimizers* $c^{\delta_k}_{\alpha_k}$ *has a convergent subsequence that converges with respect to* Ψ *towards a solution of* $F(\mathcal{F}^*c) = y$. *If the solution is unique, the whole sequence converges with respect to* Ψ, *i.e.*

$$\lim_{k\to\infty} \Psi(c^{\delta_k}_{\alpha_k} - c^\dagger) = 0. \tag{4.3}$$

Consequently, the variational Tikhonov approach with properly chosen sparsity constraints is a regularization method.

4.2 Convergence Rates for A-Priori Parameter Rules

The convergence in (4.3) can be arbitrarily slow. Therefore, conditions for establishing convergence rates need to be achieved. As the analysis that is required for nonlinear operator equations can be under several conditions on the operator F reduced to the study of the linear operator case, we limit the discussion to the linear case for which convergence rates can be shown. The results within this Section have been first published in [48].

Consider the linear and ill-posed operator equation

$$\begin{aligned} \tilde{A}x &= g \\ \tilde{A} &: X_{p,\omega} \to L_2(\Omega). \end{aligned} \tag{4.4}$$

Here, $X_{p,w}$ denotes a Banach space which is a subspace of $L_2(\Omega)$, with parameters $p \in (1,2)$ and $w = \{w_\lambda\}_{\lambda\in\Lambda}$, where Ω is a bounded open subset of $\mathbb{R}^d$, with $d \geq 1$, and Λ is an index set of (possibly tuples of) integer indices. Although one could employ more general separable Hilbert spaces than $L_2(\Omega)$, we consider here the Lebesgue space case, for simplicity.

We are in particular interested in the reconstruction of solutions of (4.4) that admit a sparse structure with respect to a given basis in the Banach space $X_{p,w}$. In these cases it is desirable to choose a regularization method that also promotes a sparse reconstruction. For instance, suitable choices for the spaces $X_{p,w}$ are the Besov spaces $B^s_{p,p}$ with $p \in (1,2)$, in case of a sufficiently smooth wavelet basis and properly chosen weights – see, e.g., [5, 37] for detailed discussions.

Instead of solving the above equation in a function space setting, we will transform it into a sequential setting. By choosing a suitable orthonormal basis $\Phi = \{\phi_\lambda : \lambda \in \Lambda\}$ for the space $L_2(\Omega)$, both x and $\tilde{A}x$ can be expressed with respect to Φ. Thus,

$$\tilde{A}x = \sum_{\lambda'} \sum_{\lambda} \langle x, \phi_\lambda \rangle \langle \tilde{A}\phi_\lambda, \phi_{\lambda'} \rangle \, \phi_{\lambda'}. \tag{4.5}$$

Defining the infinite dimensional matrix A and vectors c, y by

$$A = (\langle \tilde{A}\phi_\lambda, \phi_{\lambda'} \rangle)_{\lambda,\lambda' \in \Lambda}, \quad c = (\langle c, \phi_\lambda \rangle)_{\lambda \in \Lambda}, \quad y = (\langle g, \phi_\lambda \rangle)_{\lambda \in \Lambda}, \tag{4.6}$$

equation (4.4) can be reformulated as an (infinite) linear system

$$Ac = b. \tag{4.7}$$

To specify the spaces $X_{p,w}$, we define for a given orthonormal basis Φ and positive weights w

$$x \in X_{p,w} \iff \sum_\lambda w_\lambda |\langle x, \phi_\lambda \rangle|^p < \infty,$$

i.e., c belongs to the weighted sequence space $\ell_{p,w}$, where

$$\ell_{p,w} = \left\{ c = \{c_\lambda\}_{\lambda \in \Lambda} : \|c\|_{p,w} = \left(\sum_\lambda w_\lambda |c_\lambda|^p \right)^{\frac{1}{p}} < \infty \right\}.$$

Since $\ell_p \subset \ell_q$ with $\|c\|_q \le \|c\|_p$ for $p \le q$, one also has $\ell_{p,w} \subset \ell_{q,w'}$ for $p \le q$ and $w' \le w$. In particular, if the sequence of weights is positive and bounded from below, i.e., $0 < C \le w_\lambda$ for some $C > 0$, then $\ell_{p,w} \subset \ell_2$ for $p \le 2$.

With the above discretization, we consider the sequence space operator equation

$$\begin{aligned} Ac &= y \\ A &: \ell_{p,w} \to \ell_2, \end{aligned} \tag{4.8}$$

where A is a linear and bounded operator. Now we are prepared to investigate convergence rates for Tikhonov regularization with sparsity constraints, where the approximation of the solution is obtained as a minimizer of

$$J_\alpha(c) = \|Ac - y^\delta\|^2 + 2\alpha \Psi^p_{p,w}(c), \tag{4.9}$$

with regularization parameter $\alpha > 0$. Note that the function $\Psi^p_{p,w}$ is strictly convex since the p-powers of the norms are so. In addition, the function $\Psi^p_{p,w}$ is Fréchet differentiable. In order to obtain convergence rates we need the following source conditions:

(SC) $\Psi_{p,w}^{p}{}'(c^\dagger) = A^* v$, for some $v \in \ell_2$.

(SC I) $\Psi_{p,w}^{p}{}'(c^\dagger) = A^* A\hat{v}$, for some $\hat{v} \in \ell_{p,w}$.

For the above given source conditions we get the following convergence rates:

Proposition 4.2. *Assume that the noisy data y^δ fulfill $\|y - y^\delta\| \le \delta$ and that $p \in (1, 2)$.*

i) *If* (SC) *and $\alpha \sim \delta$, then the following error estimates hold for the minimizer c_α^δ of* (4.9)*:*
$$\|c_\alpha^\delta - c^\dagger\|_{p,w} = \mathcal{O}(\delta^{\frac{1}{2}}), \quad \|Ac_\alpha^\delta - y\| = \mathcal{O}(\delta).$$

ii) *If* (SC I) *and $\alpha \sim \delta^{\frac{2}{p+1}}$, then*
$$\|c_\alpha^\delta - c^\dagger\|_{p,w} = \mathcal{O}(\delta^{\frac{p}{p+1}}), \quad \|Ac_\alpha^\delta - y\| = \mathcal{O}(\delta).$$

Recently it is shown in [26] that under the assumption that $c^\dagger$ is sparse and (SC) holds, the convergence rate is $\mathcal{O}(\delta^{\frac{1}{p}})$ for $p \in [1, 2)$ (thus, up to $\mathcal{O}(\delta)$) with respect to the ℓ_2 norm of $c_\alpha^\delta - c^\dagger$ (which is weaker than the $\ell_{p,w}$ norm for $p < 2$). These rates are already higher, when $p < 1.5$, than the "superior limit" of $\mathcal{O}(\delta^{\frac{2}{3}})$ established for quadratic regularization. This indicates that the assumption of sparsity is a very strong source condition. Next we give a converse results for the first source condition, which shows that the above given convergence rate can only hold if the source condition is fulfilled.

Proposition 4.3. *If $\|y - y^\delta\| \le \delta$, the rate $\|Ac_\alpha^\delta - y\| = \mathcal{O}(\delta)$ holds and c_α^δ converges to $c^\dagger$ in the $\ell_{p,w}$ weak topology as $\delta \to 0$ and $\alpha \sim \delta$, then $\Psi_{p,w}^{p}{}'(c^\dagger)$ belongs to the range of the adjoint operator A^*.*

In what follows, we characterize sequences that fulfill the source condition (SC I). To this end we introduce the power of a sequence by
$$w^t = \{w_\lambda^t\}_{\lambda \in \Lambda}, \quad t \in \mathbb{R},$$
and will consider the operator
$$A : \ell_{p',w'} \to \ell_2. \tag{4.10}$$
Please note that $\|\cdot\|_{p,w}$ is still used as penalty and that p, p' and w, w' are allowed to be different, respectively. In the sequel, the dual exponents to the given p, p' will be denoted by q, q'. Consider first the case $p, p' > 1$.

Proposition 4.4. *Let $p, p' > 1$, the operator A and $\Psi_{p,w}^{p}$ be given as above, and assume that $p \le p'$, $w' \le w$ holds true. Then a solution $c^\dagger$ of $Ac = y$ fulfilling $A^*v = \Psi_{p,w}^{p}{}'(c^\dagger)$ satisfies*
$$c^\dagger \in \ell_{(p-1)q',(w')^{-q'/p'} \cdot w^{q'}}. \tag{4.11}$$

The previous result states only a necessary condition. In order to characterize the smoothness condition in terms of spaces of sequences, we relate the spaces to range(A^*):

Proposition 4.5. *Let $p, p' > 1$, the operator A and $\Psi^p_{p,w}$ be given as above, and assume that $p \le p'$, $w' \le w$ holds true. Moreover, assume*

$$\text{range}(A^*) = \ell_{\tilde{q},\tilde{w}^{-\tilde{q}/\tilde{p}}} \subset \ell_{q',w'^{-q'/p'}}$$

for some $\tilde{p}$, $\tilde{q} > 1$. Then each sequence

$$c^\dagger \in \ell_{(p-1)\tilde{q},\tilde{w}^{-\tilde{q}/\tilde{p}} \cdot w^{\tilde{q}}} \tag{4.12}$$

fulfills the smoothness condition (SC).

The above derived conditions on sequences fulfilling a source condition (SC) mean in principle that the sequence itself has to converge to zero fast enough. They can also be interpreted in terms of smoothness of an associated function: If the function system Φ in (4.5), (4.6) is formed by a wavelet basis, then the norm of a function in the Besov space $B^s_{p,p}$ coincides with a weighted ℓ_p norm of its wavelet coefficients and properly chosen weights [8]. In this sense, the source condition requires the solution to belong to a certain Besov space. The assumption on range(A^*) in Proposition 4.5 then means that the range of the dual operator equals a Besov space. Similar assumptions were used for the analysis of convergence rates for Tikhonov regularization in Hilbert scales, see [31, 30, 27].

4.3 Regularization Result for A-Posteriori Parameter Rules

We deal with Morozov's discrepancy principle as an a-posteriori parameter choice rule for Tikhonov regularization with general convex penalty terms Ψ. The results presented in this Section were first published in [1]. In this framework it can be shown that a regularization parameter α fulfilling the discrepancy principle exists, whenever the operator F satisfies some basic conditions, and that for suitable penalty terms the regularized solutions converge to the true solution in the topology induced by Ψ. It is illustrated that for this parameter choice rule it holds $\alpha \to 0$, $\delta^q/\alpha \to 0$ as the noise level δ goes to 0.

We assume the operator $F : \text{dom}(F) \subset X \to Y$ between reflexive Banach spaces X, Y, with $0 \in \text{dom}(F)$, to be weakly continuous, $q > 0$ to be fixed, and that the penalty term $\Psi(x)$ fulfills the following condition.

Condition 4.6. Let $\Psi : D(\Psi) \subset X \to \mathbb{R}^+$, with $0 \in \text{dom}(\Psi)$, be a convex functional such that

(i) $\Psi(x) = 0$ if and only if $x = 0$,

(ii) Ψ is weakly lower semicontinuous (w.r.t. the Banach space topology on X),

(iii) Ψ is weakly coercive, i.e. $\|x_n\| \to \infty \Rightarrow \Psi(x_n) \to \infty$.

We want to recover solutions $x \in X$ of $F(x) = y$, where we are given y^δ with $\|y^\delta - y\| \leq \delta$.

Definition 4.7. As before, our regularized solutions will be the minimizers x_α^δ of the Tikhonov-type variational functionals

$$J_\alpha(x) = \begin{cases} \|F(x) - y^\delta\|^q + \alpha\Psi(x) & \text{if } x \in \operatorname{dom}(\Psi) \cap \operatorname{dom}(F), \\ +\infty & \text{otherwise.} \end{cases} \tag{4.13}$$

For fixed y^δ, we denote the set of all minimizers by M_α, i.e.

$$M_\alpha = \{x_\alpha^\delta \in X \ : \ J_\alpha(x_\alpha^\delta) \leq J_\alpha(x), \ \forall x \in X\}. \tag{4.14}$$

We call a solution $x^\dagger$ of equation $F(x) = y$ an Ψ-minimizing solution if

$$\Psi(x^\dagger) = \min\{\Psi(x) : F(x) = y\},$$

and denote the set of all Ψ-minimizing solutions by $\mathcal{L}$. Throughout this paper we assume that $\mathcal{L} \neq \emptyset$.

Morozov's discrepancy principle goes now as follows.

Definition 4.8. For $1 < \tau_1 \leq \tau_2$ we choose $\alpha = \alpha(\delta, y^\delta) > 0$ such that

$$\tau_1\delta \leq \|F(x_\alpha^\delta) - y^\delta\| \leq \tau_2\delta \tag{4.15}$$

holds for some $x_\alpha^\delta \in M_\alpha$.

Condition 4.9. Assume that y^δ satisfies

$$\|y - y^\delta\| \leq \delta < \tau_2\delta < \|F(0) - y^\delta\|, \tag{4.16}$$

and that there is no $\alpha > 0$ with minimizers $x_1, x_2 \in M_\alpha$ such that

$$\|F(x_1) - y^\delta\| < \tau_1\delta \leq \tau_2\delta < \|F(x_2) - y^\delta\|.$$

Then we have the following

Theorem 4.10. *If Condition* 4.9 *is fulfilled, then there are* $\alpha = \alpha(\delta, y^\delta) > 0$ *and* $x_\alpha^\delta \in M_{\alpha(\delta,y^\delta)}$ *such that* (4.15) *holds.*

Based on this existence result, we are able to establish regularization properties.

Condition 4.11. Let $(x^n)_{n\in\mathbb{N}} \subset X$ be such that $x^n \xrightarrow{w} \bar{x} \in X$ and $\Psi(x^n) \to \Psi(\bar{x}) < \infty$, then x^n converges to $\bar{x}$ with respect to Ψ, i.e.,

$$\Psi(x^n - \bar{x}) \to 0.$$

Remark 4.12. Choosing weighted ℓ_p-norms of the coefficients with respect to some frame $\{\phi_\lambda : \lambda \in \Lambda\} \subset X$ as the penalty term, i.e.

$$\Psi_{p,w}(x) = \|x\|_{w,p} = \left(\sum_{\lambda\in\Lambda} w_\lambda |\langle x, \phi_\lambda\rangle|^p\right)^{1/p}, \quad 1 \le p \le 2, \tag{4.17}$$

where $0 < C \le w_\lambda$, satisfies Condition 4.11. Therefore the same automatically holds for $\Psi^p_{p,w}(x)$. Note that these choices also fulfill all the assumptions in Condition 4.6.

Theorem 4.13. *Let $\delta_n \to 0$ and F, Ψ satisfy the Conditions* 4.6, 4.11. *Assume that y^{δ_n} fulfills Condition* 4.9 *and choose $\alpha_n = \alpha(\delta_n, y^{\delta_n})$, $x_n \in M_{\alpha_n}$ such that* (4.15) *holds, then each sequence x_n has a subsequence that converges to an element of $\mathcal{L}$ with respect to Ψ.*

Remark 4.14. If instead of Condition 4.11 the penalty term $\Psi(x)$ satisfies the Kadec property, i.e., $x_n \xrightarrow{w} \bar{x} \in X$ and $\Psi(x_n) \to \Psi(\bar{x}) < \infty$ imply $\|x_n - \bar{x}\| \to 0$, then the convergence in Corollary 4.13 holds with respect to the norm.

Condition 4.15. For all $x^\dagger \in \mathcal{L}$ (see Definition 4.7) we assume that

$$\lim_{t\to 0+} \inf \frac{\|F((1-t)x^\dagger) - y)^q\|}{t} = 0. \tag{4.18}$$

The following lemma provides more insight as to the nature of Condition 4.15.

Lemma 4.16. *Let X be a Hilbert space and $q > 1$. If $F(x)$ is differentiable in the directions $x^\dagger \in \mathcal{L}$ and the derivatives are bounded in a neighbourhood of $x^\dagger$, then Condition* 4.15 *is satisfied.*

Theorem 4.17. *Let F, Ψ satisfy the Conditions* 4.6, 4.15. *Moreover, assume that data y^δ, $\delta \in (0, \delta^*)$, are given such that Condition* 4.9 *holds, where $\delta^* > 0$ is an arbitrary upper bound. Then the regularization parameter $\alpha = \alpha(\delta, y^\delta)$ obtained from Morozov's discrepancy principle (see Definition* 4.8*) satisfies*

$$\alpha(\delta, y^\delta) \to 0 \quad \textit{and} \quad \frac{\delta^q}{\alpha(\delta, y^\delta)} \to 0 \quad \textit{as } \delta \to 0.$$

Remark 4.18. In the proof of Theorem 4.17 we have used that $\|F(0)-y\| > 0$, which is an immediate consequence of (4.16). On the other hand, whenever $\|F(0)-y\| > 0$ we can choose

$$0 < \delta^* \leq \frac{1}{\tau_2 + 1}\|F(0) - y\|$$

and for all $0 < \delta < \delta^*$ and y^δ satisfying $\|y - y^\delta\| \leq \delta$ we obtain

$$\|F(0) - y^\delta\| \geq \|F(0) - y\| - \|y - y^\delta\| \geq \|F(0) - y\| - \delta > \tau_2\delta,$$

which is (4.16). Therefore (4.16) can be fulfilled for all δ smaller than some $\delta^* > 0$, whenever $y \neq F(0)$.

4.4 Convergence Rates for A-Posteriori Parameter Rules

Finally, we establish convergence rates with respect to the generalized Bregman distance.

Definition 4.19. Let $\partial\Psi(x)$ denote the subgradient of Ψ at $x \in X$. The generalized Bregman distance with respect to Ψ of two elements $x, z \in X$ is defined as

$$D_\Psi(x, z) = \{D_\Psi^\xi(x, z) : \xi \in \partial\Psi(z) \neq \emptyset\},$$

where

$$D_\Psi^\xi(x, z) = \Psi(x) - \Psi(z) - \langle \xi, x - z\rangle.$$

Condition 4.20. Let $x^\dagger$ be an arbitrary but fixed Ψ-minimizing solution of $F(x) = y$. Assume that the operator $F : X \to Y$ is Gâteaux differentiable and that there is $w \in Y^*$ such that

$$F'(x^\dagger)^* w \in \partial\Psi(x^\dagger). \tag{4.19}$$

Throughout the remainder of this section let $w \in Y^*$ be arbitrary but fixed fulfilling (4.19) and $\xi \in \partial\Psi(x^\dagger)$ be defined as

$$\xi = F'(x^\dagger)^* w. \tag{4.20}$$

Moreover, assume that one of the two following non-linearity conditions holds:

(i) There is $c > 0$ such that for all $x, z \in X$ it holds that

$$\langle w, F(x) - F(z) - F'(z)(x - z)\rangle \leq c\|w\|_{Y^*}\|F(x) - F(z)\|. \tag{4.21}$$

(ii) There are $\rho > 0, c > 0$ such that for all $x \in \mathrm{dom}(F) \cap \mathcal{B}_\rho(x^\dagger)$,

$$\|F(x) - F(x^\dagger) - F'(x^\dagger)(x - x^\dagger)\| \leq c\, D_\Psi^\xi(x, x^\dagger), \tag{4.22}$$

and it holds that

$$c\|w\|_{Y^*} < 1. \tag{4.23}$$

Here, $\mathcal{B}_\rho(x^\dagger)$ denotes a ball around $x^\dagger$ with radius ρ.

Theorem 4.21. *Let the operator F and the penalty term Ψ be such that Condition* 4.6 *and* 4.20 *hold. For all $0 < \delta < \delta^*$ assume that the data y^δ fulfill Condition* 4.9, *and choose $\alpha = \alpha(\delta, y^\delta)$ according to the discrepancy principle in Definition* 4.8. *Then*

$$\|F(x_\alpha^\delta) - F(x^\dagger)\| = \mathcal{O}(\delta), \quad D_\Psi^\xi(x_\alpha^\delta, x^\dagger) = \mathcal{O}(\delta). \tag{4.24}$$

5 Iterated Shrinkage for Nonlinear Ill-Posed Problems

This section is devoted to the elaboration of a numerical scheme to derive a minimizer of

$$J_\alpha(c) = \|F(\mathcal{F}^*c) - y^\delta\|^2 + \alpha\Psi(Bc), \tag{5.1}$$

for some given $\alpha > 0$ and where we have assumed (for simplicity) that $\bar{c} = 0$ motivating the shorthand notation to $\Psi(c)$ for $\Psi(c, 0)$ and where B is an isometric mapping. But, in contrast to Section 4, in which the choice of Ψ was restricted to weighted ℓ_p-norms, we allow in this section a much broader range constraints, namely the wide range of positive, one-homogeneous, lower semi–continuous and convex penalty constraints, where the ℓ_p norm is just one famous example. Further important cases such as the TV measure can be found in [3, 33, 34, 36, 49]. Since we focus here on constraints that work on the basis of frame coefficients, TV–like constraints are not directly applicable here. But there is a remarkable relation between TV penalties and frame coefficient-oriented constraints which can be explained by the inclusion $B_{1,1}^1 \subset BV \subset B_{1,1}^1 - \mathit{weak}$ (in two dimensions), see for further Harmonic analysis on BV [9, 11]. This relation yields a wavelet-based near BV reconstruction when limiting to Haar frames and using a $B_{1,1}^1$ constraint, see for further elaboration [16, 17].

One additional important condition that is necessary for our further analysis is

$$\|c\|_{\ell_2} \leq \Psi(Bc). \tag{5.2}$$

To derive a minimizer of (5.1), we follow the strategies developed for nonlinear problems with quadratic penalties suggested in [44]. These concepts seem to be also adequate when dealing with sparsity, or more general, with one-homogeneous constraints. The idea goes as follows: we replace (5.1) by a sequence of functionals from which we hope that they are easier to treat and that the sequence of minimizers converge in some sense to, at least, a critical point of (5.1). To be more concrete, for some auxiliary $a \in \ell_2$, we introduce the a surrogate functional

$$J_\alpha^s(c, a) := J_\alpha(c) + C\|c - a\|_{\ell_2}^2 - \|F(\mathcal{F}^*c) - F(\mathcal{F}^*a)\|_Y^2 \tag{5.3}$$

and create an iteration process by:

(i) Pick c^0 and some proper constant $C > 0$.

(ii) Derive a sequence $(c^n)_{n=0,1,\ldots}$ by the iteration:

$$c^{n+1} = \arg\min_c J_\alpha^s(c, c^n) \quad n = 0, 1, 2, \ldots . \tag{5.4}$$

As a minor but relatively common restriction, convergence of iterations (5.4) can only be established when the class of operators F is restricted to (twice) Fréchet differentiable operators fulfilling

$$c^n \stackrel{w}{\to} c^\star \Longrightarrow F(\mathcal{F}^* c^n) \to F(\mathcal{F}^* c^\star), \tag{5.5}$$

$$F'(\mathcal{F}^* c^n)^* y \to F'(\mathcal{F}^* c^\star)^* y \quad \text{for all } y, \quad \text{and} \tag{5.6}$$

$$\|F'(\mathcal{F}^* c) - F'(\mathcal{F}^* c')\| \le L C_2 \|c - c'\|. \tag{5.7}$$

These conditions are essentially necessary to establish weak convergence. If F is not equipped with conditions (5.5)–(5.7) as an operator from $X \to Y$, this can be achieved by assuming more regularity of x, i.e. changing the domain of F a little (hoping that the underlying application still fits with modified setting). To this end, we then assume that there exists a function space X^s, and a compact embedding operator $i^s : X^s \to X$. Then we may consider $\tilde{F} = F \circ i^s : X^s \longrightarrow Y$. Lipschitz regularity is preserved. Moreover, if now $x^n \stackrel{w}{\to} x^\star$ in X^s, then $x^n \to x^\star$ in X and, moreover, $\tilde{F}'(x^n) \to \tilde{F}'(x^\star)$ in the operator norm. This argument applies to arbitrary nonlinear continuous and Fréchet differentiable operators $F : X \to Y$ with continuous Lipschitz derivative as long as a function space X^s with compact embedding i^s into X is available.

At a first glance the made assumptions on F might seem to be somewhat restrictive. But compared to usually made assumptions in nonlinear inverse problems they are indeed reasonable and are fulfilled by numerous applications.

All what follows in the remaining section can be comprehensively retraced (including all proofs) in [46].

5.1 Properties of the Surrogate Functional

By the definition of J_α^s in (5.3) it is not clear whether the functional is positive definite or even bounded from below. This will be the case provided the constant C is chosen properly.

For given $\alpha > 0$ and c^0 we define a ball $K_r := \{c \in \ell_2 : \Psi(Bc) \le r\}$, where the radius r is given by

$$r := \frac{\|y^\delta - F(\mathcal{F}^* c^0)\|_Y^2 + 2\alpha \Psi(B c^0)}{2\alpha}. \tag{5.8}$$

This obviously ensures $c^0 \in K_r$. Furthermore, we define the constant C by

$$C := 2C_2 \max\left\{\Big(\sup_{c \in K_r} \|F'(\mathcal{F}^* c)\|\Big)^2, L\sqrt{J_\alpha(c^0)}\right\}, \tag{5.9}$$

where L is the Lipschitz constant of the Fréchet derivative of F and C_2 the upper frame bound in (3.6). We assume that c^0 was chosen such that $r < \infty$ and $C < \infty$.

Lemma 5.1. *Let r and C be chosen by* (5.8), (5.9). *Then, for all $c \in K_r$,*

$$C\|c - c^0\|_{\ell_2}^2 - \|F(\mathcal{F}^*c) - F(\mathcal{F}^*c^0)\|_Y^2 \geq 0$$

and thus, $J_\alpha(c) \leq J_\alpha^s(c, c^0)$.

In our iterative approach, this property carries over to all of the iterates.

Proposition 5.2. *Let c^0, α be given and r, C be defined by* (5.8), (5.9). *Then the functionals $J_\alpha^s(c, c^n)$ are bounded from below for all $c \in \ell_2$ and all $n \in \mathbb{N}$ and have thus minimizers. For the minimizer c^{n+1} of $J_\alpha^s(c, c^n)$ holds $c^{n+1} \in K_r$.*

The proof of the latter Proposition 5.2 directly yields

Corollary 5.3. *The sequences $(J_\alpha(c^n))_{n\in\mathbb{N}}$ and $(J_\alpha^s(c^{n+1}, c^n))_{n\in\mathbb{N}}$ are non-increasing.*

5.2 Minimization of the Surrogate Functionals

To derive an algorithm that approximates a minimizer of (5.1), we elaborate the necessary condition.

Lemma 5.4. *The necessary condition for a minimum of $J_\alpha^s(c, a)$ is given by*

$$0 \in -\mathcal{F} F'(\mathcal{F}^*c)^*(y^\delta - F(\mathcal{F}^*a)) + C\, c - C\, a + \alpha B^* \partial\Psi(Bc). \tag{5.10}$$

This result can be achieved when introducing the functional Θ via the relation $v \in \partial\Theta(c) \Leftrightarrow Bv \in \partial\Psi(Bc)$; then one obtains in the notion of subgradients,

$$\partial J_\alpha^s(c, a) = -2\mathcal{F} F'(\mathcal{F}^*c)^*(y^\delta - F(\mathcal{F}^*a)) + 2C\, c - 2C\, a + 2\alpha\partial\Theta(c).$$

Lemma 5.5. *Let $M(c, a) := \mathcal{F} F'(\mathcal{F}^*c)^*(y^\delta - F(\mathcal{F}^*a))/C + a$. The necessary condition* (5.10) *can then be casted as*

$$c = \frac{\alpha}{C} B^*(I - P_{\mathcal{C}})\left(\frac{C}{\alpha} BM(c, a)\right), \tag{5.11}$$

where $P_{\mathcal{C}}$ is an orthogonal projection onto a convex set $\mathcal{C}$.

To verify Lemma 5.5, one has to establish the relation between Ψ and $\mathcal{C}$. To this end, we consider the Fenchel or so-called dual functional of Ψ, which we will denote by Ψ^*. For a functional $\Psi : X \to \overline{\mathcal{R}}$, the dual function $\Psi^* : \mathcal{X}^* \to \overline{\mathcal{R}}$ is defined by

$$\Psi^*(x^*) = \sup_{x\in X}\{\langle x^*, x\rangle - \Psi(x)\}.$$

Since we have assumed Ψ to be a positive and one homogeneous functional, there exists a convex set $\mathcal{C}$ such that Ψ^* is equal to the indicator function $\chi_{\mathcal{C}}$ over $\mathcal{C}$. Moreover, in a Hilbert space setting, we have total duality between convex sets and positive and one homogeneous functionals, i.e. $\Psi = (\chi_{\mathcal{C}})^*$.

Consequently, with the shorthand notation $M(c,a)$ we may rewrite (5.10),

$$B\frac{M(c,a)-c}{\frac{\alpha}{C}} \in \partial\Psi(Bc),$$

and thus, by convex analysis standard arguments,

$$\frac{C}{\alpha}Bc \in \frac{C}{\alpha}\partial\Psi^*\left(B\frac{M(c,a)-c}{\frac{\alpha}{C}}\right).$$

In order to have an expression by means of projections, we expand the latter formula as follows

$$\begin{aligned} B\frac{M(c,a)}{\frac{\alpha}{C}} &\in B\frac{M(c,a)-c}{\frac{\alpha}{C}} + \frac{C}{\alpha}\partial\Psi^*\left(B\frac{M(c,a)-c}{\frac{\alpha}{C}}\right) \\ &= \left(I + \frac{C}{\alpha}\partial\Psi^*\right)\left(B\frac{M(c,a)-c}{\frac{\alpha}{C}}\right), \end{aligned}$$

which is equivalent to

$$\left(I + \frac{C}{\alpha}\partial\Psi^*\right)^{-1}\left(B\frac{M(c,a)}{\frac{\alpha}{C}}\right) = B\frac{M(c,a)-c}{\frac{\alpha}{C}}.$$

Again, by standard arguments, (for more details, see [46]) it is known that $(I + \frac{C}{\alpha}\partial\Psi^*)^{-1}$ is nothing than the orthogonal projection onto a convex set $\mathcal{C}$, and hence the assertion (5.11) follows.

Lemma 5.5 states that for minimizing (5.3) we need to solve the fixed point equation (5.11). To this end, we introduce the associated fixed point map $\Phi_{\alpha,\mathcal{C}}$ with respect to some α and $\mathcal{C}$, i.e.

$$\Phi_{\alpha,\mathcal{C}}(c,a) := \frac{\alpha}{C}B^*(I - P_{\mathcal{C}})\left(B\frac{M(c,a)}{\frac{\alpha}{C}}\right).$$

In order to ensure contractivity of $\Phi_{\alpha,\mathcal{C}}$ for some generic a we need two standard properties of convex sets, see [7].

Lemma 5.6. *Let K be a closed and convex set in some Hilbert space X, then for all $u \in X$ and all $k \in K$ the inequality $\langle u - P_K u, k - P_K u\rangle \leq 0$ holds true.*

Lemma 5.7. *Let K be a closed and convex set, then for all $u, v \in X$ the inequality*

$$\|u - v - (P_K u - P_K v)\| \leq \|u - v\|$$

holds true.

Thanks to Lemmata 5.6 and 5.7 we obtain

Lemma 5.8. *The mapping* $I - P_{\mathcal{C}}$ *is non-expansive.*

The latter statement provides contractivity of $\Phi_{\alpha,\mathcal{C}}(\cdot, a)$.

Lemma 5.9. *The operator* $\Phi_{\alpha,\mathcal{C}}(\cdot, a)$ *is a contraction, i.e.*

$$\|\Phi_{\alpha,\mathcal{C}}(c, a) - \Phi_{\alpha,\mathcal{C}}(\tilde{c}, a)\|_{\ell_2} \leq q\|c - \tilde{c}\|_{\ell_2} \quad \text{if } q := \frac{C_2 L}{C}\sqrt{J_\alpha(a)} < 1.$$

This consequently leads to

Proposition 5.10. *The fixed point map* $\Phi_{\alpha,\mathcal{C}}(c, c^n)$ *that is applied in* (5.11) *to compute c is due to definition* (5.9) *for all* $n = 0, 1, 2, \ldots$ *and all* $\alpha > 0$ *and* $\mathcal{C}$ *a contraction.*

The last proposition guarantees convergence towards a critical point of $J_\alpha^s(\cdot, c^n)$. This can be sharpened.

Proposition 5.11. *The necessary equation* (5.11) *for a minimum of the functional* $J_\alpha^s(\cdot, c^n)$ *has a unique fixed point, and the fixed point iteration converges towards the minimizer.*

By assuming more regularity on F, the latter statement can be improved a little.

Proposition 5.12. *Let* F *be a twice continuously differentiable operator. Then the functional* $J_\alpha^s(\cdot, c^n)$ *is strictly convex.*

5.3 Convergence Properties

Within this section we establish convergence properties of $(c^n)_{n\in\mathbb{N}}$. In particular, we show that $(c^n)_{n\in\mathbb{N}}$ converges strongly towards a critical point of J_α.

Lemma 5.13. *The sequence of iterates* $(c^n)_{n\in\mathbb{N}}$ *has a weakly convergent subsequence.*

This is an immediate consequence of Proposition 5.2, in which we have shown that for $n = 0, 1, 2, \ldots$ the iterates c^n remain in K_r. Moreover, with the help of Corollary 5.3, we observe the following lemma that is essentially used in the convergence proof.

Lemma 5.14. *For the iterates* c^n *holds* $\lim_{n\to\infty} \|c^{n+1} - c^n\|_{\ell_2} = 0$.

To arrive at weak convergence, we need the following preliminary lemmata. They state properties involving the general constraint Θ. To achieve strong convergence the analysis is limited to the class of constraints given by weighted ℓ_p norms.

Lemma 5.15. *Let Θ be a convex and weakly lower semi–continuous functional. For sequences $v^n \to v$ and $c^n \stackrel{w}{\to} c$, assume $v^n \in \partial\Theta(c^n)$ for all $n \in \mathbb{N}$. Then, $v \in \partial\Theta(c)$.*

Thanks to last Lemma 5.15 we therefore have weak convergence.

Lemma 5.16. *Every subsequence of $(c^n)_{n\in\mathbb{N}}$ has a weakly convergent subsequence $(c^{n_l})_{l\in\mathbb{N}}$ with weak limit $c_\alpha^\star$ that satisfies the necessary condition for a minimizer of J_α,*

$$\mathcal{F}F'(\mathcal{F}^*c_\alpha^\star)^*(y^\delta - F(\mathcal{F}^*c_\alpha^\star)) \in \alpha\partial\Theta(c_\alpha^\star). \tag{5.12}$$

Lemma 5.17. *Let $(c^{n_l})_{l\in\mathbb{N}} \subset (c^n)_{n\in\mathbb{N}}$ with $c^{n_l} \stackrel{w}{\to} c_\alpha^\star$. Then, $\lim_{l\to\infty}\Theta(c^{n_l}) = \Theta(c_\alpha^\star)$*

Combining the previous lemmata and restricting the constraints to weighted ℓ_p norms, we can achieve strong convergence of the subsequence $(c^{n_l})_{l\in\mathbb{N}}$.

Theorem 5.18. *Let $(c^{n_l})_{l\in\mathbb{N}} \subset (c^n)_{n\in\mathbb{N}}$ with $c^{n_l} \stackrel{w}{\to} c_\alpha^\star$. Assume, moreover, that*

$$\Theta(c) = \Psi(c) = \Big(\sum_j w_j|c_j|^p\Big)^{1/p}$$

with $w_j \geq r > 0$ and $1 \leq p \leq 2$. Then the subsequence $(c^{n_l})_{l\in\mathbb{N}}$ converges also in norm.

In principle, the limits of different convergent subsequences of c^n may differ. Let $c^{n_l} \to c_\alpha^\star$ be a subsequence of c^n, and let $c^{n_l\prime}$ the predecessor of c^{n_l} in c^n, i.e. $c^{n_l} = c^i$ and $c^{n_l\prime} = c^{i-1}$. Then we observe, $J_\alpha^s(c^{n_l}, c^{n_l\prime}) \to J_\alpha(c_\alpha^\star)$. Moreover, as we have $J_\alpha^s(c^{n+1}, c^n) \leq J_\alpha^s(c^n, c^{n-1})$ for all n, it turns out that the value of the Tikhonov functional for every limit $c_\alpha^\star$ of a convergent subsequence remains the same, i.e. $J_\alpha(c_\alpha^\star) = \text{const}$.

We summarize our findings and give a simple criterion that ensures strong convergence of the whole sequence $(c^n)_{n\in\mathbb{N}}$ towards a critical point of J_α.

Theorem 5.19. *Assume that there exists at least one isolated limit $c_\alpha^\star$ of a subsequence c^{n_l} of c^n. Then $c^n \to c_\alpha^\star$ as $n \to \infty$. The accumulation point $c_\alpha^\star$ is a minimizer for the functional $J_\alpha^s(\cdot, c_\alpha^\star)$ and fulfills the necessary condition for a minimizer of J_α.*

Moreover, we obtain, $J_\alpha^s(c_\alpha^\star + h, c_\alpha^\star) \geq J_\alpha^s(c_\alpha^\star, c_\alpha^\star) + \frac{C}{2}\|h\|^2$ and with Lemma 5.12 the second assertion in the theorem can be shown. The first assertion of the theorem can be directly taken from [44].

As a summary of the above reasoning we suggest the following implementation of the proposed Tikhonov-based projection iteration.

Iterated (generalized) Shrinkage for nonlinear ill-posed and inverse problems with sparsity constraints	
Given	operator F, its derivative F', matrix B, data y^δ, some initial guess c^0, and $\alpha > 0$
Initialization	$K_r = \{c \in \ell_2 : \Psi(Bc) \leq r\}$ with $r = J_\alpha(c^0)/(2\alpha)$, $C = 2C_2 \max\{\sup_{c\in K_r} \|F'(\mathcal{F}^*c)\|^2, L\sqrt{J_\alpha(c^0)}\}$
Iteration	**for** $n = 0, 1, 2, \ldots$ **until** a preassigned precision / maximum number of iterations **1.** $c^{n+1} = \frac{\alpha}{C} B^*(I - P_{\mathcal{C}})(\frac{C}{\alpha} B\, M(c^{n+1}, c^n))$ by fixed point iteration, and where $M(c^{n+1}, c^n) = c^n + \frac{1}{C}\mathcal{F} F'(\mathcal{F}^*c^{n+1})^*(y^\delta - F(\mathcal{F}^*c^n))$ **end**

5.4 Application of Sparse Recovery to SPECT

This section is devoted to the application of the developed theory to a sparse recovery problem in the field of single photon emission computed tomography (SPECT). SPECT is a medical imaging technique where one aims to reconstruct a radioactivity distribution f from radiation measurements outside the body. The measurements are described by the attenuated Radon transform (ATRT)

$$y = A(f, \mu)(s, \omega) = \int_{\mathbb{R}} f(s\omega^\perp + t\omega) e^{-\int_t^\infty \mu(s\omega^\perp + r\omega)dr} dt. \tag{5.13}$$

As the measurements depend on the (usually also unknown) density distribution μ of the tissue, we have to solve a nonlinear problem in (f, μ). A throughout analysis of the nonlinear ATRT was presented by Dicken [22], and several approaches for its solution were proposed in [4, 29, 56, 57, 43, 40, 41, 42]. If the ATRT operator is considered with

$$D(A) = H_0^{s_1}(\Omega) \times H_0^{s_2}(\Omega),$$

where $H_0^s(\Omega)$ denotes a Sobolev space over a bounded area Ω with zero boundary conditions and smoothness s, then the operator is twice continuous Fréchet differentiable with Lipschitz continuous first derivative, if s_1, s_2 are chosen large enough. A possible choice for these parameters that also reflects the smoothness properties of

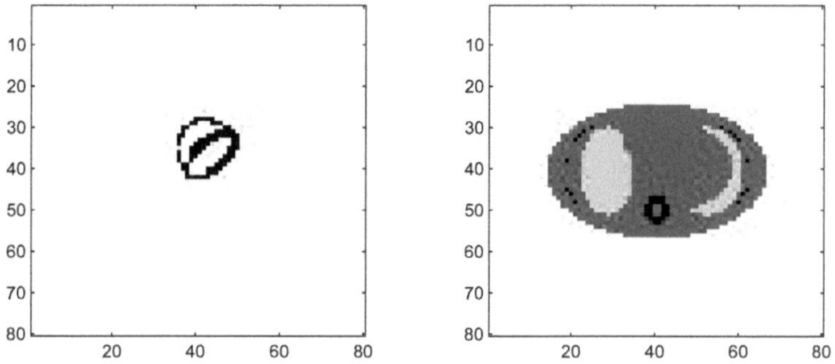

Figure 1. Activity function f^* (left) and attenuation function μ^* (right). The activity function models a cut through the heart.

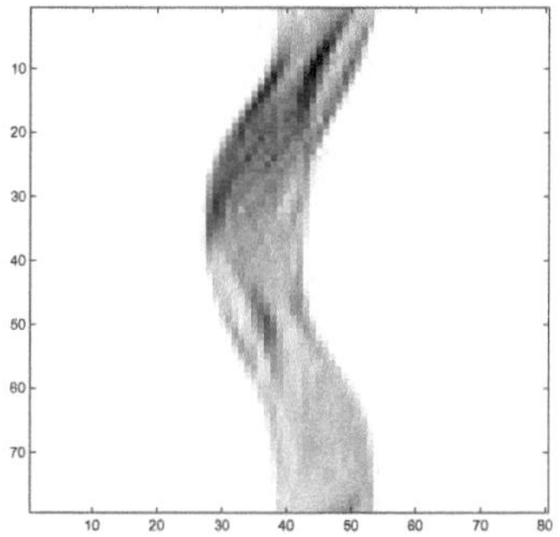

Figure 2. Generated data $g(s, \omega) = A(f^*, \mu^*)(s, \omega)$.

activity and density distribution is $s_1 > 4/9$ and $s_2 = 1/3$. For more details we refer to [41, 21]. Additionally, it has been shown that conditions (5.5), (5.7) hold [39]. For our test computations, we will use the so-called MCAT phantom [51], see Figure 1. Both functions were given as 80×80 pixel images. The sinogram data was gathered on 79 angles, equally spaced over 360 degree, and 80 samples. The sinogram belonging to the MCAT phantom is shown in Figure 2.

At first, we have to choose the underlaying frame or basis on which we put the sparsity constraint. Since a wavelet expansion might sparsely represent images/functions (better than pixel basis), we have chosen a wavelet basis (here Daubechies wavelets of order two) to represent (f, μ), i.e.

$$(f, \mu) = \left(\sum_k c(f)_k \phi_{0,k} + \sum_{j \geq 0, i, k} d(f)^i_{j,k} \psi^i_{j,k}, \sum_k c(\mu)_k \phi_{0,k} + \sum_{j \geq 0, i, k} d(\mu)^i_{j,k} \psi^i_{j,k} \right).$$

For more details we refer the reader to [12]. For our implementation we have chosen $B = I$ and $\Psi(\cdot) = \|\cdot\|_{\ell_1}$. As an observation, the speed of convergence depends heavily on the choice of the constant C in (5.3). According to our convergence analysis,

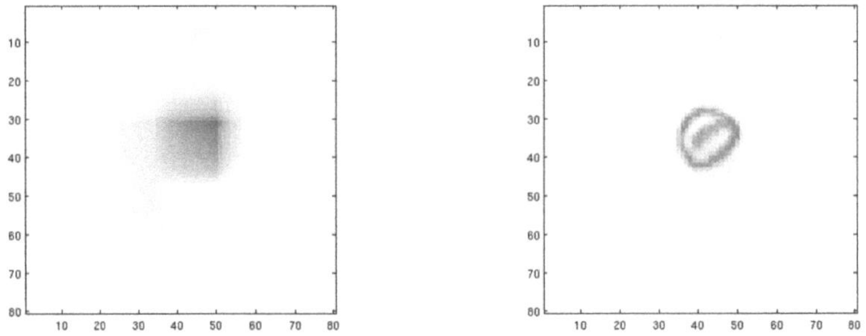

Figure 3. Reconstructions with 5% noise and $\alpha = 350$: sparsity constraint (left) and Hilbert space constraint (right).

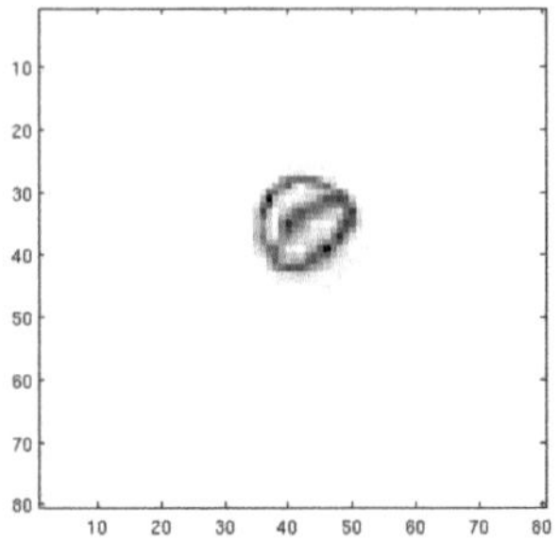

Figure 4. Reconstruction with sparsity constraint and 5% noise. The regularization parameter ($\alpha = 5$) was chosen such that $\|y^\delta - A(f^*, \mu^*)\| \approx 2\delta$.

it has to be chosen reasonably large. However, a large C speeds up the convergence of the inner iteration, but decreases the speed of convergence of the outer iteration. In our example, we needed only 2–4 inner iteration, but the outer iteration required about 5000 iterations. As the minimization in the quadratic case needed much less iterations, this suggests that the speed of convergence also increases with p.

According to (5.2), the functional Ψ will always have a bigger value than $\|\cdot\|_{\ell_2}$. If $\Psi(c)$ is not too large, then it will also dominate $\|c\|^2_{\ell_2}$, which also represents the classical L_2-norm, and we might conclude that reconstructions with the classical quadratic Hilbert space constraint and sparsity constraint will not give comparable results if the same regularization parameter is used. As Ψ is dominant, we expect a smaller (optimal) regularization parameter in the case of the penalty term Ψ. This is confirmed by our first test computations: Figure 3 shows the reconstructions from noisy data where the regularization parameter was chosen as $\alpha = 350$. The reconstruction with the quadratic Hilbert space penalty (we have used the L_2 norm) is already quite good, whereas the reconstruction for the sparsity constraint is still far off. In fact, if we

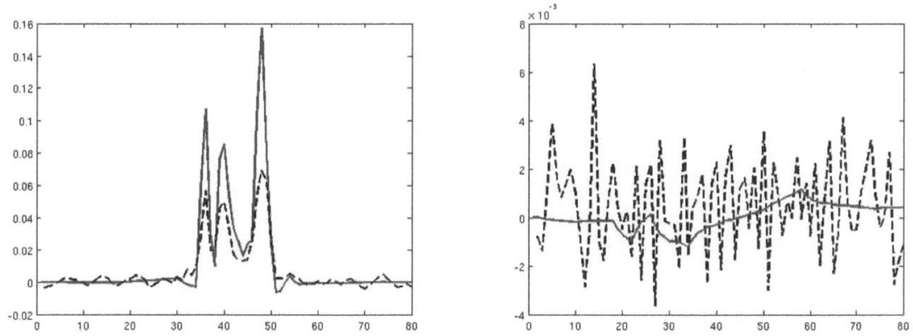

Figure 5. Values of the reconstructed activity function through the heart (left) and well below the heart (right). Solid line: reconstruction with sparsity constraint, dashed line: quadratic Hilbert space penalty.

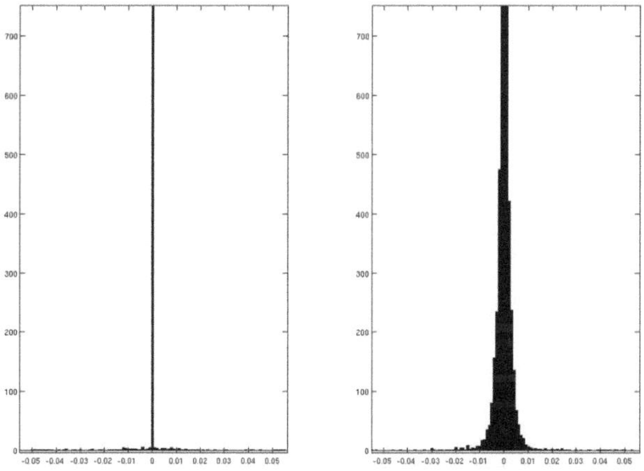

Figure 6. Histogram plot of the wavelet coefficient of the reconstructions. Left: sparsity constraint, Right: quadratic Hilbert space constraint.

consider Morozov's discrepancy principle, then the regularization parameter in the quadratic case has been chosen optimal, as we observe

$$\|y^\delta - A(f_\alpha^\delta, \mu_\alpha^\delta)\| \approx 2\delta.$$

To obtain a reasonable basis for comparison, we adjusted the regularization parameter α such that the residual had also a magnitude of 2δ in the sparsity case, which occurred for $\alpha = 5$. The reconstruction can be seen in Figure 4.

A visual inspection shows that the reconstruction with sparsity constraint yields much sharper contours. In particular, the absolute values of f in the heart are higher in the sparsity case, and the artifacts are not as bad as in the quadratic constraint case, as can be seen in Figure 5. It shows a plot of the values of the activity function for both reconstructions along a row in the image in Figures 3 and 4 respectively. The left graph shows the values at a line that goes through the heart, and right graph shows the values along a line well outside the heart, where only artifacts occur. Clearly, both reconstructions are different, but it certainly needs much more computations in order to decide in which situations a sparsity constraint has to be preferred. A histogram plot of the wavelet coefficients for both reconstructions shows that the reconstruction with sparsity constraint has much more small coefficients – it is, as we did expect, a sparse reconstruction, see Figure 6.

6 Projected Accelerated Steepest Descent for Nonlinear Ill-Posed Problems

In the previous section we have discussed the iterated generalized shrinkage method given by

$$c^{n+1} = \frac{\alpha}{C} B^*(I - P_{\mathcal{C}})\Big(\frac{C}{\alpha} B\,\{c^n + \mathcal{F} F'(\mathcal{F}^* c^{n+1})^*(y^\delta - F(\mathcal{F}^* c^n))/C\}\Big).$$

For special choices of Ψ (e.g. $\Psi(c) = \|c\|_{\ell_1}$), this iteration then allows due its simple nature an easy to implement recovery algorithm. But the convergence is rather slow and does not change substantially through different choices of Ψ. One first serious step to accelerate such types of iterations (but for linear problems) was suggested in [15], in which the authors "borrowed a leaf" from standard linear steepest descent methods by using an adaptive step length. In addition to this, the authors concluded from a detailed analysis of the characteristic dynamics of the iterated soft-shrinkage that the algorithm converges initially relatively fast, then it overshoots the ℓ_1 penalty, and it takes very long to re-correct back. The proposed way to circumvent this "external" detour is to force the iterates to remain within a particular ℓ_1 ball $B_R := \{c \in \ell_2; \|c\|_{\ell_1} \le R\}$. This has led to the constrained optimization problem

$$\min_{c \in B_R} \|A\mathcal{F}^* c - y^\delta\|^2 \tag{6.1}$$

resulting in a significantly different proceeding. The shrinkage operation is replaced by a projection P_{B_R} (where the projection $\mathrm{P}_{\mathcal{C}}(c)$ is defined for any closed convex set $\mathcal{C}$ and any c as the unique point in $\mathcal{C}$ for which the ℓ_2 distance to c is minimal) leading for properly chosen $\gamma > 0$ to the following iteration,

$$c^{n+1} = \mathrm{P}_{B_R}(c^n + \gamma \mathcal{F} A^*(y^\delta - A\mathcal{F}^* c^n)). \tag{6.2}$$

However, the speed of convergence remained very slow. Therefore, as mentioned above, the authors suggested to introduce an adaptive "descent parameter" $\gamma^n > 0$ in each iteration yielding

$$c^{n+1} = \mathrm{P}_{B_R}(c^n + \gamma^n \mathcal{F} A^*(y^\delta - A\mathcal{F}^* c^n)). \tag{6.3}$$

The authors of [15] referred to this modified algorithm as the *projected gradient iteration* or the *projected steepest descent method*. They have determined how large one can choose the successive γ^n and have shown weak as well as strong convergence of the method (with and without acceleration). Alternative approaches for sparse recovery that are closely related to the introduced method are the schemes presented in [32] and [55]. The analysis in [55] is limited to finite dimensions and the strategy provided in [32] is suited for linear inverse problems. The principle there is to reformulate the minimization problem as a bounded constrained quadratic program, and then apply iterative project gradient iterations.

In this section we show that iteration (6.3) (and also more general formulations) can be directly extended to the nonlinear situation resulting in

$$c^{n+1} = \mathrm{P}_{B_R}(c^n + \gamma^n \mathcal{F} F'(\mathcal{F}^* c^{n+1})^*(y - F(\mathcal{F}^* c^n))). \tag{6.4}$$

Again, as in the previous section, weak as well as strong convergence can only be achieved, if F is equipped with conditions (5.5)–(5.7). We also assume twice continuous Fréchet differentiability of F. But note that at the cost of more technicalities most of the results can also be achieved if F is only one time Fréchet differentiable.

Another issue that is of great importance but was neither considered in [15] nor somewhere else is to verify regularizing properties of (6.4). Elaborations on this topic, however, are not provided so far. Nevertheless, we wish to briefly mention the theory that is still provided in the literature, which is so far unfortunately limited to linear problems, see, e.g., [23, Section 5.4]. Therefore, the concepts summarized in [23] not directly apply here and need to be extended. In any case, the question arises whether the convex constraint stabilize the problem or if it is still necessary to regularize the inverse problem. In general it seems to be meaningful to assume ill-posedness. Therefore, we need to introduce an additional stabilization. The iteration (6.4) can be viewed as iteration scheme to approach the B_R*-best-approximate solution* $c_R^\dagger$, which we define as the minimizer of $\|F(\mathcal{F}^* c) - y\|^2$ on B_R, i.e.

$$\|F(\mathcal{F}^* c_R^\dagger) - y\| = \inf_c\{\|F(\mathcal{F}^* c) - y\|,\ c \in B_R\} \quad \text{and}$$

$$\|c_R^\dagger\| = \min\{\|c\|,\ \|F(\mathcal{F}^* c) - y\| = \|F(\mathcal{F}^* c_R^\dagger) - y\| \text{ and } c \in B_R\}.$$

Since $c_R^\dagger \in B_R$, it is natural to require that the regularized solutions are in B_R as well. If $c^\dagger$ denotes the generalized solution of the unconstrained problem and if $c_R^\dagger = c^\dagger$, then all "standard results" concerning stability, convergence, and convergence rates hold also for the constrained case. If $c_R^\dagger \neq c^\dagger$, one might select a different regularization method, e.g.,

$$\min_{c \in B_R} \|F(\mathcal{F}^* c) - y\|^2 + \eta\|c\|^2,$$

for some $\eta > 0$.

6.1 Preleminaries

Once a frame is selected for X, the computation of a solution x translates into finding a corresponding sequence $c \in \ell_2(\Lambda)$. Hence, the operator under consideration can be written as $F \circ \mathcal{F}^* : \ell_2(\Lambda) \to Y$. Thus, for the ease of notation we write in the remaining section (if not misleadingly used) only F instead of $F \circ \mathcal{F}^*$.

Before analyzing the proposed projected steepest descent (6.4), we provide some analysis of ℓ_2 projections onto ℓ_1 balls. The listed properties can be retraced in [15, 53], from where they are partially taken, or to some extent in [18, 19].

Lemma 6.1. $\forall a \in \ell_2(\Lambda), \forall \tau > 0 : \|\mathrm{S}_\tau(a)\|_1$ *is a piecewise linear, continuous, decreasing function of* τ; *moreover, if* $a \in \ell_1(\Lambda)$ *then* $\|\mathrm{S}_0(a)\|_1 = \|a\|_1$ *and* $\|\mathrm{S}_\tau(a)\|_1 = 0$ *for* $\tau \geq \max_i |a_i|$.

Lemma 6.2. *If* $\|a\|_1 > R$, *then the* ℓ_2 *projection of* a *on the* ℓ_1 *ball with radius* R *is given by* $\mathrm{P}_{B_R}(a) = \mathrm{S}_\mu(a)$, *where* μ *(depending on* a *and* R*) is chosen such that* $\|\mathrm{S}_\mu(a)\|_1 = R$. *If* $\|a\|_1 \leq R$ *then* $\mathrm{P}_{B_R}(a) = \mathrm{S}_0(a) = a$.

Lemma 6.1 and 6.2 provide a simple recipe for computing the projection $\mathrm{P}_{B_R}(a)$. First, sort the absolute values of the components of a (an $\mathcal{O}(m \log m)$ operation if $\#\Lambda = m$ is finite), resulting in the rearranged sequence $(a_l^*)_{l=1,\dots,m}$, with $a_l^* \geq a_{l+1}^* \geq 0; \forall l$. Next, perform a search to find k such that

$$\|\mathrm{S}_{a_k^*}(a)\|_1 = \sum_{l=1}^{k-1}(a_l^* - a_k^*) \leq R < \sum_{l=1}^{k}(a_l^* - a_{k+1}^*) = \|\mathrm{S}_{a_{k+1}^*}(a)\|_1.$$

The complexity of this step is again $\mathcal{O}(m \log m)$. Finally, set $\nu := k^{-1}(R - \|S_{a_k^*}(a)\|_1)$, and $\mu := a_k^* - \nu$. Then

$$\begin{aligned}\|\mathrm{S}_\mu(a)\|_1 &= \sum_{i\in\Lambda}\max(|a_i| - \mu, 0) = \sum_{l=1}^{k}(a_l^* - \mu)\\ &= \sum_{l=1}^{k-1}(a_l^* - a_k^*) + k\nu = \|\mathrm{S}_{a_k^*}(a)\|_1 + k\nu = R.\end{aligned}$$

In addition to the above statements, also the still provided Lemmata 5.6 (setting $K = B_R$), 5.7, and 5.8 apply to P_{B_R} and allow therewith the use of several standard arguments of convex analysis.

6.2 Projected Steepest Descent and Convergence

We have now collected some facts on the projector P_{B_R} and on convex analysis issues that allow for convergence analysis of the projected steepest descent method defined in (6.3). In what follows, we essentially proceed as in [15]. But as we shall see, several serious technical changes (including also a weakening of a few statements) but also significant extensions of the nice analysis provided in [15] need to be made. For instance, due to the nonlinearity of F, several uniqueness statements proved in [15] do not carry over in its full glory. Nevertheless, the main propositions on *weak* and *strong convergence* can be achieved (of course, at the cost of involving much more technicalities).

First, we derive the necessary condition for a minimizer of $\boldsymbol{D}(c) := \|F(c) - y\|^2$ on B_R.

Lemma 6.3. *If the vector $\tilde{c}_R \in \ell_2$ is a minimizer of $\boldsymbol{D}(c)$ on B_R then for any $\gamma > 0$,*

$$\mathrm{P}_{B_R}(\tilde{c}_R + \gamma F'(\tilde{c}_R)^*(y - F(\tilde{c}_R))) = \tilde{c}_R,$$

which is equivalent to

$$\langle F'(\tilde{c}_R)^*(y - F(\tilde{c}_R)), w - \tilde{c}_R\rangle \leq 0, \quad \textit{for all } w \in B_R.$$

This result essentially relies on the Fréchet differentiability of F (see, e.g., [59, 58]) and summarizes the following reasoning.

With the help of the first order Taylor expansion given by

$$F(c+h) = F(c) + F'(c)h + R(c,h) \quad \text{with } \|R(c,h)\| \leq \frac{L}{2}\|h\|^2$$

one has for the minimizer $\tilde{c}_R$ of $\boldsymbol{D}$ on B_R and all $w \in B_R$ and all $t \in [0, 1]$

$$\begin{aligned}
\boldsymbol{D}(\tilde{c}_R) \leq \boldsymbol{D}(\tilde{c}_R + t(w - \tilde{c}_R)) &= \|F(\tilde{c}_R + t(w - \tilde{c}_R)) - y\|^2 \\
&= \|F(\tilde{c}_R) - y + F'(\tilde{c}_R)t(w - \tilde{c}_R) + R(\tilde{c}_R, t(w - \tilde{c}_R))\|^2 \\
&= \boldsymbol{D}(\tilde{c}_R) + 2\langle F'(\tilde{c}_R)^*(F(\tilde{c}_R) - y), t(w - \tilde{c}_R)\rangle \\
&\quad + 2\langle F(\tilde{c}_R) - y, R(\tilde{c}_R, t(w - \tilde{c}_R))\rangle \\
&\quad + \|F'(\tilde{c}_R)t(w - \tilde{c}_R) + R(\tilde{c}_R, t(w - \tilde{c}_R))\|^2.
\end{aligned}$$

This implies

$$\langle F'(\tilde{c}_R)^*(y - F(\tilde{c}_R)), w - \tilde{c}_R\rangle \leq 0,$$

and therefore, for all $\gamma > 0$,

$$\langle \tilde{c}_R + \gamma F'(\tilde{c}_R)^*(y - F(\tilde{c}_R)) - \tilde{c}_R, w - \tilde{c}_R\rangle \leq 0,$$

which verifies the assertion.

Lemma 6.3 provides just a necessary condition for a minimizer $\tilde{c}_R$ of $\boldsymbol{D}$ on B_R. The minimizers of $\boldsymbol{D}$ on B_R need not be unique. Nevertheless, we have

Lemma 6.4. *If* $\tilde{c}, \tilde{\tilde{c}} \in B_R$, *if* $\tilde{c}$ *minimizes* $\boldsymbol{D}$ *and if* $\tilde{c} - \tilde{\tilde{c}} \in \ker F'(w)$ *for all* $w \in B_R$ *then* $\tilde{\tilde{c}}$ *minimizes* $\boldsymbol{D}$ *as well.*

In what follows we elaborate the convergence properties of (6.4). In a first step we establish weak convergence and in a second step we extend weak to strong convergence. To this end, we have to specify the choice of γ^n. At first, we introduce a constant r,

$$r := \max\left\{2 \sup_{c \in B_R} \|F'(c)\|^2,\ 2L\sqrt{\boldsymbol{D}(c^0)}\right\}, \tag{6.5}$$

where c^0 denotes a initial guess for the solution to be reconstructed. One role of the constant r can be seen in the following estimate which is possible by the first order Taylor expansion of F,

$$\|F(c^{n+1}) - F(c^n)\|^2 \leq \sup_{c \in B_R} \|F'(c)\|^2 \|c^{n+1} - c^n\|^2 \leq \frac{r}{2}\|c^{n+1} - c^n\|^2.$$

We define now with the help of (6.5) a sequence of real numbers which we denote by β^n that specifies the choice γ^n by setting $\gamma^n = \beta^n / r$ (as we shall see later in this section).

Definition 6.5. We say that the sequence $(\beta^n)_{n\in\mathbb{N}}$ satisfies Condition (B) with respect to the sequence $(c^n)_{n\in\mathbb{N}}$ if there exists n_0 such that:

(B1) $\quad \bar{\beta} := \sup\{\beta^n; n \in \mathbb{N}\} < \infty \quad$ and $\quad \inf\{\beta^n; n \in \mathbb{N}\} \geq 1,$

(B2) $\quad \beta^n \|F(c^{n+1}) - F(c^n)\|^2 \leq \frac{r}{2}\|c^{n+1} - c^n\|^2 \quad \forall n \geq n_0,$

(B3) $\quad \beta^n L\sqrt{\boldsymbol{D}(c^n)} \leq \frac{r}{2}.$

By condition (B1) we ensure

$$\|F(c^{n+1}) - F(c^n)\|^2 \leq \beta^n \|F(c^{n+1}) - F(c^n)\|^2.$$

The idea of adding condition (B2) is to find the largest number $\beta^n \geq 1$ such that

$$0 \leq -\|F(c^{n+1}) - F(c^n)\|^2 + \frac{r}{2\beta^n}\|c^{n+1} - c^n\|^2$$

is as small as possible. The reason can be verified below in the definition of the surrogate functional Φ_β in Lemma 6.6. The goal is to ensure that Φ_{β^n} is not too far off $\boldsymbol{D}(c^n)$. The additional restriction (B3) is introduced to ensure convexity of Φ_{β^n} and convergence of the fixed point map Ψ in Lemma 6.7 (as we will prove below).

Because the definition of c^{n+1} involves β^n and vice versa, the inequality (B2) has an implicit quality. In practice, it is not straightforward to pick β^n adequately. This issue will be discussed later in Subsection 6.3.

In the remaining part of this subsection we prove weak convergence of any subsequence of $(c^n)_{n\in\mathbb{N}}$ towards weak limits that fulfill the necessary condition for minimizers of $\boldsymbol{D}$ on B_R.

Lemma 6.6. *Assume F to be twice Fréchet differentiable and $\beta \geq 1$. For arbitrary fixed $c \in B_R$ assume $\beta L\sqrt{\mathbf{D}(c)} \leq r/2$ and define the functional $\Phi_\beta(\cdot, c)$ by*

$$\Phi_\beta(w,c) := \|F(w) - y\|^2 - \|F(w) - F(c)\|^2 + \frac{r}{\beta}\|w - c\|^2. \tag{6.6}$$

Then there exists a unique $w \in B_R$ that minimizes the restriction to B_R of $\Phi_\beta(w,c)$. We denote this minimizer by $\hat{c}$ which is given by

$$\hat{c} = \mathrm{P}_{B_R}\left(c + \frac{\beta}{r}F'(\hat{c})^*(y - F(c))\right).$$

The essential strategy of the proof goes as follows. First, since F is twice Fréchet differentiable one verifies that $\Phi_\beta(\cdot, c)$ is strictly convex in w. Therefore there exists a unique minimizer $\hat{c}$ and thus we have for all $w \in B_R$ and all $t \in [0,1]$

$$\Phi_\beta(\hat{c}, c) \leq \Phi_\beta(\hat{c} + t(w - \hat{c}), c).$$

With the short hand notation $J(\cdot) := \Phi_\beta(\cdot, c)$ it therefore follows that

$$\begin{aligned}
0 &\leq J(\hat{c} + t(w - \hat{c})) - J(\hat{c}) = tJ'(\hat{c})(w - \hat{c}) + \rho(\hat{c}, t(w - \hat{c})) \\
&= 2t\langle F(c) - y, F'(\hat{c})(w - \hat{c})\rangle + 2t\frac{r}{\beta}\langle \hat{c} - c, w - \hat{c}\rangle \\
&\quad + 2\langle F(c) - y, R(\hat{c}, t(w - \hat{c}))\rangle + \frac{r}{\beta}\|t(w - \hat{c})\|^2 \\
&\leq 2t\left\{\langle F(c) - y, F'(\hat{c})(w - \hat{c})\rangle + \frac{r}{\beta}\langle \hat{c} - c, w - \hat{c}\rangle\right\} \\
&\quad + t^2\left\{2\frac{r}{2\beta L}\frac{L}{2}\|w - \hat{c}\|^2 + \frac{r}{\beta}\|w - \hat{c}\|^2\right\}.
\end{aligned}$$

This implies for all $t \in [0,1]$

$$0 \leq \left\{\frac{\beta}{r}\langle F(c) - y, F'(\hat{c})(w - \hat{c})\rangle + \langle \hat{c} - c, w - \hat{c}\rangle\right\} + \frac{3t}{4}\|w - \hat{c}\|^2.$$

Consequently, we deduce

$$\left\langle c + \frac{\beta}{r}F'(\hat{c})^*(y - F(c)) - \hat{c}, w - \hat{c}\right\rangle \leq 0$$

which is equivalent to

$$\hat{c} = \mathrm{P}_{B_R}\left(c + \frac{\beta}{r}F'(\hat{c})^*(y - F(c))\right)$$

and the assertion is shown.

The unique minimizer $\hat{c}$ is only implicitly given. We propose to apply a simple fixed point iteration to derive $\hat{c}$. The next lemma verifies that the corresponding fixed point map is indeed contractive and can therefore be used.

Lemma 6.7. *Assume* $\beta L\sqrt{\boldsymbol{D}(x)} \le r/2$. *Then the map*

$$\Psi(\hat{c}) := \mathrm{P}_{B_R}(c + \beta/rF'(\hat{c})^*(y - F(c)))$$

is contractive and therefore the fixed point iteration $\hat{c}^{l+1} = \Psi(\hat{c}^l)$ *converges to a unique fixed point.*

The latter lemma is a consequence of the Lipschitz continuity of F' and the non-expansiveness of P_{B_R}. The last property that is needed to establish convergence is an immediate consequence of Lemma 6.6.

Lemma 6.8. *Assume* c^{n+1} *is given by*

$$c^{n+1} = \mathrm{P}_{B_R}\left(c^n + \frac{\beta^n}{r}F'(c^{n+1})^*(y - F(c^n))\right),$$

where r *is as in* (6.5) *and the* β^n *satisfy Condition* (B) *with respect to* $(c^n)_{n\in\mathbb{N}}$, *then the sequence* $\boldsymbol{D}(c^n)$ *is monotonically decreasing and* $\lim_{n\to\infty}\|c^{n+1} - c^n\| = 0$.

Now we have all ingredients for the convergence analysis together. Since for all the iterates we have by definition $c^n \in B_R$, we automatically have $\|c^n\|_2 \le R$ for all $n \in \mathbb{N}$. Therefore, the sequence $(c^n)_{n\in\mathbb{N}}$ must have weak accumulation points.

Proposition 6.9. *If* $c^\star$ *is a weak accumulation point of* $(c^n)_{n\in\mathbb{N}}$, *then it fulfills the necessary condition for a minimum of* $\boldsymbol{D}(c)$ *on* B_R, *i.e. for all* $w \in B_R$,

$$\langle F'(c^\star)^*(y - F(c^\star)), w - c^\star\rangle \le 0. \tag{6.7}$$

Since this proposition is essential and combines all the above made statements, we give the reasoning and arguments to verify (6.7) in greater detail. Since $c^{n_j} \stackrel{w}{\longrightarrow} c^\star$, we have for fixed c and a

$$\langle F'(c)c^{n_j}, a\rangle = \langle c^{n_j}, F'(c)^*a\rangle \longrightarrow \langle c^\star, F'(c)^*a\rangle = \langle F'(c)c^\star, a\rangle$$

and therefore

$$F'(c)c^{n_j} \stackrel{w}{\longrightarrow} F'(c)c^\star. \tag{6.8}$$

Due to Lemma 6.8, we also have $c^{n_j+1} \stackrel{w}{\longrightarrow} c^\star$. Now we are prepared to show the necessary condition for the weak accumulation point $c^\star$. As the iteration is given by

$$c^{n+1} = \mathrm{P}_{B_R}(c^n + \beta^n/rF'(c^{n+1})^*(y - F(c^n))),$$

we have

$$\langle c^n + \beta^n/rF'(c^{n+1})^*(y - F(c^n)) - c^{n+1}, w - c^{n+1}\rangle \le 0 \quad \text{for all } w \in B_R.$$

Specializing this inequality to the subsequence $(c^{n_j})_{j\in\mathbb{N}}$ yields

$$\langle c^{n_j} + \beta^{n_j}/rF'(c^{n_j+1})^*(y - F(c^{n_j})) - c^{n_j+1}, w - c^{n_j+1}\rangle \leq 0 \quad \text{for all } w \in B_R.$$

Therefore we obtain (due to Lemma 6.8)

$$\lim \sup_{j\to\infty} \beta^{n_j}/r\langle F'(c^{n_j+1})^*(y - F(c^{n_j})), w - c^{n_j+1}\rangle \leq 0 \quad \text{for all } w \in B_R.$$

To the latter inequality we may add

$$\beta^{n_j}/r\langle(-F'(c^{n_j+1})^* + F'(c^{n_j})^*)(y - F(c^{n_j})), w - c^{n_j+1}\rangle$$

and

$$\beta^{n_j}/r\langle F'(c^{n_j})^*(y - F(c^{n_j})), -c^{n_j} + c^{n_j+1}\rangle$$

resulting in

$$\lim \sup_{j\to\infty} \beta^{n_j}/r\langle F'(c^{n_j})^*(y - F(c^{n_j})), w - c^{n_j}\rangle \leq 0 \quad \text{for all } w \in B_R, \tag{6.9}$$

which is possible due to

$$\begin{aligned}
&|\langle(-F'(c^{n_j+1})^* + F'(c^{n_j})^*)(y - F(c^{n_j})), w - c^{n_j+1}\rangle| \\
&\qquad \leq L\|c^{n_j+1} - c^{n_j}\|\|y - F(c^{n_j})\|\|w - c^{n_j+1}\| \overset{j\to\infty}{\longrightarrow} 0
\end{aligned}$$

and

$$\begin{aligned}
&|\langle F'(c^{n_j})^*(y - F(c^{n_j})), -c^{n_j} + c^{n_j+1}\rangle| \\
&\qquad \leq \sup_{x\in B_R} \|F'(c)^*\|\|y - F(c^{n_j})\|\|c^{n_j} - c^{n_j+1}\| \overset{j\to\infty}{\longrightarrow} 0.
\end{aligned}$$

Let us now consider the inner product in (6.9) which we write as

$$\langle F'(c^{n_j})^*y, w - c^{n_j}\rangle - \langle F'(c^{n_j})^*F(c^{n_j}), w - c^{n_j}\rangle.$$

For the left summand we have by the weak convergence of $(c^{n_j})_{j\in\mathbb{N}}$ or likewise $(F'(c^\star)c^{n_j})_{j\in\mathbb{N}}$ and the assumption of F, $F'(c^{n_j})^*y \overset{j\to\infty}{\longrightarrow} F'(c^\star)^*y$,

$$\begin{aligned}
\langle F'(c^{n_j})^*y, w-c^{n_j}\rangle &= \langle(F'(c^{n_j})^* - F'(c^\star)^* + F'(c^\star)^*)y, w-c^{n_j}\rangle \\
&= \langle F'(c^{n_j})^*y - F'(c^\star)^*y, w-c^{n_j}\rangle + \langle F'(c^\star)^*y, w-c^{n_j}\rangle \\
&\overset{j\to\infty}{\longrightarrow} \langle F'(c^\star)^*y, w-c^\star\rangle \\
&= \langle F'(c^\star)^*(y-F(c^\star)), w-c^\star\rangle + \langle F'(c^\star)^*F(c^\star), w-c^\star\rangle.
\end{aligned}$$

Therefore (and since $1 \le \beta^{n_j} \le \bar{\beta}$ and again by the weak convergence of $(c^{n_j})_{j\in\mathbb{N}}$), inequality (6.9) transforms to

$$\begin{aligned}&\lim_{j\to\infty}\sup\,[\langle F'(c^\star)^*(y - F(c^\star)), w - c^\star\rangle \\ &\quad + \langle F'(c^\star)^* F(c^\star), w - c^\star + c^{n_j} - c^{n_j}\rangle - \langle F'(c^{n_j})^* F(c^{n_j}), w - c^{n_j}\rangle] \le 0\end{aligned}$$

$$\Longleftrightarrow$$

$$\begin{aligned}&\lim_{j\to\infty}\sup\,[\langle F'(c^\star)^*(y - F(c^\star)), w - c^\star\rangle \\ &\quad + \langle F'(c^\star)^* F(c^\star) - F'(c^{n_j})^* F(c^{n_j}), w - c^{n_j}\rangle] \le 0\end{aligned}$$

$$\Longleftrightarrow$$

$$\begin{aligned}&\langle F'(c^\star)^*(y - F(c^\star)), w - c^\star\rangle \\ &\quad + \lim_{j\to\infty}\sup\,\langle F'(c^\star)^* F(c^\star) - F'(c^{n_j})^* F(c^{n_j}), w - c^{n_j}\rangle \le 0.\end{aligned}$$

It remains to show that the right summand in the latter inequality is zero for all $w \in B_R$. We have by the assumptions made on F,

$$\begin{aligned}&\langle F'(c^\star)^* F(c^\star) - F'(c^{n_j})^* F(c^{n_j}), w - c^{n_j}\rangle| \\ &\quad = |\langle F'(c^\star)^* F(c^\star) - F'(c^\star)^* F(c^{n_j}) + F'(c^\star)^* F(c^{n_j}) \\ &\qquad - F'(c^{n_j})^* F(c^{n_j}), w - c^{n_j}\rangle| \\ &\quad \le |\langle F'(c^\star)^* F(c^\star) - F'(c^\star)^* F(c^{n_j}), w - c^{n_j}\rangle| \\ &\qquad + |\langle F'(c^\star)^* F(c^{n_j}) - F'(c^{n_j})^* F(c^{n_j}), w - c^{n_j}\rangle| \\ &\quad \le \sup_{x\in B_R}\|F'(c)\|\|F(c^\star) - F(c^{n_j})\|\|w - c^{n_j}\| \\ &\qquad + |\langle (F'(c^\star)^* - F'(c^{n_j})^*)(F(c^\star) - F(c^\star) + F(c^{n_j})), w - c^{n_j}\rangle| \\ &\quad \le \sup_{x\in B_R}\|F'(c)\|\|F(c^\star) - F(c^{n_j})\|\|w - c^{n_j}\| \\ &\qquad + \|(F'(c^\star)^* - F'(c^{n_j})^*)F(c^\star)\|\|w - c^{n_j}\| \\ &\qquad + L\|c^\star - c^{n_j}\|\|F(c^\star) - F(c^{n_j})\|\|w - c^{n_j}\| \\ &\quad \overset{j\to\infty}{\longrightarrow} 0.\end{aligned}$$

Consequently, for all $w \in B_R$,

$$\langle F'(c^\star)^*(y - F(c^\star)), w - c^\star\rangle \le 0.$$

After the verification of the necessary condition for weak accumulation points we show that the weak convergence of subsequences can be strengthened into convergence in norm topology. This is important to be achieved as in principle our setup is infinite dimensional.

Proposition 6.10. *With the same assumptions as in Proposition* 6.9 *and the assumptions* (5.6)–(5.7) *on the nonlinear operator* F*, there exists a subsequence* $(c^{n'_l})_{l\in\mathbb{N}} \subset (c^n)_{n\in\mathbb{N}}$ *such that* $(c^{n'_l})_{l\in\mathbb{N}}$ *converges in norm towards the weak accumulation point* $c^\star$*, i.e.*

$$\lim_{l\to\infty} \|c^{n'_l} - c^\star\| = 0.$$

The proof of this proposition is in several parts the same as in [15, Lemma 12]. Here we only mention the difference that is due to the nonlinearity of F. Denote by $(c^{n_j})_{j\in\mathbb{N}}$ the subsequence that was introduced in the proof of Proposition 6.9. Define now $u^j := c^{n_j} - c^\star$, $v^j := c^{n_j+1} - c^\star$, and $\beta^j := \beta^{n_j}$. Due to Lemma 6.8, we have $\lim_{j\to\infty} \|u^j - v^j\| = 0$. But we also have,

$$\begin{aligned}
u^j - v^j &= u^j + c^\star - \mathrm{P}_{B_R}(u^j + c^\star + \beta^j F'(v^j + c^\star)^*(y - F(u^j + c^\star)))\\
&= u^j + \mathrm{P}_{B_R}(c^\star + \beta^j F'(c^\star)^*(y - F(c^\star)))\\
&\quad - \mathrm{P}_{B_R}(u^j + c^\star + \beta^j F'(v^j + c^\star)^*(y - F(u^j + c^\star)))\\
&= u^j + \mathrm{P}_{B_R}(c^\star + \beta^j F'(c^\star)^*(y - F(c^\star)))\\
&\quad - \mathrm{P}_{B_R}(c^\star + \beta^j F'(v^j + c^\star)^*(y - F(u^j + c^\star)) + u^j) && (6.10)\\
&\quad + \mathrm{P}_{B_R}(c^\star + \beta^j F'(c^\star)^*(y - F(c^\star)) + u^j) && (6.11)\\
&\quad - \mathrm{P}_{B_R}(c^\star + \beta^j F'(c^\star)^*(y - F(c^\star)) + u^j)\\
&\quad + \mathrm{P}_{B_R}(c^\star + \beta^j F'(c^\star)^*(y - F(u^j + c^\star)) + u^j) && (6.12)\\
&\quad - \mathrm{P}_{B_R}(c^\star + \beta^j F'(c^\star)^*(y - F(u^j + c^\star)) + u^j), && (6.13)
\end{aligned}$$

where we have applied Proposition 6.9 ($c^\star$ fulfills the necessary condition) and Lemma 6.3, i.e. $c^\star = \mathrm{P}_{B_R}(c^\star + \beta^j F'(c^\star)^*(y - F(c^\star)))$. We consider now the sum of the terms (6.11)+(6.13), and obtain by the assumptions on F and since the β^j are uniformly bounded,

$$\begin{aligned}
&\|\mathrm{P}_{B_R}(c^\star + \beta^j F'(c^\star)^*(y - F(c^\star)) + u^j)\\
&\qquad - \mathrm{P}_{B_R}(c^\star + \beta^j F'(c^\star)^*(y - F(u^j + c^\star)) + u^j)\|\\
&\le \|\beta^j F'(c^\star)^*(F(u^j + c^\star) - F(c^\star))\|\\
&\le \bar{\beta} \sup_{x\in B_R} \|F'(x)\| \|F(u^j + c^\star) - F(c^\star)\| \overset{j\to\infty}{\longrightarrow} 0.
\end{aligned}$$

The sum of the terms (6.10)+(6.12) yields

$$\begin{aligned}
&\|\mathrm{P}_{B_R}(c^\star + \beta^j F'(c^\star)^*(y - F(u^j + c^\star)) + u^j) \\
&\qquad - \mathrm{P}_{B_R}(c^\star + \beta^j F'(v^j + c^\star)^*(y - F(u^j + c^\star)) + u^j)\| \\
&\le \bar{\beta}\{\|(F'(c^\star)^* - F'(v^j + c^\star)^*)(y - F(c^\star))\| \\
&\qquad + \|(F'(c^\star)^* - F'(v^j + c^\star)^*)(F(c^\star) - F(u^j + c^\star))\|\} \\
&\le \bar{\beta}\{\|(F'(c^\star)^* - F'(v^j + c^\star)^*)(y - F(c^\star))\| \\
&\qquad + L\|v^j\|\|F(c^\star) - F(u^j + c^\star)\|\} \stackrel{j\to\infty}{\longrightarrow} 0.
\end{aligned}$$

Consequently, combining $\|u^j - v^j\| \stackrel{j\to\infty}{\longrightarrow} 0$ and the two last statements, we observe that

$$\begin{aligned}
\lim_{j\to\infty} &\|\mathrm{P}_{B_R}(c^\star + \beta^j F'(c^\star)^*(y - F(c^\star)) + u^j) \\
&\qquad - \mathrm{P}_{B_R}(c^\star + \beta^j F'(c^\star)^*(y - F(c^\star))) - u^j\| = 0.
\end{aligned}$$

The remaining arguments that verify the strong convergence towards zero of a subsequence of u^j are now the same as in [15, Lemma 12].

As mentioned in [15], one can prove at the cost of more technicalities that the whole subsequence $(c^{n_j})_{j\in\mathbb{N}}$ converges in norm towards $c^\star$. We summarize the findings in the following proposition.

Proposition 6.11. *Every weak accumulation point $c^\star$ of the sequence $(c^n)_{n\in\mathbb{N}}$ defined by (6.4) fulfills the necessary condition for a minimizer of $\boldsymbol{D}$ in B_R. Moreover, there exists a subsequence $(c^{n_j})_{j\in\mathbb{N}} \subset (c^n)_{n\in\mathbb{N}}$ that converges in norm to $c^\star$.*

In the next proposition we give a condition under which norm convergence of subsequences carries over to the full sequence $(c^n)_{n\in\mathbb{N}}$.

Proposition 6.12. *Assume that there exists at least one isolated limit $c^\star$ of a subsequence $(c^{n_j})_{j\in\mathbb{N}} \subset (c^n)_{n\in\mathbb{N}}$. Then $c^n \to c^\star$ holds.*

A proof of this assertion can be directly taken from [45].

6.3 Some Algorithmic Aspects

In the previous subsection we have shown norm convergence for all β^n satisfying Condition (B). This, of course, implies also norm convergence for $\beta^n = 1$ for all $n \in \mathbb{N}$, which corresponds to the projected classical Landweber iteration. However, to accelerate the speed of convergence, we are interested in choosing, adaptively, larger values for β^n. In particular, by the reasoning made after Definition 6.5, we like to choose β^n as large as possible. The problem (even for linear operators A) is that the definition of c^{n+1} involves β^n and the inequality (B2) to restrict the choice of β^n uses c^{n+1}. This "implicit" quality does not allow for a straightforward determination of β^n.

For linear problems, conditions (B1) and (B2) are inspired by classical length-step in the steepest descent algorithm for the unconstrained functional $\|Ax - y\|^2$ leading to an accelerated Landweber iteration $x^{n+1} = x^n + \gamma^n A^*(y - Ax^n)$, for which γ^n is picked so that it gives a maximal decrease of $\|Ax - y\|^2$, i.e.

$$\gamma^n = \|A^*(y - Ax^n)\|^2 \|AA^*(y - Ax^n)\|^{-2}.$$

For nonlinear operators this condition translates into a rather non-practical suggestion for γ^n. In our situation, in which we have to fulfill Condition (B), we may derive a much simpler procedure to find a suitable γ^n (which is in our case β^n/r). Due to Lemma 6.8 we have monotonicity of $\boldsymbol{D}$ with respect to the iterates, i.e.

$$L\sqrt{\boldsymbol{D}(c^n)} \leq L\sqrt{\boldsymbol{D}(c^{n-1})} \leq \cdots \leq \frac{r}{2} = \max\Big\{ \sup_{c \in B_R} \|F'(c)\|^2, L\sqrt{\boldsymbol{D}(c^0)}\Big\}.$$

Therefore (B3), which was given by

$$L\sqrt{\boldsymbol{D}(c^n)} \leq \beta^n L\sqrt{\boldsymbol{D}(c^n)} \leq \frac{r}{2},$$

is indeed a nontrivial condition for $\beta^n \geq 1$. Namely, the smaller the decrease of $\boldsymbol{D}$, the larger we may choose β^n (when only considering (B3)). Condition (B3) can be recast as $1 \leq \beta^n \leq r/(2L\sqrt{\boldsymbol{D}(c^n)})$ and consequently, by Definition (6.5), an *explicit* (but somewhat "greedy") guess for β^n is given by

$$\beta^n = \max\left\{ \sup_{x \in B_R} \frac{\|F'(x)\|^2}{L\sqrt{\boldsymbol{D}(c^n)}}, \sqrt{\frac{\boldsymbol{D}(c^0)}{\boldsymbol{D}(c^n)}} \right\} \geq 1. \tag{6.14}$$

If this choice fulfills (B2) as well, it is retained; if it does not, it can be gradually decreased (by multiplying it with a factor slightly smaller than 1 until (B2) is satisfied.

As a summary of the above reasoning we suggest the following implementation of the proposed projected steepest descent algorithm.

Projected Steepest Descent Method for nonlinear inverse problems	
Given	operator F, its derivative $F'(c)$, data y, some initial guess c^0, and R (sparsity constraint ℓ_1-ball B_R)
Initialization	$r = \max\{2 \sup_{c \in B_R} \|F'(c)\|^2, 2L\sqrt{D(c^0)}\}$, set $q = 0.9$ (as an example)
Iteration	**for** $n = 0, 1, 2, \ldots$ **until** a preassigned precision / maximum number of iterations **1.** $\beta^n = \max\left\{\sup_{c \in B_R} \frac{\|F'(c)\|^2}{L\sqrt{D(c^n)}}, \sqrt{\frac{D(c^0)}{D(c^n)}}\right\}$ **2.** $c^{n+1} = \mathrm{P}_{B_R}(c^n + \frac{\beta^n}{r} F'(c^{n+1})^*(y - F(c^n)))$; by fixed point iteration **3.** verify (B2): $\beta^n \|F(c^{n+1}) - F(c^n)\|^2 \leq \frac{r}{2}\|c^{n+1} - c^n\|^2$ **if** (B2) is satisfied increase n and go to **1.** **otherwise** set $\beta^n = q \cdot \beta^n$ and go to **2.** **end**

6.4 Numerical Experiment: A Nonlinear Sensing Problem

The numerical experiment centers around a nonlinear sampling problem that is very closely related to the sensing problem considered in [54]. The authors of [54] have studied a sensing setup in which a continuous-time signal is mapped by a memoryless, invertible and nonlinear transformation, and then sampled in a non-ideal manner. In this context, memoryless means a static mapping that individually acts at each time instance (pointwise behavior). Such scenarios may appear in acquisition systems where the sensor introduces static nonlinearities, before the signal is sampled by a usual analog-to-digital converter. In [54] a theory and an algorithm is developed that allow a perfect recovery of a signal within a subspace from its nonlinear and non-ideal samples. In our setup we drop the invertibility requirement of the nonlinear transformation, which is indeed quite restrictive. Moreover, we focus on a subclass of problems in which the signal to be recovered is supposed to have sparse expansion.

Let us specify the sensing model. Assume we are given a reconstruction space $\mathcal{A} \subset X$ (e.g. $L_2(\mathbb{R})$) which is spanned by the frame $\{\phi_\lambda : \lambda \in \Lambda\}$ with frame bounds

$0 < C_1 \leq C_2 < \infty$. With this frame we associate two mappings, the analysis and synthesis operator,

$$\mathcal{F} : \mathcal{A} \ni f \mapsto \{\langle f, \phi_\lambda \rangle\}_{\lambda \in \Lambda} \in \ell_2(\Lambda) \quad \text{and} \quad \mathcal{F}^* : \ell_2(\Lambda) \ni x \mapsto \sum_{\lambda \in \Lambda} x_\lambda \phi_\lambda.$$

We assume that the function/signal f we wish to recover has a sparse expansion in $\mathcal{A}$. The sensing model is now determined by the nonlinear transformation $M : \mathcal{A} \to Y$ of the continuous-time function f that is point-wise given by the regularized modulus function (to have some concrete example for the nonlinear transformation)

$$M : f \mapsto M(f) = |f|_\varepsilon := \sqrt{f^2 + \varepsilon^2}.$$

This nonlinearly transformed f is then sampled in a possibly non-ideal fashion by some sampling function s yielding the following sequence of samples,

$$SM(f) = \{\langle s(\cdot - nT), M(f) \rangle_Y\}_{n \in \mathbb{Z}},$$

where we assume that the family $\{s(\cdot - nT_s), n \in \mathbb{Z}\}$ forms a frame with bounds $0 < S_1 \leq S_2 < \infty$. The goal is to reconstruct f from its samples $y = (S \circ M)(f)$. Since f belongs to $\mathcal{A}$, the reconstruction of f is equivalent with finding a sequence c such that $y = (S \circ M \circ \mathcal{F}^*)(c)$. As $\{\phi_\lambda : \lambda \in \Lambda\}$ forms a frame there might be several different sequences leading to the same function f. Among all possible solutions, we aim (as mentioned above) to find those sequences that have small ℓ_1 norm. As y might be not directly accessible (due to the presence of measurement noise) and due to the nonlinearity of the operator M, it seems more practical not to solve $y = (S \circ M \circ \mathcal{F}^*)(c)$ directly, but to find an approximation $\hat{c}$ such that

$$\hat{c} = \arg \min_{c \in B_R} \|F(c) - y\|^2 \quad \text{and,}$$

where we have used the shorthand notation $F := S \circ M \circ \mathcal{F}^*$ and where the ℓ_1 ball B_R restricts c to have a certain preassigned sparsity.

In order to apply our proposed accelerated steepest descent iteration,

$$c^{n+1} = \mathrm{P}_{B_R}\Big(c^n + \frac{\beta^n}{r} F'(c^{n+1})^*(y - F(c^n))\Big),$$

to derive an approximation to $\hat{x}$, we have to determine the constants r, see (6.5), and the Lipschitz constant L. This requires a specification of two underlying frames (the reconstruction and sampling frame). One technically motivated choice in signal sampling is the cardinal sine function. This function can be defined as the inverse Fourier transform of the characteristic function of the frequency interval $[-\pi, \pi]$, i.e.

$$\sqrt{2\pi}\,\mathrm{sinc}(\pi t) = \frac{1}{\sqrt{2\pi}} \int_{\mathbb{R}} \chi_{[-\pi,\pi]}(\omega) e^{it\omega} d\omega.$$

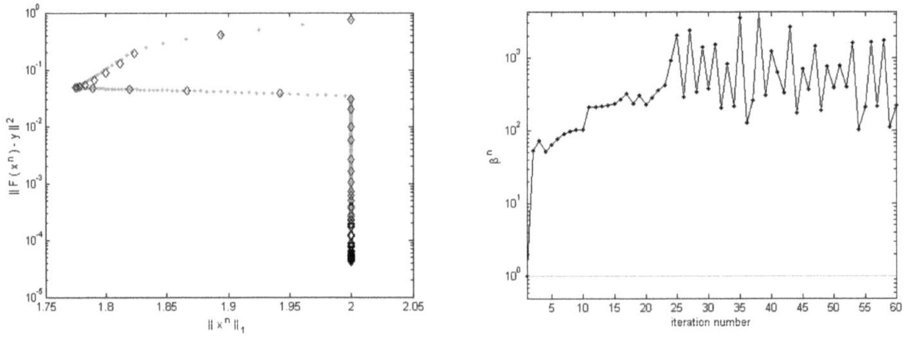

Figure 7. The left image shows the sparsity to residual plot. The black diamonds correspond to the accelerated iteration. For the non-accelerated iteration we have plotted every 20th iteration (gray dots). The right image visualizes the sequence of β^n (black) for the accelerated iteration. The gray line corresponds to $\beta = 1$.

Therefore, the resulting function spaces are spaces of bandlimited functions. The inverse Fourier transform of the L_2 normalized characteristic function $\frac{1}{\sqrt{2\Omega}}\chi_{[-\Omega,\Omega]}$ yields

$$\frac{1}{\sqrt{2\pi}}\int_{\mathbb{R}}\frac{1}{\sqrt{2\Omega}}\chi_{[-\Omega,\Omega]}(\omega)e^{it\omega}\,d\omega = \sqrt{\frac{\Omega}{\pi}}\,\mathrm{sinc}(\Omega t)$$

leading to the following definition of L_2 normalized and translated cardinal sine functions,

$$\phi_n(t) = \frac{1}{\sqrt{D_a}}\,\mathrm{sinc}\left(\frac{\pi}{D_a}(t - nT_a)\right), \quad \text{i.e. } \Omega = \frac{\pi}{D_a} \quad \text{and} \tag{6.15}$$

$$s_n(t) = \frac{1}{\sqrt{D_s}}\,\mathrm{sinc}\left(\frac{\pi}{D_s}(t - nT_s)\right), \quad \text{i.e. } \Omega = \frac{\pi}{D_s} \tag{6.16}$$

that determine the two frames. The parameters D_a and D_s are fixed and specify here the frequency cut off, whereas T_a and T_s fix the time step sizes. For all $n \in \mathbb{Z}$ we have $\|\phi_n\|_2 = \|s_n\|_2 = 1$. Moreover, it can be easily retrieved that

$$\langle\phi_n, \phi_m\rangle = \mathrm{sinc}\left(\frac{\pi}{D_a}(n-m)T_a\right) \quad \text{and} \quad \langle s_n, s_m\rangle = \mathrm{sinc}\left(\frac{\pi}{D_s}(n-m)T_s\right). \tag{6.17}$$

As long as $T_a/D_a, T_s/D_s \in \mathbb{Z}$, the frames form orthonormal systems. The inner products (6.17) are the entries of the Gramian matrices $\mathcal{F}\mathcal{F}^*$ and SS^*, respectively, for which we have $\|\mathcal{F}\mathcal{F}^*\| = \|\mathcal{F}\|^2 = \|\mathcal{F}^*\|^2 \leq C_2$ and $\|SS^*\| = \|S\|^2 = \|S^*\|^2 \leq S_2$.

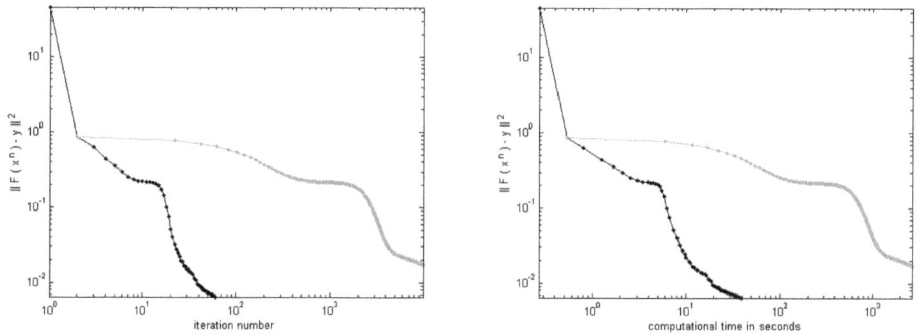

Figure 8. These images represent the residual evolution with respect to the number of iterations (left) and the computational time (right). The black dotted curves represent the residual evolution for the accelerated and the gray dotted curves for the non-accelerated scheme.

Let us now determine r and L. To this end we have to estimate $\sup_{c\in B_R}\|F'(c)\|^2$. For given $c\in B_R$, it follows that

$$\begin{aligned}\|F'(c)\| &= \sup_{h\in\ell_2,\|h\|=1}\|F'(c)h\| = \|SM'(\mathcal{F}^*c)\mathcal{F}^*h\|\\ &\le \|S\|\,\|M'(\mathcal{F}^*c)\|\,\|\mathcal{F}^*\|.\end{aligned}$$

Moreover, due to (6.15),

$$\begin{aligned}\|M'(\mathcal{F}^*c)\|^2 &= \sup_{h\in\Lambda_2,\|h\|=1}\|M'(\mathcal{F}^*c)h\|^2\\ &= \int_{\mathbb{R}}|(\mathcal{F}^*c)(t)|^2|((\mathcal{F}^*c)(t))^2+\varepsilon^2|^{-1}|h(t)|^2dt\\ &\le \frac{1}{\varepsilon^2}\int_{\mathbb{R}}\Big(\sum_n|c_n||\phi_n(t)|\Big)^2|h(t)|^2dt \le \frac{\|c\|_1^2}{\varepsilon^2 D_a}.\end{aligned}$$

Therefore, we finally obtain

$$\sup_{c\in B_R}\|F'(c)\|^2 \le \|S\|^2\,\|\mathcal{F}^*\|^2\,\frac{R^2}{\varepsilon^2 D_a} \le S_2\,C_2\,\frac{R^2}{\varepsilon^2 D_a}. \tag{6.18}$$

The Lipschitz continuity of F' is characterized by $\|F'(\tilde{c})-F'(c)\|\le L\|\tilde{c}-c\|$, for all $c,\tilde{c}\in B_R$. In order to find the Lipschitz constant L, we directly derive

$$\begin{aligned}\|F'(\tilde{c})-F'(c)\| &= \sup_{h\in\ell_2,\|h\|=1}\|F'(\tilde{c})h-F'(c)h\|\\ &= \sup_{h\in\ell_2,\|h\|=1}\|SM'(\mathcal{F}^*\tilde{c})\mathcal{F}^*h-SM'(\mathcal{F}^*c)\mathcal{F}^*h\|\\ &\le \|S\|\,\|M'(\mathcal{F}^*\tilde{c})-M'(\mathcal{F}^*c)\|\,\|\mathcal{F}^*\|\,, \end{aligned}\tag{6.19}$$

and with $M''(f) = \varepsilon^2(f^2 + \varepsilon^2)^{-3/2}$ it follows

$$\begin{aligned}
&\|M'(\mathcal{F}^*\tilde{c}) - M'(\mathcal{F}^*c)\|^2 \\
&= \sup_{h\in L_2,\|h\|=1} \int_{\mathbb{R}} |M'(\mathcal{F}^*\tilde{c}(t)) - M'(\mathcal{F}^*c(t))|^2 |h(t)|^2 dt \\
&\le \sup_{h\in L_2,\|h\|=1} \int_{\mathbb{R}} \frac{1}{\varepsilon^2} |\mathcal{F}^*\tilde{c}_n(t) - \mathcal{F}^*c(t)|^2 \, |h(t)|^2 dt \\
&\le \sup_{h\in L_2,\|h\|=1} \int_{\mathbb{R}} \frac{1}{\varepsilon^2} \Big(\sum_{n\in\mathbb{Z}} |(\tilde{c}_n - c_n)| \, |\phi_n(t)| \Big)^2 |h(t)|^2 dt \\
&\le \sup_{h\in L_2,\|h\|=1} \int_{\mathbb{R}} \sum_{n\in\mathbb{Z}} |\phi_n(t)|^2 \, |h(t)|^2 dt \, \frac{1}{\varepsilon^2} \|\tilde{c} - c\|^2.
\end{aligned}$$

To finally bound the last quantity, we have to estimate $\sum_{n\in\mathbb{Z}} |\phi_n(t)|^2$ independently on $t \in \mathbb{R}$. With definition (6.15), we observe that

$$\sum_{n\in\mathbb{Z}} |\phi_n(t)|^2 = \frac{1}{D_a} \sum_{n\in\mathbb{Z}} \operatorname{sinc}^2\Big(\frac{\pi}{D_a} t - n \frac{\pi\, T_a}{D_a} \Big) \tag{6.20}$$

is a periodic function with period T_a. Therefore it is sufficient to analyze (6.20) for $t \in [0, T_a]$. The sum in (6.20) is maximal for $t = 0$ and $t = T_a$. Consequently, with

$$\begin{aligned}
\sum_{n\in\mathbb{Z}} \operatorname{sinc}^2\Big(n \frac{\pi\, T_a}{D_a} \Big) &= 1 + \sum_{n\in\mathbb{Z}\setminus\{0\}} \operatorname{sinc}^2\Big(n \frac{\pi\, T_a}{D_a} \Big) \le 1 + \frac{2\, D_a^2}{\pi^2\, T_a^2} \sum_{n\in\mathbb{N}\setminus\{0\}} \frac{1}{n^2} \\
&= 1 + \frac{4\, D_a^2}{\pi^2\, T_a^2}
\end{aligned}$$

we obtain by combining (6.19) and (6.20),

$$\|F'(\tilde{c}) - F'(c)\| \le L\, \|\tilde{c} - c\|, \quad \text{with } L := \frac{1}{\varepsilon} \sqrt{\frac{1}{D_a} + \frac{4\, D_a}{\pi^2\, T_a^2}} \sqrt{S_2}\, \sqrt{A_2}. \tag{6.21}$$

In our concrete example (visualized in Figure 9) the ansatz space $\mathcal{A} \subset L_2(\mathbb{R})$ is spanned by functions a_n with $D_a = 0.4$ and time step size $T_a = 0.1$. The sampling map S is determined by $D_s = 0.2$ and $T_s = 0.1$. The synthetic signal which we aim to reconstruct is given by

$$f(t) = \phi_{-2}(t) - 0.5\phi_{2.5}(t).$$

For the numerical implementation we have restricted the computations to the finite interval $[-10, 10]$ which was discretized by the grid $t_k = -10 + 0.05k$ with $k =$

$0, 1, 2, \ldots$. The bounds A_2 and S_2 are estimated by the eigenvalues of adequately corresponding finite dimensional approximations of the Gramian matrices $\langle a_n, a_m \rangle$ and $\langle s_n, s_m \rangle$. For the radius of the ℓ_1 ball (determined the sparsity constraint) we have picked $R = 2$. This choice of course includes some a-priori knowledge of the solution to be reconstructed. Usually there is no a-priori information on R available. Even if not proven so far, R plays the role of an regularization parameter (so far just with numerical evidence). Therefore, we can observe in case of misspecified R a similar behavior as for inversion methods where the regularization parameter was not optimally chosen. If R is chosen too large it may easily happen that the ℓ_1 constraint has almost no impact and the solution can be arbitrarily far off the true solution. Therefore, it was suggested in [15] to choose a slowly increasing radius, i.e.

$$R^n = (n + 1)R/N,$$

where n is the iteration index and N stands for a prescribed number of iterations. This proceeding yields in all considered experiments better results. However, convergence of a scheme with varying R^n is theoretically not verified yet.

In Figure 7 (right image) one finds that β^n varies significantly from one to another iteration. This verifies the usefulness of Condition (B). From the first iteration on, the values for β^n are obviously larger than one and grow in the first phase of the iteration process (for the accelerated method only the first 60 iterations are shown). But the main impact manifests itself more in the second half of the iteration ($n > 20$) where the non-accelerated variant has a much less decay of $\sqrt{\boldsymbol{D}(x^n)}$, see Figure 8. There the values of β^n vary around 10^3 and allow that impressive fast and rapid decay of $\sqrt{\boldsymbol{D}(x^n)}$ of the accelerated descent method. For the non-accelerated method we had to compute 10^4 iterations to achieve reasonable small residuals $\sqrt{\boldsymbol{D}(x^n)}$ (but even then being far off the nice results achieved by the accelerated scheme). The right plot in Figure 8 sketches the residual decay with respect to the overall computational time that was practically necessary. Both curves (the black and the gray) were of course obtained on the same machine under same conditions. The achieved time reduction is remarkable as the accelerated iteration method has required many additional loops of the individual fixed point iterations in order to find the optimal β^n. In particular, the final residual value after $n = 10.000$ iterations for the non-accelerated method was $\sqrt{\boldsymbol{D}(x^{10000})} = 0.0172$. This value was reached by the accelerated method after $n = 28$ iteration steps (the final value after $n = 60$ iterations was $\sqrt{\boldsymbol{D}(x^{60})} = 0.0065$). The overall computational time consumption of the non-accelerated method to arrive at $\sqrt{\boldsymbol{D}(x^{10000})} = 0.0172$ was 45min and 2s, whereas the time consumption for the accelerated method for the same residual discrepancy was only 11.8s, i.e. 229 times faster. The finally resulting reconstruction including a diagram showing the nonlinearly sampled data is given in Figure 9.

Summarizing this numerical experiment, we can conclude that all the theoretical statements of the previous sections can be verified. For this particular nonlinear sensing problem we can achieve an impressive factor of acceleration. But this, however,

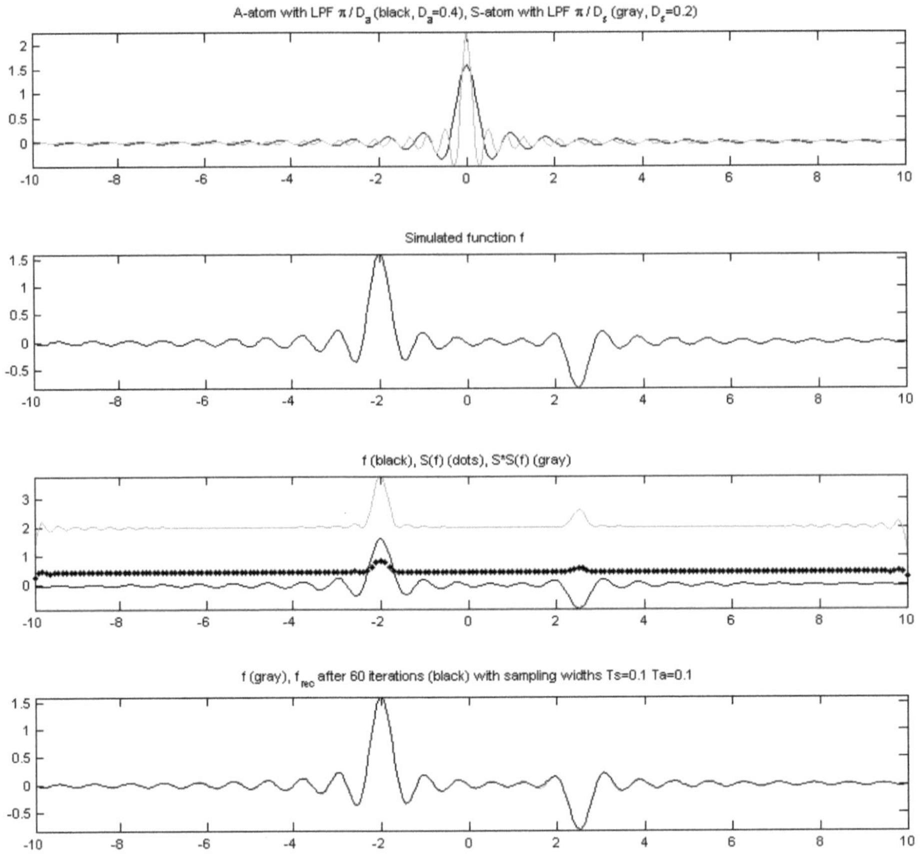

Figure 9. This overview plot shows the used atoms a_0 and s_0 (1st row), the simulated signal (2nd row), the nonlinearly and non-ideally sampled values (3rd row), and the final approximation $A^*x^{60} \in \mathcal{A}$ that was computed with accelerated iteration scheme.

holds for this concrete setting. There is no proved guaranty that the same can be achieved for other applications.

Acknowledgments. The authors would like to thank M. Fornasier for the kind invitation to contribute to the present book. G. T. especially acknowledges the vivid and fruitful interaction with the research team of R. Ramlau and the very nice and close collaboration within his own research group (thanks C. Borries, E. Herrholz, M. Nitschke, K. Pönitz, M. Rätsch, C. Reitberger). R. R. thanks in particular M. Zhariy, S. Anzengruber and E. Resmerita for their collaboration and support in producing this contribution. Doing joint work with G. T. was always enjoyable.

Bibliography

[1] S. Anzengruber and R. Ramlau, Morozovs discrepancy principle for Tikhonov-type functionals with non-linear operators. *Inverse Problems* **26**:2, (2010), 1–17.

[2] K. Bredies, D. A. Lorenz and P. Maass, A generalized conditional gradient method and its connection to an iterative shrinkage method, *Computational Optimization and Application* **42** (2009), 173–193.

[3] E. J. Candès and F. Guo, New multiscale transforms, minimum total variation synthesis: application to edge-preserving image restoration, *CalTech* (2001), preprint.

[4] Y. Censor, D. Gustafson, A. Lent and H. Tuy, A new approach to the emission computerized tomography problem: simultaneous calculation of attenuation and activity coefficients, *IEEE Trans. Nucl. Sci.* (1979), 2275–79.

[5] A. Chambolle, R. DeVore, N. Y. Lee and B. Lucier, Non-linear wavelet image processing: Variational problems, compression and noise removal through wavelet shrinkage, *IEEE Tran. Image Proc.* **7** (1998), 319–333.

[6] O. Christensen, *An Introduction to Frames and Riesz Bases*, Birkhäuser, Boston, 2003.

[7] P. G. Ciarlet, *Introduction to Numerical Linear Algebra and Optimisation*, Cambrigde Univ. Pr., Cambridge, 1995.

[8] A. Cohen, *Numerical Analysis of Wavelet Methods*, 32, North-Holland, Amsterdam, 2003.

[9] A. Cohen, W. Dahmen, I. Daubechies and R. DeVore, *Harmonic Analysis of the Space BV*, IGPM Report # 195, RWTH Aachen, 2000.

[10] A. Cohen, W. Dahmen and R. DeVore, Adaptive wavelet methods II – Beyond the elliptic case, *Found. Comput. Math.* **2** (2002), 203–245.

[11] A. Cohen, R. DeVore, P. Petrushev and H. Xu, Nonlinear approximation and the space $BV(\mathbb{R}^2)$, *American Journal of Mathematics* (1999), 587–628.

[12] I. Daubechies, *Ten Lectures on Wavelets*, SIAM, Philadelphia, 1992.

[13] I. Daubechies, M. Defrise and C. De Mol, An iterative thresholding algorithm for linear inverse problems with a sparsity constraint, *Comm. Pure Appl. Math.* **51** (2004), 1413–1541.

[14] I. Daubechies, M. Defrise and C. DeMol, An iterative thresholding algorithm for linear inverse problems with a sparsity constraint, *Comm. Pure Appl. Math* **57** (2004), 1413–1541.

[15] I. Daubechies, M. Fornasier and I. Loris, Accelerated projected gradient methods for linear inverse problems with sparsity constraints, *J. Fourier Anal. Appl.* (2008), to appear.

[16] I. Daubechies and G. Teschke, Wavelet–based image decomposition by variational functionals, *Proc. SPIE Vol. 5266, p. 94–105, Wavelet Applications in Industrial Processing; Frederic Truchetet; Ed.* (2004).

[17] ———, Variational image restoration by means of wavelets: simultaneous decomposition, deblurring and denoising, *Applied and Computational Harmonic Analysis* **19** (2005), 1–16.

[18] I. Daubechies, G. Teschke and L. Vese, Iteratively solving linear inverse problems with general convex constraints, *Inverse Problems and Imaging* **1** (2007), 29–46.

[19] ———, On some iterative concepts for image restoration, *Advances in Imaging and Electron Physics* **150** (2008), 1–51.

[20] M. Defrise and C. De Mol, A note on stopping rules for iterative methods and filtered SVD, in: *Inverse Problems: An Interdisciplinary Study* (P. C. Sabatier, ed.), Academic Press, 1987, pp. 261–268.

[21] V. Dicken, *Simultaneous Activity and Attenuation Reconstruction in Single Photon Emission Computed Tomography, a Nonlinear Ill-Posed Problem*, PhD thesis, Universität Potsdam, 5/ 1998.

[22] ———, A new approach towards simultaneous activity and attenuation reconstruction in emission tomography, *Inverse Problems* **15** (1999), 931–960.

[23] H. W. Engl, M. Hanke and A. Neubauer, *Regularization of Inverse Problems*, Kluwer, Dordrecht, 1996.

[24] M. Fornasier, Domain decomposition methods for linear inverse problems with sparsity constraints, *Inverse Problems* **23** (2007), 2505.

[25] M. Fornasier and H. Rauhut, Recovery algorithms for vector valued data with joint sparsity constraint, preprint (2006).

[26] M. Grassmair, M. Haltmeier and O. Scherzer, Sparse regularization with ℓ_q penalty term, *Inverse Problems*, **24** (2008), 1–13.

[27] E. Klann, P. Maaß and R. Ramlau, Two-step regularization methods for linear inverse problems, *J. Inverse Ill-Posed Probl.* **14** (2006), 583–609.

[28] A. K. Louis, *Inverse und schlecht gestellte Probleme*, Teubner, Stuttgart, 1989.

[29] S. H. Manglos and T. M. Young, *Constrained IntraSPECT Reconstructions from SPECT Projections*, Conf. Rec. IEEE Nuclear Science Symp. and Medical Imaging Conference, San Francisco, CA, 1993, pp. 1605–1609.

[30] P. Mathé and S. Pereverzev, Optimal discretization of inverse problems in Hilbert scales. Regularization and self-regularization of projection methods, *SIAM J. Numer. Anal.* **38** (2001), 1999–2021.

[31] A. Neubauer, When do Sobolev spaces form a Hilbert scale, *Proc. Am. Math. Soc.* **103** (1988), 557–562.

[32] R. D. Nowak, M. A. T. Figueiredo and S. J. Wright, Gradient projection for sparse reconstruction: Application to compressed sensing and other inverse problems, *IEEE J. Sel. Top. Signal Process.* **1** (2007).

[33] Y. Meyer, Oscillating patterns in image processing and nonlinear evolution equations, *University Lecture Series Volume 22, AMS* (2002).

[34] ———, Oscillating patterns in some nonlinear evolution equations, *CIME report* (2003).

[35] F. Natterer, *The Mathematics of Computerized Tomography*, Teubner, Stuttgart, 1986.

[36] S. Osher and L. Vese, *Modeling Textures with Total Variation Minimization and Oscillating Patterns in Image Processing*, University of California Los Angeles C.A.M., Report no. 02-19, 2002.

[37] R. Ramlau, *Modifizierte Landweber-Iterationen für Inverse Probleme*, PhD thesis, Universität Potsdam (1997).

[38] ______, A modified Landweber-method for inverse problems, *Journal for Numerical Functional Analysis and Optimization* **20** (1999), 79–98.

[39] R. Ramlau, Morozov's discrepancy principle for Tikhonov regularization of nonlinear operators, *Numer. Funct. Anal. and Optimiz* **23** (2002), 147–172.

[40] ______, A steepest descent algorithm for the global minimization of the Tikhonov-functional, *Inverse Problems* **18** (2002), 381–405.

[41] ______, TIGRA – an iterative algorithm for regularizing nonlinear ill-posed problems, *Inverse Problems* **19** (2003), 433–467.

[42] ______, On the use of fixed point iterations for the regularization of nonlinear ill-posed problems, *Journal for Inverse and Ill-Posed Problems* **13**:2 (2005),175–200.

[43] R. Ramlau, R. Clackdoyle, F. Noo and G. Bal, Accurate attenuation correction in SPECT imaging using optimization of bilinear functions and assuming an unknown spatially–varying attenuation distribution, *Z. Angew. Math. Mech.* **80** (2000), 613–621.

[44] R. Ramlau and G. Teschke, Tikhonov replacement functionals for iteratively solving nonlinear operator equations, *Inverse Problems* **21** (2005), 1571–1592.

[45] ______, Tikhonov replacement functionals for iteratively solving nonlinear operator equations, *Inverse Problems* **21** (2005), 1571–1592.

[46] ______, A projection iteration for nonlinear operator equations with sparsity constraints, *Numerische Mathematik* **104** (2006), 177–203.

[47] R. Ramlau, G. Teschke and M. Zhariy, Nonlinear and adaptive frame approximations schemes for ill-posed inverse problems: regularization theory and numerical experiments, *Inverse Problems* **24** (2008), 065013.

[48] R. Ramlau and E. Resmerita, Convergence rates for regularization with sparsity constraints, *ETNA* **37** (2010), 87–104.

[49] L. Rudin, S. Osher and E. Fatemi, Nonlinear total variations based noise removal algorithms, *Physica D* **60** (1992), 259–268.

[50] R. Stevenson, Adaptive solution of operator equations using wavelet frames, *SIAM J. Numer. Anal.* (2003), 1074–1100.

[51] J. A. Terry, B. M. W. Tsui, J. R. Perry, J. L. Hendricks and G. T. Gullberg, *The Design of a Mathematical Phantom of the Upper Human Torso for Use in 3-D SPECT Imaging Research*, Proc. 1990 Fall Meeting Biomed. Eng. Soc. (Blacksburg, VA), New York University Press, 1990, pp. 1467–74.

[52] G. Teschke, Multi-frame representations in linear inverse problems with mixed multi-constraints, *Applied and Computational Harmonic Analysis* **22** (2007), 43–60.

[53] G. Teschke and C. Borries, Accelerated projected steepest descent method for nonlinear inverse problems with sparsity constraints, *Inverse Problems* **26** (2010), 025007.

[54] Y. Eldar T. G. Dvorkind and E. Matusiak, Nonlinear and non-ideal sampling: theory and methods, *IEEE Trans. on Signal Processing* **56** (2008).

[55] E. van den Berg and M. P. Friedlander, In pursiut of a root, preprint (2007).

[56] A. Welch, R. Clack, P. E. Christian and G. T. Gullberg, Toward accurate attenuation correction without transmission measurements, *Journal of Nuclear Medicine* (1996), 18p.

[57] A. Welch, R. Clack, F. Natterer and G. T. Gullberg, Toward accurate attenuation correction in SPECT without transmission measurements, *IEEE Trans. Med. Imaging* (1997), 532–40.

[58] E. Zeidler, *Nonlinear Functional Analysis and its Applications III*, Springer, New York, 1985.

[59] ———, *Nonlinear Functional Analysis and its Applications I*, Springer, New York, 1986.

Author information

Ronny Ramlau, Industrial Mathematics Institute, Johannes Kepler University Linz, Altenbergerstr. 69, 4040 Linz, Austria.
E-mail: `ronny.ramlau@jku.at`

Gerd Teschke, Institute for Computational Mathematics in Science and Technology, Neubrandenburg University of Applied Sciences, Brodaer Str. 2, 17033 Neubrandenburg, Germany.
E-mail: `teschke@hs-nb.de`

Radon Series Comp. Appl. Math **9**, 263–340

An Introduction to Total Variation for Image Analysis

Antonin Chambolle, Vicent Caselles, Daniel Cremers, Matteo Novaga and Thomas Pock

Abstract. These notes address various theoretical and practical topics related to Total Variation-based image reconstruction. They focus first on some theoretical results on functions which minimize the total variation, and in a second part, describe a few standard and less standard algorithms to minimize the total variation in a finite-differences setting, with a series of applications from simple denoising to stereo, or deconvolution issues, and even more exotic uses like the minimization of minimal partition problems.

Keywords. Total variation, variational image reconstruction, functions with bounded variation, level sets, convex optimization, splitting algorithms, denoising, deconvolution, stereo.

2010 Mathematics Subject Classification. 26B30, 26B15, 49-01, 49M25, 49M29, 65-01, 65K15.

Table of Contents.

1 The Total Variation

1.1 Why Is the Total Variation Useful for Images?

The total variation has been introduced for image denoising and reconstruction in a celebrated paper of 1992 by Rudin, Osher and Fatemi [69]. Let us discuss how such a model, as well as other variational approaches for image analysis problems, arise in the context of Bayesian inference.

1.1.1 The Bayesian Approach to Image Reconstruction

Let us first consider the discrete setting, where images $g = (g_{i,j})_{1\le i,j\le N}$ are discrete, bounded ($g_{i,j} \in [0,1]$ or $\{0,\dots,255\}$) 2D-signals. The general idea for solving (linear) inverse problems is to consider

The first author is supported by the CNRS and the ANR, grant ANR-08-BLAN-0082.

- A *model*: $g = Au + n$, $u \in \mathbb{R}^{N\times N}$ is the initial "perfect" signal, A is some transformation (blurring, sampling, or more generally some linear operator, like a Radon transform for tomography). $n = (n_{i,j})$ is the noise: in the simplest situations, we consider a Gaussian norm with average 0 and standard deviation σ.
- An *a priori probability density* for "perfect" original signals, $P(u) \sim e^{-p(u)}du$. It represents the idea we have of perfect data (in other words, the model for the data).

Then, the *a posteriori* probability for u knowing g is computed from Bayes' rule, which is written as follows:

$$P(u|g)P(g) = P(g|u)P(u). \tag{BR}$$

Since the density for the probability of g knowing u is the density for $n = g - Au$, it is

$$e^{-\frac{1}{2\sigma^2}\sum_{i,j}|g_{i,j}-(Au)_{i,j}|^2}$$

and we deduce from (BR) that the density for $P(u|g)$, the probability of u knowing the observation g is

$$\frac{1}{Z(g)}e^{-p(u)}e^{-\frac{1}{2\sigma^2}\sum_{i,j}|g_{i,j}-(Au)_{i,j}|^2}$$

with $Z(g)$ a renormalization factor which is simply

$$Z(g) = \int_u e^{-\left(p(u)+\frac{1}{2\sigma^2}\sum_{i,j}|g_{i,j}-(Au)_{i,j}|^2\right)}du$$

(the integral is on all possible images u, that is, $\mathbb{R}^{N\times N}$, or $[0,1]^{N\times N}$ ...).

The idea of "maximum a posteriori" (MAP) image reconstruction is to find the "best" image as the one which maximizes this probability, or equivalently, which solves the minimum problem

$$\min_u p(u) + \frac{1}{2\sigma^2}\sum_{i,j}|g_{i,j} - (Au)_{i,j}|^2. \tag{MAP}$$

Let us observe that this is not necessarily a good idea, indeed, even if our model is perfectly well built, the image with highest probability given by the resolution of (MAP) might be very rare. Consider for instance Figure 1 where we have plotted a (of course quite strange) density on $[0,1]$, whose maximum is reached at $x = 1/20$, while, in fact, the probability that $x \in [0, 1/10]$ is $1/10$, while the probability $x \in [1/10, 1]$ is $9/10$. In particular the expectation of x is $1/2$. This shows that it might make more sense to try to compute the *expectation* of u (given g)

$$E(u|g) = \frac{1}{Z(g)}\int_u u\, e^{-\left(p(u)+\frac{1}{2\sigma^2}\sum_{i,j}|g_{i,j}-(Au)_{i,j}|^2\right)}du.$$

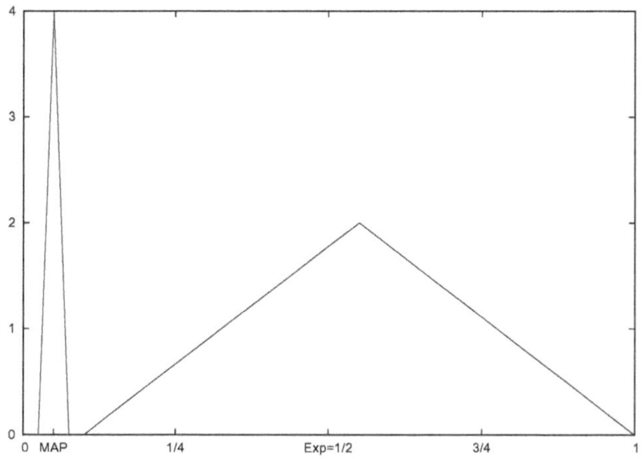

Figure 1. A strange probability density.

However, such a computation is hardly tractable in practice, and requires subtle algorithms based on complex stochastic techniques (Monte Carlo methods with Markov Chains, or MCMC). These approaches seem yet not efficient enough for complex reconstruction problems. See for instance [50, 67] for experiments in this direction.

1.1.2 Variational Models in the Continuous Setting

Now, let us forget the Bayesian, discrete model and just retain to simplify the idea of *minimizing* an energy such as in (MAP). We will now write our images in the continuous setting: as grey-level valued functions $g, u : \Omega \mapsto \mathbb{R}$ or $[0, 1]$, where $\Omega \subset \mathbb{R}^2$ will in practice be (most of the times) the square $[0, 1]^2$, but in general any (bounded) open set of $\mathbb{R}^2$, or more generally $\mathbb{R}^N$, $N \geq 1$.

The operator A will be a bounded, linear operator (for instance from $L^2(\Omega)$ to itself), but from now on, to simplify, we will simply consider $A = \mathrm{Id}$ (the identity operator $Au = u$), and return to more general (and useful) cases in the Section 3 on numerical algorithms.

In this case, the minimization problem (MAP) can be written

$$\min_{u \in L^2(\Omega)} \lambda F(u) + \frac{1}{2} \int_\Omega |u(x) - g(x)|^2 \, dx \qquad \text{(MAPc)}$$

where F is a functional corresponding to the a priori probability density $p(u)$, and which synthesizes the idea we have of the type of signal we want to recover, and $\lambda > 0$ a weight balancing the respective importance of the two terms in the problem. We consider u in the space $L^2(\Omega)$ of functions which are square-integrable, since the energy will be infinite if u is not, this might not always be the right choice (with for instance general operators A).

Now, what is the good choice for F? Standard Tychonov regularization approaches will usually consider quadratic F's, such as $F(u) = \frac{1}{2}\int_\Omega u^2\,dx$ or $\frac{1}{2}\int_\Omega |\nabla u|^2\,dx$. In this last expression,

$$\nabla u(x) = \begin{pmatrix} \frac{\partial u}{\partial x_1}(x) \\ \vdots \\ \frac{\partial u}{\partial x_N}(x) \end{pmatrix}$$

is the *gradient* of u at x. The advantage of these choices is that the corresponding problem to solve is linear, indeed, the *Euler–Lagrange equation* for the minimization problem is, in the first case,

$$\lambda u + u - g = 0,$$

and in the second,

$$-\lambda \Delta u + u - g = 0,$$

where $\Delta u = \sum_i \partial^2 u/\partial x_i^2$ is the *Laplacian* of u. Now look at Figure 2:

Figure 2. A white square on a dark background, then with noise, then restored with $F = \frac{1}{2}\int |u|^2$, then with $F = \frac{1}{2}\int |\nabla u|^2$.

In the first case, no regularization has occurred. This is simply because $F(u) = \frac{1}{2}\int u^2$ enforces no spatial regularization of any kind. Of course, this is a wrong choice, since all "natural" images show a lot of spatial regularity. On the other hand, in the second case, there is too much spatial regularization. Indeed, the image u must belong to the space $H^1(\Omega)$ of functions whose derivative is square-integrable. However, it is well known that such functions cannot present discontinuities across hypersurfaces, that is, in 2 dimension, across lines (such as the edges or boundaries of objects in an image).

A quick argument to justify this is as follows. Assume first $u : [0,1] \to \mathbb{R}$ is a 1-dimensional function which belongs to $H^1(0,1)$. Then for each $0 < s < t < 1$,

$$u(t) - u(s) = \int_s^t u'(r)\,dr \le \sqrt{t-s}\sqrt{\int_s^t |u'(r)|^2\,dr} \le \sqrt{t-s}\|u\|_{H^1}^2$$

so that u must be $1/2$-Hölder continuous (and in fact a bit better). (This computation is a bit formal, it needs to be performed on smooth functions and is then justified by density for any function of $H^1(0,1)$.)

Now if $u \in H^1((0,1) \times (0,1))$, one can check that for a.e. $y \in (0,1)$, $x \mapsto u(x,y) \in H^1(0,1)$, which essentially comes from the fact that

$$\int_0^1 \left(\int_0^1 \left| \frac{\partial u}{\partial x}(x,y) \right|^2 dx \right) dy \leq \|u\|_{H^1}^2 < +\infty.$$

It means that for a.e. y, $x \mapsto u(x,y)$ will be $1/2$-Hölder continuous in x, so that it certainly cannot jump across the vertical boundaries of the square in Figure 2. In fact, a similar kind of regularity can be shown for any $u \in W^{1,p}(\Omega)$, $1 \leq p \leq +\infty$ (although for $p = 1$ it is a bit weaker, but still "large" discontinuities are forbidden), so that replacing $\int_\Omega |\nabla u|^2 \, dx$ with $\int_\Omega |\nabla u|^p \, dx$ for some other choice of p should not produce any better result. We will soon check that the reality is a bit more complicated.

So what is a good "$F(u)$" for images? There have been essentially two types of answers, during the 80s and early 90s, to this question. As we have checked, a good F should simultaneously ensure some spatial regularity, but also preserve the edges. The first idea in this direction is due to D. Geman and S. Geman [38], where it is described in the Bayesian context. They consider an additional variable $\ell = (\ell_{i+1/2,j}, \ell_{i,j+1/2})_{i,j}$ which can take only values 0 and 1: $\ell_{i+1/2,j} = 1$ means that there is an edge between the locations (i,j) and $(i+1,j)$, while 0 means that there is no edge. Then, $p(u)$ in the a priori probability density of u needs to be replaced with $p(u,\ell)$, which typically will be of the form

$$p(u,\ell) = \lambda \sum_{i,j} ((1 - \ell_{i+\frac{1}{2},j})(u_{i+1,j} - u_{i,j})^2 + (1 - \ell_{i,j+\frac{1}{2}})(u_{i,j+1} - u_{i,j})^2)$$
$$+ \mu \sum_{i,j} (\ell_{i+\frac{1}{2},j} + \ell_{i,j+\frac{1}{2}}),$$

with λ, μ positive parameters. Hence, the problem (MAP) will now look like (taking as before $A = \mathrm{Id}$):

$$\min_{u,\ell} p(u,\ell) + \frac{1}{2\sigma^2} \sum_{i,j} |g_{i,j} - u_{i,j}|^2.$$

In the continuous setting, it has been observed by D. Mumford and J. Shah [57] that the set $\{\ell = 1\}$ could be considered as a 1-*dimensional* curve $K \subset \Omega$, while the way it was penalized in the energy was essentially proportional to its length. So that they proposed to consider the minimal problem

$$\min_{u,K} \lambda \int_{\Omega \setminus K} |\nabla u|^2 \, dx + \mu \,\mathrm{length}(K) + \int_\Omega |u - g|^2 \, dx$$

among all 1-dimensional closed subsets K of Ω and all $u \in H^1(\Omega \setminus K)$. This is the famous "Mumford–Shah" functional whose study has generated a lot of interesting mathematical tools and problems in the past 20 years, see in particular [55, 7, 30, 52].

However, besides being particularly difficult to analyse mathematically, this approach is also very complicated numerically since it requires to solve a non-convex problem, and there is (except in a few particular situations) no way, in general, to know whether a candidate is really a minimizer. The most efficient methods rely either on stochastic algorithms [56], or on variational approximations by "Γ-convergence", see [8, 9] solved by alternate minimizations. The exception is the one-dimensional setting where a dynamical programming principle is available and an exact solution can be computed in polynomial time.

1.1.3 A Convex, yet Edge-preserving Approach

In the context of image reconstruction, it was proposed first by Rudin, Osher and Fatemi in [69] to consider the "Total Variation" as a regularizer $F(u)$ for (MAPc). The precise definition will be introduced in the next section. It can be seen as an extension of the energy

$$F(u) = \int_\Omega |\nabla u(x)|\, dx$$

well defined for C^1 functions, and more generally for functions u in the Sobolev space $W^{1,1}$. The big advantage of considering such a F is that it is now convex in the variable u, so that the problem (MAPc) will now be convex and many tools from convex optimization can be used to tackle it, with a great chance of success (see Definition 3.2 and Section 3). However, as we have mentioned before, a function in $W^{1,1}$ cannot present a discontinuity across a line (in 2D) or a hypersurface (in general). Exactly as for H^1 functions, the idea is that if $u \in W^{1,1}(0,1)$ and $0 < s < t < 1$,

$$u(t) - u(s) = \int_s^t u'(r)\, dr \le \int_s^t |u'(r)|\, dr$$

and if $u' \in L^1(0,1)$, the last integral must vanish as $|t-s| \to 0$ (and, even, in fact, uniformly in s, t). We deduce that u is (uniformly) continuous on $[0,1]$, and, as before, if now $u \in W^{1,1}((0,1)\times(0,1))$ is an image in 2D, we will have that for a.e. $y \in (0,1)$, $u(\cdot, y)$ is a 1D $W^{1,1}$ function hence continuous in the variable x.

But what happens when one tries to resolve

$$\min_u \lambda \int_0^1 |u'(t)|\, dt + \int_0^1 |u(t) - g(t)|^2\, dt \; ? \qquad (1.1)$$

Consider the simple case where $g = \chi_{(1/2,1)}$ (that is 0 for $t < 1/2$, 1 for $t > 1/2$). First, there is a "maximum principle": if u is a candidate (which we assume in $W^{1,1}(0,1)$, or to simplify continuous and piecewise C^1) for the minimization, then also $v = \min\{u, 1\}$ is. Moreover, $v' = u'$ whenever $u < 1$ and $v' = 0$ a.e. on $\{v = 1\}$, that is, where $u \ge 1$. So that clearly, $\int_0^1 |v'| \le \int_0^1 |u'|$ (and the inequality is strict if

$v \neq u$). Moreover, since $g \leq 1$, also $\int_0^1 |v-g|^2 \leq \int_0^1 |u-g|^2$. Hence,

$$\mathcal{E}(v) := \lambda \int_0^1 |v'(t)|\,dt + \int_0^1 |v(t)-g(t)|^2\,dt \leq \mathcal{E}(u)$$

(with a strict inequality if $v \neq u$). This tells us that a minimizer, if it exists, must be ≤ 1 a.e. (1 here is the maximum value of g). In the same way, one checks that it must be ≥ 0 a.e. (the minimum value of g). Hence we can restrict ourselves to functions between 0 and 1.

Then, by symmetry, $t \mapsto 1-u(1-t)$ has the same energy as u, and by convexity,

$$\mathcal{E}\left(\frac{1-u(1-\cdot)+u}{2}\right) \leq \frac{1}{2}\mathcal{E}(1-u(1-\cdot)) + \frac{1}{2}\mathcal{E}(u) = \mathcal{E}(u)$$

so that $v : t \mapsto (1-u(1-t)+u(t))/2$ has also lower energy (and again, one can show that the energy is "strictly" convex so that this is strict if $v \neq u$): but $v = u$ iff $u(1-t) = 1-u(t)$, so that any solution must have this symmetry.

Let now $m = \min u = u(a)$, and $M = 1-m = \max u = u(b)$: it must be that (assuming for instance that $a < b$)

$$\int_0^1 |u'(t)|\,dt \geq \int_a^b |u'(t)|\,dt \geq \int_a^b u'(t)\,dt = M - m = 1-2m$$

(and again all this is strict except when u is nondecreasing, or nonincreasing).

To sum up, a minimizer u of $\mathcal{E}$ should be between two values $0 \leq m \leq M = 1-m \leq 1$ (hence $m \in [0,1/2]$), and have the symmetry $u(1-t) = 1-u(t)$. In particular, we should have

$$\mathcal{E}(u) \geq \lambda(M-m) + \int_0^{\frac{1}{2}} m^2 + \int_{\frac{1}{2}}^1 (1-M)^2 = \lambda(1-2m) + m^2$$

which is minimal for $m = \lambda > 0$ provided $\lambda \leq 1/2$, and $m = 1/2$ if $\lambda \geq 1/2$ (remember $m \in [0,1/2]$). In particular, in the latter case $\lambda \geq 1/2$, we deduce that the only possible minimizer is the function $u(t) \equiv 1/2$.

Assume then that $\lambda < 1/2$, so that for any u,

$$\mathcal{E}(u) \geq \lambda(1-\lambda)$$

and consider for $n \geq 2$, $u_n(t) = \lambda$ if $t \in [0, 1/2 - 1/n]$, $u_n(t) = 1/2 + n(t - 1/2)(1/2-\lambda)$ if $|t-1/2| \leq 1/n$, and $1-\lambda$ if $t \geq 1/2 + 1/n$. Then, since u_n is nondecreasing, $\int_0^1 |u'| = \int_0^1 u' = 1-2\lambda$ so that

$$\mathcal{E}(u_n) \leq \lambda(1-2\lambda) + \left(1-\frac{2}{n}\right)\lambda^2 + \frac{2}{n} \to \lambda(1-\lambda)$$

as $n \to \infty$. Hence: $\inf_u \mathcal{E}(u) = \lambda(1-\lambda)$. Now, for a function u to be a minimizer, we see that: it must be nondecreasing and grow from λ to $1-\lambda$ (otherwise the term $\int_0^1 |u'|$ will be too large), and it must satisfy as well

$$\int_0^1 |u(t) - g(t)|^2 \, dt = \lambda^2,$$

while from the first condition we deduce that $|u-g| \geq \lambda$ a.e.: hence we must have $|u-g| = \lambda$, that is, $u = \lambda$ on $[0, 1/2)$ and $1-\lambda$ on $(1/2, 1]$. But this u, which is actually the limit of our u_n's, is not classically derivable: this shows that one must extend in an appropriate way the notion of derivative to give a solution to problem (1.1) of minimizing $\mathcal{E}$: otherwise it cannot have a solution. In particular, we have seen that for all the functions u_n, $\int_0^1 |u_n'| = 1 - 2\lambda$, so that for our discontinuous limit u it is reasonable to assume that $\int |u'|$ makes sense. This is what we will soon define properly as the "total variation" of u, and we will see that it makes sense for a whole category of non necessarily continuous functions, namely, the "functions with bounded variation" (or BV functions). Observe that we could define, in our case, for any $u \in L^1(0,1)$,

$$F(u) = \inf\left\{ \lim_{n\to\infty} \int_0^1 |u_n'(t)| \, dt \; : \; u_n \to u \text{ in } L^1(0,1) \text{ and } \lim_n \int_0^1 |u_n'| \text{ exists.}\right\}.$$

In this case, we could check easily that our discontinuous solution is the (unique) minimizer of

$$\lambda F(u) + \int_0^1 |u(t) - g(t)|^2 \, dt.$$

It turns out that this definition is consistent with the more classical definition of the total variation which we will introduce hereafter, in Definition 1.1 (see inequality (1.2) and Theorem 1.3).

What have we learned from this example? If we introduce, in Tychonov's regularization, the function $F(u) = \int_\Omega |\nabla u(x)| \, dx$ as a regularizer, then in general the problem (MAPc) will have no solution in $W^{1,1}(\Omega)$ (where F makes sense). But, there should be a way to appropriately extend F to more general functions which can have (large) discontinuities and not be in $W^{1,1}$, so that (MAPc) has a solution, and this solution can have edges! This was the motivation of Rudin, Osher and Fatemi [69] to introduce the *Total Variation* as a regularizer $F(u)$ for inverse problems of type (MAPc). We will now introduce more precisely, from a mathematical point of view, this functional, and give its main properties.

1.2 Some Theoretical Facts: Definitions, Properties

The material in this part is mostly extracted from the textbooks [41, 75, 35, 7], which we invite the reader to consult for further details.

1.2.1 Definition

Definition 1.1. The *total variation* of an image is defined by duality: for $u \in L^1_{\mathrm{loc}}(\Omega)$ it is given by

$$J(u) = \sup\left\{ -\int_\Omega u \operatorname{div}\phi \, dx \; : \; \phi \in C_c^\infty(\Omega;\mathbb{R}^N), |\phi(x)| \le 1 \;\forall x \in \Omega \right\}. \qquad \text{(TV)}$$

A function is said to have Bounded Variation whenever $J(u) < +\infty$. Typical examples include:

- A smooth function $u \in C^1(\Omega)$ (or in fact a function $u \in W^{1,1}(\Omega)$): in this case,

$$-\int_\Omega u \operatorname{div}\phi \, dx = \int_\Omega \phi \cdot \nabla u \, dx$$

and the sup over all ϕ with $|\phi| \le 1$ is $J(u) = \int_\Omega |\nabla u| \, dx$.

- The characteristic function of a set with smooth (or $C^{1,1}$) boundary: $u = \chi_E$, in this case

$$-\int_\Omega u \operatorname{div}\phi \, dx = -\int_{\partial E} \phi \cdot \nu_E \, d\sigma$$

and one can reach the sup (which corresponds to $\phi = -\nu_E$, the outer normal to ∂E, on $\partial E \cap \Omega$, while $\phi = 0$ on $\partial E \cap \partial\Omega$) by smoothing, in a neighborhood of the boundary, the gradient of the signed distance function to the boundary. We obtain that $J(u) = \mathcal{H}^{N-1}(\partial E \cap \Omega)$, the *perimeter* of E in Ω.

Here, $\mathcal{H}^{N-1}(\cdot)$ is the $(N-1)$-dimensional Hausdorff measure, see for instance [36, 55, 7] for details.

1.2.2 An Equivalent Definition

It is well known (see for instance [70]) that any $u \in L^1_{\mathrm{loc}}(\Omega)$ defines a *distribution*

$$\begin{array}{rccl} T_u \; : & \mathcal{D}(\Omega) & \to & \mathbb{R} \\ & \phi & \mapsto & \displaystyle\int_\Omega \phi(x)u(x)\, dx \end{array}$$

where here $\mathcal{D}(\Omega)$ is the space of smooth functions with compact support ($C_c^\infty(\Omega)$) endowed with a particular topology, and T_u is a continuous linear form on $\mathcal{D}(\Omega)$, that is, $T_u \in \mathcal{D}'(\Omega)$. We denote by $\langle T, \phi \rangle_{\mathcal{D}',\mathcal{D}} \in \mathbb{R}$ the duality product between a linear form $T \in \mathcal{D}'$ and a vector $\phi \in \mathcal{D}$. The derivative of T_u is then defined as ($i = 1, \dots, N$)

$$\left\langle \frac{\partial T_u}{\partial x_i}, \phi \right\rangle_{\mathcal{D}',\mathcal{D}} := -\left\langle T_u, \frac{\partial \phi}{\partial x_i} \right\rangle_{\mathcal{D}',\mathcal{D}} = -\int_\Omega u(x) \frac{\partial \phi}{\partial x_i}(x)\, dx$$

(which clearly extends the integration by parts: if u is smooth, then $\partial T_u/\partial x_i = T_{\partial u/\partial x_i}$). We denote by Du the (vectorial) distribution $(\partial T_u/\partial x_i)_{i=1}^N$.

Then, if $J(u) < +\infty$, it means that for all vector field $\phi \in C_c^\infty(\Omega; \mathbb{R}^N)$

$$\langle Du, \phi \rangle_{\mathcal{D}', \mathcal{D}} \leq J(u) \sup_{x \in \Omega} |\phi(x)|.$$

This means that Du defines a linear form on the space of continuous vector fields, and by Riesz' representation theorem it follows that it defines a Radon measure (precisely, a vector-valued (or signed) Borel measure on Ω which is finite on compact sets), which is globally bounded, and its norm (or *variation* $|Du|(\Omega) = \int_\Omega |Du|$) is precisely the total variation $J(u)$. See for instance [75, 35, 7] for details.

1.2.3 Main Properties of the Total Variation

Lower Semi-continuity Definition 1.1 has a few advantages. It can be introduced for *any* locally integrable function (without requiring any regularity or derivability). But also, $J(u)$ is written as a sup of linear forms

$$L_\phi : u \mapsto -\int_\Omega u(x) \operatorname{div} \phi(x)\, dx$$

which are continuous with respect to very weak topologies (in fact, with respect to the "distributional convergence" related to the space $\mathcal{D}'$ introduced in the previous section).

For instance, if $u_n \rightharpoonup u$ in $L^p(\Omega)$ for any $p \in [1, +\infty)$ (or weakly-$*$ for $p = \infty$), or even in $L^p(\Omega')$ for any $\Omega' \subset\subset \Omega$, then $L_\phi u_n \to L_\phi u$. But it follows that

$$L_\phi u = \lim_n L_\phi u_n \leq \liminf_n J(u_n)$$

and taking then the sup over all smooth fields ϕ with $|\phi(x)| \leq 1$ everywhere, we deduce that

$$J(u) \leq \liminf_{n \to \infty} J(u_n), \tag{1.2}$$

that is, J is (sequentially) *lower semi-continuous* (l.s.c.) with respect to all the above mentioned topologies. [The idea is that a sup of continuous functions is l.s.c.]

In particular, it becomes obvious to show that with $F = J$, problem (MAPc) has a solution. Indeed, consider a minimizing sequence for

$$\min_u \mathcal{E}(u) := J(u) + \|u - g\|^2_{L^2(\Omega)},$$

which is a sequence $(u_n)_{n \geq 1}$ such that $\mathcal{E}(u_n) \to \inf_u \mathcal{E}(u)$.

As $\mathcal{E}(u_n) \leq \mathcal{E}(0) < +\infty$ for n large enough (we assume $g \in L^2(\Omega)$), and $J \geq 0$), we see that (u_n) is bounded in $L^2(\Omega)$ and it follows that up to a subsequence (still denoted (u_n)), it converges weakly to some u, that is, for any $v \in L^2(\Omega)$,

$$\int_\Omega u_n(x) v(x)\, dx \to \int_\Omega u(x) v(x)\, dx.$$

But then it is known that

$$\|u - g\|_{L^2} \le \liminf_n \|u_n - g\|_{L^2},$$

and since we also have (1.2), we deduce that

$$\mathcal{E}(u) \le \liminf_n \mathcal{E}(u_n) = \inf \mathcal{E}$$

so that u is a minimizer. □

Convexity Now, is u unique? The second fundamental property of J which we deduce from Definition 1.1 is its *convexity*: for any u_1, u_2 and $t \in [0, 1]$,

$$J(tu_1 + (1-t)u_2) \le tJ(u_1) + (1-t)J(u_2). \tag{1.3}$$

It follows, again, because J is the supremum of the linear (hence convex) functions L_ϕ: indeed, one clearly has

$$L_\phi(tu_1 + (1-t)u_2) = tL_\phi(u_1) + (1-t)L_\phi(u_2) \le tJ(u_1) + (1-t)J(u_2)$$

and taking the sup in the left-hand side yields (1.3).

Hence in particular, if u and u' are two solutions of (MAPc), then

$$\begin{aligned}\mathcal{E}\left(\frac{u+u'}{2}\right) &\le \frac{\lambda}{2}(J(u) + J(u')) + \int_\Omega \left|\frac{u+u'}{2} - g\right|^2 dx \\ &= \frac{1}{2}(\mathcal{E}(u) + \mathcal{E}(u')) - \frac{1}{4}\int_\Omega (u - u')^2\, dx\end{aligned}$$

which would be strictly less than the inf of $\mathcal{E}$, unless $u = u'$: hence the minimizer of (MAPc) exists, and is unique.

Homogeneity It is obvious for the definition that for each u and $t > 0$,

$$J(tu) = tJ(u), \tag{1.4}$$

that is, J is positively *one-homogeneous*.

1.2.4 Functions with Bounded Variation

We introduce the following definition:

Definition 1.2. The space $BV(\Omega)$ of *functions with bounded variation* is the set of functions $u \in L^1(\Omega)$ such that $J(u) < +\infty$, endowed with the norm $\|u\|_{BV(\Omega)} = \|u\|_{L^1(\Omega)} + J(u)$.

This space is easily shown to be a Banach space. It is a natural (weak) "closure" of $W^{1,1}(\Omega)$. Let us state a few essential properties of this space.

Meyers–Serrin's Approximation Theorem We first state a theorem which shows that BV function may be "well" approximated with smooth functions. This is a refinement of a classical theorem of Meyers and Serrin [54] for Sobolev spaces.

Theorem 1.3. *Let $\Omega \subset \mathbb{R}^N$ be an open set and let $u \in BV(\Omega)$: then there exists a sequence $(u_n)_{n\geq 1}$ of functions in $C^\infty(\Omega) \cap W^{1,1}(\Omega)$ such that*

(i) $u_n \to u$ *in* $L^1(\Omega)$,

(ii) $J(u_n) = \int_\Omega |\nabla u_n(x)|\, dx \to J(u) = \int_\Omega |Du|$ *as* $n \to \infty$.

Before sketching the proof, let us recall that in Sobolev's spaces $W^{1,p}(\Omega)$, $p < \infty$, the thesis of this classical theorem is stronger, since one proves that $\|\nabla u_n - \nabla u\|_{L^p} \to 0$, while here one cannot expect $J(u_n - u) = \int_\Omega |Du_n - Du| \to 0$ as $n \to \infty$. This is easily illustrated by the following example: let $\Omega = (-1, 1)$, and $u(t) = -1$ if $t < 0$, $u(t) = 1$ if $t \geq 0$. Then, the sequence $u_n(t) = \tanh(n \times t)$ clearly converges to u, with

$$\int_{-1}^{1} u_n'(t)\, dt = 2\tanh(n) \ \to\ 2 = J(u)$$

as $n \to \infty$, but clearly $J(u_n - u) \approx 4$ for large n. In fact, it is clear that if v is any smooth approximation of u such as shown in Figure 3, then clearly the variation $J(u - v)$ of $w = u - v$ is given by

$$\begin{aligned} &|w(0^-) - w(-1)| + |w(0^+) - w(0^-)| + |w(1) - w(0^+)| = \\ &|v(0) - v(-1)| + \qquad 2 \qquad + \ |v(1) - v(0)| \ \approx 4 \end{aligned}$$

and cannot be made arbitrarily small.

Proof. Let us now explain how Theorem 1.3 is proven. The idea is to smooth u with a "mollifier" (or a "smoothing kernel"): as usual one considers a function $\eta \in C_c^\infty(B(0,1))$ with $\eta \geq 0$ and $\int_{B(0,1)} \eta(x)\, dx = 1$. For each $\varepsilon > 0$, one considers $\eta_\varepsilon(x) := (1/\varepsilon)^N \eta(x/\varepsilon)$: then, η_ε has support in the ball $B(0,\varepsilon)$, and $\int_{\mathbb{R}^N} \eta_\varepsilon\, dx = 1$. If $u \in L^1(\mathbb{R}^N)$, it is then classical that the functions

$$u_\varepsilon(x) = u * \eta_\varepsilon(x) := \int_{\mathbb{R}^N} u(y)\eta_\varepsilon(x-y)\, dy = \int_{B(0,\varepsilon)} u(x-y)\eta_\varepsilon(y)\, dy$$

are smooth (because the first expression of the convolution product can be differentiated infinitely many times under the integral), and converge to u, in $L^1(\mathbb{R}^N)$, as $\varepsilon \to 0$ (the convergence is easily shown for continuous function with compact support, and can be shown by density for L^1 functions).

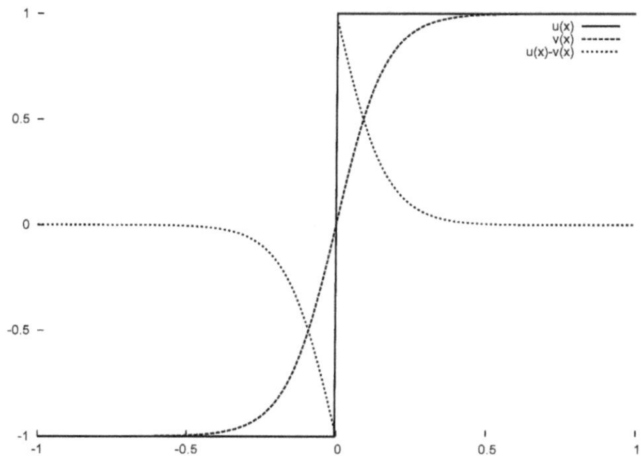

Figure 3. Smooth approximation of a step function.

Then, if $u \in BV(\mathbb{R}^N)$, one also have that for any $\phi \in C_c^\infty(\mathbb{R}^N;\mathbb{R}^N)$ with $|\phi| \le 1$ a.e., (to simplify we assume η is even)

$$\begin{aligned} -\int_{\mathbb{R}^N} \phi(x)\cdot\nabla u_\varepsilon(x)\,dx &= \int_{\mathbb{R}^N} u_\varepsilon(x)\mathrm{div}\,\phi(x)\,dx \\ &= \int_{\mathbb{R}^N}\int_{\mathbb{R}^N} \eta_\varepsilon(x-y)u(y)\mathrm{div}\,\phi(x)\,dy\,dx \\ &= \int_{\mathbb{R}^N} u(y)\mathrm{div}\,(\phi_\varepsilon)(y)\,dy \end{aligned}$$

where we have used Fubini's theorem, and the fact that $(\mathrm{div}\,\phi)_\varepsilon = \mathrm{div}\,(\phi_\varepsilon)$. By Definition 1.1, this is less than $J(u)$. Taking then the sup on all admissible ϕ's, we end up with

$$J(u_\varepsilon) = \int_{\mathbb{R}^N} |\nabla u_\varepsilon|\,dx \le J(u)$$

for all $\varepsilon > 0$. Combined with (1.2), it follows that

$$\lim_{\varepsilon\to 0} J(u_\varepsilon) = J(u).$$

This shows the theorem, when $\Omega = \mathbb{R}^N$.

When $\Omega \neq \mathbb{R}^N$, this theorem is shown by a subtle variant of the classical proof of Meyers–Serrin's theorem [54], see for instance [41] or [7, Theorem 3.9] for details. Let us insist that the result is not straightforward, and, in particular, that in general the function u_n can not be supposed to be smooth up to the boundary. □

Rellich's Compactness Theorem The second important property of BV functions is the following compactness theorem:

Theorem 1.4. *Let $\Omega \subset \mathbb{R}^N$ be a bounded domain with Lipschitz boundary, and let $(u_n)_{n\geq 1}$ be a sequence of functions in $BV(\Omega)$ such that $\sup_n \|u_n\|_{BV} < +\infty$. Then there exists $u \in BV(\Omega)$ and a subsequence $(u_{n_k})_{k\geq 1}$ such that $u_{n_k} \to u$ (strongly) in $L^1(\Omega)$ as $k \to \infty$.*

Proof. If we assume that the theorem is known for functions in $W^{1,1}(\Omega)$, then the extension to BV functions simply follows from Theorem 1.3. Indeed, for each n, we can find $u'_n \in C^\infty(\Omega) \cap W^{1,1}(\Omega)$ with $\|u_n - u'_n\|_{L^1} \leq 1/n$ and $\|u'_n\|_{BV(\Omega)} \leq \|u_n\|_{BV(\Omega)} + 1/n$. Then, we apply Rellich's compactness theorem in $W^{1,1}(\Omega)$ to the sequence u'_n: it follows that there exists $u \in L^1(\Omega)$ and a subsequence $(u'_{n_k})_k$ with $u'_{n_k} \to u$ as $k \to \infty$. Clearly, we have $\|u_{n_k} - u\|_{L^1} \leq 1/n_k + \|u'_{n_k} - u\|_{L^1} \to 0$ as $k \to \infty$. Moreover, $u \in BV(\Omega)$, since its variation is bounded as follows from (1.2).

A complete proof (including the proof of Rellich's theorem) is found in [7], proof of Theorem 3.23. The regularity of the domain Ω is crucial here, since the proof relies on an extension argument outside of Ω: it needs the existence of a linear "extension" operator $T : BV(\Omega) \to BV(\Omega')$ for any $\Omega' \supset\supset \Omega$, such that for each $u \in BV(\Omega)$, Tu has compact support in Ω', $Tu(x) = u(x)$ for a.e. $x \in \Omega$, and $\|Tu\|_{BV(\Omega')} \leq C\|u\|_{BV(\Omega)}$. Then, the proof follows by mollifying the sequence Tu_n, introducing the smooth functions $\eta_\varepsilon * Tu_n$, applying Ascoli–Arzelà's theorem to the mollified functions, and a diagonal argument. □

Sobolev's Inequalities We observe here that the classical inequalities of Sobolev

$$\|u\|_{L^{\frac{N}{N-1}}(\mathbb{R}^N)} \leq C \int_{\mathbb{R}^N} |Du| \tag{1.5}$$

if $u \in L^1(\mathbb{R}^N)$, and Poincaré–Sobolev

$$\|u - m\|_{L^{\frac{N}{N-1}}(\Omega)} \leq C \int_{\mathbb{R}^N} |Du| \tag{1.6}$$

where Ω is bounded with Lipschitz boundary, and m is the average of u on Ω, valid for $W^{1,1}$ functions, clearly also hold for BV function as can be deduced from Theorem 1.3.

1.3 The Perimeter. Sets with Finite Perimeter

1.3.1 Definition, and an Inequality

Definition 1.5. A measurable set $E \subset \Omega$ is a *set of finite perimeter* in Ω (or *Caccioppoli set*) if and only if $\chi_E \in BV(\Omega)$. The total variation $J(\chi_E)$ is the *perimeter* of E in Ω, denoted by $\mathrm{Per}(E;\Omega)$. If $\Omega = \mathbb{R}^N$, we simply denote $\mathrm{Per}(E)$.

We observe that a "set" here is understood as a measurable set in $\mathbb{R}^N$, and that this definition of the perimeter makes it depend on E only up to sets of zero Lebesgue measure. In general, in what follows, the sets we will consider will be rather equivalence classes of sets which are equal up to Lebesgue negligible sets.

The following inequality is an essential property of the perimeter: for any $A, B \subseteq \Omega$ sets of finite perimeter, we have

$$\mathrm{Per}(A \cup B; \Omega) + \mathrm{Per}(A \cap B; \Omega) \leq \mathrm{Per}(A; \Omega) + \mathrm{Per}(B; \Omega). \tag{1.7}$$

Proof. The proof is as follows: we can consider, invoking Theorem 1.3, two sequences u_n, v_n of smooth functions, such that $u_n \to \chi_A$, $v_n \to \chi_B$, and

$$\int_\Omega |\nabla u_n(x)|\, dx \to \mathrm{Per}(A; \Omega) \quad \text{and} \quad \int_\Omega |\nabla v_n(x)|\, dx \to \mathrm{Per}(B; \Omega) \tag{1.8}$$

as $n \to \infty$. Then, it is easy to check that $u_n \vee v_n := \max\{u_n, v_n\} \to \chi_{A\cup B}$ as $n \to \infty$, while $u_n \wedge v_n := \min\{u_n, v_n\} \to \chi_{A\cap B}$ as $n \to \infty$. We deduce, using (1.2), that

$$\mathrm{Per}(A \cup B; \Omega) + \mathrm{Per}(A \cap B; \Omega) \leq \liminf_{n\to\infty} \int_\Omega |\nabla(u_n \vee v_n)| + |\nabla(u_n \wedge v_n)|\, dx. \tag{1.9}$$

But for almost all $x \in \Omega$, $|\nabla(u_n \vee v_n)(x)| + |\nabla(u_n \wedge v_n)(x)| = |\nabla u_n(x)| + |\nabla v_n(x)|$, so that (1.7) follows from (1.9) and (1.8). □

1.3.2 The Reduced Boundary, and a Generalization of Green's Formula

It is shown that if E is a set of finite perimeter in Ω, then the derivative $D\chi_E$ can be expressed as

$$D\chi_E = \nu_E(x)\mathcal{H}^{N-1} \llcorner \partial^* E \tag{1.10}$$

where $\nu_E(x)$ and $\partial^* E$ can be defined as follows: $\partial^* E$ is the set of points x where the "blow-up" sets

$$E_\varepsilon = \{y \in B(0,1) \,:\, x + \varepsilon y \in E\}$$

(see Figure 4) converge as ε to 0 to a semi-space $P_{\nu_E(x)} = \{y \,:\, y \cdot \nu_E(x) \geq 0\} \cap B(0,1)$ in $L^1(B(0,1))$, in the sense that their characteristic functions converge, or in other words

$$|E_\varepsilon \setminus P_{\nu_E(x)}| + |P_{\nu_E(x)} \setminus E_\varepsilon| \to 0$$

as $\varepsilon \to 0$. Here $|E|$ denotes the Lebesgue measure of the set E. This definition of the boundary simultaneously defines the (inner) normal vector $\nu_E(x)$.

The set $\partial^* E$ is called the "reduced" boundary of E (the "true" definition of the reduced boundary is a bit more precise and the precise set slightly smaller than ours, but still (1.10) is true with our definition, see [7, Chapter 3]).

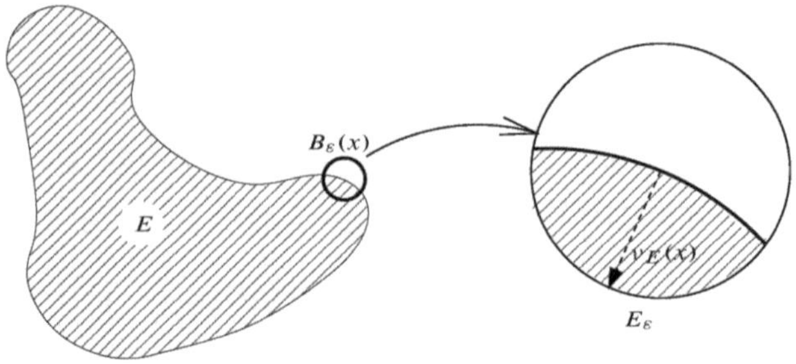

Figure 4. The blow-up of E near the point x.

Equation (1.10) means that for any C^1 vector field ϕ, one has

$$\int_E \operatorname{div} \phi(x)\, dx = -\int_{\partial^* E} \phi \cdot \nu_E(x)\, d\mathcal{H}^{N-1}(x) \tag{1.11}$$

which is a sort of generalization of Green's formula to sets of finite perimeter.

This generalization is useful as shows the following example: let $x_n \in (0,1)^2$, $n \geq 1$, be the sequence of rational points (in $\mathbb{Q}^2 \cap (0,1)^2$), and let $E = \bigcup_{n\geq 1} B(x_n, \varepsilon 2^{-n})$, for some $\varepsilon > 0$ fixed.

Then, one sees that E is an open, dense set in $(0,1)^2$. In particular its "classical" (topological) boundary ∂E is very big, it is $[0,1]^2 \setminus E$ and has Lebesgue measure equal to $1 - |E| \geq 1 - \pi\varepsilon^2/3$. In particular its length is infinite.

However, one can show that E is a finite perimeter set, with perimeter less than $\sum_n 2\pi\varepsilon 2^{-n} = \pi\varepsilon$. Its "reduced boundary" is, up to the intersections, the set

$$\partial^* E \approx \bigcup_{n\geq 1} \partial B(x_n, \varepsilon 2^{-n}).$$

One shows that this "reduced boundary" is always, as in this simple example, a *rectifiable set*, that is, a set which can be almost entirely covered with a countable union of C^1 hypersurfaces, up to a set of Hausdorff $\mathcal{H}^{N-1}$ measure zero: there exist $(\Gamma_i)_{i\geq 1}$, hypersurfaces of regularity C^1, such that

$$\partial^* E \subset \mathcal{N} \cup \left(\bigcup_{i=1}^{\infty} \Gamma_i \right), \quad \mathcal{H}^{N-1}(\mathcal{N}) = 0. \tag{1.12}$$

In particular, $\mathcal{H}^{N-1}$ a.e., the normal $\nu_E(x)$ is a normal to the surface(s) Γ_i such that $x \in \Gamma_i$.

1.3.3 The Isoperimetric Inequality

For $u = \chi_E$, equation (1.5) becomes the celebrated isoperimetric inequality:

$$|E|^{\frac{N-1}{N}} \le C\,\mathrm{Per}(E) \tag{1.13}$$

for all finite-perimeter set E of bounded volume, with the best constant C reached by balls:

$$C^{-1} = N(\omega_N)^{1/N}$$

where $\omega_N = |B(0,1)|$ is the volume of the unit ball in $\mathbb{R}^N$.

1.4 The Co-area Formula

We now can state a fundamental property of BV functions, which will be the key of our analysis in the next sections dealing with applications. This is the famous "co-area" formula of Federer and Fleming:

Theorem 1.6. *Let $u \in BV(\Omega)$: then for a.e. $s \in \mathbb{R}$, the set $\{u > s\}$ is a finite-perimeter set in Ω, and one has*

$$J(u) = \int_\Omega |Du| = \int_{-\infty}^{+\infty} \mathrm{Per}(\{u > s\}; \Omega)\, ds. \tag{CA}$$

It means that the total variation of a function is also the accumulated surfaces of all its level sets. (Of course, is these levels are smooth enough, it corresponds to their actual classical surface, or length in two dimension.) The proof of this result is quite complicated (we refer to [37, 35, 75, 7]) but let us observe that:

- It is relatively simple if $u = p \cdot x$ is an affine function, defined for instance on a simplex T (or in fact any open set). Indeed, in this case, $J(u) = |T|\,|p|$, and $\partial\{u > s\}$ are hypersurfaces $\{p \cdot x = s\}$, and it is not too difficult to compute the integral $\int_s \mathcal{H}^{N-1}(\{p \cdot x = s\})$.
- For a general $u \in BV(\Omega)$, we can approximate u with piecewise affine functions u_n with $\int_\Omega |\nabla u_n|\, dx \to J(u)$. Indeed, one can first approximate u with the smooth functions provided by Theorem 1.3, and then these smooth functions by piecewise affine functions using the standard finite elements theory. Then, we will obtain using (1.2) and Fatou's lemma that $\int_\mathbb{R} \mathrm{Per}(\{u > s\}; \Omega)\, ds \le J(u)$.
- The reverse inequality $J(u) \le \int_\mathbb{R} \mathrm{Per}(\{u > s\}; \Omega)\, ds = \int_\mathbb{R} J(\chi_{\{u>s\}})\, ds$, can easily be deduced by noticing that if $\phi \in C_c^\infty(\Omega)$ with $\|\phi\| \le 1$, one has

$\int_\Omega \operatorname{div} \phi \, dx = 0$, so that (using Fubini's theorem)

$$
\begin{aligned}
&\int_\Omega u \operatorname{div} \phi \, dx \\
&= \int_{\{u>0\}} \int_0^{u(x)} ds \operatorname{div} \phi(x) \, dx - \int_{\{u<0\}} \int_{u(x)}^0 ds \operatorname{div} \phi(x) \, dx \\
&= \int_0^\infty \int_\Omega \chi_{\{u>s\}}(x) \operatorname{div} \phi(x) \, dx \, ds - \int_{-\infty}^0 \int_\Omega (1 - \chi_{\{u>s\}}(x)) \operatorname{div} \phi(x) \, dx \, ds \\
&= \int_{-\infty}^\infty \int_{\{u>s\}} \operatorname{div} \phi \, dx \, ds \\
&\leq \int_{-\infty}^\infty \operatorname{Per}(\{u > s\}; \Omega) \, ds
\end{aligned}
$$

and taking then the sup over all admissible ϕ's in the leftmost term.

Remark. Observe that (1.7) also follows easily from (CA), indeed, let $u = \chi_A + \chi_B$, then $J(u) \leq J(\chi_A) + J(\chi_B) = \operatorname{Per}(A; \Omega) + \operatorname{Per}(B; \Omega)$, while from (CA) we get that

$$
J(u) = \int_0^2 \operatorname{Per}(\{\chi_A + \chi_B > s\}; \Omega) \, ds = \operatorname{Per}(A \cup B; \Omega) + \operatorname{Per}(A \cap B; \Omega).
$$

1.5 The Derivative of a BV Function

To end up this theoretical section on BV functions, we mention an essential result on the measure Du, defined for any $u \in BV(\Omega)$ by

$$
\int \phi(x) \cdot Du(x) = -\int u(x) \operatorname{div} \phi(x) \, dx
$$

for any smooth enough vector field ϕ with compact support. As mentioned in Section 1.2.2, it is a bounded Radon measure. A derivation theorem due to Radon and Nikodym (and a refined version due to Besicovitch) shows that such a measure can be decomposed with respect to any positive radon measure μ into

$$
Du = f(x) \, d\mu + \nu \tag{1.14}
$$

where μ a.e.,

$$
f(x) = \lim_{\rho \to 0} \frac{Du(B(x, \rho))}{\mu(B(x, \rho)}
$$

(and in particular the theorem states that the limit exists a.e.), $f \in L^1_\mu(\Omega)$, that is, $\int_\Omega |f| \, d\mu < +\infty$, and $\nu \perp \mu$, which means that there exists a Borel set $E \subset \Omega$ such that $|\nu|(\Omega \setminus E) = 0$, $\mu(E) = 0$.

If the function $u \in W^{1,1}(\Omega)$, then $Du = \nabla u(x)\,dx$, with ∇u the "weak gradient" a vector-valued function in $L^1(\Omega;\mathbb{R}^N)$. Hence, the decomposition (1.14) with $\mu = dx$ (the Lebesgue measure), holds with $f = \nabla u$ and $\nu = 0$, and one says that Du is "absolutely continuous" with respect to Lebesgue's measure. This is not true anymore for a generic function $u \in BV(\Omega)$. One has

$$Du = \nabla u(x)\,dx + D^s u$$

where the "singular part" $D^s u$ vanishes if and only if $u \in W^{1,1}$, and $\nabla u \in L^1(\Omega;\mathbb{R}^N)$ is the "approximate gradient" of u.

The singular part can be further decomposed. Let us call J_u the "jump set" of u, defined as follows:

Definition 1.7. Given $u \in BV(\Omega)$, we say that $x \in J_u$ if and only if there exist $u_-(x), u_+(x) \in \mathbb{R}$ with $u_-(x) \neq u_+(x)$, and $\nu_u(x) \in \mathbb{R}^N$ a unit vector such that the functions, defined for $y \in B(0,1)$ for $\varepsilon > 0$ small enough

$$y \mapsto u(x + \varepsilon y)$$

converge as $\varepsilon \to 0$, in $L^1(B(0,1))$, to the function

$$y \mapsto u_-(x) + (u_+(x) - u_-(x))\chi_{\{y\cdot\nu_u(x)\geq 0\}}$$

which takes value $u_+(x)$ in the half-space $\{y \cdot \nu_u(x) \geq 0\}$, and $u_-(x)$ in the other half-space $\{y \cdot \nu_u(x) < 0\}$.

In particular, this is consistent with our definition of $\partial^* E$ in Section 1.3.2: $\partial^* E = J_{\chi_E}$, with $(\chi_E)_+(x) = 1$, $(\chi_E)_-(x) = 0$, and $\nu_{\chi_E}(x) = \nu_E$. The triple (u_-, u_+, ν_u) is almost unique: it is unique up to the permutation $(u_+, u_-, -\nu_u)$. For a scalar function u, the canonical choice is to take $u_+ > u_-$, whereas for vectorial BV functions, one must fix some arbitrary rule.

One can show that J_u is a rectifiable set (see Section 1.3.2, eq. (1.12)), in fact, it is a countable union of rectifiable sets since one can always write

$$J_u \subseteq \bigcup_{n \neq m} \partial^*\{u > s_n\} \cap \partial^*\{u > s_m\},$$

for some countable, dense sequence $(s_n)_{n\geq 1}$: the jump set is where two different level sets meet.

One then has the following fundamental result (see for instance [7]):

Theorem 1.8 (Federer–Volpert). *Let $u \in BV(\Omega)$. Then one has*

$$Du = \nabla u(x)\,dx + Cu + (u_+(x) - u_-(x))\nu_u(x)\,d\mathcal{H}^{N-1} \llcorner J_u$$

where Cu is the "Cantor part" of Du, which is singular with respect to the Lebesgue measure, and vanishes on any set E with $\mathcal{H}^{N-1}(E) < +\infty$. In other words, for any $\phi \in C_c^1(\Omega; \mathbb{R}^N)$,

$$-\int_\Omega u(x) \operatorname{div} \phi(x)\, dx = \int_\Omega \nabla u(x) \cdot \phi(x)\, dx + \int_\Omega \phi(x) \cdot Cu(x) + \int_{J_u} (u_+(x) - u_-(x))\phi(x) \cdot \nu_u(x)\, d\mathcal{H}^{N-1}(x). \quad (1.15)$$

Observe that (1.15) is a generalized version of (1.11).

As we have seen, an example of a function with absolutely continuous derivative is given by any function $u \in W^{1,1}(\Omega)$ (or more obviously $u \in C^1(\overline{\Omega})$).

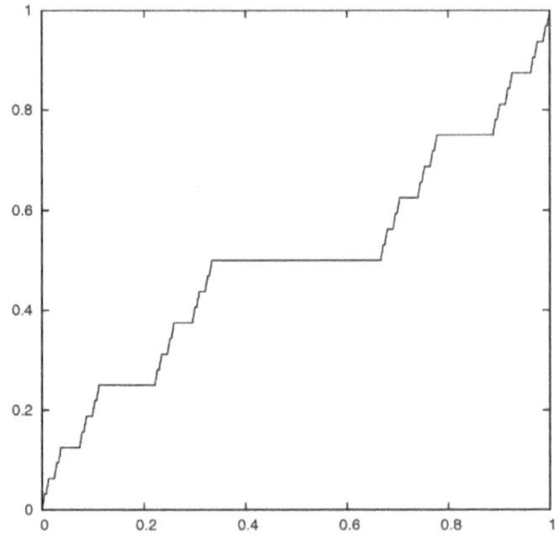

Figure 5. The "devil's staircase" or Cantor–Vitali function.

An example of a function whose derivative is a pure jump is given by $u = \chi_E$, E a Caccioppoli set (see Section 1.3). A famous example of a function with derivative purely Cantorian is the Cantor–Vitali function, obtained as follows: $\Omega = (0,1)$ and we let $u_0(t) = t$, and for any $n \geq 0$,

$$u_{n+1}(t) = \begin{cases} \frac{1}{2} u_n(3t) & 0 \leq t \leq \frac{1}{3} \\ \frac{1}{2} & \frac{1}{3} \leq t \leq \frac{2}{3} \\ \frac{1}{2}(u_n(3t-2)+1) & \frac{2}{3} \leq t \leq 1. \end{cases}$$

Then, one checks that

$$\sup_{(0,1)} |u_{n+1} - u_n| = \frac{1}{2} \sup_{(0,1)} |u_n - u_{n-1}| = \frac{1}{2^n} \times \frac{1}{6}$$

so that $(u_n)_{n \geq 1}$ is a Cauchy sequence and converges uniformly to some function u. This function (see Figure 5) is constant on each interval of the complement of the triadic Cantor set, which has zero measure in $(0,1)$. Hence, almost everywhere, its classical derivative exists and is zero. One can deduce that the derivative Du is singular with respect to Lebesgue's measure. On the other hand, it is continuous as a uniform limit of continuous functions, hence Du has no jump part. In fact, $Du = Cu$, which, in this case, is the measure $\mathcal{H}^{\ln 2/\ln 3} \llcorner C / \mathcal{H}^{\ln 2/\ln 3}(C)$.

2 Some Functionals Where the Total Variation Appears

2.1 Perimeter Minimization

In quite a few applications it is important to be able to solve the following problem:

$$\min_{E \subset \Omega} \lambda \mathrm{Per}(E;\Omega) - \int_E g(x)\,dx. \tag{2.1}$$

The intuitive idea is as follows: if $\lambda = 0$, then this will simply choose $E = \{g \geq 0\}$: that is, we find the set E by thresholding the values of g at 0. Now, imagine that this is precisely what we would like to do, but that g has some noise, so that a brutal thresholding of its value will produce a very irregular set. Then, choosing $\lambda > 0$ in (2.1) will start regularizing the set $\{g > 0\}$, and the high values of λ will produce a very smooth, but possibly quite approximate, version of that set.

We now state a proposition which is straightforward for people familiar with linear programming and "LP"-relaxation:

Proposition 2.1. *Problem* (2.1) *is convex. In fact, it can be relaxed as follows:*

$$\min_{u \in BV(\Omega;[0,1])} \lambda J(u) - \int_\Omega u(x)g(x)\,dx \tag{2.2}$$

and given any solution u of the convex problem (2.2), *and any value $s \in [0,1)$, the set $\{u > s\}$ (or $\{u \geq s\}$ for $s \in (0,1]$) is a solution of* (2.1).

Proof. This is a consequence of the co-area formula. Denote by m^* the minimum value of Problem (2.1). One has

$$\begin{aligned} \int_\Omega u(x)g(x)\,dx &= \int_\Omega \left(\int_0^{u(x)} ds \right) g(x)\,dx \\ &= \int_\Omega \int_0^1 \chi_{\{u>s\}}(x) g(x)\,ds\,dx = \int_0^1 \int_{\{u>s\}} g(x)\,dx\,ds \end{aligned} \tag{2.3}$$

so that, if we denote $\mathcal{E}(E) = \lambda \mathrm{Per}(E;\Omega) - \int_E g(x)\,dx$, it follows from (2.3) and (CA) that for any $u \in BV(\Omega)$ with $0 \leq u \leq 1$ a.e.,

$$\lambda J(u) - \int_\Omega u(x)g(x)\,dx = \int_0^1 \mathcal{E}(\{u > s\})\,ds \geq m^*. \tag{2.4}$$

Hence the minimum value of (2.2) is larger than m^*, on the other hand, it is also less since (2.1) is just (2.2) restricted to characteristic functions. Hence both problems have the same values, and it follows from (2.4) that if $\lambda J(u) - \int_\Omega ug\,dx = m^*$, that is, if u is a minimizer for (2.2), then for a.e. $s \in (0,1)$, $\{u > s\}$ is a solution to (2.1). Denote by S the set of such values of s. Now let $s \in [0,1)$, and let $(s_n)_{n\geq 1}$ be a decreasing sequence of values such that $s_n \in S$ and $s_n \to s$ as $n \to \infty$. Then, $\{u > s\} = \bigcup_{n\geq 1}\{u > s_n\}$, and, in fact, $\lim_{n\to\infty}\{u > s_n\} = \{u > s\}$ (the limit is in the L^1 sense, that is, $\int_\Omega |\chi_{\{u>s_n\}} - \chi_{\{u>s\}}|dx \to 0$ as $n \to \infty$). Using (1.2), it follows

$$m^* \leq \mathcal{E}(\{u > s\}) \leq \liminf_{n\to\infty} \mathcal{E}(\{u > s_n\}) = m^*$$

so that $s \in S$: it follows that $S = [0,1)$. □

The meaning of this result is that it is always possible to solve a problem such as (2.1) despite it apparently looks non-convex, and despite the fact the solution might be nonunique (although we will soon see that it is quite "often" unique). This has been observed several times in the past [27], and probably the first time for numerical purposes in [14]. In Section 3 we will address the issues of algorithms to tackle this kind of problems.

2.2 The Rudin–Osher–Fatemi Problem

We now concentrate on problem (MAPc) with $F(u) = \lambda J(u)$ as a regularizer, that is, on the celebrated "Rudin–Osher–Fatemi" problem (in the "pure denoising case": we will not consider any operator A as in (MAP)):

$$\min_u \lambda J(u) + \frac{1}{2}\int_\Omega |u(x) - g(x)|^2\,dx. \tag{ROF}$$

As mentioned in Section 1.2.3, this problem has a unique solution (it is strictly convex).

Let us now show that as in the previous section, the level sets $E_s = \{u > s\}$ solve a particular variational problem (of the form (2.1), but with $g(x)$ replaced with some s-dependent function). This will be of particular interest for our further analysis.

2.2.1 The Euler–Lagrange Equation

Formally:

$$-\lambda \mathrm{div}\,\frac{Du}{|Du|} + u - g = 0 \tag{2.5}$$

but this is hard to interpret. In particular because one can show there is always "staircasing", as soon as $g \in L^\infty(\Omega)$, so that there always are large areas where "$Du = 0$".

On can interpret the equation in the viscosity sense. Or try to derive the "correct" Euler–Lagrange equation in the sense of convex analysis. This requires to define properly the "subgradient" of J.

Definition 2.2. For X a Hilbert space, the subgradient of a convex function $F : X \to (-\infty, +\infty]$ is the operator ∂F which maps $x \in X$ to the (possibly empty set)

$$\partial F(x) = \{p \in X \,:\, F(y) \geq F(x) + \langle v, y - x\rangle \;\forall y \in X\}.$$

We introduce the set

$$\mathcal{K} = \{-\operatorname{div}\phi \,:\, \phi \in C_c^\infty(\Omega;\mathbb{R}^N) \,:\, |\phi(x)| \leq 1 \;\forall x \in \Omega\}$$

and the closure K if $\mathcal{K}$ in $L^2(\Omega)$, which is shown to be

$$K = \{-\operatorname{div} z \,:\, z \in L^\infty(\Omega;\mathbb{R}^N) \,:\, -\operatorname{div}(z\chi_\Omega) \in L^2(\mathbb{R}^N)\}$$

where the last condition means that

(i) $-\operatorname{div} z \in L^2(\Omega)$, i.e., there exists $\gamma \in L^2(\Omega)$ such that $\int_\Omega \gamma u\,dx = \int_\Omega z\cdot\nabla u\,dx$ for all smooth u with compact support;

(ii) the above also holds for $u \in H^1(\Omega)$ (not compactly supported), in other words $z \cdot \nu_\Omega = 0$ on $\partial\Omega$ in the weak sense.

Definition 1.1 defines J as

$$J(u) = \sup_{p\in\mathcal{K}} \int_\Omega u(x)p(x)\,dx,$$

so that if $u \in L^2(\Omega)$ it is obvious that, also,

$$J(u) = \sup_{p\in K} \int_\Omega u(x)p(x)\,dx. \tag{2.6}$$

In fact, one shows that K is the largest set in $L^2(\Omega)$ such that (2.6) holds for any $u \in L^2(\Omega)$, in other words

$$K = \left\{p \in L^2(\Omega) \,:\, \int_\Omega p(x)u(x)\,dx \leq J(u) \;\forall u \in L^2(\Omega)\right\}. \tag{2.7}$$

Then, if we consider J as a functional over the Hilbert space $X = L^2(\Omega)$, we have:

Proposition 2.3. *For $u \in L^2(\Omega)$,*

$$\partial J(u) = \left\{p \in K \,:\, \int_\Omega p(x)u(x)\,dx = J(u)\right\}. \tag{2.8}$$

Proof. It is not hard to check that if $p \in K$ and $\int_\Omega pu\,dx = J(u)$, then $p \in \partial J(u)$, indeed, for any $v \in L^2(\Omega)$, using (2.6),

$$J(v) \geq \int_\Omega p(x)v(x)\,dx = J(u) + \int_\Omega (v(x) - u(x))p(x)\,dx.$$

The converse inclusion can be proved as follows: if $p \in \partial J(u)$, then for any $t > 0$ and $v \in \mathbb{R}^N$, as J is one-homogeneous (1.4),

$$tJ(v) = J(tv) \geq J(u) + \int_\Omega p(x)(tv(x) - u(x))\,dx,$$

dividing by t and sending $t \to \infty$, we get $J(v) \geq \int_\Omega pv\,dx$, hence $p \in K$, by (2.7). On the other hand, sending t to 0 shows that $J(u) \leq \int_\Omega pu\,dx$ which shows our claim. □

Remark. We see that $K = \partial J(0)$.

The Euler–Lagrange Equation for (ROF) We can now derive the equation satisfied by u which minimizes (ROF): for any v in $L^2(\Omega)$, we have

$$\begin{aligned}\lambda J(v) &\geq \lambda J(u) + \frac{1}{2}\int_\Omega (u-g)^2 - (v-g)^2 dx \\ &= \lambda J(u) + \int_\Omega (u-v)\left(\frac{u+v}{2} - g\right)dx \\ &= \lambda J(u) + \int_\Omega (v-u)(g-u)\,dx - \frac{1}{2}\int_\Omega (u-v)^2 dx. \qquad (2.9)\end{aligned}$$

In particular, for any $t \in \mathbb{R}$,

$$\lambda(J(u + t(v-u)) - J(u)) - t\int_\Omega (v-u)(g-u)\,dx \geq -\frac{t^2}{2}\int_\Omega (v-u)^2\,dx.$$

The left-hand side of the last expression is a convex function of $t \in \mathbb{R}$, one can show quite easily that a convex function which is larger than a concave parabola and touches at the maximum point ($t = 0$) must be everywhere larger than the maximum of the parabola (here zero).

We deduce that

$$\lambda(J(u + t(v-u)) - J(u)) - t\int_\Omega (v-u)(g-u)\,dx \geq 0$$

for any t, in particular for $t = 1$, which shows that $\frac{g-u}{\lambda} \in \partial J(u)$. Conversely, if this is true, then obviously (2.9) holds so that u is the minimizer of J. It follows that the Euler–Lagrange equation for (ROF) is

$$\lambda \partial J(u) + u - g \ni 0 \qquad (2.10)$$

which, in view of (2.8) and the characterization of K, is equivalent to the existence of $z \in L^\infty(\Omega;\mathbb{R}^N)$ with:

$$\begin{cases} -\lambda \operatorname{div} z(x) + u(x) = g(x) & \text{a.e. } x \in \Omega \\ |z(x)| \le 1 & \text{a.e. } x \in \Omega \\ z \cdot \nu = 0 & \text{on } \partial\Omega \text{ (weakly)} \\ z \cdot Du = |Du|, \end{cases} \tag{2.11}$$

the last equation being another way to write that $\int (-\operatorname{div} z) u \, dx = J(u)$.

If u is smooth and $\nabla u \neq 0$, the last condition ensures that $z = \nabla u / |\nabla u|$ and we recover (2.5).

In all cases, we see that z must be orthogonal to the level sets of u (from $|z| \le 1$ and $z \cdot Du = |Du|$), so that $-\operatorname{div} z$ is still the curvature of the level sets.

In 1D, z is a scalar and the last condition is $z \cdot u' = |u'|$, so that $z \in \{-1, +1\}$ whenever u is not constant, while $\operatorname{div} z = z'$: we see that the equation becomes $u = g$ or $u =$ constant (and in particular the "staircasing" always occurs if g is not monotonous).

2.2.2 The Problem Solved by the Level Sets

We introduce the following problems, parameterized by $s \in \mathbb{R}$:

$$\min_E \lambda \operatorname{Per}(E;\Omega) + \int_E s - g(x)\, dx. \tag{ROF$_s$}$$

Given $s \in \mathbb{R}$, let us denote by E_s a solution of (ROF$_s$) (whose existence follows in a straightforward way from Rellich's Theorem 1.4 and (1.2)).

Then the following holds

Lemma 2.4. *Let $s' > s$: then $E_{s'} \subseteq E_s$.*

This lemma is found for instance in [4, Lemma 4]. Its proof is very easy.

Proof. We have (to simplify we let $\lambda = 1$):

$$\operatorname{Per}(E_s) + \int_{E_s} s - g(x)\, dx \le \operatorname{Per}(E_s \cup E_{s'}) + \int_{E_s \cup E_{s'}} s - g(x)\, dx,$$

$$\operatorname{Per}(E_{s'}) + \int_{E_{s'}} s' - g(x)\, dx \le \operatorname{Per}(E_s \cap E_{s'}) + \int_{E_s \cap E_{s'}} s' - g(x)\, dx$$

and summing both inequalities we get:

$$\operatorname{Per}(E_s) + \operatorname{Per}(E_{s'}) + \int_{E_s} s - g(x)\, dx + \int_{E_{s'}} s' - g(x)\, dx$$

$$\le \operatorname{Per}(E_s \cup E_{s'}) + P(E_s \cap E_{s'}) + \int_{E_s \cup E_{s'}} s - g(x)\, dx + \int_{E_s \cap E_{s'}} s' - g(x)\, dx.$$

Using (1.7), it follows that

$$\int_{E_{s'}} s' - g(x)\,dx - \int_{E_s \cap E_{s'}} s' - g(x)\,dx \le \int_{E_s \cup E_{s'}} s - g(x)\,dx - \int_{E_s} s - g(x)\,dx,$$

that is,

$$\int_{E_{s'} \setminus E_s} s' - g(x)\,dx \le \int_{E_{s'} \setminus E_s} s - g(x)\,dx$$

hence

$$(s' - s)|E_{s'} \setminus E_s| \le 0.$$

It shows that $E_s' \subseteq E_s$, up to a negligible set, as soon as $s' > s$. □

In particular, it follows that E_s is unique, except for at most countably many values of s. Indeed, we can introduce the sets $E_s^+ = \bigcap_{s'<s} E_{s'}$ and $E_s^- = \bigcup_{s'>s} E_{s'}$, then one checks that E_s^+ and E_s^- are respectively the largest and smallest solutions of $(\mathrm{ROF_s})$[1]. There is uniqueness when the measure $|E_s^+ \setminus E_s^-| = 0$. But the sets $(E_s^+ \setminus E_s^-)$, $s \in \mathbb{R}$, are all disjoint, so that their measure must be zero except for at most countably many values.

Let us introduce the function

$$u(x) = \sup\{s \in \mathbb{R} : x \in E_s\}.$$

We have that $u(x) > s$ if there exists $t > s$ with $x \in E_t$, so that in particular, $x \in E_s^-$; conversely, if $x \in E_s^-$, $x \in E_{s'}$ for some $s' > s$, so that $u(x) > s$: $\{u > s\} = E_s^-$. (In the same way, we check $E_s^+ = \{u \ge s\}$.)

Lemma 2.5. *The function u is the minimizer of* (ROF).

Proof. First of all, we check that $u \in L^2(\Omega)$. This is because

$$\lambda \mathrm{Per}(E_s;\Omega) + \int_{E_s} s - g(x)\,dx \le 0$$

(the energy of the empty set), hence

$$s|E_s| \le \int_{E_s} g(x)\,dx.$$

It follows that

$$\int_0^M s|E_s|\,ds \le \int_0^M \int_{E_s} g(x)\,dx\,ds,$$

[1] Observe that since the set E_s are normally defined up to negligible sets, the intersections $E_s^+ = \bigcap_{s'<s} E_{s'}$ might not be well-defined (as well as the unions E_s^-). A rigorous definition requires, actually, either to consider only countable intersections/unions, or to first choose a precise representative of each E_s, for instance the set $\{x \in E_s : \lim_{\rho\to 0} |E_s \cap B(x,\rho)|/|B(x,\rho)| = 1\}$ of points of Lebesgue density 1, which can be shown, for E_s minimizing $(\mathrm{ROF_s})$, to be an open set.

but $\int_0^M s|E_s|\,ds = \int_{E_0}\int_0^{u(x)\wedge M} s\,ds\,dx = \int_{E_0}(u(x)\wedge M)^2/2\,dx$ (using Fubini's theorem), while in the same way $\int_0^M \int_{E_s} g(x)\,dx\,ds = \int_{E_0}(u(x)\wedge M)g(x)\,dx$. We recall that here $u(x)\wedge M = \min\{u(x), M\}$.

Hence

$$\frac{1}{2}\int_{E_0}(u\wedge M)^2\,dx \le \int_{E_0}(u\wedge M)g\,dx \le \left(\int_{E_0}(u\wedge M)^2\,dx\int_{E_0} g^2\,dx\right)^{\frac{1}{2}}$$

so that

$$\int_{E_0}(u(x)\wedge M)^2\,dx \le 4\int_{E_0} g(x)^2\,dx$$

and sending $M \to \infty$ it follows that

$$\int_{\{u>0\}} u(x)^2\,dx \le 4\int_{\{u>0\}} g(x)^2\,dx. \tag{2.12}$$

In the same way, we can show that

$$\int_{\{u<0\}} u(x)^2\,dx \le 4\int_{\{u<0\}} g(x)^2\,dx. \tag{2.13}$$

This requires the observation that the set $\{-u > -s\} = \{u < s\} = \Omega \setminus E_s^+$ is a minimizer of the problem

$$\min_E \operatorname{Per}(E;\Omega) + \int_E g(x) - s\,dx$$

which easily follows from the fact that $\operatorname{Per}(E;\Omega) = \operatorname{Per}(\Omega\setminus E;\Omega)$ for any set of finite perimeter $E \subset \Omega$: it follows that if we replace g with $-g$ in ($\mathrm{ROF_s}$), then the function u is replaced with $-u$. We deduce from (2.12) and (2.13) that $u \in L^2(\Omega)$.

Let now $v \in BV(\Omega)\cap L^2(\Omega)$: we have for any $M > 0$,

$$\begin{aligned}\int_{-M}^{M}\left(\lambda\operatorname{Per}(E_s^-;\Omega) + \int_{E_s^-} s - g(x)\,dx\right)ds \\ \le \int_{-M}^{M}\left(\lambda\operatorname{Per}(\{v>s\};\Omega) + \int_{\{v>s\}} s - g(x)\,dx\right)ds\end{aligned} \tag{2.14}$$

since E_s^- is a minimizer for ($\mathrm{ROF_s}$). Notice that (using Fubini's theorem again)

$$\begin{aligned}\int_{-M}^{M}\int_{\{v>s\}} s - g(x)\,dx\,ds &= \int_\Omega\int_{-M}^{M}\chi_{\{v>s\}}(x)(s-g(x))\,ds\,dx \\ &= \frac{1}{2}\int_\Omega((v(x)\wedge M) - g(x))^2 - ((v(x)\wedge(-M)) - g(x))^2\,dx,\end{aligned}$$

hence

$$\int_{-M}^{M}\left(\int_{\{v>s\}} s-g(x)\,dx\right)ds+\int_{\Omega}(M+g(x))^2\,dx$$
$$=\frac{1}{2}\int_{\Omega}(v(x)-g(x))^2\,dx+\mathcal{R}(v,M) \quad (2.15)$$

where

$$\mathcal{R}(v,M)=\frac{1}{2}\bigg(\int_{\Omega}((v(x)\wedge M)-g(x))^2-(v(x)-g(x))^2\,dx$$
$$+\int_{\Omega}(-M-g(x))^2-((v(x)\wedge(-M))-g(x))^2\,dx\bigg).$$

It suffices now to check that for any $v\in L^2(\Omega)$,

$$\lim_{M\to\infty}\mathcal{R}(v,M)=0.$$

We leave it to the reader. From (2.14) and (2.15), we get that

$$\lambda\int_{-M}^{M}\operatorname{Per}(\{u>s\};\Omega)\,ds+\frac{1}{2}\int_{\Omega}(u-g)^2\,dx+\mathcal{R}(u,M)$$
$$\le\lambda\int_{-M}^{M}\operatorname{Per}(\{v>s\};\Omega)\,ds+\frac{1}{2}\int_{\Omega}(v-g)^2\,dx+\mathcal{R}(v,M),$$

sending $M\to\infty$ (and using the fact that both u and v are in L^2, so that $\mathcal{R}(\cdot,M)$ goes to 0) we deduce

$$\lambda J(u)+\frac{1}{2}\int_{\Omega}(u-g)^2\,dx\le\lambda J(v)+\frac{1}{2}\int_{\Omega}(v-g)^2\,dx,$$

that is, the minimality of u for (ROF). □

We have proved the following result:

Proposition 2.6. *A function u solves* (ROF) *if and only if for any $s\in\mathbb{R}$, the set $\{u>s\}$ solves* (ROF_s).

Normally, we should write "for almost any s", but as before by approximation it is easy to show that if it is true for almost all s, then it is true for all s.

This is interesting for several applications. It provides another way to solve problems such as (2.1) (through an unconstrained relaxation – but in fact both problems are relatively easy to solve). But most of all, it gives a lot of information on the level sets of u, as problems such as (2.1) have been studied thoroughly in the past 50 years.

The link between minimal surfaces and functions minimizing the total variation was first identified by De Giorgi and Bombieri, as a tool for the study of minimal surfaces.

Proposition 2.6 can be generalized easily to the following case: we should have that u is a minimizer of

$$\min_u J(u) + \int_\Omega G(x, u(x))\, dx$$

for some G measurable in x, and convex, C^1 in u, if and only if for any $s \in \mathbb{R}$, $\{u > s\}$ minimizes

$$\min_E \operatorname{Per}(E;\Omega) + \int_E g(x,s)\, dx$$

where $g(x,s) = \partial_s G(x,s)$. The case of nonsmooth G is also interesting and has been studied by Chan and Esedoglu [26].

2.2.3 A few Explicit Solutions

The results in this section are simplifications of results which are found in [5, 4] (see also [3] for more results of the same kind).

Let us show how Proposition 2.6 can help build explicit solutions of (ROF), in a few very simple cases. We consider the two following cases: $\Omega = \mathbb{R}^2$, and

(i) $g = \chi_{B(0,R)}$ the characteristic of a ball;

(ii) $g = \chi_{[0,1]^2}$ the characteristic of the unit square.

In both cases, $g = \chi_C$ for some convex set C. First observe that obviously, $0 \le u \le 1$. As in the introduction, indeed, one easily checks that $u \wedge 1 = \min\{u, 1\}$ has less energy than u. Another way is to observe that $E_s = \{u > s\}$ solves

$$\min_E \lambda \operatorname{Per}(E) + \int_E s - \chi_C(x)\, dx,$$

but if $s > 1$, $s - \chi_C$ is always positive so that $E = \emptyset$ is clearly optimal, while if $s < 0$, $s - \chi_C$ is always negative so that $E = \mathbb{R}^2$ is optimal (and in this case the value is $-\infty$).

Now, if $s \in (0,1)$, E_s solves

$$\min_E \lambda \operatorname{Per}(E) - (1-s)|E \cap C| + s|E \setminus C|. \tag{2.16}$$

Let P be a half-plane containing C: observe that $\operatorname{Per}(E \cap P) \le \operatorname{Per}(E)$, since we replace the part of a boundary outside of P with a straight line on ∂P, while $|(E \cap P) \setminus C| \le |E \setminus C|$: hence, the set $E \cap P$ has less energy than E, see Figure 6. Hence $E_s \subset P$. As C, which is convex, is the intersection of all half-planes containing it, we deduce that $E_s \subset C$. (In fact, this is true in any dimension for any convex set C.)

We see that the problem for the level sets E_s (2.16) becomes:

$$\min_{E \subset C} \lambda \operatorname{Per}(E) - (1-s)|E|. \tag{2.17}$$

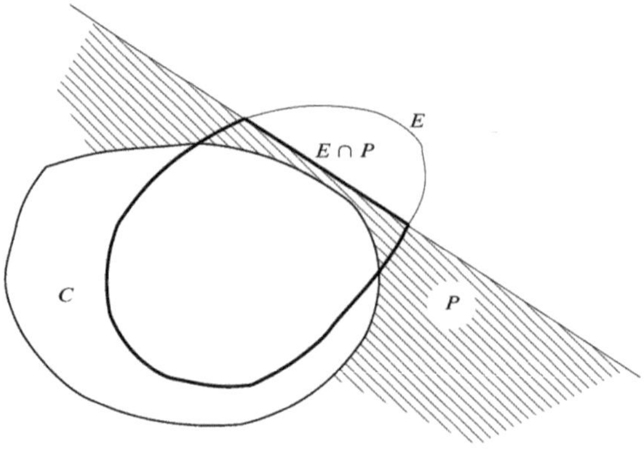

Figure 6. $E \cap P$ has less energy than E.

The Characteristic of a Ball If $g = \chi_{B(0,R)}$, the ball of radius R (in $\mathbb{R}^2$), we see that thanks to the isoperimetric inequality (1.13),

$$\lambda \mathrm{Per}(E) - (1-s)|E| \geq \lambda 2\sqrt{\pi}\sqrt{|E|} - (1-s)|E|,$$

and for $|E| \in [0, |B(0,R)|]$, the right-hand side is minimal only if $|E| = 0$ or $|E| = |B(0,R)|$, with value 0 in the first case, $2\lambda\pi R - (1-s)\pi R^2$ in the second. Since for these two choices, the isoperimetric inequality is an equality, we deduce that the min in (2.17) is actually attained by $E = \emptyset$ if $s \geq 1 - 2\lambda/R$, and $E = B(0,R)$ if $s \leq 1 - 2\lambda/R$. Hence the solution is

$$u = \left(1 - \frac{2\lambda}{R}\right)^+ \chi_{B(0,R)}$$

for any $\lambda > 0$ (here $x^+ = \max\{x, 0\}$ is the positive part of the real number x). In fact, this result holds in all dimension (with 2 replaced with N in the expression). See also [53].

The Characteristic of a Square The case of the characteristic of a square is a bit different. The level sets E_s need to solve (2.17). From the Euler–Lagrange equation (see also (2.11)) it follows that the curvature of $\partial E_s \cap C$ is $(1-s)/\lambda$. An accurate study (see also [5, 4, 43]) shows that, if we define (for $C = [0,1]^2$)

$$C_R = \bigcup_{x : B(x,R) \subset C} B(x, R)$$

Figure 7. Solution u for $g = \chi_{[0,1]^2}$.

and let R^* be the value of R for which $\mathrm{Per}(C_R)/|C_R| = 1/R$, then for any $s \in [0,1]$,

$$E_s = \begin{cases} \emptyset & \text{if } s \geq 1 - \frac{\lambda}{R^*} \\ C_{\lambda/(1-s)} & \text{if } s \leq 1 - \frac{\lambda}{R^*} \end{cases}$$

while letting $v(x) = 1/\mathbb{R}^*$ if $x \in C_{R^*}$, and $1/R$ if $x \in \partial C_R$, we find

$$u = (1 - \lambda v(x))^+,$$

see Figure 7.

2.2.4 The Discontinuity Set

We now show that the jump set of u_λ is always contained in the jump set of g. More precisely, we will describe shortly the proof of the following result, which was first proved in [20]:

Theorem 2.7. *Let $g \in BV(\Omega) \cap L^\infty(\Omega)$ and u solve* (ROF). *Then $J_u \subseteq J_g$ (up to a set of zero $\mathcal{H}^{N-1}$-measure).*

Hence, if g is already a BV function, the Rudin–Osher–Fatemi denoising will never produce new discontinuities.

First, we use the following important regularity result, from the theory of minimal surfaces [41, 6]:

Proposition 2.8. *Let $g \in L^\infty(\Omega)$, $s \in \mathbb{R}$, and E_s be a minimizer of* (ROF_s). *Then $\Sigma = \partial E_s \setminus \partial^* E_s$ is a closed set of Hausdorff dimension at most $N - 8$, while near each $x \in \partial^* E_s$, $\partial^* E_s$ is locally the graph of a function of class $W^{2,q}$ for all $q < +\infty$ (and, in dimension $N = 2$, $W^{2,\infty} = C^{1,1}$).*

It means that outside of a very small set (which is empty if $N \le 7$), then the boundary of E_s is C^1, and the normal is still differentiable but in a weaker sense. We now can show Theorem 2.7.

Proof. The jump set J_u is where several level sets intersect: if we choose $(s_n)_{n\ge 1}$ a dense sequence in $\mathbb{R}^N$ of levels such that $E_n = E_{s_n} = \{u > s_n\}$ are finite perimeter sets each solving (ROF_{s_n}), we have

$$J_u = \bigcup_{n\neq m} (\partial^* E_n \cap \partial^* E_m).$$

Hence it is enough to show that for any n, m with $n \neq m$,

$$\mathcal{H}^{N-1}((\partial^* E_n \cap \partial^* E_m) \setminus J_g) = 0$$

which precisely means that $\partial^* E_n \cap \partial^* E_m \subseteq J_g$ up to a negligible set. Consider thus a point $x^0 \in \partial^* E_n \cap \partial^* E_m$ such that (without loss of generality we let $x^0 = 0$):

(i) Up to a change of coordinates, in a small neighborhood $\{x = (x_1, \dots, x_N) = (x', x_N) : |x'| < R, |x_N| < R\}$ of $x^0 = 0$, the sets E_n and E_m coincide respectively to $\{x_N < v_n(x')\}$ and $\{x_N < v_m(x')\}$, with v_n and v_m in $W^{2,q}(B')$ for all $q < +\infty$, where $B' = \{|x'| < R\}$.

(ii) The measure of the contact set $\{x' \in B' : v_n(x') = v_m(x')\}$ is positive.

We assume without loss of generality that $s_n < s_m$, so that $E_m \subseteq E_n$, hence $v_n \ge v_m$ in B'.

From $(\mathrm{ROF_s})$, we see that the function v_l, $l \in \{n, m\}$, must satisfy

$$\int_{B'} \left[\sqrt{1 + |\nabla v_l(x')|^2} + \int_0^{v_l(x')} (s_l - g(x', x_N))\, dx_N \right] dx'$$
$$\le \int_{B'} \left[\sqrt{1 + |\nabla v_l(x') + t\phi(x')|^2} + \int_0^{v_l(x')+t\phi(x')} (s_l - g(x', x_N))\, dx_N \right] dx'$$

for any smooth $\phi \in C_c^\infty(B')$ and any $t \in \mathbb{R}$ small enough, so that the perturbation remains in the neighborhood of x^0 where E_{s_n} and E_{s_m} are subgraphs. Here, the first integral corresponds to the perimeter of the set $\{x_N < v_l(x') + t\phi(x')\}$ and the second to the volume integral in $(\mathrm{ROF_s})$.

We consider $\phi \ge 0$, and compute

$$\lim_{\substack{t\to 0,\\ t>0}} \frac{1}{t} \Bigg(\int_{B'} \left[\sqrt{1 + |\nabla v_l(x') + t\phi(x')|^2} + \int_0^{v_l(x')+t\phi(x')} (s_l - g(x', x_N))\, dx_N \right] dx'$$
$$- \int_{B'} \left[\sqrt{1 + |\nabla v_l(x')|^2} + \int_0^{v_l(x')} (s_l - g(x', x_N))\, dx_N \right] dx' \Bigg) \ge 0,$$

we find that v_l must satisfy

$$\int_{B'} \frac{\nabla v_l(x') \cdot \nabla \phi(x')}{\sqrt{1 + |\nabla v_l(x')|^2}} + (s_l - g(x', v_l(x') + 0))\phi(x')\, dx' \geq 0.$$

Integrating by parts, we find

$$-\text{div}\, \frac{\nabla v_l}{\sqrt{1 + |\nabla v_l|^2}} + s_l - g(x', v_l(x') + 0) \geq 0, \tag{2.18}$$

and in the same way, taking this time the limit for $t < 0$, we find that

$$-\text{div}\, \frac{\nabla v_l}{\sqrt{1 + |\nabla v_l|^2}} + s_l - g(x', v_l(x') - 0) \leq 0. \tag{2.19}$$

Both (2.18) and (2.19) must hold almost everywhere in B'. At the contact points x' where $v_n(x') = v_m(x')$, since $v_n \geq v_m$, we have $\nabla v_n(x') = \nabla v_m(x')$, while $D^2 v_n(x') \geq D^2 v_m(x')$ at least at a.e. contact point x'. In particular, it follows

$$-\text{div}\, \frac{\nabla v_n}{\sqrt{1 + |\nabla v_n|^2}}(x') \leq -\text{div}\, \frac{\nabla v_m}{\sqrt{1 + |\nabla v_m|^2}}(x'). \tag{2.20}$$

If the contact set $\{v_n = v_m\}$ has positive ($(N-1)$-dimensional) measure, we can find a contact point x' such that (2.20) holds, as well as both (2.18) and (2.19), for both $l = n$ and $l = m$. It follows

$$g(x', v_n(x') + 0) - s_n \leq g(x', v_m(x') - 0) - s_m$$

and denoting x_N the common value $v_n(x') = v_m(x')$, we get

$$0 < s_m - s_n \leq g(x', x_N - 0) - g(x', x_N + 0).$$

It follows that $x = (x', x_N)$ must be a jump point of g (with the values below x_N larger than the values above x_N, so that the jump occurs in the same direction for g and u). This concludes the proof: the possible jump points outside of J_g are negligible for the $(N-1)$-dimensional measure. The precise meaning of $g(x', x_N \pm 0)$ and the relationship to the jump of g is rigorous for a.e. x', see the "slicing properties of BV functions" in [7]. □

Remark. We have also proved that for almost all points in J_u, we have $\nu_u = \nu_g$, that is, the orientation of the jumps of u and g are the same, which is intuitively obvious.

2.2.5 Regularity

The same idea, based on the control on the curvature of the level sets which is provided by (ROF_s), yields further continuity results for the solutions of (ROF). The following theorems are proved in [21]:

Theorem 2.9. *Assume $N \le 7$ and let u be a minimizer. Let $A \subset \Omega$ be an open set and assume that $g \in C^{0,\beta}(A)$ for some $\beta \in [0,1]$. Then, also $u \in C^{0,\beta}(A')$ for any $A' \subset\subset A$.*

Here, $A' \subset\subset A$ means that $\overline{A'} \subset A$. The proof of this result is quite complicated and we refer to [21]. The reason for restriction on the dimension is clear from Proposition 2.8, since we need here in the proof that ∂E_s is globally regular for all s. The next result is proved more easily:

Theorem 2.10. *Assume $N \le 7$ and Ω is convex. Let u solve* (ROF)*, and suppose g is uniformly continuous with modulus of continuity ω (that is, $|g(x)-g(y)| \le \omega(|x-y|)$ for all x, y, with ω continuous, nondecreasing, and $\omega(0) = 0$). Then u has the same modulus of continuity.*

For instance, if g is globally Lipschitz, then also u is with same constant.

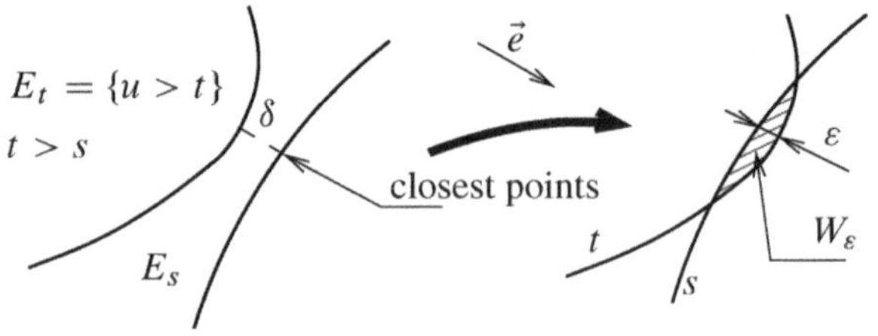

Figure 8. The set W_ε is obtained by moving E_t towards E_s.

Proof. We only sketch the proof: by approximation we can assume that Ω is smooth and uniformly convex, and, as well, that g is smooth up to the boundary.

We consider two levels s and $t > s$ and the corresponding level sets E_s and E_t. Let $\delta = \text{dist}(\Omega \cap \partial E_s, \Omega \cap \partial E_t)$ be the distance between the level sets s and t of u. The strict convexity of Ω and the fact that ∂E_s are smooth, and orthogonal to the boundary (because $z \cdot \nu = 0$ on $\partial\Omega$ in (2.11)), imply that this minimal distance cannot be reached by points on the boundary $\partial\Omega$, but that there exists $x_s \in \partial E_s \cap \Omega$ and $x_t \in \partial E_t \cap \Omega$, with $|x_s - x_t| = \delta$. We can also exclude the case $\delta = 0$, by arguments similar to the previous proof.

Let $e = (x_s - x_t)/\delta$. This must be the outer normal to both ∂E_t and ∂E_s, respectively at x_t and x_s.

The idea is to “slide” one of the sets until it touches the other, in the direction e (see Figure 8). We consider, for $\varepsilon > 0$ small, $(E_t + (\delta + \varepsilon)e) \setminus \overline{E}_s$, and a connected component W_ε which contains x_s on its boundary. We then use (ROF_t), comparing E_t with $E_t \setminus (W_\varepsilon - (\delta + \varepsilon)e)$ and E_s with $E_s \cup W_\varepsilon$:

$$
\begin{aligned}
&\mathrm{Per}(E_t; \Omega) + \int_{E_t} (t - g(x))\, dx \\
&\qquad \le \mathrm{Per}(E_t \setminus (W_\varepsilon - (\delta + \varepsilon)e); \Omega) + \int_{E_t \setminus (W_\varepsilon - (\delta+\varepsilon)e)} (t - g(x))\, dx \\
&\mathrm{Per}(E_s; \Omega) + \int_{E_s} (s - g(x))\, dx \\
&\qquad \le \mathrm{Per}(E_s \cup W_\varepsilon; \Omega) + \int_{E_s \cup W_\varepsilon} (s - g(x))\, dx.
\end{aligned} \tag{2.21}
$$

Now, if we let $L_t = \mathcal{H}^{N-1}(\partial W_\varepsilon \setminus \partial E_s)$ and $L_s = \mathcal{H}^{N-1}(\partial W_\varepsilon \cap \partial E_s)$, we have that

$$
\mathrm{Per}(E_t \setminus (W_\varepsilon - (\delta + \varepsilon)e), \Omega) = \mathrm{Per}(E_t, \Omega) - L_t + L_s
$$

and

$$
\mathrm{Per}(E_s \cup W_\varepsilon, \Omega) = \mathrm{Per}(E_s, \Omega) + L_t - L_s,
$$

so that, summing both equations in (2.21), we deduce

$$
\int_{W_\varepsilon - (\delta+\varepsilon)e} (t - g(x))\, dx \le \int_{W_\varepsilon} (s - g(x))\, dx.
$$

Hence,

$$
(t - s)|W_\varepsilon| \le \int_{W_\varepsilon} (g(x + (\delta + \varepsilon)e) - g(x))\, dx \le |W_\varepsilon| \omega(\delta + \varepsilon).
$$

Dividing both sides by $|W_\varepsilon| > 0$ and sending then ε to zero, we deduce

$$
t - s \le \omega(\delta).
$$

The regularity of u follows. Indeed, if $x, y \in \Omega$, with $u(x) = t > u(y) = s$, we find $|u(x) - u(y)| \le \omega(\delta) \le \omega(|x - y|)$ since $\delta \le |x - y|$, δ being the minimal distance between the level surface of u through x and the level surface of u through y. □

3 Algorithmic Issues

3.1 Discrete Problem

To simplify, we let now $\Omega = (0, 1)^2$ and we will consider here a quite straightforward discretization of the total variation in 2D, as

$$
TV_h(u) = h^2 \sum_{i,j} \frac{\sqrt{|u_{i+1,j} - u_{i,j}|^2 + |u_{i,j+1} - u_{i,j}|^2}}{h}. \tag{3.1}
$$

Here in the sum, the differences are replaced by 0 when one of the points is not on the grid. The matrix $(u_{i,j})$ is our discrete image, defined for instance for $i, j = 1, \dots, N$, and $h = 1/N$ is the discretization step: TV_h is therefore an approximation of the 2D total variation of a function $u \in L^1((0,1)^2)$ at scale $h > 0$.

It can be shown in many ways that (3.1) is a "correct" approximation of the Total Variation J introduced previously and we will skip this point. The most simple result is as follows:

Proposition 3.1. *Let $\Omega = (0,1)^2$, $p \in [1, +\infty)$, and $G : L^p(\Omega) \to \mathbb{R}$ a continuous functional such that $\lim_{c\to\infty} G(c+u) = +\infty$ for any $u \in L^p(\Omega)$ (this coerciveness assumption is just to ensure the existence of a solution to the problem, and other situations could be considered). Let $h = 1/N > 0$ and $u^h = (u_{i,j})_{1\le i,j\le N}$, identified with $u^h(x) = \sum_{i,j} u_{i,j}\chi_{((i-1)h,ih)\times((j-1)h,jh)}(x)$, be the solution of*

$$\min_{u^h} TV_h(u^h) + G(u^h).$$

Then, there exists $u \in L^p(\Omega)$ such that some subsequence $u^{h_k} \to u$ as $k \to \infty$ in $L^1(\Omega)$, and u is a minimizer in $L^p(\Omega)$ of

$$J(u) + G(u).$$

But more precise results can be shown, including with error bounds, see for instance recent works [51, 46].

In what follows, we will choose $h = 1$ (introducing h yields in general to straightforward changes in the other parameters). Given $u = (u_{i,j})$ a discrete image $(1 \ge i, j, \ge N)$, we introduce the discrete gradient

$$(\nabla u)_{i,j} = \begin{pmatrix} (D_x^+ u)_{i,j} \\ (D_y^+ u)_{i,j} \end{pmatrix} = \begin{pmatrix} u_{i+1,j} - u_{i,j} \\ u_{i,j+1} - u_{i,j} \end{pmatrix}$$

except at the boundaries: if $i = N$, $(D_x^+ u)_{N,j} = 0$, and if $j = N$, $(D_y^+ u)_{i,N} = 0$. Let $X = \mathbb{R}^{N\times N}$ be the vector space where u lives, then ∇ is a linear map from X to $Y = X \times X$, and (if we endow both spaces with the standard Euclidean scalar product), its adjoint ∇^*, denoted by $-\mathrm{div}$, is defined by

$$\langle \nabla u, p\rangle_Y = \langle u, \nabla^* p\rangle_X = -\langle u, \mathrm{div}\, p\rangle_X$$

for any $u \in X$ and $p = (p^x_{i,j}, p^y_{i,j}) \in Y$, is given by the following formulas

$$(\mathrm{div}\, p)_{i,j} = p^x_{i,j} - p^x_{i-1,j} + p^y_{i,j} - p^y_{i,j-1}$$

for $2 \le i, j \le N-1$, and the difference $p^x_{i,j} - p^x_{i-1,j}$ is replaced with $p^x_{i,j}$ if $i = 1$, and with $-p^x_{i-1,j}$ if $i = N$, while $p^y_{i,j} - p^y_{i,j-1}$ is replaced with $p^y_{i,j}$ if $j = 1$ and with $-p^y_{i,j-1}$ if $j = N$.

We will focus on algorithms for solving the discrete problem

$$\min_{u \in X} \lambda \|\nabla u\|_{2,1} + \frac{1}{2}\|u - g\|^2, \tag{3.2}$$

where $\|v\|^2 = \sum_{i,j} v_{i,j}^2$ and $\|p\|_{2,1} = \sum_{i,j} \sqrt{(p_{i,j}^x)^2 + (p_{i,j}^y)^2}$. In this section and all what follows, $J(\cdot)$ will now denote the discrete total variation $J(u) = \|\nabla u\|_{2,1}$ (and we will not speak anymore of the continuous variation introduced in the previous section). The problem (3.2) can also be written in the more general form

$$\min_{u \in X} F(Au) + G(u) \tag{3.3}$$

where $F : Y \to \mathbb{R}_+$ and $G : X \to \mathbb{R}$ are convex functions and $A : X \to Y$ is a linear operator (in the discretization of (ROF), $A = \nabla$, $F(p) = \|p\|_{2,1}$, $G(u) = \|u - g\|^2/(2\lambda)$).

It is essential here that F, G are convex, since we will focus on techniques of convex optimization which can produce quite efficient algorithms for problems of the form (3.3), provided F and G have a simple structure.

3.2 Basic Convex Analysis – Duality

Before detailing a few numerical methods to solve (3.2) let us recall the basics of convex analysis in finite-dimensional spaces (all the results we state now are true in a more general setting, and the proofs in the Hilbertian framework are the same). We refer of course to [68, 32] for more complete information.

3.2.1 Convex Functions – Legendre–Fenchel Conjugate

Let X be a finite-dimensional, Euclidean space (or a Hilbert space). Recall that a subset $C \subset X$ of X is said to be *convex* if and only if for any $x, x' \in C$, the segment $[x, x'] \subset C$, that is, for any $t \in [0, 1]$,

$$tx + (1 - t)x' \in C.$$

Let us now introduce a similar definition for functions:

Definition 3.2. We say that the function $F : X \to [-\infty, +\infty]$ is

- *convex* if and only if for any $x, x' \in X$, $t \in [0, 1]$,

$$F(tx + (1 - t)x') \le tF(x) + (1 - t)F(x'), \tag{3.4}$$

- *proper* if and only if F is not identically $-\infty$ or $+\infty$[2]

[2] Notice that if F is convex, $F \equiv -\infty$ if and only if there exists $x \in X$ s.t. $F(x) = -\infty$, so that a proper convex function has values in $\mathbb{R} \cup \{+\infty\}$.

- *lower-semicontinuous* (l.s.c.) if and only for any $x \in X$ and $(x_n)_n$ a sequence converging to x,
$$F(x) \leq \liminf_{n\to\infty} F(x_n). \tag{3.5}$$

We let $\Gamma^0(X)$ be the set of all convex, proper, l.s.c. functions on X.

It is well known, and easy to show, that if F is twice differentiable at any $x \in X$, then it is convex if and only if $D^2F(x) \geq 0$ at any $x \in X$, in the sense that for any x, y, $\sum_{i,j} \partial^2_{i,j} F(x) y_i y_j \geq 0$. If F is of class C^1, one has that F is convex if and only if
$$\langle \nabla F(x) - \nabla F(y), x - y\rangle \geq 0 \tag{3.6}$$
for any $x, y \in X$.

For any function $F : X \to [-\infty, +\infty]$, we define the *domain*
$$\operatorname{dom} F = \{x \in X \ :\ F(x) < +\infty\}$$
(which is always a convex set if F is) and the *epigraph*
$$\operatorname{epi} F = \{(x,t) \in X \times \mathbb{R} \ :\ t \geq F(x)\},$$
which is convex if and only if F is.

Then, it is well known that F is l.s.c. if and only if for any $\lambda \in \mathbb{R}$, $\{F \leq \lambda\}$ is closed, if and only if epi F is closed in $X \times \mathbb{R}$. [Indeed, if F is l.s.c. and $(x_n) \in \{F \leq \lambda\}$ converges to x, then $F(x) \leq \liminf_n F(x_n) \leq \lambda$, so that $\{F \leq \lambda\}$ is closed; if this set is closed and $(x_n, t_n)_n$ is a sequence of epi F converging to (x,t), then for any $\lambda > t$, $(x,t) \in \{F \leq \lambda\}$, that is $F(x) \leq \lambda$, so that $F(x) \leq t$ and $(x,t) \in \operatorname{epi} F$; and if epi F is closed and $(x_n)_n \to x$, for any $t > \liminf_n F(x_n)$ there exists a subsequence (x_{n_k}) such that $(x_{n_k}, t) \in \operatorname{epi} F$ for each k, so that $(x,t) \in \operatorname{epi} F$ hence $F(x) \leq t$. We deduce (3.5).]

Hence, we have $F \in \Gamma^0$ if and only if epi F is closed, convex, nonempty and differs from $X \times \mathbb{R}$.

Another standard fact is that in finite dimension, any convex F is locally Lipschitz in the interior of its domain.

Definition 3.3 (Legendre–Fenchel conjugate). We define the Legendre–Fenchel conjugate F^* of F for any $p \in X$ by
$$F^*(p) = \sup_{x\in X} \{\langle p, x\rangle - F(x)\}.$$

It is obvious that F^* is convex, l.s.c. (as a supremum of linear, continuous functions). We will soon see that it is also proper as soon as F is convex and proper: hence it maps Γ^0 into itself (and in fact, onto). The following is the most classical and fundamental result of convex duality:

Theorem 3.4. *Let $F \in \Gamma^0$: then $F^{**} = F$.*

Before proving the theorem, let us state an important separation result on which it relies:

Theorem 3.5. *Let X be an Euclidean space, $C \subset X$ be a closed convex set, and $x \notin C$. Then there exists a closed hyperplane separating strictly x and C, that is: there exists $p \in X$ and $\alpha \in \mathbb{R}$ such that*

$$\langle p, x\rangle > \alpha \geq \langle p, z\rangle$$

for any $z \in C$.

This result is in its most general form a consequence of the Hahn–Banach theorem, however, in the finite dimensional, or Hilbert setting, its proof is much easier than in the general setting:

Proof. Let $\bar{x} \in C$ be the projection of x onto C:

$$\bar{x} = \arg\min\{\|x - x'\|^2 : x' \in C\}.$$

The proof that it exists is standard and relies on the "parallelogram identity" $\|a + b\|^2 + \|a - b\|^2 = 2(\|a\|^2 + \|b\|^2)$, and the fact that X is complete.

Let $p = x - \bar{x}$. We know that for any $z \in C$,

$$\langle x - \bar{x}, z - \bar{x}\rangle \leq 0$$

as is deduced from a straightforward first variation argument. Hence

$$\langle p, z\rangle \leq \alpha = \langle p, \bar{x}\rangle.$$

On the other hand,

$$\langle p, x\rangle = \langle p, \bar{x}\rangle + \|p\|^2 > \alpha.$$ □

Now we can prove Theorem (3.4):

Proof. First, for any x and p, $F^*(p) \geq \langle p, x\rangle - F(x)$, so that $\langle p, x\rangle - F^*(p) \leq F(x)$, and taking the sup over p we get that $F^{**}(x) \leq F(x)$. Hence we need to show the opposite inequality.

To simplify, we prove it only in case dom $F = X$. If dom $F \neq X$, then it is easy to approximate F as a sup of functions $(F_\delta)_{\delta>0}$ in Γ^0 with dom $F_\delta = X$ (for instance, the Moreau–Yosida regularizations $F_\delta(x) = \min_y F(y) + \|x - y\|^2/(2\delta) \leq F(x)$), and recover the result.

Let now $(x, t) \notin$ epi F, which is closed and convex. That is, $t < F(x)$. By the separation theorem, there exists p, s, α such that

$$\langle p, x\rangle + st > \alpha \geq \langle p, z\rangle + su$$

for any $(z, u) \in \text{epi}\, F$. Sending $u \to +\infty$ we see that $s \leq 0$. On the other hand, since we have assumed $\text{dom}\, F = X$, we have that $x \in \text{dom}\, F$ and $t < F(x)$, so that

$$\langle p, x \rangle + st > \alpha \geq \langle p, x \rangle + sF(x),$$

and it follows $s < 0$.

Now for any z,

$$\left\langle \frac{p}{s}, x \right\rangle + t < \frac{\alpha}{s} \leq \left\langle \frac{p}{s}, z \right\rangle + F(z)$$

we deduce that $\langle \frac{p}{s}, x \rangle + t < -F^*(-p/s)$, so that $t < F^{**}(x)$. It follows that $F(x) \leq F^{**}(x)$. □

3.2.2 Subgradient

The definition is already given in Definition 2.2. Now F is a convex function defined on a finite dimensional space X.

Definition 3.6. For $x \in X$,

$$\partial F(x) = \{p \in X \;:\; F(y) \geq F(x) + \langle p, y - x \rangle \;\forall y \in \text{dom}\, F\}$$

and $\text{dom}\, \partial F = \{x \;:\; \partial F(x) \neq \emptyset\} \subset \text{dom}\, F$.

If F is differentiable at x, then $\partial F(x) = \{\nabla F(x)\}$.

Now, for any p, x, $\langle p, x \rangle \leq F(x) + F^*(p)$. But $p \in \partial F(x)$ implies that

$$\langle p, x \rangle - F(x) \geq \langle p, y \rangle - F(y)$$

for all y, hence

$$\langle p, x \rangle - F(x) \geq F^*(p).$$

Hence, one deduces the Legendre–Fenchel identity:

Proposition 3.7. *For any F convex, $p \in \partial F(x)$ if and only if*

$$\langle p, x \rangle = F(x) + F^*(p).$$

Moreover if $F \in \Gamma^0$, so that $F^{**} = F$, then this is equivalent to $x \in \partial F^*(p)$.

We have the following obvious proposition:

Proposition 3.8. *Let F be convex, then $x \in \arg\min_X F$ if and only if $0 \in \partial F(x)$.*

Proof. Indeed, this is equivalent to $F(y) \geq F(x) + \langle 0, y - x \rangle$ for all y. □

The following is trickier to show, but we skip the proof:

Proposition 3.9. *Let F, G be convex and assume* $\mathrm{int}(\mathrm{dom}\, G) \cap \mathrm{dom}\, F \neq \emptyset$*: then* $\partial(F+G) = \partial F + \partial G$.

The inclusion $\partial(F+G) \supset \partial F + \partial G$ always holds. For a proof, see [32]. The condition that $\mathrm{int}(\mathrm{dom}\, G) \cap \mathrm{dom}\, F \neq \emptyset$ (which is always true, for instance, if $\mathrm{dom}\, G = X$), is way too strong in finite dimension, where one may just assume that the relative interiors of the domains of G and F have a nonempty intersection [68]. If not, the result might not be true, as shows the example where $F(x) = +\infty$ if $x < 0$, $-\sqrt{x}$ if $x \geq 0$, and $G(x) = F(-x)$.

Monotonicity The subgradient of a convex function F is an example of "monotone" operator (in the sense of Minty): clearly from the definition we have for any x, y and any $p \in \partial F(x)$, $q \in \partial F(y)$,

$$\langle p - q, x - y \rangle \geq 0. \tag{3.7}$$

(Just sum the two inequalities $F(y) \geq F(x) + \langle p, y - x \rangle$ and $F(x) \geq F(y) + \langle q, x - y \rangle$.) Compare with (3.6). This is an essential property, as numerous algorithms have been designed (mostly in the 70s) to find the zeroes of monotone operators and their numerous variants. Also, a quite complete theory is available for defining and describing the flow of such operators [71, 19].

3.2.3 The Dual of (ROF)

We now can easily derive the "dual" problem of (3.2) (and, in fact, also of (ROF) since everything we will write here also holds in the Hilbertian setting).

Recall that J denotes now the discrete total variation introduced in (3.2). Let u be a minimizer: from Propositions 3.8 and 3.9, we have

$$0 \in \partial\Big(\lambda J + \frac{1}{2}\|\cdot - g\|^2\Big)(u) = \lambda \partial J(u) + u - g.$$

(We recover in the discrete setting the same Euler–Lagrange equation as in the continuous setting, see (2.10).)

Hence: $(g-u)/\lambda \in \partial J(u)$, hence from Proposition 3.7, $u \in \partial J^*((g-u)/\lambda)$, hence $v = (g-u)/\lambda$ solves

$$0 \in v - \frac{g}{\lambda} + \frac{1}{\lambda}\partial J^*(v)$$

which is exactly the equation which characterizes the minimality for

$$\min_v \frac{1}{2}\left\| v - \frac{g}{\lambda} \right\|^2 + \frac{1}{\lambda} J^*(v). \tag{3.8}$$

Now, what is J^*? This is quite simple: J can (as in the continuous setting) be defined by duality, indeed,

$$\begin{aligned} J(u) = \|\nabla u\|_{2,1} &= \sup\{\langle \xi, \nabla u\rangle_Y \ : \ |\xi_{i,j}| \le 1 \ \forall i,j\} \\ &= \sup\{-\langle \operatorname{div} \xi, u\rangle_X \ : \ |\xi_{i,j}| \le 1 \ \forall i,j\} \\ &= \sup_p \langle p, x\rangle_X - H(p) \end{aligned}$$

where, letting

$$K = \{p = -\operatorname{div} \xi \in X \ : \ \|\xi_{i,j}\| \le 1 \ \forall i,j\}, \tag{3.9}$$

we have $H(p) = 0$ for $p \in K$ and $H(p) = +\infty$ if $p \notin K$ (H is called the "characteristic function of K"). Hence J is the Legendre–Fenchel conjugate H^* of this function H.

Since K is closed and convex, $H \in \Gamma^0$, so that $J^* = H^{**} = H$. Hence the dual problem (3.8) is also

$$\min_{v \in K} \frac{1}{2}\left\| v - \frac{g}{\lambda}\right\|^2 = \min_{|\xi|_{i,j} \le 1} \frac{1}{2}\left\| \operatorname{div} \xi + \frac{g}{\lambda}\right\|^2 \tag{3.10}$$

and we recover u by letting $u = g - \lambda v = g + \lambda \operatorname{div} \xi$.

We will see that this problem has a structure which makes it nicer (easier) to solve than the primal problem (3.2).

3.2.4 "Proximal" Operator

We end this section by introducing more generally the "proximal" operator associated to a function $F \in \Gamma^0$. For any $F \in \Gamma^0$ it is not hard to show that for any $\delta > 0$, problem

$$\min_y \delta F(y) + \frac{1}{2}\|y - x\|^2$$

always have a solution, which is unique. The equation for this solution y is

$$\delta \partial F(y) + y - x \ni 0$$

hence

$$y = (I + \delta \partial F)^{-1}(x) \tag{3.11}$$

is well-defined and uniquely defined. The mapping $(I + \delta\partial F)^{-1}$ is called the "proximal map" of δF and sometimes denoted $\operatorname{prox}_{\delta F}$. The following identity, which is exactly our derivation of the dual problem in the previous section, is due to Moreau:

$$x = (I + \delta \partial F)^{-1}(x) + \delta\left(I + \frac{1}{\delta}\partial F^*\right)^{-1}\left(\frac{x}{\delta}\right), \tag{3.12}$$

and for $\delta = 1$ it reduces to

$$x = (I + \partial F)^{-1}(x) + (I + \partial F^*)^{-1}(x).$$

Examples. If $F(x) = \alpha x^2/2$ for some $\alpha > 0$, we check that

$$(I + \delta\partial F)^{-1}(x) = \frac{x}{1+\delta\alpha}.$$

The reader may check that this is coherent with (3.12) and the fact that $F^*(p) = p^2/(2\alpha)$.

If $F(x)$ is the characteristic function of a closed, convex set $C \subset X$, that is $F(x)=0$ if $x \in C$ and $+\infty$ else, then

$$(I + \delta\partial F)^{-1}(x) = \Pi_C(x),$$

the Euclidean projection on C of x, which actually minimizes $\|x-y\|^2$ over all $y \in C$. On the other hand, it follows from (3.12) that

$$\bar{y} = \left(I + \frac{1}{\delta}\partial F^*\right)^{-1}(y) = y - \frac{1}{\delta}\Pi_C(\delta y)$$

which is some kind of "shrinkage" or "soft-thresholding" of y from which one removes the projection on $(1/\delta)C$. The point $\bar{y}$ is the minimizer of

$$\min_z \frac{\|z-y\|^2}{2} + \frac{1}{\delta}h_C(z)$$

where $F^* = h_C$ is the *support function* of C, defined by $h_C(z) = \sup_{x\in C}\langle z, x\rangle$.

In the sequel we introduce a few possible algorithms to solve (3.2) or (3.10).

3.3 Gradient Descent

Consider first an elementary problem which is to minimize over X a function $F \in \Gamma^0$, which is differentiable and such that ∇F is Lipschitz with some constant L (one says that F is $C^{1,1}$).

It is not the case for (3.2), but it is the case for the approximation

$$F_\varepsilon(u) = \sum_{i,j}\sqrt{\varepsilon^2 + |(\nabla u)_{i,j}|^2} + \frac{1}{2}\|u-g\|^2$$

for any $\varepsilon > 0$ (and it is clear that as $\varepsilon \to 0$, this problem will approximate the other one). In this case, ∇F_ε is Lipschitz with a constant of order $1/\varepsilon$.

Then, the most straightforward approach is the "gradient descent": choose $h > 0$ a step, any $x^0 \in X$ and let for any $n \geq 0$

$$u^{n+1} = u^n - h\nabla F(u^n).$$

(Of course the step h needs not, and should not, be constant, but for simplicity we stick to this case).

This method is not very efficient (and should *not* be used!). A complexity bound can be derived:

Theorem 3.10 (Nesterov [60]). *Assume $h \in (0, 2/L)$. Then $F(u^k) \to \min F = F(x^*)$ as $k \to \infty$. The best rate of convergence is obtained for $h = 1/L$, and is*

$$F(u^k) - F(x^*) \le \frac{2L\|u^0 - u^*\|^2}{k+4}.$$

Observe that the estimate depends on the quality of the initial guess. For a proof, see [60], Theorem 2.1.14 and Corollary 2.1.2.

For solving the approximation F_ε, one sees that the step h should be taken of order ε. This approach cannot be used to solve the dual problem (3.10): indeed, although the objective function is quite smooth in this case (and the gradient, $\nabla(\operatorname{div} p + g/\lambda)$, is Lipschitz with constant $L \le 8$), it has to be minimized on a convex set (hence with a constraint). For this reason, we need to introduce constrained variants of the Gradient Descent algorithm.

3.3.1 Splitting, and Projected Gradient Descent

We follow in this presentation the paper of Beck and Teboulle [13]. Assume we want to solve

$$\min_{x \in X} F(x) + G(x)$$

where: F is $C^{1,1}$ (∇F is L-Lipschitz), and G is "simple", meaning that the "prox" $(I + h\partial G)^{-1}$ is easy to compute. This is for instance the case for the dual problem (3.10): in this case

$$F(p) = \frac{1}{2}\|\operatorname{div} p + g\|^2 \quad \text{and} \quad G(p) = \begin{cases} 0 & \text{if } \|p_{i,j}\| \le 1\ \forall i,j \\ +\infty & \text{else.} \end{cases}$$

We see that (see Section 3.2.4)

$$(I + h\partial G)^{-1}(p) = \arg\min_q \frac{1}{2}\|q - p\|^2 + hG(q) = \Pi_C(p)$$

where Π_C denotes the orthogonal projection onto $\{p : \|p_{i,j}\| \le 1\ \forall i,j\}$ and is straightforward to compute.

A good idea to solve the problem is to do descent steps alternatively in F and G, as we will now describe. This is an example of "splitting" (introduced first by Douglas and Rachford, see for instance [49] for a general form): in practice, we solve successively one step of the gradient descent of F (in an explicit way), and one step of the gradient descent of G (in an implicit way), in order to obtain a "full" gradient descent of $F + G$. Hence the term "forward-backwards" splitting, see [28].

In case G is, like here, a characteristic function and $(I + h\partial G)^{-1}$ is a projection), then it reduces to a "projected gradient algorithm": we do one explicit step of descent in F, and then reproject the point on the constraint.

The resulting algorithm is hence as follows: we choose $x^0 \in X$, and let

$$x^{n+1} = (I + h\partial G)^{-1}(x^n - h\nabla F(x^n)) \tag{3.13}$$

for a given, fixed step $h > 0$. One can also write the iteration $x^{n+1} \in x^n - h(\nabla F(x^n) + \partial G(x^{n+1}))$ which makes apparent the forward-backwards splitting.

Again, this algorithm is quite slow, but it is interesting to understand the intuitive idea behind it. In fact, if ∇F is L-Lipschitz, we can write for any $x, y \in X$

$$\begin{aligned} F(y) &= F(x) + \left\langle \int_0^1 \nabla F(x + t(y - x))\, dt, y - x \right\rangle \\ &\le F(x) + \langle \nabla F(x), y - x\rangle + \frac{L}{2}\|y - x\|^2 \end{aligned} \tag{3.14}$$

so that the parabola $y \mapsto Q_L(y, x) = F(x) + \langle \nabla F(x), y - x\rangle + \frac{L}{2}\|y - x\|^2$ approximates from above the function F. Now, assume $x = x^n$ and we replace the minimization of F, at step n, with the minimization of $Q_L(y, x^n)$ w.r. y. Then, we find $y = x^n - (1/L)\nabla F(x^n)$, that is, a step of the gradient descent algorithms with step $1/L$. This is a way to interpret that algorithm, and provides a natural way to extend it to the minimization of $F + G$: indeed, we can now let

$$Q_L(y, x) = F(x) + \langle \nabla F(x), y - x\rangle + \frac{L}{2}\|y - x\|^2 + G(y)$$

and we see, as before, that $F(y) + G(y) \le Q_L(y, x)$ for any x, y.

Now, consider the problem $\min_y Q_L(y, x^n)$. The equation is

$$\nabla F(x^n) + L(y - x^n) + \partial G(y) \ni 0$$

and we find that the solution is nothing else than the iterate x^{n+1} given by (3.13), provided $h = 1/L$.

The following is Lemma 2.3 in [13]:

Lemma 3.11. *Let $x \in X$, and $h > 0$, and let $y = \arg\min Q_{1/h}(\cdot, x)$ be such that $F(y) + G(y) \le Q_{1/h}(y, x)$ (which is true as soon as $h < 1/L$, by* (3.14)*). Then for any $z \in X$*

$$\begin{aligned} (F(z) + G(z)) - (F(y) + G(y)) &\ge \frac{1}{2h}\|x - y\|^2 + \frac{1}{h}\langle x - z, y - x\rangle \\ &= \frac{1}{2h}(\|y - z\|^2 - \|x - z\|^2). \end{aligned} \tag{3.15}$$

Proof. By assumption,

$$(F(z) + G(z)) - (F(y) + G(y)) \ge F(z) + G(z) - Q_{1/h}(y, x). \tag{3.16}$$

Also, $F(z) \geq F(x) + \langle \nabla F(x), z - x \rangle$, while $G(z) \geq G(y) + \langle p, z - y \rangle$ where $p = (x - h\nabla F(x) - y)/h \in \partial G(y)$. Hence

$$F(z) + G(z) \geq F(x) + G(y) + \langle \nabla F(x), z - x \rangle + \langle p, z - y \rangle.$$

We deduce from (3.16) that

$$\begin{aligned}
(F(z) + G(z)) - (F(y) + G(y)) & \\
\geq F(x) + G(y) + \langle \nabla F(x), z - x \rangle + \langle p, z - y \rangle & \\
- F(x) - \langle \nabla F(x), y - x \rangle - \frac{1}{2h}\|y - x\|^2 - G(y) & \\
= \langle \nabla F(x) + p, z - y \rangle - \frac{1}{2h}\|y - x\|^2 & \\
= \frac{1}{h}\langle x - y, z - y \rangle - \frac{1}{2h}\|y - x\|^2 = \frac{1}{2h}\|y - x\|^2 + \frac{1}{h}\langle x - y, z - x \rangle. & \quad \square
\end{aligned}$$

This allows to prove the following result (see Beck and Teboulle [13], Theorem 3.1):

Theorem 3.12. *Let* $(x^n)_n$ *satisfy* (3.13), *and* $h = 1/L$. *Then*

$$(F(x^k) + G(x^k)) - (F(x^*) + G(x^*)) \leq \frac{L\|x^0 - x^*\|^2}{2k} \tag{3.17}$$

for any $k \geq 1$, *and for any solution* x^* *of the problem.*

Hence the rate of convergence is essentially the same as for the standard gradient descent, when $G \equiv 0$.

Proof. We follow the very elegant proof of Beck and Teboulle [13]. First use (3.15) with $z = x^*$, $y = x^{n+1}$, $x = x^n$, $h = 1/L$:

$$\frac{2}{L}((F(x^*) + G(x^*)) - (F(x^{n+1}) + G(x^{n+1}))) \geq \|x^{n+1} - x^*\|^2 - \|x^n - x^*\|^2,$$

which we sum from $n = 0$ to $k - 1$:

$$\frac{2}{L}\left(k(F(x^*)+G(x^*)) - \sum_{n=1}^{k}(F(x^n)+G(x^n))\right) \geq \|x^k - x^*\|^2 - \|x^0 - x^*\|^2. \tag{3.18}$$

We use (3.15) again with $z = x = x^n$, $h = 1/L$:

$$\frac{2}{L}((F(x^n) + G(x^n)) - (F(x^{n+1}) + G(x^{n+1}))) \geq \|x^{n+1} - x^n\|^2,$$

which we multiply by n before summing from 0 to $k-1$:

$$\frac{2}{L}\Big(\sum_{n=0}^{k-1} n(F(x^n)+G(x^n)) - \sum_{n=1}^{k}(n-1)(F(x^n)+G(x^n))\Big)$$
$$= \frac{2}{L}\Big(\sum_{n=1}^{k-1}(F(x^n)+G(x^n)) - (k-1)(F(x^k)+G(x^k))\Big)$$
$$\geq \sum_{n=0}^{k-1} n\|x^{n+1}-x^n\|^2.$$

We add this last equation to (3.18) and find:

$$\frac{2}{L}(k(F(x^*)+G(x^*)) - k(F(x^k)+G(x^k)))$$
$$\geq \|x^k - x^*\|^2 - \|x^0 - x^*\|^2 + \sum_{n=0}^{k-1} n\|x^{n+1}-x^n\|^2 \geq -\|x^0 - x^*\|^2$$

from which (3.17) follows. □

Hence: this provides a convergent (but slow) way to minimize the dual problem (and many variants).

3.3.2 Improvements: Optimal First-order Methods

The rate of convergence of theses methods are slow. It is shown by Nesterov [60] that first order method can theoretically not achieve a better rate of convergence than C/k^2 (after k iterations). A few variants of the previous methods achieve such a rate of convergence and are recommended in the implementations.

Nesterov/Beck and Teboulle's Acceleration In [13] the following iteration is proposed, as a variant of an acceleration for the gradient descent proposed by Nesterov in [59]: let $x^0 \in X = \mathbb{R}^N$, $y^1 = x^0$, $t_1 = 1$, and

$$x^k = \Big(I + \frac{1}{L}\partial G\Big)^{-1}\Big(y^k - \frac{1}{L}\nabla F(y^k)\Big),$$
$$t_{k+1} = \frac{1+\sqrt{1+4t_k^2}}{2}, \quad y^{k+1} = x^k + \frac{t_k - 1}{t_{k+1}}(x^k - x^{k-1}).$$

Then the following theorem holds.

Theorem 3.13 (Beck and Teboulle [13, Theorem 4.1]). *For any minimizer* x^*,

$$F(x^k) + G(x^k) - (F(x^*) + G(x^*)) \leq \frac{2L\|x^0 - x^*\|^2}{(k+1)^2}.$$

Yu. Nesterov himself also proposed an improvement of his earlier algorithm for non-smooth problems in [61], which is similar in spirit but a bit more complex to describe. See also [15] for similar accelerations for a smaller class of problems.

3.4 Augmented Lagrangian Approaches

Another class of methods for solving (3.2) are the "augmented Lagrangian" methods, also known as "split Bregman" iterations, or "Alternating directions method of multipliers"

The basic idea is as follows: instead of solving (3.2), we solve the *constrained* problem

$$\min_{p=\nabla u} \lambda\|p\|_{2,1} + \frac{1}{2}\|u-g\|^2.$$

Then, to enforce the constraint, we use an augmented Lagrangian approach, which consists in introducing

$$L_\alpha(p,u,\mu) = \lambda\|p\|_{2,1} + \frac{1}{2}\|u-g\|^2 + \langle \mu, p - \nabla u\rangle + \frac{\alpha}{2}\|p - \nabla u\|^2$$

where here, $\mu \in Y$ is a Lagrange multiplier for the constraint $p = \nabla u$. The method consists then in minimizing alternatingly L_α w.r. p, u, and updating the Lagrange multiplier μ:

$$\begin{aligned} u^{k+1} &= \arg\min_u L_\alpha(p^k, u, \mu^k) \\ p^{k+1} &= \arg\min_p L_\alpha(p, u^{k+1}, \mu^k) \\ \mu^{k+1} &= \mu^k + \alpha(p^{k+1} - \nabla u^{k+1}) \end{aligned}$$

and this method is shown to converge. It seems it was first studied by Gabay and Mercier, 1976, Glowinski and Marrocco, 1975. Then, it was analyzed in a more general framework in [31]. See also [33] and the references therein for a recent study on these approaches.

3.5 Primal-dual Approaches

The last interesting family of algorithms to solve (3.2) are maybe the primal-dual approaches, or "Arrow–Hurwicz" type methods. The idea goes back to [11], but seems to be first found in this framework in a paper of Appleton and Talbot [10]. Consider the formulation (3.3), using Legendre–Fenchel's duality, it can also be written as:

$$\min_{x\in X}\max_{y\in Y}\langle y, Ax\rangle - F^*(y) + G(x) \tag{3.19}$$

and the idea is to alternate gradient descent in x and ascent in y. Take care that in some cases, it provides iterations which are almost identical to the iterations provided by the previous splitting approaches. An important observation is that standard convex analysis shows that under very weak assumptions, the min and max may be swapped in (3.19). This is another approach to the dual problem of (3.3), which can be found by writing

$$\begin{aligned}\min_{x\in X} F(Ax) + G(x) &= \min_{x\in X}\max_{y\in Y}\langle y, Ax\rangle - F^*(y) + G(x)\\ &= \max_{y\in Y}\min_{x\in X}\langle A^*y, x\rangle + G(x) - F^*(y)\\ &= \max_{y\in Y} -(G^*(-A^*y) + F^*(y)).\end{aligned}$$

Moreover, one deduces immediately that the quantity

$$\mathcal{G}(x, y) = F(Ax) + G(x) + G^*(-A^*y) + F^*(y),$$

known as the *primal-dual gap*, is always nonnegative, and vanishes only if $(\hat{x}, \hat{y})$ is a saddle-point of (3.19), hence satisfying

$$\langle y, A\hat{x}\rangle - F^*(y) + G(\hat{x}) \le \langle \hat{y}, A\hat{x}\rangle - F^*(\hat{y}) + G(\hat{x}) \le \langle \hat{y}, Ax\rangle - F^*(\hat{y}) + G(x) \tag{3.20}$$

for all $(x, y) \in X \times Y$.

This suggests the following approach, which consists in performing simultaneously an approximate gradient descent in x and gradient ascent in y: choose x^0, y^0, $\tau, \sigma > 0$ two time-steps, and let

$$\begin{aligned}y^{n+1} &= (I + \sigma\partial F^*)^{-1}(y^n + \sigma Ax^n)\\ x^{n+1} &= (I + \tau\partial G)^{-1}(x^n - \tau A^*y^{n+1}).\end{aligned}$$

The scheme, as is, is proposed in a paper of Zhu and Chan [74], with an interesting (and very efficient) acceleration which is obtained by varying the time-steps, but unfortunately no proof of convergence exists. A global study of such schemes is found in a recent preprint [34].

We have provided recently in [64] the following variant, inspired by a paper of L. Popov [66], where the convergence of a variant of the "extragradient" method of G. Korpelevich [45] is studied. The algorithm is as follows: we choose $x^0 = \bar{x}^0$, y^0, and let for each $n \geq 0$:

$$\begin{aligned} y^{n+1} &= (I + \sigma\partial F^*)^{-1}(y^n + \sigma A\bar{x}^n) \\ x^{n+1} &= (I + \tau\partial G)^{-1}(x^n - \tau A^* y^{n+1}) \\ \bar{x}^{n+1} &= 2x^{n+1} - x^n. \end{aligned} \tag{3.21}$$

(Observe that the algorithm could also be written with a variable $\bar{y}$ instead of $\bar{x}$, and iterating first in x, then in y.) Our approach can also be seen as a slight variant of the extragradient algorithm [45], or its generalization in [58], but it seems original.

This algorithm, first mentioned in [64], has been recently presented as a particular case of a more general algorithm in [34], and a proof of convergence is also given there. A detailed study is found in [25], with accelerations in case some of the convex functions are smoother. It reduces trivially to the standard Douglas–Rachford splitting in case $A = \mathrm{Id}$ (and probably if A is invertible), just perform the change of variable $v^n = y^n - x^n/\tau$ and check against the formula proposed in [49, eq. (10)]. We give here an alternate proof of convergence, which is inspired from [66] and [58]. Actually, we can show convergence of the iterates to a solution in finite dimension, while in the general case, mimicking the proof of [58] where A. Nemirovski computes rates of convergence for a general version of the extragradient algorithm, we find a convergence of a primal-dual gap to zero, in $O(1/n)$. For practical use, we introduce the partial primal-dual gap

$$\mathcal{G}_{B_1\times B_2}(x,y) = \max_{y'\in B_2} \langle y', Ax\rangle - F^*(y') + G(x) - \min_{x'\in B_1} \langle y, Ax'\rangle - F^*(y) + G(x'),$$

($\mathcal{G} = \mathcal{G}_{X\times Y}$). Then, as soon as $B_1 \times B_2$ contains a saddle-point $(\hat{x}, \hat{y})$, defined by (3.20), we have

$$\mathcal{G}_{B_1\times B_2}(x,y) \geq \langle \hat{y}, Ax\rangle - F^*(\hat{y}) + G(x) - \langle y, A\hat{x}\rangle - F^*(y) + G(\hat{x}) \geq 0$$

and it vanishes only if (x, y) is itself a saddle-point.

Theorem 3.14. *Let $L = \|A\|$ and assume problem* (3.19) *has a saddle-point $(\hat{x}, \hat{y})$. Then, if $\tau\sigma L^2 < 1$ and $(x_n, \bar{x}_n, y_n)$ are defined by* (3.21)*:*

(a) *For any n,*

$$\frac{\|y^n - \hat{y}\|^2}{2\sigma} + \frac{\|x^n - \hat{x}\|^2}{2\tau} \leq C\left(\frac{\|y^0 - \hat{y}\|^2}{2\sigma} + \frac{\|x^0 - \hat{x}\|^2}{2\tau}\right) \tag{3.22}$$

where the constant $C \leq (1 - \tau\sigma L^2)^{-1}$.

(b) *If we let $x_N = (\sum_{=1}^N x^n)/N$ and $y_N = (\sum_{n=1}^N y^n)/N$, for any bounded $B_1 \times B_2 \subset X \times Y$ the restricted gap has the following bound:*

$$\mathcal{G}_{B_1\times B_2}(x_N, y_N) \leq \frac{C(B_1, B_2)}{n}. \tag{3.23}$$

Moreover, the weak cluster points of (x_N, y_N) are saddle-points of (3.19).

(c) *If the dimension of the spaces X and Y is finite, then there exists a saddle-point (x^*, y^*) such that $x^n \to x^*$ and $y^n \to y^*$ (of course, then, also $(x_n, y_n) \to (x^*, y^*)$).*

The proof of Theorem 3.14 is found in Appendix A.

The estimate (3.23) is relatively weak but seems to show that the method is in some sense optimal (but slow). However, we'll see in the next Section 3.7 that the convergence can be apparently improved by varying the time-steps and relaxation parameters, as suggested in [74], although we do not have a clear explanation for this (and it is possible that this acceleration is problem-dependent, as opposed to the result of Theorem 3.13).

Observe that if $F^*(y)/|y| \to \infty$ as $|y| \to \infty$, then for any $R > 0$, $F^*(y) \geq R|y|$ for y large enough which yields that $\operatorname{dom} F \subset B(0, R)$. Hence F has full domain. It is classical that in this case, F is locally Lipschitz in Y.

One checks, then, that

$$\max_{y \in Y} \langle y, Ax_n \rangle - F^*(y) + G(x_n) = F(Ax_n) + G(x_n)$$

is reached at some $y \in \partial F(Ax_n)$, which is globally bounded thanks to (3.22). It follows from (3.23) that $F(Ax_n) + G(x_n) - (F(A\bar{x}) + G(\bar{x})) \leq C/n$ for some constant depending on the starting point (x^0, y^0), F and L. In the same way, if $\lim_{|x| \to \infty} G(x)/|x| \to \infty$, we have $F^*(y_n) + G^*(-A^* y_n) - (F^*(\hat{y}) + G^*(-A^* \hat{y})) \leq C/n$. If both $F^*(y)/|y|$ and $G(x)/|x|$ diverge as $|y|$ and $|x|$ go to infinity, then the global gap $\mathcal{G}(x_n, y_n) \leq C/n$.

It is easy to check that this approach is useful for many variants of (3.2) or similar discretizations of (2.1). We leave this as an exercise to the reader, see also the examples in Section 3.7.

3.6 Graph-cut Techniques

A last approach to solve (3.2) (or, in fact, a slight variant) is to use "graph-cuts" or "maximal flow" algorithms [1]. It has been noticed long ago [63] that maximal flow/minimum cut techniques could be used to solve discrete problems of the form (2.1), that is, to compute finite sets minimizing a discrete variant of the perimeter and an additional external field term.

This approach has gained in popularity in the past ten years, mostly because of incredibly fast algorithms, specially coined for image processing applications, see in particular [16].

Combined with the discrete counterpart of Proposition 2.6, it leads to efficient techniques for solving (only) the denoising problem (ROF) in the discrete setting. In fact, although this approach is strictly limited to problems of the form

$$\min_{u \in \mathbb{R}^{N \times N}} J(u) + \sum_{i,j} \Psi_{i,j}(u_{i,j})$$

with each function $\Psi_{i,j}$ convex, and J involving pairwise interactions,[3] such as

$$J(u) = \sum_{i,j} |u_{i+1,j} - u_{i,j}| + |u_{i,j+1} - u_{i,j}|,$$

an algorithm described in a paper by D. Hochbaum [42] provides a way to solve *exactly* (up to machine precision) this minimization for simple terms $\Psi_{i,j}(u_{i,j})$ such as $(u_{i,j} - g_{i,j})^2$.

We will not describe this approach in these notes, and refer to [23], and the references therein, for details.

3.7 Comparisons of the Numerical Algorithms

It is of course naive to believe that one single algorithm will fulfill the needs of all inverse problems which can be regularized by the total variation. However, the method (3.21) has the advantage of being easy to adapt to many different situations, and provide good results in most cases. Clearly, research should then focus on improving the numerical methods specifically for each particular applications, once it has been checked that the approach was efficient. In the following, we will present a comparison of the algorithms discussed in the previous sections for computing the minimizer of the discrete (ROF) problem (3.2). Besides this, we will also compare different discrete approximations of the total variation including anisotropic approximations and upwind schemes.

In order to evaluate the performance of each algorithm, we used the following procedure. First, we ran the proposed modified extragradient method (3.21) for a large number of iterations (10 000) to generate the ground truth solution (the primal dual gap was always less then 10^{-6}). Then, we ran the different algorithms until the root mean squared error (RMSE) of the current iterate to the ground truth solution was less than $tol = 10^{-3}$.

P-GD	Primal, gradient descend, $\varepsilon = 0.001$
D-PGD	Dual, projected gradient descend
D-BT	Dual, fast iterative shrinkage thresholding algorithm [13]
PD-AL	Primal-dual, augmented Lagrangian approach [33], $\alpha = 20$
PD-ME	Primal-dual, modified extragradient, $\tau = 0.01$
PD-MEG	Primal-dual, modified extragradient, GPU version, $\tau = 0.01$
PD-ZC	Primal-dual, Arrow–Hurwitz method, varying steps [74]
GC-8(16)	Graph-cut, 8(16)-connected graph, 8-bit accuracy [23]

Table 1. Explanation of the algorithms of the performance evaluation.

[3] In fact, slightly more complex situations can be considered, see for instance [44].

Table 1 gives an overview of the algorithms and parameter settings we used in our performance evaluation. All algorithms were implemented in pure Matlab, except for the graph-cut algorithm, whose main routine was implemented in optimized C/C++ (see [23] for more details). In addition, we implemented an parallel version of the modified extragradient algorithm on the (graphics processing unit) GPU using the CUDA framework. Executed on a Nvidia Tesla C1060 GPU, this results in a dramatic speedup of approximately 200 compared to the pure Matlab implementation. Note that on the other hand, an efficient parallel implementation of graph-cut algorithms is still an open problem. It can therefore be expected that algorithms that can be executed in parallel will play an important role in future. Finally, we note that since graph-cut algorithms compute the exact solution with respect to a discretized solution space, their results are not directly comparable to those of the continuous optimization algorithms.

Figure 9 shows the input image and results of our evaluation. The first row shows the clean and noisy input images and the second row shows results for different values of the regularization parameter λ. In the last row we provide a comparison between the graph-cut algorithms and the continuous minimization algorithms for $\lambda = 1$. The graph cut algorithm was executed using a 8- and a 16-connected graph. The modified extragradient method was executed using the simple forward differences approximation (see (3.1)) and a more sophisticated upwind scheme proposed in [24]. One can see that for a 8-connected graph, the so-called metrication errors of graph-cut algorithms are clearly visible. For a 16-connected graph, the results are very close to those of the continuous algorithms. Comparing the results of the continuous algorithms, one can observe that the upwind scheme delivers sharp edges almost independent of the edge orientation, whereas the simple forward differences approximation exhibits some blurring at certain edge orientations.

λ	1/16	1/8	1/4	1/2	1
P-GD	800/20.6	–/–	–/–	–/–	–/–
D-PGD	110/2.57	350/8.93	1120/27.93	3340/86.78	–/–
D-BT	40/1.28	80/2.54	140/4.72	270/8.31	460/15.41
PD-AL	50/1.40	90/2.65	150/4.55	250/7.55	420/12.5
PD-ME	40/1.09	70/2.06	120/3.64	220/6.04	410/10.99
PD-MEG	40/0.005	70/0.009	120/0.016	220/0.029	410/0.053
PD-ZC	20/0.56	50/1.30	90/2.43	150/3.84	300/8.02
GC-8	–/0.59	–/0.67	–/0.79	–/0.95	–/1.31
GC-16	–/1.10	–/1.27	–/1.58	–/2.06	–/2.96

Table 2. Performance evaluation of various minimization algorithms. The entries in the table refer to [iterations/time (sec)].

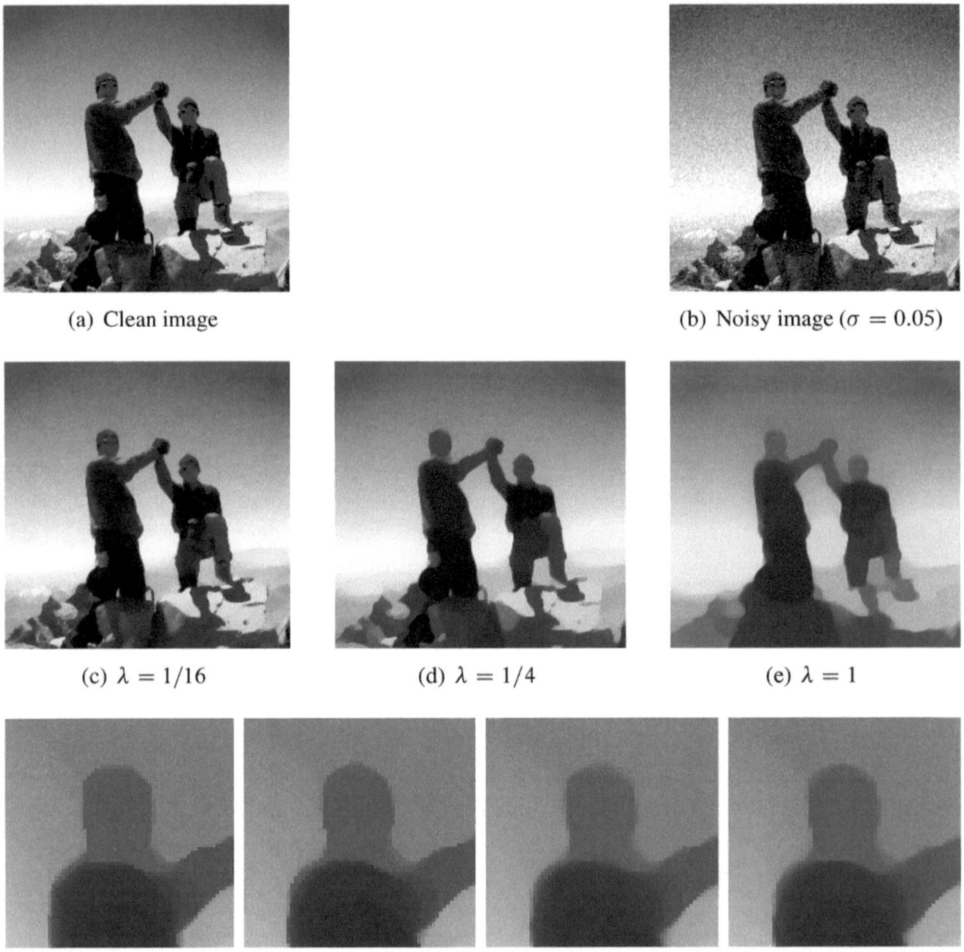

(a) Clean image
(b) Noisy image ($\sigma = 0.05$)
(c) $\lambda = 1/16$
(d) $\lambda = 1/4$
(e) $\lambda = 1$
(f) GC-8
(g) GC-16
(h) PD-ME, simple
(i) PD-ME, upwind

Figure 9. Results and comparison of the proposed minimization algorithms. (a) and (b) show the clean input image of size (256×256) and a degraded version generated by adding white Gaussian noise of standard deviation $\sigma = 0.05$. (c)–(e) show some results of the minimization algorithms for different values of the regularization parameter λ. (f)–(i) show a comparison of different discrete approximations of the total variation. (f) and (g) are anisotropic polygonal approximations of different order used by the graph cut techniques. (h) and (i) are more isotropic approximations used by the primal-dual minimization algorithm.

Table 2 shows the results of our performance evaluation. The table entries refer to the number of iterations the algorithm needed until the RMSE was less than *tol*. The second numbers refer to the respective execution time in seconds. If no entry is present, the algorithm did not meet the convergence criterion within 10 000 iterations. At first, one can observe that for larger values of λ the problem gets harder for all algorithms. Actually, simple gradient based methods even fail to converge within 10 000 iterations for larger values of λ. The PD-ZC algorithm is the fastest iterative method. It is slightly better than the proposed PD-ME algorithm. However, while the PD-ME method is proven to converge with a certain rate, a proof of convergence for the PD-ZC method (if there is one) is still an open problem. Furthermore, while the PD-ZC method is tailored for minimizing the (ROF) model, we will see in the next sections that the PD-ME method is applicable for a much larger class of problems. Interestingly, the PD-AL algorithm, which is often considered to be the most competing algorithm for minimizing the (ROF) model, is clearly outperformed by PD-ZC and PD-ME. The graph-cut algorithms are fast and deliver exact discrete solutions but an efficient parallelization is still an open problem. In contrast, the continuous PD-ME algorithm is easy to parallelize and its GPU-based variant yields the overall fastest method.

4 Applications

4.1 Total Variation Based Image Deblurring and Zooming

(a) Original image (b) Degraded image (c) Wiener filter (d) TV-deblurring

Figure 10. Motion deblurring using total variation regularization. (a) and (b) show the clean image and a degraded version containing motion blur of approximately 30 pixels and Gaussian noise of standard deviation $\sigma = 0.02$. (c) is the result of standard Wiener filtering. (d) is the result of the total variation based deblurring method. Note that the TV-based method yields visually much more appealing results.

The standard (ROF) model can be easily extended for image deblurring and digital zooming.

$$\min_u \left\{ \int_\Omega |Du| + \frac{\lambda}{2} \int_\Omega (Au - f)^2 dx \right\} \tag{4.1}$$

where $\Omega \subset \mathbb{R}^2$ is the domain of the image and A is a linear operator. In the case of image deblurring, A is the blurring kernel. In the case of image zooming, A describes the downsampling procedure, which is often assumed to be a blurring kernel followed by a subsampling operator. This problem can be easily rewritten in terms of a saddle-point problem (3.19).

$$\min_u \max_{p,q} \langle p, Du \rangle + \langle q, Au - f \rangle - I_{\|p\|_\infty \leq 1} - \frac{1}{2\lambda}\|q\|^2, \tag{4.2}$$

which can then be solved by the iterates (3.21). Here, I_S denotes the indicator function of the set S.

Figure 10 shows the application of the energy (4.1) to motion deblurring. While the classical Wiener filter is not able to restore the image the total variation based approach yields a far better result. Figure 11 shows the application of (4.1) to zooming. On can observe that total variation based zooming leads to a superresolved image with sharp boundaries whereas standard bicubic interpolation does not preserve sharp boundaries.

(a) Original images (b) Bicubic interpolation (c) TV-zooming

Figure 11. Image zooming using total variation regularization. (a) shows the original image and a by a factor of 4 downsampled version. (b) is the result of zooming by a factor of 4 using bicubic interpolation. (c) is the result of the total variation based zooming model. One can see that total variation based zooming yields much sharper image edges.

4.2 Total Variation with L^1 Data Fidelity Term

Similar to the (ROF) model, the TV-L^1 model [62, 26, 12] is defined as the variational problem

$$\min_u \left\{ \int_\Omega |Du| + \lambda \int_\Omega |u - f| dx \right\}. \tag{4.3}$$

The difference compared to the (ROF) model is that the squared L^2 data fidelity term has been replaced by the L^1 norm. Although the change is small, the TV-L^1 model offers some desirable properties. First, it turns out that the TV-L^1 model is more effective than the (ROF) model in removing impulse noise (e.g. salt and pepper noise) [62].

Second, the TV-L^1 model is contrast invariant. This means that, if u is a solution of (4.3) for a certain input image f, then cu is also a solution for cf for $c \in \mathbb{R}^+$. Therefore the TV-L^1 model has a strong geometrical meaning which makes it useful for scale-driven feature selection and denoising of shapes.

Being not strictly convex, computing a minimizer of the TV-L^1 model is a hard task. Several methods have been proposed to compute an approximate minimizer using fixed point iterations based on smoothing [72] or quadratic splitting techniques [12]. Recently, an augmented Lagrangian method has been proposed to compute the exact minimizer of the TV-L^1 model [33]. In order to apply the proposed modified extragradient method to solve the TV-L^1 model, we again rewrite it in terms of a saddle-point problem (3.19)

$$\min_u \max_{p,q} \langle p, Du \rangle + \langle q, u - f \rangle - I_{\{\|p\|_\infty \le 1\}} - I_{\{|q|_\infty \le \lambda\}}, \tag{4.4}$$

which can then be solved using the iterates (3.21).

Figure 12 shows the restoration of an image containing impulse noise (e.g. salt and pepper noise). As expected the (ROF) model can not restore the image without loosing fine scale details. On the other hand, the TV-L^1 model does not give too much weight to the outliers and hence leads to much better results. This shows that it is important to use a data term which matches the expected noise model.

4.3 Variational Models with Possibly Nonconvex Data Terms

The approach in this section is described with more details in the paper [65]. Let us consider the problem of finding the minimizer of an energy functional $F : L^1(\Omega) \to [0, \infty]$ of the form

$$\min_u \left\{ F(u) = \int_\Omega f(x, u(x), \nabla u(x))\, dx \right\}, \tag{4.5}$$

where Ω is a d-dimensional bounded open subset of $\mathbb{R}^N$ and $u : \Omega \to \mathbb{R}$ is an unknown scalar function. For $d = 2$, Ω is usually assumed to be a rectangular image domain. The *Lagrangian* $f(x, t, p)$ is the "core" of the energy functional and is used to model the characteristics of the energy functional. We will assume here that $f(x, t, p)$ is continuous in (x, t), and convex in p, but not necessarily in t.

4.3.1 Convex Representation

We can introduce a general theoretical framework which is quite classical in the calculus of variations, although not so well-known. The basic concept is the idea of *cartesian currents* [39, 40], which consists in taking the whole graph $(x, u(x))$ of a function as the "object" to optimize upon, rather than the function u itself. It is related to the so-called theory of *calibration*, which was recently brought back to light by Alberti et al. in [2], as an approach to characterize the minimizers of the Mumford–Shah

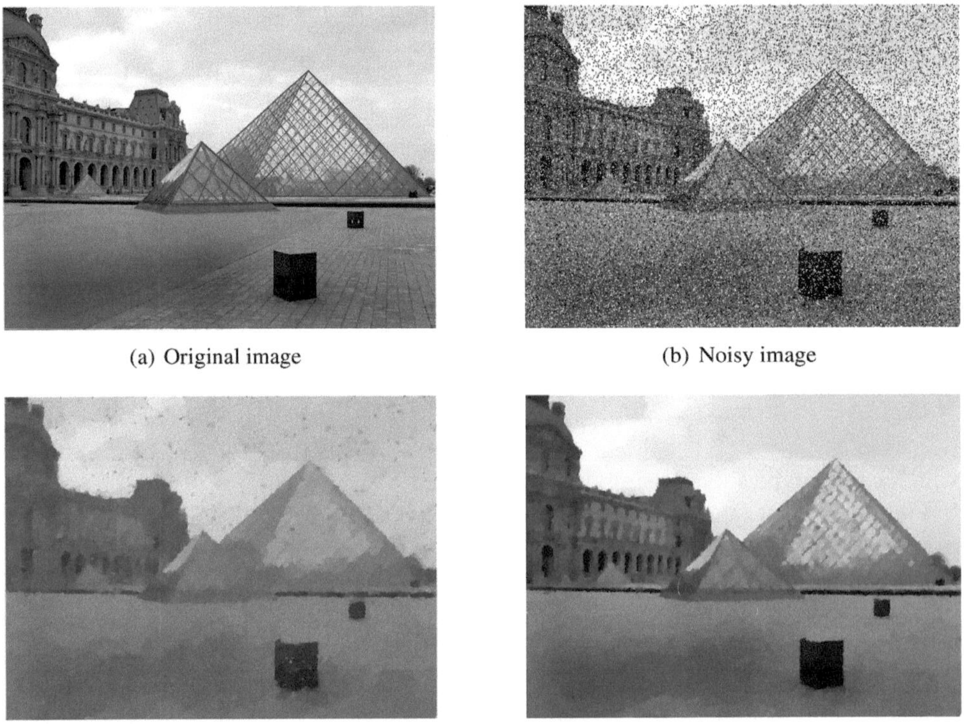

(a) Original image

(b) Noisy image

(c) (ROF)

(d) TV-L^1

Figure 12. Image denoising in the case of impulse noise. (a) shows the clean image and (b) is a noisy version which has been corrupted by 25% salt and pepper noise. (c) is the result of the (ROF) model. (d) is the result of the TV-L^1 model. Note that the TV-L^1 model is able to remove the noise while still preserving some small details.

functional [57] by an implicit (and new) convex representation: it allows to actually characterize (some) minimizers of the Mumford–Shah functional by means of divergence free vector in higher dimensions.

Let us start by considering the subgraph of the function $u(x)$, which is the collection of all points lying below the function value $u(x)$. Figure 13 shows an example for a one-dimensional function $u(x)$ where the subgraph is represented as the gray area. We also introduce the function $\mathbf{1}_u(x,t) : \Omega \times \mathbb{R} \to \{0, 1\}$ which is the characteristic function of the subgraph of $u(x)$:

$$\mathbf{1}_u(x,t) = \begin{cases} 1 & \text{if } u(x) > t \\ 0 & \text{otherwise.} \end{cases} \tag{4.6}$$

Furthermore let us denote by Γ_u the boundary of $\mathbf{1}_u(x,t)$. For the sake of simplicity, we assume first that u is smooth.

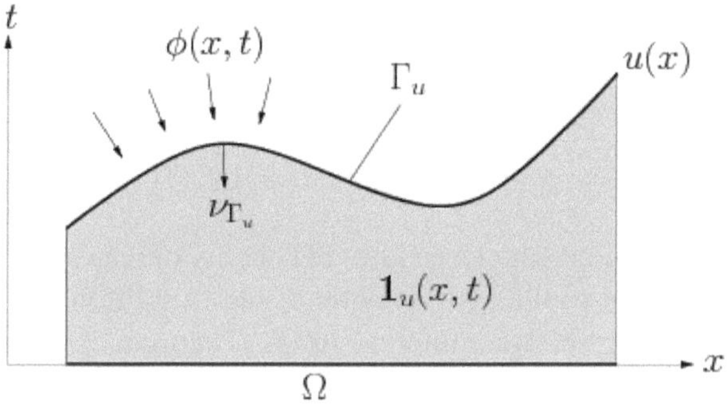

Figure 13. A one-dimensional function $u(x)$, its two-dimensional subgraph $\mathbf{1}_u$ and the vector field $\phi(x,t)$. The function $\mathbf{1}_u$ is supposed to be equal to 1 in the gray area and 0 outside.

The key idea is now to consider the flux of a vector field $\phi = (\phi^x, \phi^t) : \Omega \times \mathbb{R} \to \mathbb{R}^N \times \mathbb{R}$ through the boundary Γ_u

$$\Phi = \int_{\Gamma_u} \phi \cdot \nu_{\Gamma_u} d\,\mathcal{H}^N, \tag{4.7}$$

where $\mathcal{H}^N$ denotes the N-dimensional Hausdorff measure. ν_{Γ_u} denotes the inner unit normal to Γ_u which is given by

$$\nu_{\Gamma_u} = \frac{1}{\sqrt{1+|\nabla u(x)|^2}} \begin{pmatrix} \nabla u(x) \\ -1 \end{pmatrix}. \tag{4.8}$$

Alternatively, since we have $D\mathbf{1}_u = \nu_{\Gamma_u} \cdot \mathcal{H}^N$ on Γ_u, the flux can be written as

$$\Phi = \int_{\Gamma_u} \phi \cdot \nu_{\Gamma_u} d\,\mathcal{H}^N = \int_{\Omega\times\mathbb{R}} \phi \cdot D\mathbf{1}_u, \tag{4.9}$$

where the expression $D\mathbf{1}_u$ denotes the distributional derivative of $\mathbf{1}_u$, which is, in an integral sense, also well defined for characteristic functions. In the following, it will turn out that by choosing an appropriate vector field ϕ, $F(u)$ can be expressed as the maximal flux of ϕ through Γ_u.

Theorem 4.1. *For any function $u \in W^{1,1}(\Omega;\mathbb{R})$ the functional*

$$F(u) = \int_\Omega f(x, u(x), \nabla u(x))\, dx, \tag{4.10}$$

with $f(x,t,p)$ being continuous and positive in t and convex in p, can be written as the higher dimensional convex functional

$$\mathcal{F}(\mathbf{1}_u) := \sup_{\phi\in\mathcal{K}} \int_{\Omega\times\mathbb{R}} \phi \cdot D\mathbf{1}_u, \tag{4.11}$$

where the convex set $\mathcal{K}$ is given by

$$\mathcal{K} = \{\phi = (\phi^x, \phi^t) \in C_0(\Omega \times \mathbb{R}; \mathbb{R}^N \times \mathbb{R}) : \\ \phi^t(x,t) \ge f^*(x,t,\phi^x(x,t)) , \ \forall x,t \in \Omega \times \mathbb{R}\}. \quad (4.12)$$

Here, $f^*(x,t,p^*)$ denotes the Legendre–Fenchel conjugate (or convex conjugate) of $f(x,t,p)$ with respect to the last variable p, see Definition 3.3. Now, $f(x,t,p)$ being convex and lower semi-continuous in p, Theorem 3.4 yields the equality $f^{**} = f$.

Theorem 4.1 essentially states, that the potentially (but not necessarily) non convex functional (4.5) of a scalar function in dimension N can be rewritten as a convex functional in dimension $N+1$. Moreover, it is seen from the definition that this convex functional is "a sort of" total variation, and essentially has the same structure. To sum up, we have recast problem (4.5) as the minimization of a modified (nonuniform, anisotropic) perimeter, and the new problem is similar to (2.1). It is remarkable that this works for functions $f(x,u(x),\nabla u(x))$ with a quite arbitrary behavior in $u(x)$ (although continuous, in that variable). On the other hand, this comes along with an increased computational complexity, since we added a dimension to the problem.[4]

Proof. Let us sketch the proof of Theorem 4.1. We first check that for any $\phi \in \mathcal{K}$, we have

$$F(u) \ge \int_{\Omega\times\mathbb{R}} \phi \cdot D\mathbf{1}_u. \quad (4.13)$$

Indeed, using (4.9) and the definition of the inner unit normal (4.8), the flux can be rewritten as

$$\begin{aligned}\int_{\Omega\times\mathbb{R}} \phi \cdot D\mathbf{1}_u &= \int_{\Gamma_u} \phi(x,t) \cdot \begin{pmatrix} \nabla u(x) \\ -1 \end{pmatrix} \frac{d\mathcal{H}^N(x,t)}{\sqrt{1+|\nabla u(x)|^2}} \\ &= \int_\Omega \phi^x(x,u(x)) \cdot \nabla u(x) - \phi^t(x,u(x))\, dx, \end{aligned} \quad (4.14)$$

as $\sqrt{1+|\nabla u(x)|^2}$ is nothing else as the Jacobian of the change of variable $\Gamma_u \ni (x,t) \mapsto x \in \Omega$. Since $\phi \in \mathcal{K}$, it follows

$$\int_{\Omega\times\mathbb{R}} \phi \cdot D\mathbf{1}_u \le \int_\Omega \phi^x(x,u(x)) \cdot \nabla u(x) - f^*(x,t,\phi^x(x,u(x)))\, dx,$$

which is less than $F(u)$ by definition of the convex conjugate f^*. This shows (4.13).

The proof that the supremum is actually $F(u)$, that is, of (4.11), is more technical. Essentially, one would need to choose $\phi^x(x,u(x)) = \nabla_p f(x,u(x),\nabla u(x))$ at the point $(x,u(x))$ (since $p^* = \nabla_p f(x,t,p)$ reaches the maximum in $\max_p \langle q,p\rangle -$

[4] And, in fact, it is not completely surprising if one thinks first of the case $N = 0$....

$f(x,t,p)$, at least when f is differentiable at p, see Proposition 3.7), and $\phi^t(x,u(x)) = f^*(x,t,\phi^x(x,u(x)))$. If f, u are smooth enough (essentially, C^1), then such a choice can be performed. In other cases, it is shown that one can build a continuous field $\phi \in \mathcal{K}$ such that the flux (4.9) is arbitrarily close to $F(u)$. □

Remark 4.2. In fact, the theorem still holds for $u \in BV(\Omega)$ a bounded variation function, and a Lagrangian $f(x,t,p)$ with linear growth (in p) at ∞, with a similar proof. It can also be extended to Lagrangians which take the value $+\infty$, such as illustrated in Figure 14(c), with some additional regularity assumptions in x, t. See [65] for details.

We have now transformed the problem of computing the minimizer of (4.5) into computing the minimizer of

$$\min_{\mathbf{1}_u} \left\{ \mathcal{F}(\mathbf{1}_u) = \sup_{\phi \in \mathcal{K}} \int_{\Omega \times \mathbb{R}} \phi \cdot D\mathbf{1}_u \right\}. \tag{4.15}$$

Minimization in (4.15) is carried out over binary functions, which comprise a non convex set. Therefore we replace the function $\mathbf{1}_u$ in (4.15) by a more general function $v \in \mathcal{C}$, where the convex set $\mathcal{C}$ is given by

$$\mathcal{C} = \left\{ v \in BV(\Omega \times \mathbb{R}; [0,1]) \ : \ \lim_{t \to -\infty} v(x,t) = 1 \ , \ \lim_{t \to +\infty} v(x,t) = 0 \right\}. \tag{4.16}$$

Hence we consider the relaxed problem

$$\min_{v \in \mathcal{C}} \left\{ \mathcal{F}(v) := \sup_{\phi \in \mathcal{K}} \int_{\Omega \times \mathbb{R}} \phi \cdot Dv \right\}. \tag{4.17}$$

Using this relaxation we essentially minimize the convex envelope $\mathcal{F}(v)$ of $\mathcal{F}(\mathbf{1}_u)$.

4.3.2 Convex Relaxation

Our intention is still to solve the binary problem. Hence, the question remains in which sense the minimizers of (4.17) and (4.15) are related? Indeed, one can show that a simple thresholding produces a solution of (4.15) from one of (4.17). This is summarized by the following result, which generalizes Proposition 2.1.

Proposition 4.3. *Let* v^* *by a global minimizer of* (4.17). *Then for any* $s \in [0,1)$ *the characteristic function* $\mathbf{1}_{\{v^* > s\}}$ *is also a global minimizer of* (4.15).

Proof. The proof is the same as the proof of Proposition 2.1, as soon as one has observed that $\mathcal{F}$ satisfies the generalized co-area formula:

$$\mathcal{F}(v) = \int_{-\infty}^{+\infty} \mathcal{F}(\mathbf{1}_{\{v > s\}})\, ds. \tag{4.18}$$

This follows from the fact that $\mathcal{F}$ can be represented as

$$\mathcal{F}(v) = \int_{\Omega\times\mathbb{R}} h(x,t,Dv) = \int_{\Omega\times\mathbb{R}} h\Big(x,t,\frac{Dv}{|Dv|}\Big)|Dv| \tag{4.19}$$

where h is the convex, l.s.c. and one-homogeneous function of $Dv = (D^x v, D^t v)$ defined as the support function of the convex set $\{\phi = (\phi^x,\phi^t) \,:\, \phi^t \geq f^*(x,t,\phi^x)\}$, for any (x,t):

$$h(x,t,Dv) := \sup_{\phi^t \geq f^*(x,t,\phi^x)} \phi\cdot Dv. \tag{4.20}$$

This function is shown to be nothing else as:

$$h(x,t,Dv) = \begin{cases} |D^t v| f(x,t,D^x v/|D^t v|) & \text{if } D^t v < 0,\\ f^\infty(x,t,D^x v) & \text{if } D^t v = 0,\\ +\infty & \text{if } D^t v > 0, \end{cases} \tag{4.21}$$

where $f^\infty(x,t,p^x) := \lim_{\lambda\to+\infty} f(x,t,\lambda p^x)/\lambda$ is the *recession function of* f, see for instance [29, 40].

Hence, for any v, if we let $\nu_v = Dv/|Dv|$ (the Besicovitch derivative of the measure Dv with respect to its variation $|Dv|$), we have, using the standard co-area formula for BV functions (in a form which is more general than (CA) [37, 35, 75, 7]):

$$\begin{aligned} \mathcal{F}(v) &= \int_{\Omega\times\mathbb{R}} h(x,t,\nu_v(x,t))|Dv| \\ &= \int_{-\infty}^{+\infty}\int_{\Omega\times\mathbb{R}} h(x,t,\nu_v(x,t))|D\mathbf{1}_{\{v>s\}}|\,ds \\ &= \int_{-\infty}^{+\infty}\int_{\Omega\times\mathbb{R}} h(x,t,D\mathbf{1}_{\{v>s\}}/|D\mathbf{1}_{\{v>s\}}|)|D\mathbf{1}_{\{v>s\}}|\,ds \\ &= \int_{-\infty}^{+\infty} \mathcal{F}(\mathbf{1}_{\{v>s\}})\,ds, \end{aligned}$$

where we have used the fact that $\mathcal{H}^{N-1}$ a.e. on the boundary of $\{v > s\}$, $\nu_v = \nu_{\{v>s\}} = D\mathbf{1}_{\{v>s\}}/|D\mathbf{1}_{\{v>s\}}|$, that is, the gradient of v is normal to its level lines. □

4.3.3 Numerical Resolution

After a suitable discretization (as described in Section 3.1), problem (4.17), which is of the form (3.19), can be solved for instance by algorithm (3.21).

The most complicated step requires to project onto the (discretized version of the) set $\mathcal{K}$ defined in (4.12). Let us describe now a few examples. We will assume that $f(x,t,p)$ has the form $g(x,t) + h(p)$. Then, $\mathcal{K}$ is reduced to

$$\{\phi \,:\, \phi^t(x,t) \geq h^*(\phi^x(x,t)) - g(x,t)\}$$

and we essentially need to know how to project onto the convex set $\{q = (q^x, q^t) \,:\, q^t \geq h^*(q^x)\}$.

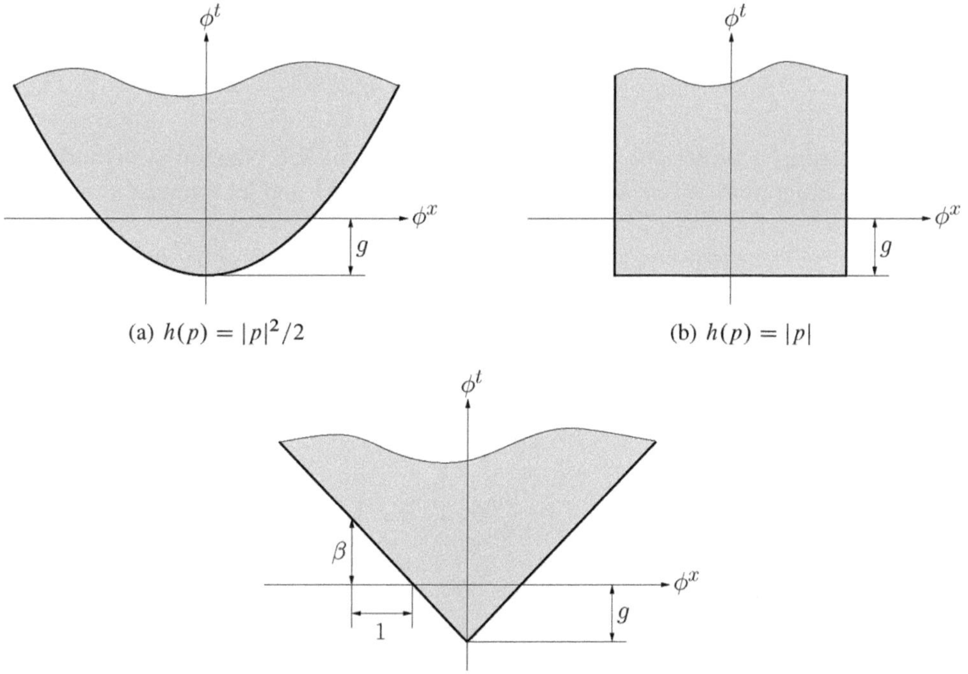

(a) $h(p) = |p|^2/2$ (b) $h(p) = |p|$

(c) $h(p) = 0$ if $|p| \le \beta$ and $+\infty$ else

Figure 14. The sets $\{q^t \ge h^*(q^x)\}$ for various h: (a) Quadratic, (b) Total Variation, (c) Lipschitz regularization.

Quadratic Regularization If $h(p) = |p|^2/2$, then $h^*(q) = |q|^2/2$ as well, and we need to know how to project some $q_0 = (q_0^x, q_0^t)$ onto $K = \{q = (q^x, q^t) : q^t \ge |q^x|^2/2\}$, see Figure 14(a).

If q_0 does not satisfy the constraint, that is, $q_0^t < |q_0^x|^2/2$, we need to project q_0 onto the paraboloid $q^t = |q^x|^2/2$. Hence we must solve the following unconstrained optimization problem

$$\min_q \left\{ \frac{|q - q_0|^2}{2} - \lambda\left(q^t - \frac{|q^x|^2}{2}\right)\right\}, \tag{4.22}$$

where λ is a Lagrange multiplier for the equality constraint $q^t - |q^x|^2/2 = 0$. The optimal conditions of (4.22) are given by

$$\begin{aligned} q^x - q_0^x + \lambda q^x &= 0 \\ q^t - q_0^t - \lambda &= 0 \\ q^t - \frac{|q^x|^2}{2} &= 0. \end{aligned} \tag{4.23}$$

After eliminating q^t and q^x, we arrive at the following cubic equation for λ:

$$\lambda^3 + \lambda^2(q_0^t + 2) + \lambda(2q_0^t + 1) + q_0^t - \frac{|q_0^x|^2}{2} = 0. \tag{4.24}$$

Instead of using a direct cubic solver for (4.24) we utilize Newton's method. We choose a starting point $\lambda^0 = \max\{0, -(2q_0^t + 1)/3)\} + 1$ and let for each $n \geq 0$

$$\lambda^{n+1} = \lambda^n - \frac{(\lambda^n)^3 + (\lambda^n)^2(q_0^t + 2) + (\lambda^n)(2q_0^t + 1) + q_0^t - \frac{|q_0^x|^2}{2}}{3(\lambda^n)^2 + 2(\lambda^n)(q_0^t + 2) + 2q_0^t + 1}. \tag{4.25}$$

We found this scheme to have a quite fast convergence. We never experienced more than 10–20 iterations to achieve a reasonable accuracy. Then, after computing the solution of (4.24), the solution of the projection is given by

$$q = \left(\frac{q_0^x}{1+\lambda}, q_0^t + \lambda\right). \tag{4.26}$$

(a) Left input image (b) Right input image (c) "True" disparity

Figure 15. Rectified stereo image pair and the ground truth disparity, where black pixels correspond to unknown disparity values.

Total Variation Regularization In case $h(p) = |p|$, then $h^*(q) = 0$ for $|q| \leq 1$, and $+\infty$ else, and the projection of q_0 onto the convex $\{q = (q^x, q^t) : q^t \geq 0, |q^x| \leq 1\}$, see Figure 14(b), is simply given by:

$$q = \left(\frac{q_0^x}{\max\{1, |q_0^x|\}}, \max\{0, q_0^t\}\right). \tag{4.27}$$

Lipschitz Constraint One advantage of this approach is that a Lipschitz constraint is quite easy to enforce. We consider $h(p) = 0$ if $|p| \leq L$, $+\infty$ else. Then, the convex conjugate of h is simply $h^*(q) = L|q|$, and we just need to know how to project q_0

onto the convex cone $\{q^t \geq L|q^x|\}$, see Figure 14(c). This projection is of course straightforward: is given by

$$q = \left(\mu \frac{q_0^x}{|q_0^x|}, \mu L\right), \tag{4.28}$$

where μ is given by

$$\mu = \frac{\max\{0, |q_0^x| + Lq_0^t\}}{1 + L^2}. \tag{4.29}$$

Example. Figure 16 shows three examples of stereo reconstruction (Figure 15 shows the input images and the true disparities) using the three different regularizers described above. Of course, only the total variation performs well in this context. As expected, the Lipschitz constraint limits the slope of the solution, while the quadratic contstraint also oversmooths. The best compromise would be to take h a Huber function, quadratic near zero and linear for large values, see [65] for this example.

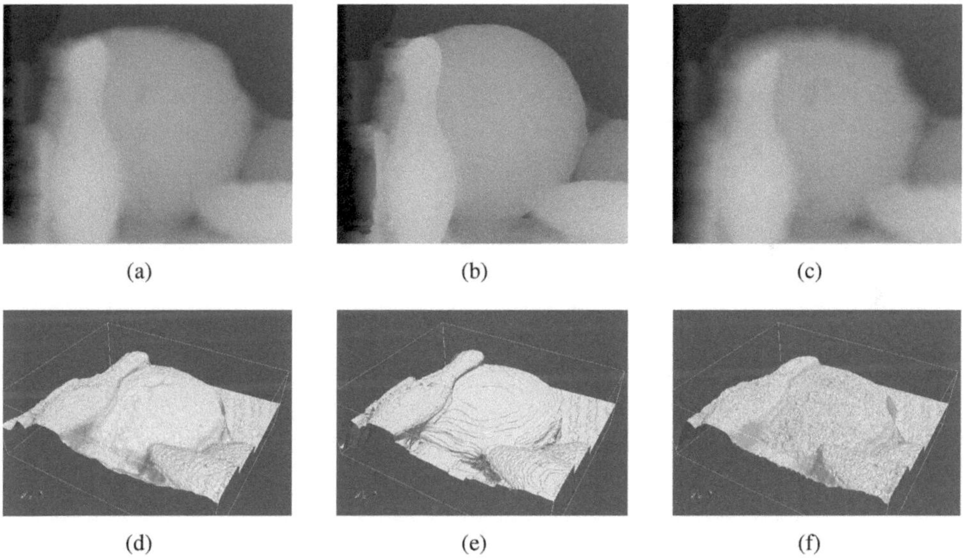

(a) (b) (c)

(d) (e) (f)

Figure 16. Application of different convex regularity terms to disparity estimation. First column: Quadratic regularization, second column: Total Variation regularization, third column: Lipschitz regularization.

4.4 The Minimal Partition Problem

Consider now the problem of finding a segmentation into k sets of a domain $\Omega \subset \mathbb{R}^N$, which minimizes the total interface between the sets, as in the piecewise constant

Mumford–Shah problem:

$$\min_{(R_i)_{i=1}^k,(c_i)_{i=1}^k} \frac{\lambda}{2}\sum_{i=1}^k \mathrm{Per}(R_i;\Omega) + \frac{1}{2}\sum_{i=1}^k \int_{R_i} |g(x)-c_i|^2\,dx \tag{4.30}$$

where $c_i \in \mathbb{R}$ and where the regions $(R_i)_{i=1}^k$ form a partition of Ω, that is, $R_i \cap R_j = \emptyset$ if $i \neq j$ and $\bigcup_{i=1}^k R_i = \Omega$. It corresponds to the best least square approximation of g with a piecewise constant function $u = \sum_i c_i \chi_{R_i}$, with the total perimeter of the partition as the cost of a particular approximation (the total number of set k should in theory be free, however it is always possible to bound it from above if g is bounded, so that there is no loss of generality in keeping it fixed). Of course, given the partition $(R_i)_{i=1}^k$, the optimal constant $c_i = (1/|R_i|)\int_{R_i} g\,ds$ is the average value of g on R_i for each $i = 1,\dots,k$. On the other hand, finding the minimum of (4.30) with respect to the partition $(R_i)_{i=1}^k$ is a hard task, even for fixed values $(c_i)_{i=1}^k$. It is known that its discrete counterpart (the *Pott's model*) is NP-hard, so that it is unlikely that (4.30) has a simple convex representation, at least without increasing drastically the number of variables.

We will show that one can consider simple convex approximations of the interface term in (4.30) which can be actually minimized, and in many cases provide a solution of the original problem (although nothing of this kind is known in general).

Let us introduce $v_i = \chi_{R_i} \in BV(\Omega)$: the functions v_i satisfy $v_i(x) \in \{0,1\}$ and $\sum_{i=1}^k v_i(x) = 1$ a.e. in Ω. Moreover, letting $\mathbf{v} = (v_1,\dots,v_k) \in BV(\Omega;\mathbb{R}^k)$, we see that

$$J_{\mathbf{v}} = \bigcup_{i=1}^k J_{v_i} = \bigcup_{i=1}^k \Omega \cap \partial^* R_i$$

and the total surface of the interface is

$$\mathcal{H}^{N-1}(J_{\mathbf{v}}) = \frac{1}{2}\sum_{i=1}^k \mathrm{Per}(R_i;\Omega) \tag{4.31}$$

since in the right-hand side, the common boundary of R_i and R_j is counted twice for all $i \neq j$.

We can therefore define the partition functional as

$$\mathcal{J}(\mathbf{v}) = \begin{cases} \mathcal{H}^{N-1}(J_{\mathbf{v}}) & \text{if } \mathbf{v} \in BV(\Omega;\{0,1\}^k) \text{ with } \sum_{i=1}^k v_i = 1 \text{ a.e.,} \\ +\infty & \text{else.} \end{cases}$$

Then, the best convex approximation of $\mathcal{J}$ should be its convex l.s.c. envelope $\mathcal{J}^{**}$ (in $L^2(\Omega;\mathbb{R}^k)$), as defined in Definition 3.3:

$$\mathcal{J}^{**}(\mathbf{v}) = \sup_{\mathbf{w}\in L^2(\Omega;\mathbb{R}^k)} \langle \mathbf{w},\mathbf{v}\rangle - \mathcal{J}^*(\mathbf{w})$$

where

$$\mathcal{J}^*(\mathbf{w}) = \sup_{\mathbf{v}\in L^2(\Omega;\mathbb{R}^k)} \langle \mathbf{w}, \mathbf{v}\rangle - \mathcal{J}(\mathbf{v}).$$

However, this provides an abstract definition of $\mathcal{J}^{**}$ but does not say how to actually compute it or minimize it. It is possible, though, to show that the domain of $\mathcal{J}^{**}$ is

$$\operatorname{dom} \mathcal{J}^{**} = \left\{ \mathbf{v} \in BV(\Omega; [0,1]^k),\ \sum_{i=1}^k v_i = 1 \text{ a.e. in } \Omega \right\},$$

see [22] where a different representation is used but this can be deduced by a simple change of variable.

To be able to numerically solve the problem, one should find a convex, l.s.c. functional $J \le \mathcal{J}$[5] with a particular form, which can be handled and provide a problem that can be actually solved. A typical form is $J(\mathbf{v}) = \int_\Omega F(x, D\mathbf{v})$ for $F(x,p)$ some function, convex in p and measurable in x. Moreover, one should at least require that $J = \mathcal{J}$ on its domain (that is, on binary functions $\mathbf{v} \in \{0,1\}^k$ with $\sum_i v_i = 1$). Eventually, one should try to find the largest possible J in this class, so that it becomes more likely that a solution of

$$\min_{\mathbf{v}} \lambda J(v) + \frac{1}{2}\int_\Omega \sum_{i=1}^k v_i(x)|g(x) - c_i|^2\, dx \tag{4.32}$$

is itself binary (in the domain of $\mathcal{J}$), and therefore provides a minimizer of (4.30) for fixed $(c_i)_{i=1}^k$.

Several choices have been proposed in the literature. In [73], it is proposed to use simple LP-relaxation exactly as for the multiway cut problem in the discrete literature [1]. Hence one just lets

$$J(v) = \frac{1}{2}\sum_{i=1}^k \int_\Omega |Dv_i| \quad \text{if } \mathbf{v} \in BV(\Omega; [0,1]^k), \tag{4.33}$$

and $+\infty$ else, which naturally extends (4.31) to the domain of $\mathcal{J}^{**}$. However, it can be shown that this relaxation is too small, see [22, Proposition A.1.] and Figure 17(b).

On the other hand, [48] propose to use the vectorial total variation (appropriately rescaled), which is defined exactly like in (TV), that is,

$$\begin{aligned} J(v) &= \frac{1}{\sqrt{2}}\int_\Omega |D\mathbf{v}| \\ &= \frac{1}{\sqrt{2}} \sup\left\{ -\int \mathbf{v}\cdot \operatorname{div}\phi \,:\, \phi \in C_c^\infty(\Omega; \mathbb{R}^{N\times k}),\ \sum_{i,j} \phi_{i,j}^2 \le 1 \right\}. \end{aligned} \tag{4.34}$$

[5] Hence $J \le \mathcal{J}^{**}$, since it is the largest convex, l.s.c. functional below $\mathcal{J}$.

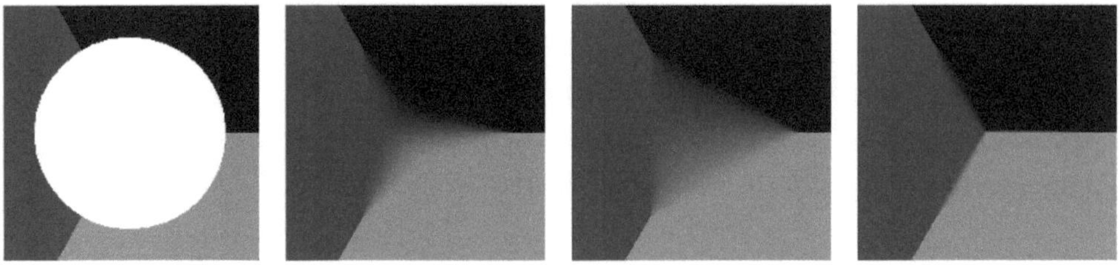

(a) Input image (b) Relaxation (4.33) (c) Relaxation (4.34) (d) Relaxation (4.35)

Figure 17. Triple junction experiment with $k = 3$. (a) shows the input image with given boundary datum. (b) shows the result using the relaxation of [73], and (c) the relaxation of [48]. (d) shows the result of the proposed relaxation.

Examples show that this does not perform much better than the previous choice for recovering triple junctions, see Figure 17(c).

The "best" possible choice, if of the form $\int_\Omega F(x, D\mathbf{v})$ for a convex, even function $F(x, p)$ (w.r. p), can be shown to be

$$J(\mathbf{v}) = \int_\Omega \Psi(D\mathbf{v}) \tag{4.35}$$

for $\Psi : \mathbb{R}^{N\times k} \to [0, +\infty]$ the convex, 1-homogeneous function given by

$$\Psi(p) = \sup\left\{ \sum_{i=1}^{k} \langle p_i, q_i \rangle \ : \ |q_i - q_j| \le 1 \ \forall 1 \le i < j \le k \right\}.$$

(Note in particular that $\Psi(p) = +\infty$ if $\sum_{i=1}^k p_i \neq 0$, which is not an issue since if $\mathbf{v} \in \operatorname{dom} \mathcal{J}^{**}$, $\sum v_i = 1$ hence $\sum_i Dv_i = 0$.) In our notation, for p a vector in $\mathbb{R}^{N\times k}$, $p_j = (p_{i,j})_{i=1}^N$ is a N-dimensional vector for each $j = 1, \dots, k$. Letting

$$\mathcal{K} = \{-\operatorname{div} \phi : \phi \in C_c^\infty(\Omega; \mathbb{R}^{N\times k}), |\phi_i(x) - \phi_j(x)| \le 1 \ \forall x \in \Omega, 1 \le i < j \le k\},$$

we can also define J as the support function of $\mathcal{K}$:

$$J(\mathbf{v}) = \sup_{\mathbf{w}\in\mathcal{K}} \langle \mathbf{w}, \mathbf{v} \rangle,$$

which is a variant of (TV): it is again a sort of total variation and we can hope to minimize it with the techniques described in Section 3. This construction is related to the theory of "paired calibration" introduced in the 1990s by Lawlor–Morgan and Brakke [47, 18].

(a) Input image

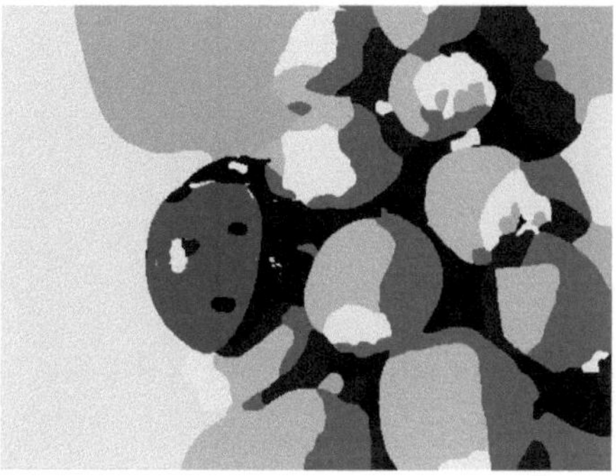

(b) Segmentation, $k = 8$

Figure 18. Piecewise constant Mumford–Shah segmentation of a natural image. (a) shows the input image and (b) is the minimizer of energy (4.30). The mean color values c_i of the partitions have been initialized using k-means clustering.

In order to minimize (4.32) using the techniques described in these notes, one first chooses to minimize alternatingly with respect to the partition $(R_i)_{i=1}^k$, described by the function $\mathbf{v}$, and with respect to the constants $(c_i)_{i=1}^k$ (which, in general, should lead to a local minimizer, or even a mere critical point of the problem). Minimizing with respect to the c_i's is straightforward (the solution is simply the average of the data in the region R_i). The minimization with respect to $\mathbf{v}$ is performed, for instance, using Algorithm (3.21). A crucial step is the projection onto $K = \{p \in \mathbb{R}^{N\times k} : |p_i - p_j| \leq 1 \ \forall i, j\}$, whose Ψ is the support function. This is performed by alternate projections, following Dykstra's algorithm for projecting onto intersection of convex sets [17].

Figure 17 shows the minimization of J with a boundary datum which enforces the completion of three regions: we find a triple point with 120° angles, as the theory expects. Observe that lower relaxations provide wrong results (with a relaxed energy strictly below the interfacial energy of the original problem). The next Figure 18 shows an example of a minimization of (4.32) done following this approach. (The numerical computations have been performed on a GPU.)

A A Proof of Convergence

We prove in this section Theorem 3.14, that is, a convergence estimate for the modified Douglas–Rachford (or extragradient) algorithm (3.21). The assumption that $x^0 = \bar{x}^0$ can also be written $x^{-1} = x^0$ and $\bar{x}^0 = 2x^0 - x^{-1}$, which is consistent with the definition of $\bar{x}^{n+1}$ for $n \geq 0$. The proof which follows is heavily inspired by [58] (for the estimate) and [66] (for the convergence proof). See also [25] for extensions and accelerations in smoother cases.

Proof. We have

$$\partial F^*(y^{n+1}) \ni \frac{y^n - y^{n+1}}{\sigma} + A\bar{x}^n$$
$$\partial G(x^{n+1}) \ni \frac{x^n - x^{n+1}}{\tau} - A^* y^{n+1}$$

so that for any $(x, y) \in X \times Y$,

$$F^*(y) \geq F^*(y^{n+1}) + \left\langle \frac{y^n - y^{n+1}}{\sigma}, y - y^{n+1} \right\rangle + \langle A\bar{x}^n, y - y^{n+1} \rangle$$
$$G(x) \geq G(x^{n+1}) + \left\langle \frac{x^n - x^{n+1}}{\tau}, x - x^{n+1} \right\rangle - \langle y^{n+1}, A(x - x^{n+1}) \rangle.$$

Summing both inequalities, it follows:

$$\begin{aligned} F^*(y) + G(x) &+ \frac{\|y - y^n\|^2}{2\sigma} + \frac{\|x - x^n\|^2}{2\tau} \\ &\geq F^*(y^{n+1}) + G(x^{n+1}) + \langle A(x^{n+1} - \bar{x}^n), y^{n+1} - y \rangle \\ &\quad + \langle Ax^{n+1}, y \rangle - \langle y^{n+1}, Ax \rangle \\ &\quad + \frac{\|y - y^{n+1}\|^2}{2\sigma} + \frac{\|x - x^{n+1}\|^2}{2\tau} + \frac{\|y^n - y^{n+1}\|^2}{2\sigma} + \frac{\|x^n - x^{n+1}\|^2}{2\tau}. \end{aligned} \tag{A.1}$$

Now:

$$\begin{aligned} &\langle A(x^{n+1} - \bar{x}^n), y^{n+1} - y \rangle \\ &\quad = \langle A((x^{n+1} - x^n) - (x^n - x^{n-1})), y^{n+1} - y \rangle \\ &\quad = \langle A(x^{n+1} - x^n), y^{n+1} - y \rangle - \langle A(x^n - x^{n-1}), y^n - y \rangle \\ &\qquad - \langle A(x^n - x^{n-1}), y^{n+1} - y^n \rangle \\ &\quad \geq \langle A(x^{n+1} - x^n), y^{n+1} - y \rangle - \langle A(x^n - x^{n-1}), y^n - y \rangle \\ &\qquad - L\|x^n - x^{n-1}\| \|y^{n+1} - y^n\|. \end{aligned} \tag{A.2}$$

For any $\delta > 0$, we have that (using $2ab \leq \delta a^2 + b^2/\delta$ for any a, b)

$$L\|x^n - x^{n-1}\| \|y^{n+1} - y^n\| \leq \frac{L\delta\tau}{2\tau}\|x^n - x^{n-1}\|^2 + \frac{L\sigma}{2\delta\sigma}\|y^{n+1} - y^n\|^2$$

and we choose $\delta = \sqrt{\sigma/\tau}$, so that $L\delta\tau = L\sigma/\delta = \sqrt{\sigma\tau}L < 1$.

Summing the last inequality together with (A.1) and (A.2), we get that for any $x \in X$ and $y \in Y$,

$$
\begin{aligned}
&\frac{\|y-y^n\|^2}{2\sigma} + \frac{\|x-x^n\|^2}{2\tau} \\
&\geq [\langle Ax^{n+1}, y\rangle - F^*(y) + G(x^{n+1})] - [\langle Ax, y^{n+1}\rangle - F^*(y^{n+1}) + G(x)] \\
&\quad + \frac{\|y-y^{n+1}\|^2}{2\sigma} + \frac{\|x-x^{n+1}\|^2}{2\tau} + (1-\sqrt{\sigma\tau}L)\frac{\|y^n-y^{n+1}\|^2}{2\sigma} \\
&\quad + \frac{\|x^n-x^{n+1}\|^2}{2\tau} - \sqrt{\sigma\tau}L\frac{\|x^{n-1}-x^n\|^2}{2\tau} \\
&\quad + \langle A(x^{n+1}-x^n), y^{n+1}-y\rangle - \langle A(x^n-x^{n-1}), y^n-y\rangle.
\end{aligned} \tag{A.3}
$$

Let us now sum (A.3) from $n=0$ to $N-1$: it follows that for any x and y,

$$
\begin{aligned}
&\sum_{n=1}^{N} [\langle Ax^n, y\rangle - F^*(y) + G(x^n)] - [\langle Ax, y^n\rangle - F^*(y^n) + G(x)] \\
&\qquad + \frac{\|y-y^N\|^2}{2\sigma} + \frac{\|x-x^N\|^2}{2\tau} + (1-\sqrt{\sigma\tau}L)\sum_{n=1}^{N}\frac{\|y^n-y^{n-1}\|^2}{2\sigma} \\
&\qquad + (1-\sqrt{\sigma\tau}L)\sum_{n=1}^{N-1}\frac{\|x^n-x^{n-1}\|^2}{2\tau} + \frac{\|x^N-x^{N-1}\|^2}{2\tau} \\
&\quad \leq \frac{\|y-y^0\|^2}{2\sigma} + \frac{\|x-x^0\|^2}{2\tau} + \langle A(x^N-x^{N-1}), y^N-y\rangle
\end{aligned}
$$

where we have used $x^{-1} = x^0$. Now, as before,

$$
\langle A(x^N-x^{N-1}), y^N-y\rangle \leq \|x^N-x^{N-1}\|^2/(2\tau) + (\tau\sigma L^2)\|y-y^N\|^2/(2\sigma),
$$

and it follows

$$
\begin{aligned}
&\sum_{n=1}^{N} [\langle Ax^n, y\rangle - F^*(y) + G(x^n)] - [\langle Ax, y^n\rangle - F^*(y^n) + G(x)] \\
&\quad + (1-\sigma\tau L^2)\frac{\|y-y^N\|^2}{2\sigma} + \frac{\|x-x^N\|^2}{2\tau} + (1-\sqrt{\sigma\tau}L)\sum_{n=1}^{N}\frac{\|y^n-y^{n-1}\|^2}{2\sigma} \\
&\quad + (1-\sqrt{\sigma\tau}L)\sum_{n=1}^{N-1}\frac{\|x^n-x^{n-1}\|^2}{2\tau} \leq \frac{\|y-y^0\|^2}{2\sigma} + \frac{\|x-x^0\|^2}{2\tau}.
\end{aligned} \tag{A.4}
$$

First we choose $(x, y) = (\hat{x}, \hat{y})$ a saddle-point in (A.4). Then, it follows from (3.20) that the first summation in (A.4) is non-negative, and point *(a)* in Theorem 3.14 follows. We then deduce from (A.4) and the convexity of G and F^* that, letting $x_N = (\sum_{n=1}^N x^n)/N$ and $y_N = (\sum_{n=1}^N y^n)/N$,

$$[\langle Ax_N, y\rangle - F^*(y) + G(x_N)] - [\langle Ax, y_N\rangle - F^*(y_N) + G(x)] \\ \leq \frac{1}{N}\left(\frac{\|y - y^0\|^2}{2\sigma} + \frac{\|x - x^0\|^2}{2\tau}\right) \tag{A.5}$$

for any $(x, y) \in X \times Y$, which yields (3.23). Consider now a weak cluster point (x^*, y^*) of (x_N, y_N) (which is a bounded sequence, hence weakly compact). Since G and F^* are convex and l.s.c. they also are weakly l.s.c., and it follows from (A.5) that

$$[\langle Ax^*, y\rangle - F^*(y) + G(x^*)] - [\langle Ax, y^*\rangle - F^*(y^*) + G(x)] \leq 0$$

for any $(x, y) \in X \times Y$: this shows that (x^*, y^*) satisfies (3.20) and therefore is a saddle-point. We have shown point *(b)* in Theorem 3.14.

It remains to prove the convergence to a saddle point if the spaces X and Y are finite-dimensional. Point *(a)* establishes that (x^n, y^n) is a bounded sequence, so that some subsequence (x^{n_k}, y^{n_k}) converges to some limit (x^*, y^*), strongly since we are in finite dimension. Observe that (A.4) implies that $\lim_n(x^n - x^{n-1}) = \lim_n(y^n - y^{n-1}) = 0$, in particular also x^{n_k-1} and y^{n_k-1} converge respectively to x^* and y^*. It follows that the limit (x^*, y^*) is a fixed point of the algorithm (3.21), hence a saddle point of our problem.

We can then take $(x, y) = (x^*, y^*)$ in (A.3), which we sum from $n = n^k$ to $N - 1$, $N > n_k$. We obtain

$$\frac{\|y^* - y^N\|^2}{2\sigma} + \frac{\|x^* - x^N\|^2}{2\tau} + (1 - \sqrt{\sigma\tau}L) \sum_{n=n_k+1}^{N} \frac{\|y^n - y^{n-1}\|^2}{2\sigma} \\ - \frac{\|x^{n_k} - x^{n_k-1}\|^2}{2\tau} + (1 - \sqrt{\sigma\tau}L) \sum_{n=n_k}^{N-1} \frac{\|x^n - x^{n-1}\|^2}{2\tau} + \frac{\|x^N - x^{N-1}\|^2}{2\tau} \\ + \langle A(x^N - x^{N-1}), y^N - y^*\rangle - \langle A(x^{n_k} - x^{n_k-1}), y^{n_k} - y^*\rangle \\ \leq \frac{\|y^* - y^{n_k}\|^2}{2\sigma} + \frac{\|x^* - x^{n_k}\|^2}{2\tau}$$

from which we easily deduce that $x^N \to x^*$ and $y^N \to y^*$ as $N \to \infty$. □

Acknowledgments. These lecture notes have been prepared for the summer school on sparsity organized in Linz, Austria, by Massimo Fornasier and Ronny Romlau, during the week August 31 to September 4, 2009. They contain work done in collaboration with Vicent Caselles and Matteo Novaga (for the first theoretical parts), and with Thomas Pock and Daniel Cremers, for the algorithmic parts. All are obviously warmly thanked for the work we have done together so far – and the work still to be done! I thank particularly Thomas Pock for pointing out very recent references on primal-dual algorithms and help me clarify the jungle of algorithms.

I also thank, obviously, the organizers of the summer school for inviting me to give these lectures. It was a wonderful scientific event, and we had a great time in Linz.
Antonin Chambolle, November 2009

Bibliography

[1] R. K. Ahuja, T. L. Magnanti and J. B. Orlin, *Network Flows: Theory, Algorithms, and Applications*, Prentice Hall Inc., Englewood Cliffs, NJ, 1993.

[2] G. Alberti, G. Bouchitté and D. Dal Maso, The calibration method for the Mumford–Shah functional and free-discontinuity problems., *Calc. Var. Partial Differential Equations* **16** (2003), 299–333.

[3] W. K. Allard, Total variation regularization for image denoising, I. Geometric theory, *SIAM J. Math. Anal.* **39** (2007), 1150–1190.

[4] F. Alter, V. Caselles and A. Chambolle, A characterization of convex calibrable sets in $\mathbb{R}^N$, *Math. Ann.* **332** (2005), 329–366.

[5] ______, Evolution of characteristic functions of convex sets in the plane by the minimizing total variation flow, *Interfaces Free Bound.* **7** (2005), 29–53.

[6] L. Ambrosio, *Corso Introduttivo alla Teoria Geometrica della Misura ed alle Superfici Minime*, Appunti dei Corsi Tenuti da Docenti della Scuola. [Notes of Courses Given by Teachers at the School], Scuola Normale Superiore, Pisa, 1997.

[7] L. Ambrosio, N. Fusco and D. Pallara, *Functions of Bounded Variation and Free Discontinuity Problems*, The Clarendon Press Oxford University Press, New York, 2000.

[8] L. Ambrosio and V. M. Tortorelli, Approximation of functionals depending on jumps by elliptic functionals via Γ-convergence, *Comm. Pure Appl. Math.* **43** (1990), 999–1036.

[9] ______, On the approximation of free discontinuity problems, *Boll. Un. Mat. Ital. B (7)* **6** (1992), 105–123.

[10] B. Appleton and H. Talbot, Globally optimal geodesic active contours, *J. Math. Imaging Vision* **23** (2005), 67–86.

[11] K. J. Arrow, L. Hurwicz and H. Uzawa, *Studies in Linear and Non-linear Programming*, With contributions by H. B. Chenery, S. M. Johnson, S. Karlin, T. Marschak, R. M. Solow. Stanford Mathematical Studies in the Social Sciences, vol. II, Stanford University Press, Stanford, Calif., 1958.

[12] J.-F. Aujol, G. Gilboa, T. Chan and S. Osher, Structure-texture image decomposition–modeling, algorithms, and parameter selection, *Int. J. Comput. Vis.* **67** (2006), 111–136.

[13] A. Beck and M. Teboulle, A fast iterative shrinkage-thresholding algorithm for linear inverse problems, *SIAM J. Imaging Sci.* **2** (2009), 183–202.

[14] G. Bellettini, M. Paolini and C. Verdi, Convex approximations of functionals with curvature, *Atti Accad. Naz. Lincei Cl. Sci. Fis. Mat. Natur. Rend. Lincei (9) Mat. Appl.* **2** (1991), 297–306.

[15] J. Bioucas-Dias and M. Figueiredo, A new TwIST: two-step iterative shrinkage / thresholding algorithms for image restoration, *IEEE Trans. on Image Processing* **16** (2007), 2992–3004.

[16] Y. Boykov and V. Kolmogorov, An experimental comparison of Min-Cut/Max-Flow algorithms for energy minimization in vision, *IEEE Trans. Pattern Analysis and Machine Intelligence* **26** (2004), 1124–1137.

[17] J. P. Boyle and R. L. Dykstra, *A Method for Finding Projections onto the Intersection of Convex Sets in Hilbert Spaces*, Advances in order restricted statistical inference (Iowa City, Iowa, 1985), Lecture Notes in Statist. 37, Springer, Berlin, 1986, pp. 28–47.

[18] K. A. Brakke, Soap films and covering spaces, *J. Geom. Anal.* **5** (1995), 445–514.

[19] H. Brézis, *Opérateurs Maximaux Monotones et Semi-groupes de Contractions dans les Espaces de Hilbert*, North-Holland Publishing Co., Amsterdam, 1973, North-Holland Mathematics Studies, No. 5. Notas de Matemática (50).

[20] V. Caselles, A. Chambolle and M. Novaga, The discontinuity set of solutions of the TV denoising problem and some extensions, *Multiscale Model. Simul.* **6** (2007), 879–894.

[21] V. Caselles, A. Chambolle and M. Novaga, Regularity for solutions of the total variation denoising problem, *Revista Matematica Iberoamericana*, to appear.

[22] A. Chambolle, D. Cremers and T. Pock, *A Convex Approach for Computing Minimal Partitions*, CMAP, Ecole Polytechnique, France, Report no. 649, 2008.

[23] A. Chambolle and J. Darbon, On total variation minimization and surface evolution using parametric maximum flows, *Int J Comput Vis* (2009).

[24] A. Chambolle, S. Levine and B. Lucier, *Some Variations on Total Variation–Based Image Smoothing*, CMAP, Ecole Polytechnique, Report, 2009, http://hal.archives-ouvertes.fr/hal-00370195/fr/.

[25] A. Chambolle and T. Pock, *A First-order Primal-dual Algorithm for Convex Problems with Applications to Imaging*, Ecole Polytechnique/T.U. Graz, Report, 2010.

[26] T. F. Chan and S. Esedoḡlu, Aspects of total variation regularized L^1 function approximation, *SIAM J. Appl. Math.* **65** (2005), 1817–1837 (electronic).

[27] T. F. Chan, S. Esedoḡlu and M. Nikolova, Algorithms for finding global minimizers of image segmentation and denoising models, *SIAM J. Appl. Math.* **66** (2006), 1632–1648 (electronic).

[28] P. Combettes and V. Wajs, Signal recovery by proximal forward-backward splitting, *SIAM Multiscale Modelling and Simulation* (2006).

[29] G. Dal Maso, Integral representation on $BV(\Omega)$ of Γ-limits of variational integrals, *Manuscr. Math.* **30** (1979), 387–416 (English).

[30] G. David, *Global Minimizers of the Mumford–Shah Functional*, Current developments in mathematics, 1997 (Cambridge, MA), Int. Press, Boston, MA, 1999, pp. 219–224.

[31] J. Eckstein and D. Bertsekas, On the Douglas–Rachford splitting method and the proximal point algorithm for maximal monotone operators, *Mathematical Programming* **55** (1992).

[32] I. Ekeland and R. Témam, *Convex Analysis and Variational Problems*, english ed, Classics in Applied Mathematics 28, Society for Industrial and Applied Mathematics (SIAM), Philadelphia, PA, 1999, Translated from the French.

[33] E. Esser, *Applications of Lagrangian-Based Alternating Direction Methods and Connections to Split Bregman*, UCLA, Center for Applied Math., CAM Reports no. 09-31, 2009.

[34] E. Esser, X. Zhang and T. Chan, *A General Framework for a Class of First Order Primal-Dual Algorithms for TV Minimization*, UCLA, Center for Applied Math., CAM Reports no. 09-67, 2009.

[35] L. C. Evans and R. F. Gariepy, *Measure Theory and Fine Properties of Functions*, CRC Press, Boca Raton, FL, 1992.

[36] K. J. Falconer, *The Geometry of Fractal Sets*, Cambridge Tracts in Mathematics 85, Cambridge University Press, Cambridge, 1986.

[37] H. Federer, *Geometric Measure Theory*, Springer, New York, 1969.

[38] S. Geman and D. Geman, Stochastic relaxation, Gibbs distributions, and the Bayesian restoration of images, *IEEE Trans. PAMI* PAMI-6 (1984), 721–741.

[39] M. Giaquinta, G. Modica and J. Souček, *Cartesian Currents in the Calculus of Variations. I*, Ergebnisse der Mathematik und ihrer Grenzgebiete 37, Springer, Berlin, 1998.

[40] ———, *Cartesian Currents in the Calculus of Variations. II*, Ergebnisse der Mathematik und ihrer Grenzgebiete 38, Springer, Berlin, 1998.

[41] E. Giusti, *Minimal Surfaces and Functions of Bounded Variation*, Birkhäuser, Basel, 1984.

[42] D. S. Hochbaum, An efficient algorithm for image segmentation, Markov random fields and related problems, *J. ACM* **48** (2001), 686–701 (electronic).

[43] B. Kawohl and T. Lachand-Robert, Characterization of Cheeger sets for convex subsets of the plane, *Pacific J. Math.* **225** (2006), 103–118.

[44] V. Kolmogorov and R. Zabih, What energy functions can be minimized via graph cuts?, *IEEE Trans. Pattern Analysis and Machine Intelligence* **2** (2004), 147–159.

[45] G. M. Korpelevich, Extrapolational gradient methods and their connection with modified Lagrangians, *Ehkon. Mat. Metody* **19** (1983), 694–703 (Russian).

[46] M.-J. Lai, B. J. Lucier and J. Wang, The convergence of a central-difference discretization of Rudin–Osher–Fatemi model for image denoising, in: *Scale Space and Variational Methods in Computer Vision*, Lecture Notes in Computer Science 5567, pp. 514–526, Springer, 2009.

[47] G. Lawlor and F. Morgan, Paired calibrations applied to soap films, immiscible fluids, and surfaces or networks minimizing other norms, *Pacific J. Math.* **166** (1994), 55–83.

[48] J. Lellmann, J. Kappes, J. Yuan, F. Becker and C. Schnörr, Convex multi-class image labeling by simplex-constrained total variation, in: *Scale Space and Variational Methods in Computer Vision*, Lecture Notes in Computer Science 5567, pp. 150–162, Springer, 2009.

[49] P.-L. Lions and B. Mercier, Splitting algorithms for the sum of two nonlinear operators, *SIAM J. Numer. Anal.* **16** (1979), 964–979 (English).

[50] C. Louchet, *Modèles Variationnels et Bayésiens pour le d'Ébruitage d'Images: de la Variation Totale vers les Moyennes Non-locales*, Ph.D. thesis, Université Paris-Descartes, 2008.

[51] B. J. Lucier and J. Wang, *Error Bounds For Finite-Difference Methods For Rudin–Osher–Fatemi Image Smoothing*, UCLA, Center for Applied Math., CAM Reports no. 09-70, 2009.

[52] F. Maddalena and S. Solimini, Lower semicontinuity properties of functionals with free discontinuities, *Arch. Ration. Mech. Anal.* **159** (2001), 273–294.

[53] Y. Meyer, *Oscillating Patterns in Image Processing and Nonlinear Evolution Equations*, University Lecture Series 22, American Mathematical Society, Providence, RI, 2001.

[54] N. Meyers and J. Serrin, $H = W$, *Proc. Nat. Acad. Sci. USA* **51** (1964), 1055–1056.

[55] J.-M. Morel and S. Solimini, *Variational Methods in Image Segmentation*, Birkhäuser, Boston, 1995.

[56] D. Mumford, Pattern theory: the mathematics of perception, in: *Proceedings of the International Congress of Mathematicians*, 3, 2002.

[57] D. Mumford and J. Shah, Optimal Approximation by Piecewise Smooth Functions and Associated Variational Problems, *Comm. Pure Appl. Math.* **42** (1989), 577–685.

[58] A. Nemirovski, Prox-method with rate of convergence $O(1/t)$ for variational inequalities with Lipschitz continuous monotone operators and smooth convex-concave saddle point problems, *SIAM J. Optim.* **15** (2004), 229–251 (English).

[59] Yu. E. Nesterov, A method for solving the convex programming problem with convergence rate $O(1/k^2)$, *Dokl. Akad. Nauk SSSR* **269** (1983), 543–547.

[60] ———, *Introductory Lectures on Convex Optimization*, Applied Optimization 87, Kluwer Academic Publisher, 2004.

[61] ———, Smooth minimization of nonsmooth functions, *Mathemetical programmind Series A* **103** (2005), 127–152.

[62] M. Nikolova, A variational approach to remove outliers and impulse noise, *J. Math. Image Vis.* **20** (2004), 99–120.

[63] J. C. Picard and H. D. Ratliff, Minimum cuts and related problems, *Networks* **5** (1975), 357–370.

[64] T. Pock, D. Cremers, H. Bischof and A. Chambolle, An algorithm for minimizing the Mumford–Shah functional, in: *ICCV Proceedings*, LNCS, Springer, 2009.

[65] T. Pock, D. Cremers, H. Bischof and A. Chambolle, Global solutions of variational models with convex regularization, *SIAM J. Imaging Sci.*, to appear.

[66] L. Popov, A modification of the Arrow–Hurwicz method for search of saddl e points, *Mathematical Notes* **28** (1980), 845–848.

[67] M. Protter, I. Yavneh and M. Elad, *Closed-Form MMSE Estimation for Signal Denoising Under Sparse Representation Modelling Over a Unitary Dictionary*, Technion, Haifa, Report, 2009.

[68] R. T. Rockafellar, *Convex Analysis*, Princeton Landmarks in Mathematics, Princeton University Press, Princeton, NJ, 1997, Reprint of the 1970 original, Princeton Paperbacks.

[69] L. Rudin, S. J. Osher and E. Fatemi, Nonlinear total variation based noise removal algorithms, *Physica D.* **60** (1992), 259–268, [also in *Experimental Mathematics: Computational Issues in Nonlinear Science* (Proc. Los Alamo Conf. 1991)].

[70] L. Schwartz, *Théorie des Distributions*, Hermann, 1966, (2 volumes).

[71] R. E. Showalter, *Monotone Operators in Banach Space and Nonlinear Partial Differential Equations*, Mathematical Surveys and Monographs 49, American Mathematical Society, Providence, RI, 1997.

[72] C. R. Vogel and M. E. Oman, Iterative methods for total variation denoising, *SIAM J. Sci. Comput* **17** (1996), 227–238, Special issue on iterative methods in numerical linear algebra (Breckenridge, CO, 1994).

[73] C. Zach, D. Gallup, J. M. Frahm and M. Niethammer, Fast global labeling for real-time stereo using multiple plane sweeps, in: *Vision, Modeling, and Visualization 2008*, pp. 243–252, IOS Press, 2008.

[74] M. Zhu and T. Chan, *An Efficient Primal-Dual Hybrid Gradient Algorithm for Total Variation Image Restoration*, UCLA, Center for Applied Math., CAM Reports no. 08-34, 2008.

[75] W. P. Ziemer, *Weakly Differentiable Functions*, Springer, New York, 1989, Sobolev spaces and functions of bounded variation.

Author information

Antonin Chambolle, CMAP, Ecole Polytechnique, CNRS, 91128, Palaiseau, France.
E-mail: antonin.chambolle@polytechnique.fr

Vicent Caselles, Departament de Tecnologia, Universitat Pompeu-Fabra, Barcelona, Spain.
E-mail: vicent.caselles@tecn.upf.es

Daniel Cremers, Department of Computer Science, University of Bonn, Römerstraße 164, 53117 Bonn, Germany.
E-mail: dcremers@cs.uni-bonn.de

Matteo Novaga, Dipartimento di Matematica, Università di Padova, Via Trieste 63, 35121 Padova, Italy.
E-mail: novaga@math.unipd.it

Thomas Pock, Institute for Computer Graphics and Vision, Graz University of Technology, 8010 Graz, Austria.
E-mail: pock@icg.tugraz.at

Bei Fragen zur Produktsicherheit wenden Sie sich bitte an:
If you have any questions regarding product safety,
please contact:

Walter de Gruyter GmbH
Genthiner Straße 13
10785 Berlin
productsafety@degruyterbrill.com